Studies in Fuzziness and Soft Computing

Volume 440

The series "Studies in Fuzziness and Soft Computing" contains publications on various topics in the area of soft computing, which include fuzzy sets, rough sets, neural networks, evolutionary computation, probabilistic and evidential reasoning, multi-valued logic, and related fields. The publications within "Studies in Fuzziness and Soft Computing" are primarily monographs and edited volumes. They cover significant recent developments in the field, both of a foundational and applicable character. An important feature of the series is its short publication time and world-wide distribution. This permits a rapid and broad dissemination of research results.

Indexed by SCOPUS, DBLP, WTI Frankfurt eG, zbMATH, SCImago.

All books published in the series are submitted for consideration in Web of Science.

Ali Mahmoodirad · Harish Garg ·
Sadegh Niroomand

Recent Advances on Fuzzy Sets

Theory and Applications

Springer

Ali Mahmoodirad
Department of Mathematics, Bab. C.
Islamic Azad University
Babol, Iran

Harish Garg
Department of Mathematics
Thapar Institute of Engineering
and Technology
Patiala, Punjab, India

Sadegh Niroomand
Department of Industrial Engineering
Firouzabad Higher Education Center
Shiraz University of Technology
Shiraz, Iran

ISSN 1434-9922 ISSN 1860-0808 (electronic)
Studies in Fuzziness and Soft Computing
ISBN 978-3-032-08775-1 ISBN 978-3-032-08776-8 (eBook)
https://doi.org/10.1007/978-3-032-08776-8

This Springer imprint is published by the registered company Springer Nature Switzerland AG
The registered company address is: Gewerbestrasse 11, 6330 Cham, Switzerland

Competing Interests The authors have no competing interests to declare that are relevant to the content of this manuscript.

Contents

Abbreviations

BFS	Bipolar fuzzy set
CFFN	Cubic Fermatean fuzzy number
CPFWGA	Cubic picture fuzzy weighted geometric aggregation
DEA	Data envelopment analysis
DMU	Decision making unit
FFNIS	Fermatean fuzzy negative ideal solutions
FFPIS	Fermatean fuzzy positive ideal solutions
FFS	Fermatean fuzzy sets
GSTrT2IFN	Generalized symmetric trapezoidal type-2 intuitionistic fuzzy number
GTIFN	Generalized triangular intuitionistic fuzzy number
GTrT2IFN	Generalized trapezoidal type-2 intuitionistic fuzzy number
HBFS	Hesitant bipolar fuzzy set
HFE	Hesitant fuzzy element
HFN	Hesitant fuzzy number
HFS	Hesitant fuzzy sets
IFPFN	Interval-valued Pythagorean fuzzy number
IHFS	Intuitionistic hesitant fuzzy set
INHFE	Interval neutrosophic hesitant fuzzy element
INHFS	Interval neutrosophic hesitant fuzzy set
IVFFN	Interval-valued Fermatean fuzzy number
IVFFS	Interval-valued Fermatean fuzzy sets
IVHFE	Interval-valued hesitant fuzzy element
IVHFS	Interval-valued hesitant fuzzy set
IVIHFE	Interval-valued intuitionistic fuzzy element
IVIHFS	Interval-valued intuitionistic hesitant fuzzy set
IVNN	Interval-valued neutrosophic fuzzy number
IVPFS	Interval-valued Pythagorean fuzzy set
IVPHFS	Interval-valued Pythagorean fuzzy set
IVPIHFE	Interval-valued picture hesitant fuzzy element
IVPIHFS	Interval-valued picture hesitant fuzzy set

IVSFN	Interval-valued spherical fuzzy number
IVSFS	Interval-valued spherical fuzzy set
IVSWGM	Interval-valued spherical weighted geometric mean
IVTrNN	Interval-valued trapezoidal neutrosophic fuzzy number
NBFN	Neutrosophic bipolar fuzzy number
NFN	Neutrosophic fuzzy number
NFS	Neutrosophic fuzzy sets
PFS	Pythagorean fuzzy sets
PFWA	Pythagorean fuzzy weighted arithmetic mean
PFWAA	Picture fuzzy weighted arithmetic average
PFWG	Pythagorean fuzzy weighted geometric mean
PFWGM	Picture fuzzy weighted geometric mean
PFWIA	Picture fuzzy weighted interaction aggregation
PHFS	Pythagorean hesitant fuzzy set
PIFS	Picture fuzzy sets
PIHFS	Picture hesitant fuzzy set
SFN	Spherical fuzzy number
SFS	Spherical fuzzy sets
SHFS	Spherical hesitant fuzzy set
SVNBFN	Single-valued neutrosophic bipolar fuzzy number
SVNHFS	Single-valued neutrosophic hesitant fuzzy set
SVNN	Single-valued neutrosophic fuzzy number
SVNNWA	Single-valued neutrosophic weighted arithmetic
SVNNWG	Single-valued neutrosophic weighted geometric
SWAM	Spherical weighted arithmetic mean
SWGM	Spherical weighted geometric mean
T2IFS	Type-2 intuitionistic fuzzy set
T2NN	Type-2 neutrosophic number
T2PFN	Type-2 Pythagorean fuzzy numbers
TBFN	Triangular bipolar fuzzy number
TIFN	Triangular intuitionistic fuzzy number
TrBFN	Trapezoidal bipolar fuzzy number
TrFFN	Trapezoidal Fermatean fuzzy number
TrIFN	Trapezoidal intuitionistic fuzzy number
TrNFN	Trapezoidal neutrosophic fuzzy number
TrPHFN	Trapezoidal Pythagorean hesitant fuzzy number
TrSFN	Trapezoidal spherical fuzzy number
TT2PFN	Triangular type-2 Pythagorean fuzzy number
ZHD	Hellinger Z-number distance

List of Figures

List of Tables

Chapter 1
Fuzzy Set Theory

1.1 Introduction

The fuzzy theory was introduced by Professor Zadeh [20] for the first time. This theory is useful for modelling the vague and uncertain systems and made a way for introducing other theories of the literature such as uncertainty theory, etc. for modelling uncertain systems. According to Prof. Zadeh, the membership of the fuzzy subsets is inexact and many of the sets in real-world have such property. Since 1965 that this theorem was introduced, many mathematical structures and methods have been introduced for dealing with uncertain systems that more or less are based on the fuzzy theory. In this section, some definitions and theorems about fuzzy theory are reviewed, then the fuzzy numbers and their ranking functions are introduced. In continue, the presented contents are developed for type-2 fuzzy sets and numbers. An application of transportation problem is also provided for this chapter.

1.2 Fuzzy Set Theory

In this section some concepts of the fuzzy set theory are described in terms of some definitions and theorems. The contents of this section are taken from the studies of Zadeh [20], Dubois and Prade [2] and Lai and Hwang [8].

Definition 1.1 Suppose X as an unempty set (here is called the reference set). The set $\widetilde{A} = \{(x, \mu_{\tilde{A}}(x))|x \in X\}$, where $\mu_{\tilde{A}} : X \to [0, 1]$, is called a fuzzy set in X. Here, $\mu_{\tilde{A}}$ is the membership function of the fuzzy set $\tilde{A}$. For any $x \in X$ the value of $\mu_{\tilde{A}}(x)$ represents its associated membership in the set $\tilde{A}$.

Definition 1.2 The fuzzy set $\tilde{A}$ is a subset of $\tilde{B}$ (shown as $\tilde{A} \subseteq \tilde{B}$) if and only if $\mu_{\tilde{A}}(x) \leq \mu_{\tilde{B}}(x)$ for any $x \in X$.

A. Mahmoodirad et al., *Recent Advances on Fuzzy Sets*, Studies in Fuzziness and Soft Computing 440, https://doi.org/10.1007/978-3-032-08776-8_1

Definition 1.3 The fuzzy sets $\tilde{A}$ and $\tilde{B}$ are equal (shown as $\tilde{A} = \tilde{B}$) if and only if $\mu_{\tilde{A}}(x) = \mu_{\tilde{B}}(x)$ for any $x \in X$.

Definition 1.4 The support of the fuzzy set $\tilde{A}$ is shown by $Supp(\tilde{A})$ or $S(\tilde{A})$ and determined as $S(\tilde{A}) = \{x \in X \mid \mu_{\tilde{A}}(x) > 0\}$.

Definition 1.5 The height of the fuzzy set $\tilde{A}$ is shown by $hgt(\tilde{A})$ and determined as $hgt(\tilde{A}) = \sup \mu_{\tilde{A}}(x)$.

Definition 1.6 The fuzzy set $\tilde{A}$ is normal if and only if $hgt(\tilde{A}) = 1$.

Definition 1.7 The core of the fuzzy set $\tilde{A}$ is shown by $core(\tilde{A})$ and determined as $core(\tilde{A}) = \{x \in X \mid \mu_{\tilde{A}}(x) = 1\}$.

Definition 1.8 For two fuzzy sets $\tilde{A}$ and $\tilde{B}$,

$$\tilde{A}^c = \{ (x, \mu_{\tilde{A}^c}(x)) \mid x \in X \}, \mu_{\tilde{A}^c}(x) = 1 - \mu_{\tilde{A}}(x), \tag{1.1}$$

$$\tilde{A} \cup \tilde{B} = \{ (x, \mu_{\tilde{A}\cup\tilde{B}}(x)) \mid x \in X \}, \mu_{\tilde{A}\cup\tilde{B}}(x) = \max\{\mu_{\tilde{A}}(x), \mu_{\tilde{B}}(x)\}, \tag{1.2}$$

$$\tilde{A} \cap \tilde{B} = \{ (x, \mu_{\tilde{A}\cap\tilde{B}}(x)) \mid x \in X \}, \mu_{\tilde{A}\cap\tilde{B}}(x) = \min\{\mu_{\tilde{A}}(x), \mu_{\tilde{B}}(x)\}. \tag{1.3}$$

where $\tilde{A}^c$ is the supplement of the fuzzy set $\tilde{A}$.

Definition 1.9 An α-cut of the fuzzy set $\tilde{A}$ is a crisp set and defined as $A_\alpha = \{x \in X \mid \mu_{\tilde{A}}(x) \geq \alpha\}$.

Definition 1.10 The fuzzy set $\tilde{A}$ is convex if and only if for $x_1, x_2 \in X$ and $\lambda \in [0, 1]$ the below condition $\mu_{\tilde{A}}(\lambda x_1 + (1 - \lambda)x_2) \geq \min\{\mu_{\tilde{A}}(x_1), \mu_{\tilde{A}}(x_2)\}$, is held (see Fig. 1.1).

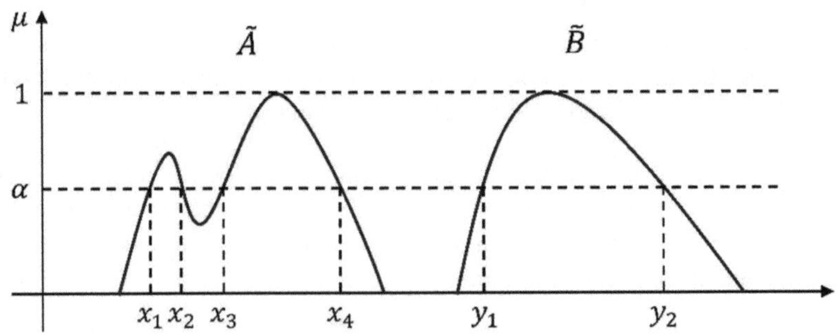

Fig. 1.1 A non-convex fuzzy set ($\tilde{A}$) and a convex fuzzy set ($\tilde{B}$).

Definition 1.11 Suppose $\tilde{A}_1, \tilde{A}_2, \ldots, \tilde{A}_n$ be the fuzzy subsets from $X_1, X_2, \ldots, X_n$ respectively. The Cartesian multiplication of the subsets $\tilde{A}_1, \tilde{A}_2, \ldots, \tilde{A}_n$ on the set $X = X_1 \times X_2 \times \cdots \times X_n$ is defined as,

$$\begin{aligned} &\tilde{A}_1 \times \tilde{A}_2 \times \cdots \times \tilde{A}_n \\ &= \left\{ \left((x_1, x_2, \ldots, x_n), \mu_{\tilde{A}_1 \times \tilde{A}_2 \times \cdots \times \tilde{A}_n}(x_1, x_2, \ldots, x_n) \right) \middle| x_i \in X, i = 1, 2, \ldots, n \right\}, \end{aligned} \tag{1.4}$$

where $\mu_{\tilde{A}_1 \times \tilde{A}_2 \times \cdots \times \tilde{A}_n}(x_1, x_2, \ldots, x_n) = \min\left\{ \mu_{\tilde{A}_i}(x_i), i = 1, 2, \ldots, n \right\}$.

Definition 1.12 (*Extension principle*) Suppose $f : X \to Y$ be a function of $X = X_1 \times X_2 \times \cdots \times X_n$ to Y, and $\tilde{A}_1, \tilde{A}_2, \ldots, \tilde{A}_n$ be the fuzzy subsets from $X_1, X_2, \ldots, X_n$, respectively. The fuzzy set $\tilde{B}$ from Y is defined as $\tilde{B} = f(\tilde{A}_1, \ldots, \tilde{A}_n) = \left\{ (y, \mu_{\tilde{B}}(y)) \middle| y = f(x_1, \ldots, x_n), x_i \in X_i, i = 1, 2, \ldots, n \right\}$, where,

$$\mu_{\tilde{B}}(y) = \begin{cases} \sup\limits_{(x_1, \ldots, x_n) \in f^{-1}(y)} \min\left\{ \mu_{\tilde{A}_1}(x_1), \ldots, \mu_{\tilde{A}_n}(x_n) \right\}, & y \in \mathit{range}(f), \\ 0, & y \notin \mathit{range}(f). \end{cases} \tag{1.5}$$

Definition 1.13 The normal and convex fuzzy set $\tilde{A}$ is a fuzzy number if its membership function is piecewise continuous on X.

Definition 1.14 The fuzzy number $\tilde{A}$ is non-negative if for any $x < 0$ the membership function value of $\mu_{\tilde{A}}(x) = 0$ is obtained.

Definition 1.15 Assume L and R as two non-ascending functions where $L, R : \mathbb{R}^+ \to [0, 1]$, and $L(0) = R(0) = 1$. The fuzzy number $\tilde{A}$ with below membership function is called an *LR* fuzzy number.

$$\mu_{\tilde{A}}(x) = \begin{cases} L\left(\frac{a_2 - x}{a_2 - a_1}\right), & a_1 \le x \le a_2, \\ 1, & a_2 \le x \le a_3, \\ R\left(\frac{x - a_3}{a_4 - a_3}\right), & a_3 \le x \le a_4. \end{cases} \tag{1.6}$$

This fuzzy number is shown by $\tilde{A} = (a_1, a_2, a_3, a_4)_{LR}$ where $a_1 \le a_2 \le a_3 \le a_4$. In addition, the functions L and R are called the reference functions. This number is shown by Fig. 1.2.

Definition 1.16 The fuzzy number $\tilde{A} = (a_1, a_2, a_3, a_4)_{LR}$, is non-negative, if $a_1 \ge 0$, and is negative, if $a_4 < 0$.

Definition 1.17 The *LR* fuzzy number $\tilde{A} = (a_1, a_2, a_3, a_4)_{LR}$ with the reference functions of $L(x) = R(x) = \max\{0, 1 - x\}$, is called trapezoidal fuzzy number shown as $\tilde{A} = (a_1, a_2, a_3, a_4)$. Its membership function is obtained as below and its represented by Fig. 1.3.

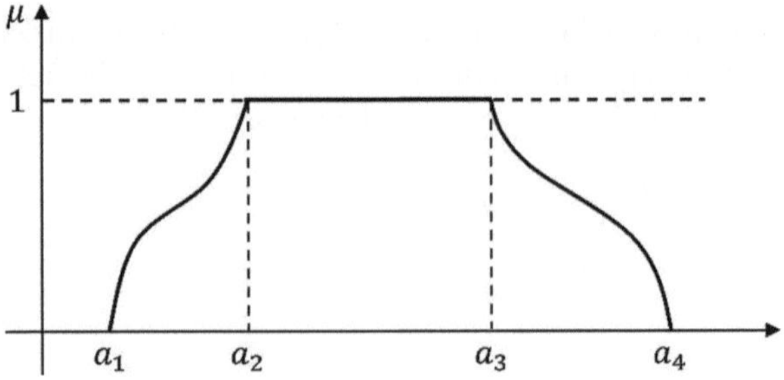

Fig. 1.2 The schematic of the *LR* fuzzy number $\tilde{A} = (a_1, a_2, a_3, a_4)_{LR}$

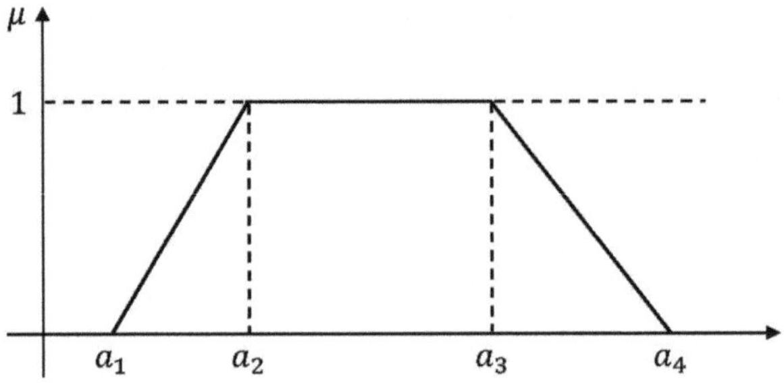

Fig. 1.3 The schematic of the trapezoidal fuzzy number $\tilde{A} = (a_1, a_2, a_3, a_4)$

$$\mu_{\tilde{A}}(x) = \begin{cases} \frac{x-a_1}{a_2-a_1}, & a_1 \le x \le a_2 \\ 1, & a_2 \le x \le a_3, \\ \frac{a_4-x}{a_4-a_3}, & a_3 \le x \le a_4, \end{cases} \tag{1.7}$$

Definition 1.18 The *LR* fuzzy number $\tilde{A} = (a_1, a_2, a_3)_{LR}$ with the reference functions of $L(x) = R(x) = 1 - x$ is called triangular fuzzy number shown as $\tilde{A} = (a_1, a_2, a_3)$. Its membership function is obtained as below and it is represented by Fig. 1.4.

$$\mu_{\tilde{A}}(x) = \begin{cases} \frac{x-a_1}{a_2-a_1}, & a_1 \le x \le a_2, \\ 1, & x = a_2, \\ \frac{a_3-x}{a_3-a_2}, & a_2 \le x \le a_3. \end{cases} \tag{1.8}$$

Remark 1.1 In the trapezoidal fuzzy number $\tilde{A} = (a_1, a_2, a_3, a_4)$ if $a_2 = a_3$, its is converted to a triangular fuzzy number.

Definition 1.19 As another representation, the fuzzy number $\tilde{A} = (m, n, \alpha, \beta)$ is a trapezoidal fuzzy number if α and β are considered as the left and right ranges of this fuzzy number. Its membership function is as below,

Fig. 1.4 The schematic of the triangular fuzzy number $\tilde{A} = (a_1, a_2, a_3)$

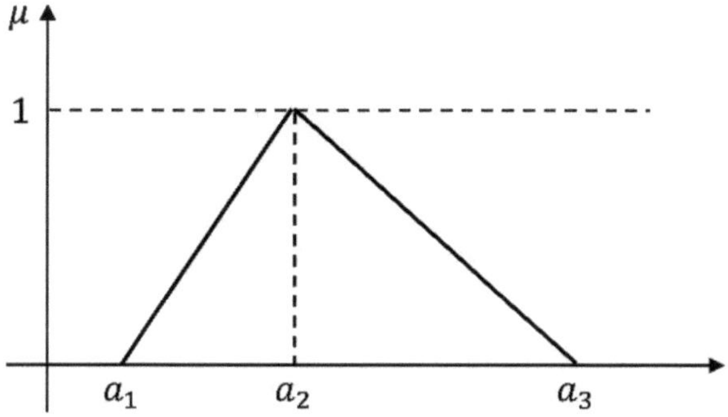

$$\mu_{\tilde{A}}(x) = \begin{cases} 1 - \frac{m-x}{\alpha}, & m - \alpha \le x \le m, \\ 1, & m \le x \le n, \\ 1 - \frac{x-n}{\beta}, & n \le x \le n + \beta, \\ 0, & O.W. \end{cases} \tag{1.9}$$

Definition 1.20 Considering the triangular *LR* fuzzy numbers $\tilde{A} = (a_1, a_2, a_3)_{LR}$, $\tilde{B} = (b_1, b_2, b_3)_{LR}$, and $\tilde{C} = (c_1, c_2, c_3)_{RL}$, the below operations can be defined.

- $\tilde{A} + \tilde{B} = (a_1 + b_1, a_2 + b_2, a_3 + b_3)_{LR}$,
- $\tilde{A} - \tilde{C} = (a_1 - c_3, a_2 - c_2, a_3 - c_1)_{LR}$,
- $\tilde{A} \otimes \tilde{B} \approx (a_1 b_1, a_2 b_2, a_3 b_3)_{LR}$, for the case of $a_1, b_1 > 0$,
- $\tilde{A} \odot \tilde{B} \approx \left(\frac{a_1}{b_3}, \frac{a_2}{b_2}, \frac{a_3}{b_1}\right)_{LR}$, for the case of $a_1, b_1 > 0$,
- $r\tilde{A} = \begin{cases} (ra_1, ra_2, ra_3)_{LR}, & r > 0, \\ (ra_3, ra_2, ra_1)_{RL}, & r < 0, \end{cases}$ for the case of $r \in \mathbb{R}$.

Definition 1.21 Considering the trapezoidal *LR* fuzzy numbers $\tilde{A} = (a_1, a_2, a_3, a_4)_{LR}$, $\tilde{B} = (b_1, b_2, b_3, b_4)_{LR}$, and $\tilde{C} = (c_1, c_2, c_3, c_4)_{RL}$, the below operations can be defined.

- $\tilde{A} + \tilde{B} = (a_1 + b_1, a_2 + b_2, a_3 + b_3, a_4 + b_4)_{LR}$,
- $\tilde{A} - \tilde{C} = (a_1 - b_4, a_2 - b_3, a_3 - b_2, a_4 - b_1)_{LR}$,
- $\tilde{A} \otimes \tilde{B} \approx (a_1 b_1, a_2 b_2, a_3 b_3, a_4 b_4)_{LR}$, for the case of $a_1, b_1 > 0$,
- $\tilde{A} \odot \tilde{B} \approx \left(\frac{a_1}{b_4}, \frac{a_2}{b_3}, \frac{a_3}{b_2}, \frac{a_4}{b_1}\right)_{LR}$, for the case of $a_1, b_1 > 0$,
- $r\tilde{A} = \begin{cases} (ra_1, ra_2, ra_3, ra_4)_{LR}, & r > 0, \\ (ra_4, ra_3, ra_2, ra_1)_{RL}, & r < 0, \end{cases}$ for the case of $r \in \mathbb{R}$.

Definition 1.22 An α-cut of the fuzzy numbers $\tilde{A} = (a_1, a_2, a_3)$, and $\tilde{B} = (a_1, a_2, a_3, a_4)$, are defined by the below relationships respectively.

$$A_\alpha = \{x | \mu_A(x) \ge \alpha\} = [A_1(\alpha), A_2(\alpha)] = [(a_2 - a_1)\alpha + a_1, a_3 - (a_3 - a_2)\alpha], \tag{1.10}$$

$$B_\alpha = \{x|\mu_A(x) \geq \alpha\} = [A_1(\alpha), A_2(\alpha)] = [(a_2 - a_1)\alpha + a_1, a_4 - (a_4 - a_3)\alpha]. \tag{1.11}$$

1.3 Ranking Function

Ranking of the fuzzy numbers is interesting for researchers because of its important role in approximation, decision making, estimation, etc. [3, 6]. In this method a function with real values is defined for each fuzzy number from a set of fuzzy numbers and the order of the fuzzy values is obtained according to their equivalent real values obtained by the mentioned function. Converting a fuzzy number to a real value may facilitate their comparison and further analysis by traditional mathematical methods. For example, in fuzzy decision making, as the alternatives with fuzzy number are evaluated in a vague environment, in fact, the comparison of the fuzzy numbers is the comparison of the alternatives. The ranking methods of fuzzy numbers are classified in three groups of defuzzification based methods, ranking factors methods, and pairwise comparison methods [11, 15]. The defuzzification based methods are easier than the others but as a disadvantage, in these methods the fuzzy information of a fuzzy number or event is lost.

Definition 1.23 Consider the ranking function of $\Re : F(R) \to R$, that converts a fuzzy number to a real number. Considering two fuzzy numbers $\tilde{A}$ and $\tilde{B}$ the below comparisons can be made.

- $\Re(\tilde{A}) > \Re(\tilde{B}) \Rightarrow \tilde{A} \succ \tilde{B}$,
- $\Re(\tilde{A}) < \Re(\tilde{B}) \Rightarrow \tilde{A} \prec \tilde{B}$,
- $\Re(\tilde{A}) = \Re(\tilde{B}) \Rightarrow \tilde{A} \approx \tilde{B}$.

There are various ranking functions for fuzzy numbers in the literature. For example, Yager [18] has proposed the ranking function of $\Re(\tilde{A}) = m+n+\frac{1}{2}(\beta - \alpha)$, for the trapezoidal fuzzy number $\tilde{A} = (m, n, \alpha, \beta)$. For the triangular fuzzy number $\tilde{A} = (a, b, c)$, where $a \leq b \leq c$, the ranking function of $\Re(\tilde{A}) = \frac{a+b+c}{3}$, is defined. In addition, the center of gravity ranking function of $\Re(\tilde{A}) = \frac{\int x\mu_{\tilde{A}}(x)dx}{\int \mu_{\tilde{A}}(x)dx}$, is defined for the continuous fuzzy number $\tilde{A}$ [19]. Its discrete form can be shown as $\Re(\tilde{A}) = \frac{\sum x\mu_{\tilde{A}}(x)}{\sum \mu_{\tilde{A}}(x)}$.

1.4 Possibility Theory

The possibility theory is a new method in probability theory inspired from mathematics to deal with some uncertainties. This approach for the first time was introduced by Zadeh [22] to measure a fuzzy event. Then, this approach was extended by Dubois and Prade [2]. In this theorem, not only the probable events are analyzed, all possible and contradictory events with their possibility degrees are considered. In continue, this theory and its related definitions and concepts are presented (see [4, 9]).

1.4.1 Possibility Measure

Definition 1.24 Considering the non-empty set X as a sample space and its power set $P(X)$ the function $Pos : \mathrm{P}(X) \rightarrow [0, 1]$ is the possibility measure of X if it respects to the conditions below.

- $Pos\{X\} = 1$,
- $Pos\{\emptyset\} = 0$, where $\emptyset$ is the empty set,
- $Pos\left\{\bigcup_i A_i\right\} = sup_i\, Pos\{A_i\}$, for any collection $\{A_i\}$ in $P(X)$.

Notably, the triplet $(X, \mathrm{P}(X), Pos)$ is called the possibility space.

Theorem 1.1 *For the possibility space* $(X, \mathrm{P}(X), Pos)$, *the below points are true.*

- *For any* $A \in P(X) : 0 \leq Pos\{A\} \leq 1$,
- *For* $A \subset B$, $Pos\{A\} \leq Pos\{B\}$,
- *For any* $A, B \in P(X) : Pos\{A \cup B\} \leq Pos\{A\} + Pos\{B\}$.

Definition 1.25 For the possibility space $(X, \mathrm{P}(X), Pos)$, the Kernel of the possibility set is defined by the set $X^+ = \{x \in X | Pos\{x\} > 0\}$.

Theorem 1.2 *Considering the possibility spaces* $(X_i, \mathrm{P}(X_i), Pos_i)$ (where $i = 1, 2, \ldots, n$), *the set* $X = X_1 \times X_2 \times \cdots \times X_n$, *and the measure* $Pos = Pos_1 \wedge Pos_2 \wedge \cdots \wedge Pos_n$ *(where the notation* $\wedge$ *is the minimum operator), then the function* Pos *is a possibility measure of* X *and* $(X, \mathrm{P}(X), Pos)$ *is a possibility space.*

Definition 1.26 Considering the possibility spaces $(X_i, \mathrm{P}(X_i), Pos_i)$ (where $i = 1, 2, \ldots, n$), the set $X = X_1 \times X_2 \times \cdots \times X_n$, and the measure $Pos = Pos_1 \wedge Pos_2 \wedge \cdots \wedge Pos_n$ (where the notation $\wedge$ is the minimum operator), then $(X, \mathrm{P}(X), Pos)$ is called the product possibility space of the spaces $(X_i, \mathrm{P}(X_i), Pos_i)$ (where $i = 1, 2, \ldots, n$).

Theorem 1.3 *Considering the possibility spaces* $(X_i, \mathrm{P}(X_i), Pos_i)$ *(where* $i = 1, 2, \ldots$) *the set* $X = X_1 \times X_2 \times \cdots$, *and the measure* $Pos\{A\} = sup_{(X_1, X_2, \ldots) \in X}\, Pos_1\{X_1\} \wedge Pos_2\{X_2\} \wedge \cdots$ *(where the notation* $\wedge$ *is the minimum operator), then the function* Pos *is a possibility measure of* X *and* $(X, \mathrm{P}(X), Pos)$ *is a possibility space.*

Definition 1.27 *Considering the possibility spaces* $(X_i, \mathrm{P}(X_i), Pos_i)$ *(where* $i = 1, 2, \ldots$), *the set* $X = X_1 \times X_2 \times \cdots$, *and the measure* $Pos = Pos_1 \wedge Pos_2 \wedge \cdots$ *(where the notation* $\wedge$ *is the minimum operator), then* $(X, \mathrm{P}(X), Pos)$ *is called the infinite product possibility space of the spaces* $(X_i, \mathrm{P}(X_i), Pos_i)$ *(where* $i = 1, 2, \ldots$).

1.4.2 Necessity Measure

The necessity measure of event A ($Nec\{A\}$) is equal to one minus the possibility of occurrence of its complement event. This measure shows the certainty and necessity amount of its occurrence.

Definition 1.28 Considering the possibility space $(X, \mathrm{P}(X), Pos)$ and the arbitrary set A from the power set $\mathrm{P}(X)$, then the necessity measure of A is defined as below, where A^c is the complement of A.

$$Nec\{A\} = 1 - Pos\{A^c\}, \tag{1.12}$$

Therefore, an event occurs necessarily only when there is no possibility of occurring its complement event.

Theorem 1.4 *Assuming the possibility space* $(X, \mathrm{P}(X), Pos)$, *then,*

- $Nec\{X\} = 1$,
- $Nec\{\emptyset\} = 0$,
- *If* $Pos\{A\} < 1$, *then* $Nec\{A\} = 0$,
- *If* $A \subset B$, *then* $Nec\{A\} \leq Nec\{B\}$,
- *For any* $A \in P(X) : Nec\{A\} + Pos\{A^c\} = 1$.

1.4.3 Credibility Measure

The possibility theory mentioned above, has not the self-duality property which is disadvantage for this theory. For example, consider the case that investors know the possibility measure of reaching a portfolio to the goal benefit (say $Pos\{A\}$), but they cannot measure (or do not know) the possibility for not reaching the portfolio to the goal benefit (say $Pos\{A^c\}$). This may be a concern for them as according to the possibility theory $Pos\{A\} + Pos\{A^c\}$ may not be equal to 1. Considering this disadvantage, Liu and Liu [10] introduced a theory with self-duality property called fuzzy credibility theory (where its measure is called the credibility measure shown by $Cr\{A\}$ for event A) to be used instead of the possibility theory. The fuzzy credibility theory has been used in many studies such as Mirzaei et al. [13] and Mahmoodirad et al. [12], in order to tackle the fuzziness of the parameters and variables.

Definition 1.29 Considering the possibility space $(X, \mathrm{P}(X), Pos)$ and the arbitrary set A from the power set $\mathrm{P}(X)$, then the credibility measure of A is defined as below,

$$Cr\{A\} = \frac{1}{2}(Nec\{A\} + Pos\{A\}). \tag{1.13}$$

Theorem 1.5 *Considering the possibility space* $(X, \mathrm{P}(X), Pos)$ *and the arbitrary set* A *from the power set* $\mathrm{P}(X)$, *then the below relation is held.*

$$Nec\{A\} \leq Cr\{A\} \leq Pos\{A\}. \tag{1.14}$$

Theorem 1.6 *Considering the possibility space* $(X, \mathrm{P}(X), Pos)$ *and the arbitrary sets A and B from the power set* $\mathrm{P}(X)$, *then the below relation is held.*

- $Cr\{X\} = 1$,
- $Cr\{\emptyset\} = 0$,
- *If* $A \subset B$, *then* $Cr\{A\} \leq Cr\{B\}$,
- $Cr\{A\} + Cr\{A^c\} = 1$,
- $Cr\{A \cup B\} \leq Cr\{A\} + Cr\{B\}$.

Definition 1.30 A fuzzy variable is defined as a function from the possibility space $(X, \mathrm{P}(X), Pos)$ to real numbers.

Definition 1.31 Assuming the fuzzy variable ξ from the possibility space $(X, \mathrm{P}(X), Pos)$, then the β-level set of ξ is defined as $\xi_\beta = \{\xi(x)|x \in X, Pos\{x\} \geq \beta\}$, and the set $\{\xi(x)|x \in X, Pos\{x\} > 0\} = \{\xi(x)|x \in X^+\}$, is called the support of ξ where X^+ is the kernel of the possibility space $(X, \mathrm{P}(X), Pos)$.

Definition 1.32 The fuzzy variable ξ,

- is non-negative if and only if $Pos\{\xi < 0\} = 0$,
- is positive if and only if $Pos\{\xi \leq 0\} = 0$,
- is continuous if and only if $Pos\{\xi = x\}$ be a continuous function of x.

Theorem 1.7 *If the fuzzy variable* ξ *is continuous, then the functions* $Pos\{\xi \geq x\}$ *and* $Pos\{\xi \leq x\}$ *are continuous for* x. *In addition, the function* $Pos\{x \leq \xi \leq y\}$ *is continuous for* $\{(x, y)|x < y\}$.

Definition 1.33 Assuming the fuzzy variable ξ from the possibility space $(X, \mathrm{P}(X), Pos)$, then its membership function is obtained as below,

$$\mu(t) = Pos\{x \in X|\xi(x) = t\}, t \in \mathbb{R}. \tag{1.15}$$

Definition 1.8 The fuzzy variable ξ with the membership function of μ,

- is non-negative if and only if $\mu(x) = 0$ for $x < 0$,
- is positive if and only if $\mu(x) = 0$ for $x \leq 0$,
- is continuous if and only if μ be a continuous function.

Definition 1.34 Assuming ξ_1 and ξ_2 as fuzzy variables from the possibility space $(X, \mathrm{P}(X), Pos)$, then $\xi_1 = \xi_2$ if and only if $\xi_1(x) = \xi_2(x)$ for all $x \in X$.

Definition 1.35 Assuming the function $f : \mathbb{R}^n \to \mathbb{R}$ and the fuzzy variable ξ_i from the possibility space $(X_i, P(X_i), Pos_i)$ (where $i = 1, 2, \ldots, n$), then the fuzzy variable $\xi = f(\xi_1, \xi_2, \ldots, \xi_n)$ is defined on the product possibility set $(X, \mathrm{P}(X), Pos)$ as below.

$$\xi(x_1, x_2, \ldots, x_n) = f(\xi_1(x_1), \xi_2(x_2), \ldots, \xi_n(x_n)), (x_1, x_2, \ldots, x_n) \in X. \quad (1.16)$$

Definition 1.36 Assuming the fuzzy variable ξ with the membership function of μ and the real number r, then $Pos\{\xi \le r\} = sup_{x \le r}\, \mu(x)$, and $Nec\{\xi \le r\} = 1 - sup_{x>r}\, \mu(x)$.

Example 1.1 According to Definition 1.36, for the triangular fuzzy variable $\xi = (a, b, c)$ and real number r,

$$Pos\{\xi \le r\} = \begin{cases} 0, & r \le a, \\ \frac{r-a}{b-a}, & a \le r \le b, \\ 1, & b \le r, \end{cases} \quad (1.17)$$

$$Pos\{\xi \ge r\} = \begin{cases} 1, & r \le b, \\ \frac{c-r}{c-b}, & c \le r \le b, \\ 0, & r \ge c, \end{cases} \quad (1.18)$$

$$Nec\{\xi \le r\} = \begin{cases} 0, & r \le b, \\ \frac{r-b}{c-b}, & b \le r \le c, \\ 1, & r \ge c, \end{cases} \quad (1.19)$$

$$Nec\{\xi \ge r\} = \begin{cases} 1, & r \le a, \\ \frac{b-r}{b-a}, & a \le r \le b, \\ 0, & r \ge c, \end{cases} \quad (1.20)$$

$$Cr\{\xi \le r\} = \begin{cases} 0, & r \le a, \\ \frac{r-a}{2(b-a)}, & a \le r \le b, \\ \frac{r+c-2b}{2(c-b)}, & b \le r \le c, \\ 1, & c \le r, \end{cases} \quad (1.21)$$

$$Cr\{\xi \ge r\} = \begin{cases} 1, & r \le a, \\ \frac{2b-a-r}{2(b-a)}, & a \le r \le b, \\ \frac{c-r}{2(c-b)}, & b \le r \le c, \\ 0, & r \ge c, \end{cases} \quad (1.22)$$

Example 1.2 According to Definition 1.36, for the trapezoidal fuzzy variable $\zeta = (a, b, c, d)$ and real number t,

$$Pos\{\zeta \ge t\} = \begin{cases} 1, & t \le c, \\ \frac{d-t}{d-c}, & c \le t \le d, \\ 0, & t \ge d, \end{cases}$$

$$Pos\{\zeta \le t\} = \begin{cases} 0, & t \le a, \\ \frac{t-a}{b-a}, & a \le t \le b, \\ 1, & t \ge b, \end{cases}$$

$$Nec\{\zeta \ge t\} = \begin{cases} 1, & t \le a \\ \frac{b-t}{b-a}, & a \le t \le b \\ 0, & b \le t \end{cases}$$

$$Nec\{\zeta \le t\} = \begin{cases} 0, & t \le c, \\ \frac{t-c}{d-c}, & c \le t \le d, \\ 1, & t \ge d, \end{cases}$$

$$Cr\{\zeta \le t\} = \begin{cases} 0, & t \le a, \\ \frac{t-a}{2(b-a)}, & a \le t \le b, \\ 0.5, & b \le t \le c, \\ \frac{t-2c+d}{2(d-c)}, & c \le t \le d, \\ 1, & t \ge d, \end{cases}$$

$$Cr\{\zeta \ge t\} = \begin{cases} 1, & t \le a \\ \frac{2b-a-t}{2(b-a)}, & a \le t \le b \\ 0.5, & b \le t \le c \\ \frac{d-t}{2(d-c)}, & c \le t \le d \\ 0, & t \ge d, \end{cases}$$

Theorem 1.9 *Considering the triangular fuzzy variable* $\xi = (a, b, c)$, *the trapezoidal fuzzy variable* $\zeta = (a, b, c, d)$, *and the confidence level* β *where* $0 < \beta \le 1$, *the below possibility measures are obtained.*

- $Pos\{\xi \le r\} \ge \beta \Leftrightarrow (1-\beta)a + \beta b \le r$,
- $Pos\{\xi \ge r\} \ge \beta \Leftrightarrow \beta b + (1-\beta)c \ge r$,
- $Pos\{\zeta \le r\} \ge \beta \Leftrightarrow (1-\beta)a + \beta b \le r$,
- $Pos\{\zeta \ge r\} \ge \beta \Leftrightarrow \beta c + (1-\beta)d \ge r$.

Corollary 1.1 *The possibility measure shows the most optimistic occurrence confidence level of a fuzzy event. Therefore, this measure is used for the cases that decision maker has an optimistic viewpoint.*

Theorem 1.10 *Considering the triangular fuzzy variable* $\xi = (a, b, c)$, *the trapezoidal fuzzy variable* $\zeta = (a, b, c, d)$, *and the confidence level* β *where* $0 < \beta \le 1$, *the below necessity measures are obtained.*

- $Nec\{\xi \le r\} \ge \beta \Leftrightarrow (1-\beta)b + \beta c \le r$,
- $Nec\{\xi \ge r\} \ge \beta \Leftrightarrow \beta a + (1-\beta)b \ge r$,
- $Nec\{\zeta \le r\} \ge \beta \Leftrightarrow (1-\beta)c + \beta d \le r$,
- $Nec\{\zeta \ge r\} \ge \beta \Leftrightarrow \beta a + (1-\beta)b \ge r$.

Corollary 1.2 *The necessity measure shows the most pessimistic occurrence confidence level of a fuzzy event. Therefore, this measure is used for the cases that decision maker has a pessimistic viewpoint.*

Theorem 1.11 *Considering the triangular fuzzy variable* $\xi = (a, b, c)$*, the trapezoidal fuzzy variable* $\zeta = (a, b, c, d)$*, and the confidence level* β *where* $0 < \beta \leq 1$*, the below credibility measures are obtained.*

- $Cr\{\xi \leq r\} \geq \beta \Leftrightarrow \begin{cases} (1-2\beta)a + 2\beta b \leq r, & \beta \leq 0.5, \\ (2-2\beta)b + (2\beta - 1)c \leq r, & \beta > 0.5, \end{cases}$
- $Cr\{\xi \geq r\} \geq \beta \Leftrightarrow \begin{cases} 2\beta b + (1-2\beta)c \geq r, & \beta \leq 0.5, \\ (2\beta - 1)a + (2-2\beta)b \geq r, & \beta > 0.5, \end{cases}$
- $Cr\{\zeta \leq r\} \geq \beta \Leftrightarrow \begin{cases} (1-2\beta)a + 2\beta b \leq r, & \beta \leq 0.5, \\ (2-2\beta)c + (2\beta - 1)d \leq r, & \beta > 0.5, \end{cases}$
- $Cr\{\zeta \geq r\} \geq \beta \Leftrightarrow \begin{cases} 2\beta c + (1-2\beta)d \geq r, & \beta \leq 0.5, \\ (2\beta - 1)a + (2-2\beta)b \geq r, & \beta > 0.5. \end{cases}$

For unnormal fuzzy variables, instead of the above-mentioned credibility measures, the extended credibility measure defined by the below definition is used.

Definition 1.37 The extended credibility measure $\left(\widetilde{Cr}\right)$ of a fuzzy event (say where $B \subset \mathbb{R}$) is defined as below,

$$\widetilde{Cr}\{\tilde{\xi} \in B\} = \frac{1}{2}\left(\sup_{x \in \mathbb{R}} \mu_{\tilde{\xi}}(x) + \sup_{x \in B} \mu_{\tilde{\xi}}(x) - \sup_{x \in B^c} \mu_{\tilde{\xi}}(x) \right), \tag{1.23}$$

1.4.4 General Fuzzy Measure

Considering the possibility space $(X, \mathrm{P}(X), Pos)$ and the arbitrary set A from the power set $\mathrm{P}(X)$, then the general fuzzy measure of A which is shown by $Me\{A\}$, is defined as below [14, 17],

$$Me\{A\} = \gamma Pos\{A\} + (1-\gamma)Nec\{A\}, \tag{1.24}$$

where, the parameter γ with the range of $0 \leq \gamma \leq 1$ can determine the pessimistic/optimistic degree of the measure. Therefore,

- if $\gamma = 0$, then, $Me\{A\} = Nec\{A\}$,
- if $\gamma = 0.5$, then, $Me\{A\} = Cr\{A\}$,
- and if $\gamma = 1$, then,$Me\{A\} = Pos\{A\}$.

Theorem 1.12 *Considering the possibility space* $(X, \mathrm{P}(X), Pos)$ *and the arbitrary set A from the power set* $P(X)$*, then the below relationships are obtained for the general fuzzy measure.*

- $Me\{X\} = 1$,
- $Me\{\emptyset\} = 0$,
- For any $A \in P(X) : Nec\{A\} \le Me\{A\} \le Pos\{A\}$,
- For any $A \in P(X) : 0 \le Me\{A\} \le 1$,
- If $A \subset B : Me\{A\} \le M\{B\}$,
- For any $A \in P(X)$, and $\gamma \le 0.5, 0 \le Me\{A\} + Me\{A^c\} \le 1$,
- For any $A \in P(X)$, and $\gamma \ge 0.5, 1 \le Me\{A\} + Me\{A^c\} \le 2$,
- For any $A, B \in P(X)$, and $\gamma \ge 0.5,\ Me\{A \cup B\} \le Me\{A\} + Me\{B\}$.

Example 1.3 For the triangular fuzzy variable $\xi = (a, b, c)$ and real number r,

$$Me\{\xi \le r\} = \begin{cases} 0, & r \le a \\ \gamma\left(\frac{r-a}{b-a}\right), & a \le r \le b \\ \gamma + (1-\gamma)\left(\frac{r-b}{c-b}\right), & b \le r \le c \\ 1, & c \le r \end{cases} \tag{1.25}$$

$$Me\{\xi \ge r\} = \begin{cases} 1, & r \le a \\ \gamma + (1-\gamma)\left(\frac{b-r}{b-a}\right), & a \le r \le b \\ \gamma\left(\frac{c-r}{c-b}\right), & b \le r \le c \\ 1, & c \le r \end{cases} \tag{1.26}$$

Example 1.4 For the trapezoidal fuzzy variable $\zeta = (a, b, c, d)$ and real number r,

$$Me\{\zeta \le r\} = \begin{cases} 0, & r \le a, \\ \gamma\left(\frac{r-a}{b-a}\right), & a \le r \le b, \\ \gamma, & b \le r \le c, \\ \gamma + (1-\gamma)\left(\frac{r-c}{d-c}\right), & c \le r \le d, \\ 1, & r \ge d, \end{cases} \tag{1.27}$$

$$Me\{\zeta \ge r\} = \begin{cases} 1, & r \le a, \\ \gamma + (1-\gamma)\left(\frac{b-r}{b-a}\right), & a \le r \le b, \\ \gamma, & b \le r \le c, \\ \gamma\left(\frac{d-r}{d-c}\right), & c \le r \le d, \\ 0, & d \le r, \end{cases} \tag{1.28}$$

Theorem 1.13 *Considering the triangular fuzzy variable $\xi = (a, b, c)$, the trapezoidal fuzzy variable $\zeta = (a, b, c, d)$, and the confidence level β where $0 < \beta \le 1$, the below general measures are obtained.*

$$Me\{\xi \leq r\} \geq \beta \Leftrightarrow \begin{cases} \left(\frac{\gamma-\beta}{\gamma}\right)a + \left(\frac{\beta}{\gamma}\right)b \leq r, & \beta \leq \gamma, \\ \left(\frac{1-\beta}{1-\gamma}\right)b + \left(\frac{\beta-\gamma}{1-\gamma}\right)c \leq r, & \beta > \gamma, \end{cases} \tag{1.29}$$

$$Me\{\xi \geq r\} \geq \beta \Leftrightarrow \begin{cases} \left(\frac{\beta}{\gamma}\right)b + \left(\frac{\gamma-\beta}{\gamma}\right)c \geq r, & \beta \leq \gamma, \\ \left(\frac{\beta-\gamma}{1-\gamma}\right)a + \left(\frac{1-\beta}{1-\gamma}\right)b \geq r, & \beta > \gamma, \end{cases} \tag{1.30}$$

$$Me\{\zeta \leq r\} \geq \beta \Leftrightarrow \begin{cases} \left(\frac{\gamma-\beta}{\gamma}\right)a + \left(\frac{\beta}{\gamma}\right)b \leq r, & \beta \leq \gamma, \\ \left(\frac{1-\beta}{1-\gamma}\right)c + \left(\frac{\beta-\gamma}{1-\gamma}\right)d \leq r, & \beta > \gamma, \end{cases} \tag{1.31}$$

$$Me\{\zeta \geq r\} \geq \beta \Leftrightarrow \begin{cases} \left(\frac{\beta}{\gamma}\right)c + \left(\frac{\gamma-\beta}{\gamma}\right)d \geq r, & \beta \leq \gamma, \\ \left(\frac{\beta-\gamma}{1-\gamma}\right)a + \left(\frac{1-\beta}{1-\gamma}\right)b \geq r, & \beta > \gamma, \end{cases} \tag{1.32}$$

1.4.5 The Optimistic and Pessimistic Values

Definition 1.38 Assuming ξ as a fuzzy variable and $\alpha \in [0, 1]$, the α-optimistic value of ξ which is shown by $\xi(\alpha)$, is defined as $\xi_{sup}(\alpha) = \sup\{r | Cr\{\xi \geq r\} \geq \alpha\}$.

This value is equal to the supremum value that the fuzzy variable ξ can obtain at the credibility value of α.

Definition 1.39 Assuming ξ as a fuzzy variable and $\alpha \in [0, 1]$, the α-pessimistic value of ξ which is shown by $\xi(\alpha)$, is defined as $\xi_{inf}(\alpha) = \inf\{r | Cr\{\xi \leq r\} \geq \alpha\}$.

This value is equal to the infimum value that the fuzzy variable ξ can obtain at the credibility value of α.

This is notable to mention that the α-optimistic and α-pessimistic values are used to rank fuzzy variables in ranking-based problems.

Example 1.5 For the triangular fuzzy variable $\xi = (a, b, c)$, the α-optimistic and α-pessimistic values are obtained as below.

$$\xi(\alpha)\begin{cases} 2\alpha b + (1 - 2\alpha)c, & \alpha \leq 0.5, \\ (2\alpha - 1)a + (2 - 2\alpha)b, & \alpha > 0.5, \end{cases}_{sup}$$

$$\xi(\alpha)\begin{cases} (1 - 2\alpha)a + 2\alpha b, & \alpha \leq 0.5, \\ (2 - 2\alpha)b + (2\alpha - 1)c, & \alpha > 0.5, \end{cases}_{inf}$$

Example 1.6 For the trapezoidal fuzzy variable $\zeta = (a, b, c, d)$, the α-optimistic and α-pessimistic values are obtained as below.

$$\zeta(\alpha)\begin{cases} 2\alpha c + (1-2\alpha)d, & \alpha \le 0.5, \\ (2\alpha - 1)a + (2-2\alpha)b, & \alpha > 0.5, \end{cases}_{sup}$$

$$\zeta(\alpha)\begin{cases} (1-2\alpha)a + 2\alpha b, & \alpha \le 0.5, \\ (2-2\alpha)c + (2\alpha - 1)d, & \alpha > 0.5, \end{cases}_{inf}$$

1.4.6 Expected Value Operator

In this section the expected value operator of fuzzy variables is defined, and some properties of this operator are discussed [10].

Definition 1.40 The expected value of the fuzzy variable ξ is defined as below if at least one of the integrals is finite.

$$E[\xi] = \int_0^{\infty} Cr\{\xi \ge r\}dr - \int_{-\infty}^{0} Cr\{\xi \le r\}dr \tag{1.33}$$

Example 1.7 For the triangular fuzzy variable $\xi = (a, b, c)$, the expected value of $E[\xi] = \frac{1}{4}(a + 2b + c)$, is obtained.

Example 1.8 For the trapezoidal fuzzy variable $\zeta = (a, b, c, d)$, the expected value of $E[\zeta] = \frac{1}{4}(a + b + c + d)$, is obtained.

Theorem 1.14 *Considering the fuzzy variable ξ (where its expected value exists) and the real values a* and *b, the relationship $E[a\xi + b] = aE[\xi] + b$ is held.*

Definition 1.41 *The variance of the fuzzy variable ξ is defined as $V[\xi] = E\left[(\xi - E)^2\right]$.*

Theorem 1.15 *Considering the fuzzy variable ξ (where its variance exists) and the real values α and β the relationship $V[\alpha\xi + \beta] = \alpha^2 V[\xi]$ is held.*

1.5 Fuzzy Decision

In a fuzzy optimization problem, the fuzzy objective function(s) and fuzzy constraints are considered by their membership functions. In such problem, a decision is defined as selecting a set of values for the variables to satisfy the objective function(s) and constraints simultaneously. Therefore, "decision" in such environment is the subscription of the fuzzy objective functions and the fuzzy constraints. According to Belman and Zadeh [1] a decision in a fuzzy environment is described by the below definition.

Definition 1.42 It is assumed that fuzzy objective functions $\tilde{G}_1, \ldots, \tilde{G}_n$ and fuzzy constraints $\tilde{C}_1, \ldots, \tilde{C}_m$ with their associated membership functions construct a fuzzy problem. In this problem, the decision $(\tilde{D})$ is the subscription of the objective functions and constraints and is defined as $\tilde{D} = \tilde{G}_1 \cap \cdots \cap \tilde{G}_n \cap \tilde{C}_1 \cap \cdots \cap \tilde{C}_m$ with the membership function of $\mu_{\tilde{D}} = \min\{\mu_{\tilde{G}_1}, \ldots, \mu_{\tilde{G}_n}, \mu_{\tilde{C}_1}, \mu_{\tilde{C}_m}\}$.

Example 1.9 [23] The fuzzy objective function "x should be significantly greater than 10" has the membership function of $\mu_{\tilde{G}}(x) = \begin{cases} 0 & x \le 10 \\ \left(1 + (x-10)^{-2}\right)^{-1} & x > 10 \end{cases}$. The constraint "$x$ should be close to 11" has the membership function of $\mu_{\tilde{C}}(x) = \begin{cases} 0 & x \le 11 \\ \left(1 + (x-11)^{4}\right)^{-1} & x > 11 \end{cases}$. The membership function of $\mu_{\tilde{D}} = \min\{\mu_{\tilde{G}}, \mu_{\tilde{C}}\}$ is obtained as below.

$$\mu_{\tilde{D}}(x) = \begin{cases} 0 & x \le 10 \\ \min\left\{\left(1 + (x-10)^{-2}\right)^{-1}, \left(1 + (x-11)^{4}\right)^{-1}\right\} & x > 10 \end{cases}$$
$$= \begin{cases} 0 & x \le 10, \\ \left(1 + (x-11)^{4}\right)^{-1} & x > 10. \end{cases}$$

1.6 Type-2 Fuzzy Sets

Type-2 fuzzy sets are an extension of type-1 fuzzy sets which were introduced by Zadeh [21] for the first time. In this type of fuzzy sets, the membership function values are of fuzzy type. Therefore, type-1 fuzzy sets have real type membership function value of the interval [0, 1] while type-2 fuzzy sets have type-1 fuzzy membership function value with the support value from the interval [0, 1]. Clearly, the applications of type-2 fuzzy sets are for the cases that the membership function values cannot be determined exactly as real number. In continue some concepts of type-2 fuzzy sets are presented [7].

Definition 1.43 In possibility space (θ, P, Pos), the regular fuzzy variable ξ is defined as a measurable function of θ to $[0, 1]$ where for any $t \in [0, 1]$, $\{\gamma \in \theta | \xi(\gamma) \le t\} \in P$.

A discrete fuzzy variable is shown as $\xi \sim \begin{pmatrix} r_1 & r_2 & \ldots & r_n \\ \mu_1 & \mu_2 & \ldots & \mu_n \end{pmatrix}$, where for any i, $r_i \in [0, 1]$, $\mu_i > 0$, and $\max\{\mu_i\} = 1$.

If $\xi = (r_1, r_2, r_3, r_4)$ and $0 \le r_1 < r_2 < r_3 \le r_4 \le 1$, then ξ is called a regular trapezoidal fuzzy variable and if $\xi = (r_1, r_2, r_3)$ and $0 \le r_1 < r_2 < r_3 \le 1$, then ξ is called a regular triangular fuzzy variable.

Example 1.10 The fuzzy variable ξ where its values 0.3, 0.5, 0.7, 0.9, and 1, have the possibility distribution values of 0.2, 0.4, 1, 0.6, and 0.8, respectively, is shown by $\xi \sim \begin{pmatrix} 0.3\;0.5\;0.7 & 0.91 \\ 0.2\;0.4\;\;1 & 0.60.8 \end{pmatrix}$.

Definition 1.44 A type-2 fuzzy set $\tilde{A}$ in X is a set in which the membership function of each member is a fuzzy set. This means that, the membership degree of each element is no longer an exact value, but a fuzzy set. This membership function is called type-2 membership function. A type-2 fuzzy set is defined as follow.

$$\tilde{A} = \{((x, u), \mu_{\tilde{A}}(x, u)) : \forall x \in X, \forall u \in J_x \subseteq [0, 1]\}, \tag{1.34}$$

Here, for $x \in X$, the term $\mu_{\tilde{A}}(x, u)$ (where $0 \leq \mu_{\tilde{A}}(x, u) \leq 1$) is the type-2 membership function, $J_x \subseteq [0, 1]$ is the primary membership of $x \in X$ which is the domain of the secondary membership function $\mu_{\tilde{A}}(x)$ so that all $u \in J_x$ are the primary membership grades of x.

Type-2 fuzzy set $\tilde{A}$ can be shown by the forms below,

$$\tilde{A} = \{(x, \tilde{\mu}_{\tilde{A}}(x)) | \forall x \in X\}, \tag{1.35}$$

$$\tilde{A} = \int\limits_{x \in X} \frac{\mu_{\tilde{A}}(x)}{x} = \int\limits_{x \in X} \frac{\left[\int_{u \in J_x} \frac{f_x(u)}{u}\right]}{x} \tag{1.36}$$

where, $0 \leq \mathrm{f_x(u)} \leq 1$, and for special case of $x = x'$ and $u = u' \in J_{x'}$, there is $f_{x'}(u') = \mu_{\tilde{A}}(x', u')$. Here, each of the signs $\int_x$ and $\int_u$, operates as a union operator.

If for any pair of (x, u), there is $f_x(u) = 1$, then this type-2 fuzzy set is interval-valued type-2 fuzzy set. Actually, and interval-valued type-2 fuzzy set is identified by the impact of uncertainty which is the union of all primary membership functions. In other words, we can say that $FOU(\tilde{A}) = \bigcup_{x \in X} J_x$. Here, FOU is the footprint of uncertainty.

Like type-1 fuzzy variables, a type-2 fuzzy variable also is defined as a function from the fuzzy possibility space to the set of real numbers. If (θ, P, Pos) be the fuzzy possibility space, then, type-2 fuzzy variable is defined as a function from θ to $\mathbb{R}$, where for any $t \in \mathbb{R}$ there is $\{\gamma \in \theta | \tilde{\xi}(\gamma) \leq t\} \in P$. Therefore, $\tilde{\mu}_{\tilde{\xi}}(x)$ which is called the secondary possibility distribution of type-2 fuzzy variable $\tilde{\xi}$, is a function with the characteristics of $\mathbb{R} \to \mathbb{R}[0, 1]$ and $\tilde{\mu}_{\tilde{\xi}}(x) = Pos\{\gamma \in \theta | \tilde{\xi}(\gamma) = x\}, x \in \mathbb{R}$. The term $\mu_{\tilde{\xi}}(x, u)$ which is called type-2 possibility distribution is the domain or support of $\tilde{\mu}_{\tilde{\xi}}(x)$. It is a function with the property of $\mathbb{R} \times J_x \to [0, 1]$ and is defined as $\mu_{\tilde{\xi}}(x, u) = Pos\{\tilde{\mu}_{\tilde{\xi}}(x) = u\}$, $(x, u) \in \mathbb{R} \times J_x, J_x \subseteq [0, 1]$. Therefore,$J_x = \{u \in [0, 1] | \mu_{\tilde{\xi}}(x) > 0\}$. Here, J_x is the primary possibility distribution of x, and for the case of $x = x'$, $\tilde{\mu}_{\tilde{\xi}}(x') = \mu_{\tilde{\xi}}(x', u)$ obtains the secondary possibility distribution for any $u \in J_x$.

Example 1.11 Assume the fuzzy variable $\tilde{A} = \{(x, \tilde{\mu}_{\tilde{A}}(x)) : x \in X\}$ and its reference set X $= \{4, 5, 6\}$. The primary memberships of $J_4 = \{0.3, 0.4, 0.6\}$, $J_5 = \{0.6, 0.8, 0.9\}$, and $J_6 = \{0.5, 0.6, 0.7, 0.8\}$, are considered. Now, the secondary membership values for point 4 are as below.

$$\tilde{\mu}_{\tilde{A}}(4) = \tilde{\mu}_{\tilde{A}}(4, u) = (0.6/0.3) + (1.0/0.4) + (0.7/0.6) \sim \begin{pmatrix} 0.3 & 0.4 & 0.6 \\ 0.6 & 1.0 & 0.7 \end{pmatrix},$$

This means that, $\mu_{\tilde{A}}(4, 0.3) = 0.6$, $\mu_{\tilde{A}}(4, 0.4) = 1$, and $\mu_{\tilde{A}}(4, 0.6) = 0.7$. As more details, the value of $\mu_{\tilde{A}}(4, 0.3) = 0.6$, means that for point 4, the secondary membership value of 0.6 is assigned for the primary membership value of 0.3. Therefore, the set $\tilde{A}$ selects the value of 4 with membership values of $\begin{pmatrix} 0.3 & 0.4 & 0.6 \\ 0.6 & 1.0 & 0.7 \end{pmatrix}$. For other values the below membership values are considered.

$$\begin{aligned} \tilde{\mu}_{\tilde{A}}(5) = \tilde{\mu}_{\tilde{A}}(5, u) &= (0.7/0.6) + (1.0/0.8) \\ &+ (0.8/0.9) \sim \begin{pmatrix} 0.6 & 0.8 & 0.9 \\ 0.7 & 1.0 & 0.8 \end{pmatrix}, \end{aligned}$$

$$\begin{aligned} \tilde{\mu}_{\tilde{A}}(6) = \tilde{\mu}_{\tilde{A}}(6, u) &= (0.3/0.5) + (0.4/0.6) \\ &+ (1.0/0.7) + (0.5/0.8) \\ &\sim \begin{pmatrix} 0.5 & 0.6 & 0.7 & 0.8 \\ 0.3 & 0.4 & 1.0 & 0.5 \end{pmatrix}. \end{aligned}$$

Therefore, the discrete type-2 fuzzy variable $\tilde{A}$ is represented in the below forms.

$$\begin{aligned} \tilde{A} = &(0.6/0.3)/4 + (1.0/0.4)/4 + (0.7/0.6)/4 \\ &+ (0.7/0.6)/5 + (1/0.8)/5 + (0.8/0.9)/5 \\ &+ (0.3/0.5)/6 + (0.4/0.6)/6 + (1.0/0.7)/6 \\ &+ (0.5/0.8)/6, \end{aligned}$$

Also, $\tilde{A}$ can be written as,

$$\tilde{A} = \begin{cases} 4, \ \mu_{\tilde{A}}(4), \\ 5, \ \mu_{\tilde{A}}(5), \\ 6, \ \mu_{\tilde{A}}(6). \end{cases}$$

Example 1.12 Assume the type-2 triangular fuzzy variable $\tilde{\xi} = (r_1, r_2, r_3; \theta_l, \theta_r)$ where r_1, r_2, r_3 are real numbers and $\theta_l, \theta_r \in [0, 1]$ are the parameters to show the degree of uncertainty to assign the value of x to $\tilde{\xi}$. The secondary possibility distribution of $\tilde{\mu}_{\tilde{A}}(x)$ from the variable $\tilde{\xi}$ is defined as,

$$\tilde{\mu}_{\xi}(x) = \left(\frac{x-r_1}{r_2-r_1} - \theta_l \min\left\{\frac{x-r_1}{r_2-r_1}, \frac{r_2-x}{r_2-r_1}\right\},\right.$$
$$\left.\frac{x-r_1}{r_2-r_1}, \frac{x-r_1}{r_2-r_1} + \theta_r \min\left\{\frac{x-r_1}{r_2-r_1}, \frac{r_2-x}{r_2-r_1}\right\}\right) \quad (1.37)$$

for any $x \in [r_1, r_2]$, and

$$\mu_{\xi}(x) = \left(\frac{r_3-x}{r_3-r_2} - \theta_l \min\left\{\frac{r_3-x}{r_3-r_2}, \frac{x-r_2}{r_3-r_2}\right\},\right.$$
$$\frac{r_3-x}{r_3-r_2}, \frac{r_3-x}{r_3-r_2} + \theta_r \min\left\{\frac{r_3-x}{r_3-r_2}, \frac{x-r_2}{r_3-r_2}\right\}. \quad (1.38)$$

for any $x \in r_2, r_3$.

In a type-1 triangular fuzzy variable such as (r_1, r_2, r_3), the membership value (possibility value) of any point is a constant value from the interval [0, 1]. But in a type-2 triangular fuzzy variable such as $(r_1, r_2, r_3; \theta_l, \theta_r)$ the primary membership values are not fixed but those are from the interval [0, 1]. Here, the primary distributions are extended by the parameters θ_l and θ_r. For the special case of $\theta_l, \theta_r = 0$, a type-2 triangular fuzzy variable is converted to a type-1 triangular fuzzy variable. Now, according to (1.37) and (1.38), the below membership distribution is obtained.

$$\tilde{\mu}_{\xi}(x) = \begin{cases} \left(\frac{x-r_1}{r_2-r_1} - \theta_l \frac{x-r_1}{r_2-r_1}, \frac{x-r_1}{r_2-r_1}, \frac{x-r_1}{r_2-r_1} + \theta_r \frac{x-r_1}{r_2-r_1}\right) & \text{if } x \in \left[r_1, \frac{r_1+r_2}{2}\right], \\ \left(\frac{x-r_1}{r_2-r_1} - \theta_l \frac{r_2-x}{r_2-r_1}, \frac{x-r_1}{r_2-r_1}, \frac{x-r_1}{r_2-r_1} + \theta_r \frac{r_2-x}{r_2-r_1}\right) & \text{if } x \in \frac{r_1+r_2}{2}, r_2, \\ \left(\frac{r_3-x}{r_3-r_2} - \theta_l \frac{x-r_2}{r_3-r_2}, \frac{r_3-x}{r_3-r_2}, \frac{r_3-x}{r_3-r_2} + \theta_r \frac{x-r_2}{r_3-r_2}\right) & \text{if } x \in r_2, \frac{r_2+r_3}{2}, \\ \left(\frac{r_3-x}{r_3-r_2} - \theta_l \frac{r_3-x}{r_3-r_2}, \frac{r_3-x}{r_3-r_2}, \frac{r_3-x}{r_3-r_2} + \theta_r \frac{r_3-x}{r_3-r_2}\right) & \text{if } x \in \frac{r_2+r_3}{2}, r_3, \end{cases} \quad (1.39)$$

As a numerical example, for the type-2 triangular fuzzy variable $\tilde{\xi} = (2, 3, 4; 0.5, 0.8)$, the below secondary membership distribution is obtained.

$$\tilde{\mu}_{\xi}(x) = \begin{array}{ll} (0.5x-1, x-2, 1.8x-3.6) & \text{if } x \in [2, 2.5], \\ (1.5x-3.5, x-2, 0.2x+0.4) & \text{if } x \in 2.5, 3 \\ (5.5-1.5x, 4-x, -0.2x-1.6) & \text{if } x \in 3, 3.5 \\ (2-0.5x, 4-x, 7.2-1.8x) & \text{if } x \in 3.5, 4 \end{array} \quad (1.40)$$

Here, the secondary possibility distribution for any x is a triangular fuzzy value. Therefore, $\tilde{\mu}_{\xi}(2.5) = (0.25, 0.5, 0.9)$, $\tilde{\mu}_{\xi}(3.2) = (0.7, 0.8, 0.96)$, etc. It means that the secondary possibility distribution of $\tilde{\mu}_{\xi}(2.5)$ is from 0.25 to 0.9. The footprint of uncertainty (FOU) for $\tilde{\xi} = (2, 3, 4; 0.5, 0.8)$ is shown by Fig. 1.5.

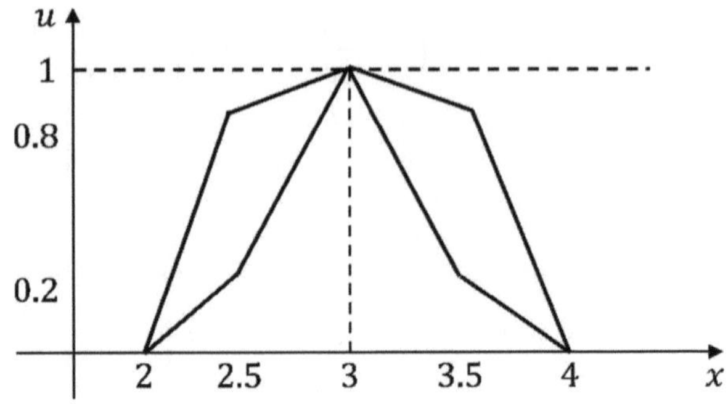

Fig. 1.5 The footprint of uncertainty for the type-2 triangular fuzzy variable $\tilde{\xi} = (2, 3, 4; 0.5, 0.8)$.

1.6.1 Critical Values for Regular Fuzzy Variables

There are here types of critical values of the regular fuzzy variable $\tilde{\xi}$ such as $CV^*[\tilde{\xi}]$ (optimistic), $CV_*[\tilde{\xi}]$ (pessimistic), and $CV[\tilde{\xi}]$ explained below [16].

$$CV^*[\tilde{\xi}] = \sup_{\alpha \in [0,1]} \{\min\{\alpha, Pos\{\tilde{\xi} \geq \alpha\}\}\}, \tag{1.41}$$

$$CV_*[\tilde{\xi}] = \sup_{\alpha \in [0,1]} \{\min\{\alpha, Nec\{\tilde{\xi} \geq \alpha\}\}\}, \tag{1.42}$$

$$CV[\tilde{\xi}] = \sup_{\alpha \in [0,1]} \{\min\{\alpha, Cr\{\tilde{\xi} \geq \alpha\}\}\}, \tag{1.43}$$

Theorem 1.16 [16] *For the regular trapezoidal fuzzy variable* $\tilde{\xi} = (r_1, r_2, r_3, r_4)$, *the optimistic critical value of* $CV^*[\tilde{\xi}] = \frac{r_4}{1+r_4-r_3}$, *the pessimistic critical value of* $CV_*[\tilde{\xi}] = \frac{r_2}{1+r_2-r_1}$, *and the below critical value is obtained.*

$$CV[\tilde{\xi}] = \begin{cases} \frac{2r_2-r_1}{1+2(r_2-r_1)} r_2 > 0.5 \\ 0.5 r_2 \leq 0.5 < r_3 \\ \frac{r_4}{1+2(r_4-r_3)} r_3 \leq 0.5 \end{cases} \tag{1.44}$$

Example 1.13 According to Theorem 1.16, for the fuzzy variable $\tilde{\xi} = (0.1, 0.3, 0.4, 0.6)$, the critical values of $CV^*[\tilde{\xi}] = \frac{r_4}{1+r_4-r_3} = \frac{1}{2}$, $CV_*[\tilde{\xi}] = \frac{r_2}{1+r_2-r_1} = \frac{1}{4}$, and $CV[\tilde{\xi}] = \frac{3}{7}$ are calculated.

Theorem 1.17 [16] *For the regular triangular fuzzy variable* $\tilde{\xi} = (r_1, r_2, r_3)$, *the optimistic critical value of* $CV^*[\tilde{\xi}] = \frac{r_3}{1+r_3-r_2}$, *the pessimistic critical value of* $CV_*[\tilde{\xi}] = \frac{r_2}{1+r_2-r_1}$, *and the below critical value is obtained.*

$$CV[\tilde{\xi}] = \begin{cases} \frac{2r_2-r_1}{1+2(r_2-r_1)} r_2 > 0.5 \\ \frac{r_3}{1+2(r_3-r_2)} r_3 \leq 0.5 \end{cases} \tag{1.45}$$

1.6.2 Reducing Type-2 Fuzzy Variables Using Critical Values

In order to decrease the degree of difficulty when dealing with type-2 fuzzy variables, those can be reduced to type-1 fuzzy variables (which can be normal or not). A method proposed by Qin et al. [16] is to apply the above-mentioned critical values. In this method, the critical values of a type-2 fuzzy variable are used to show its equivalent type-1 fuzzy variable. This method is described by the below example.

Example 1.14 Consider the type-2 fuzzy variable of Example 1.11 where $\tilde{\mu}_{\tilde{A}}(4)$, $\tilde{\mu}_{\tilde{A}}(5)$, and $\tilde{\mu}_{\tilde{A}}(6)$ are discrete regular fuzzy variables. For the case of $\tilde{\mu}_{\tilde{A}}(4)$, the below possibility measure is defined.

$$Pos\{\tilde{\mu}_{\tilde{A}}(4) \geq \alpha\} = \begin{cases} 1\alpha \leq 0.4 \\ 0.7\alpha < 0.4 \leq 0.6 \\ 00.6 < \alpha \leq 1 \end{cases} \tag{1.46}$$

Therefore,

$$\begin{aligned} CV^*[\tilde{\mu}_{\tilde{A}}(4)] &= \sup_{\alpha\in[0,0.4]} \min\{\alpha, 1\} \vee \sup_{\alpha\in 0.4,0.6} \min\{\alpha, 0.7\} \vee \sup_{\alpha\in 0.6,1} \min\{\alpha, 0\} \\ &= 0.4 \vee 0.6 \vee 0 = 0.6 \end{aligned}$$

Similarly, the below critical values are obtained.

$$\begin{aligned} &CV^*[\tilde{\mu}_{\tilde{A}}(4)] = 0.6, CV^*[\tilde{\mu}_{\tilde{A}}(5)] = 0.8, CV^*[\tilde{\mu}_{\tilde{A}}(6)] = 0.6, \\ &CV^*[\tilde{\mu}_{\tilde{A}}(4)] = 0.4, CV^*[\tilde{\mu}_{\tilde{A}}(5)] = 0.6, CV^*[\tilde{\mu}_{\tilde{A}}(6)] = 0.6, \\ &CV^*[\tilde{\mu}_{\tilde{A}}(4)] = 0.4, CV^*[\tilde{\mu}_{\tilde{A}}(5)] = 0.65, CV^*[\tilde{\mu}_{\tilde{A}}(6)] = 0.6, \end{aligned}$$

Then, according to the above critical values, the below equivalent type-1 fuzzy variables are obtained respectively.

$$\begin{pmatrix} 4 & 5 & 6 \\ 0.6 & 0.8 & 0.6 \end{pmatrix}, \begin{pmatrix} 4 & 5 & 6 \\ 0.4 & 0.6 & 0.6 \end{pmatrix}, \begin{pmatrix} 4 & 5 & 6 \\ 0.4 & 0.65 & 0.6 \end{pmatrix},$$

This method is summarized by the below theorem.

Theorem 1.18 [16] *Consider the triangular type-2 fuzzy variable* $\tilde{\xi} = (r_1, r_2, r_3; \theta_l, \theta_r)$. *Then, according to its optimistic critical value, pessimistic critical value, and critical value, the below possibility distributions are obtained respectively for its equivalent type-1 fuzzy variable* (ξ_1).

$$\mu_{\xi_1}(x) = \begin{cases} \frac{(1+\theta_r)(x-r_1)}{r_2-r_1+\theta_r(x-r_1)} x \in \left[r_1, \frac{r_1+r_2}{2}\right] \\ \frac{(1-\theta_r)x+\theta_r r_2-r_1}{r_2-r_1+\theta_r(r_2-x)} x \in \frac{r_1+r_2}{2}, r_2 \\ \frac{(-1+\theta_r)x-\theta_r r_2+r_3}{r_3-r_2+\theta_r(x-r_2)} x \in r_2, \frac{r_2+r_3}{2} \\ \frac{(1+\theta_r)(r_3-x)}{r_3-r_2+\theta_r(r_3-x)} x \in \frac{r_2+r_3}{2}, r_3 \end{cases} \tag{1.47}$$

$$\mu_{\xi_2}(x) = \begin{cases} \frac{x-r_1}{r_2-r_1+\theta_l(x-r_1)} x \in \left[r_1, \frac{r_1+r_2}{2}\right] \\ \frac{x-r_1}{r_2-r_1+\theta_l(r_2-x)} x \in \frac{r_1+r_2}{2}, r_2 \\ \frac{r_3-x}{r_3-r_2+\theta_l(x-r_2)} x \in r_2, \frac{r_2+r_3}{2} \\ \frac{r_3-x}{r_3-r_2+\theta_l(r_3-x)} x \in \frac{r_2+r_3}{2}, r_3 \end{cases} \tag{1.48}$$

$$\mu_{\xi_3}(x) = \begin{cases} \frac{(1+\theta_r)(x-r_1)}{r_2-r_1+2\theta_r(x-r_1)} x \in \left[r_1, \frac{r_1+r_2}{2}\right] \\ \frac{(1-\theta_l)x+\theta_r r_2-r_1}{r_2-r_1+2\theta_l(r_2-x)} x \in \frac{r_1+r_2}{2}, r_2 \\ \frac{(-1+\theta_l)x-\theta_l r_2+r_3}{r_3-r_2+2\theta_l(x-r_2)} x \in r_2, \frac{r_2+r_3}{2} \\ \frac{(1+\theta_r)(r_3-x)}{r_3-r_2+2\theta_r(r_3-x)} x \in \frac{r_2+r_3}{2}, r_3 \end{cases} \tag{1.49}$$

As the type-1 fuzzy variables obtained by Theorem 1.17 are not always normal, for such variables, instead on credibility measure, generalized credibility measure is used. The below theorem explains this issue.

Theorem 1.19 [16] *Consider type-2 fuzzy variable* $\tilde{\xi}_i = \left(r_1^i, r_2^i, r_3^i; \theta_{l,i}, \theta_{r,i}\right)$ *(where* $i = 1, \ldots, n$*) and its equivalent type-1 fuzzy variable* ξ_i *(where* $i = 1, \ldots, n$*) obtained by above-mentioned critical value method. Assume any pair of the fuzzy variables* $\xi_1, \ldots, \xi_n$ *are independent and further assume* $k_i \geq 0$ *(for* $i = 1, \ldots, n$*).*

- *Considering the generalized credibility level* $\alpha \in 0, 0.5$*, if* $\alpha \in 0, 0.25$ *then* $\tilde{C}r\left\{\sum_{i=1}^n k_i\xi_i \leq t\right\} \geq \alpha$ *is equivalent to* $\sum_{i=1}^n \frac{\left(1-2\alpha+(1-4\alpha)\theta_{r,i}\right)k_i r_1^i+2\alpha k_i r_2^i}{1+(1-4\alpha)\theta_{r,i}} \leq t$*. If* $\alpha \in 0.25, 0.5$ *then* $\tilde{C}r\left\{\sum_{i=1}^n k_i\xi_i \leq t\right\} \geq \alpha$ *is equivalent to* $\sum_{i=1}^n \frac{(1-2\alpha)k_i r_1^i+\left(2\alpha+(4\alpha-1)\theta_{l,i}\right)k_i r_2^i}{1+(4\alpha-1)\theta_{l,i}} \leq t$*.*
- *Considering the generalized credibility level* $\alpha \in 0.5, 1$*, if* $\alpha \in 0.5, 0.75$ *then* $\tilde{C}r\left\{\sum_{i=1}^n k_i\xi_i \leq t\right\} \geq \alpha$ *is equivalent to* $\sum_{i=1}^n \frac{(2\alpha-1)k_i r_3^i+\left(2(1-\alpha)+(3-4\alpha)\theta_{l,i}\right)k_i r_2^i}{1+(3-4\alpha)\theta_{l,i}} \leq t$*. If* $\alpha \in 0.75, 1$ *then* $\tilde{C}r\left\{\sum_{i=1}^n k_i\xi_i \leq t\right\} \geq \alpha$ *is equivalent to* $\sum_{i=1}^n \frac{\left(2\alpha-1+(4\alpha-3)\theta_{r,i}\right)k_i r_3^i+2(1-\alpha)k_i r_2^i}{1+(4\alpha-3)\theta_{r,i}} \leq t$*.*

Corollary 1.3 *According to the above theorem, the equivalent crisp form of* $\tilde{C}r\left\{\sum_{i=1}^n k_i\xi_i \geq t\right\} \geq \alpha$ *is obtained easily. For example, considering the generalized credibility level* $\alpha \in 0, 0.5$*, if* $\alpha \in 0, 0.25$ *then* $\tilde{C}r\left\{\sum_{i=1}^n k_i\xi_i \geq t\right\} \geq \alpha$ *is equivalent to* $\sum_{i=1}^n \frac{\left(1-2\alpha+(1-4\alpha)\theta_{l,i}\right)k_i r_3^i+2\alpha k_i r_2^i}{1+(1-4\alpha)\theta_{l,i}} \geq t$*. If* $\alpha \in 0.25, 0.5$ *then* $\tilde{C}r\left\{\sum_{i=1}^n k_i\xi_i \leq t\right\} \geq \alpha$ *is equivalent to* $\sum_{i=1}^n \frac{(1-2\alpha)k_i r_3^i+\left(2\alpha+(4\alpha-1)\theta_{r,i}\right)k_i r_2^i}{1+(4\alpha-1)\theta_{r,i}} \geq t$*.*

1.6.3 Defuzzification of Type-2 Fuzzy Variables

In order to defuzzify a type-2 fuzzy variable, first it is converted to a type-1 fuzzy variable, then applying and defuzzification method of type-1 fuzzy variables, its equivalent crisp form is obtained.

Example 1.15 At the end of Example 1.14 three equivalent type-1 fuzzy variables are obtained for the type-2 fuzzy variable of Example 1.11. Applying the gravity center method, we obtain crisp values 5, 5.125, and 5.121 respectively.

1.7 Trapezoidal Interval Type-2 Fuzzy Numbers

Interval type-2 fuzzy sets are of the most applicable type-2 fuzzy sets as they cover more uncertainty than other fuzzy sets. In this sub-section the concepts of interval type-2 fuzzy sets are described.

Assume $\tilde{A} = (a_1, a_2, a_3, a_4; H_1(\tilde{A}), H_2(\tilde{A}))$ as a type-1 trapezoidal fuzzy number shown by Fig. 1.6. In this number, $H_1(\tilde{A})$ shows the membership function value of a_2, and $H_2(\tilde{A})$ shows the membership function value of a_3, where $0 \leq H_1(\tilde{A}) \leq 1$ and $0 \leq H_2(\tilde{A}) \leq 1$. For the case of $a_2 = a_3$, it is converted to a triangular type-1 fuzzy number.

An interval type-2 fuzzy set contains two type-1 membership functions such as upper membership function ($\tilde{A}^U$) which is limited by the upper bound of $FOU(\tilde{A})$, and lower membership function ($\tilde{A}^L$) which is limited by the lower bound of $FOU(\tilde{A})$. For the case with trapezoidal type-1 upper and lower membership functions, the trapezoidal interval type-2 fuzzy number is obtained. Such a number is represented as $\tilde{A} = \left(\tilde{A}^U, \tilde{A}^l\right) = \left(\left(a_1^U, a_2^U, a_3^U, a_4^U; H_1\left(\tilde{A}^U\right), H_2\left(\tilde{A}^U\right)\right), \left(a_1^L, a_2^L, a_3^L, a_4^L; H_1\left(\tilde{A}^L\right), H_2\left(\tilde{A}^L\right)\right)\right)$ (see also Fig. 1.7). Here, $H_i\left(\tilde{A}^U\right)$ (for $i = 1, 2$) shows the membership value of the point a_{i+1}^U in the upper trapezoidal membership function ($\tilde{A}^U$), and $H_i\left(\tilde{A}^L\right)$ (for

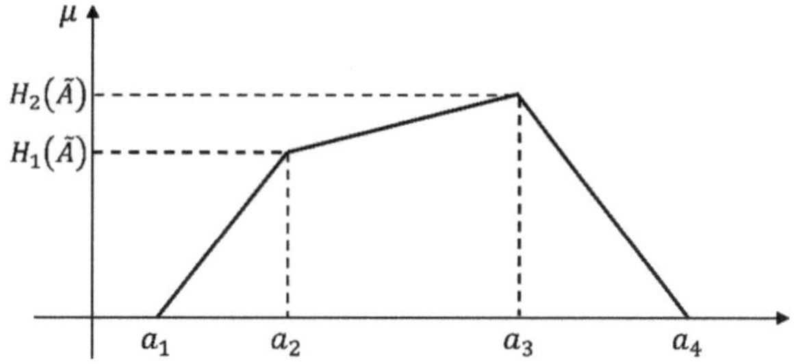

Fig. 1.6 Trapezoidal type-1 fuzzy number $\tilde{A} = (a_1, a_2, a_3, a_4; H_1(\tilde{A}), H_2(\tilde{A}))$

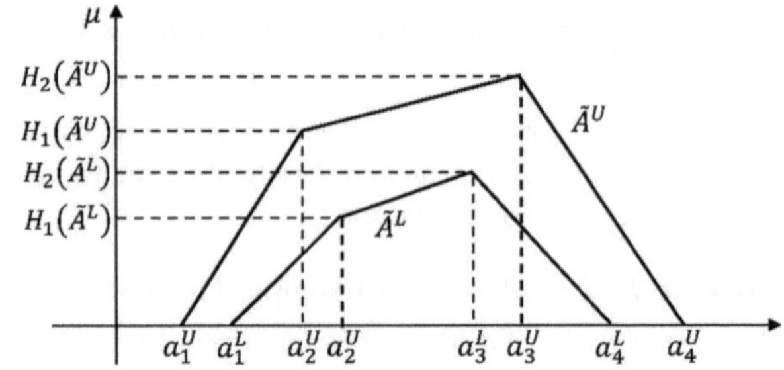

Fig. 1.7 Trapezoidal interval type-2 fuzzy number $\tilde{A} = (\tilde{A}^U, \tilde{A}^l)$

$i = 1, 2)$ shows the membership value of the point $\boldsymbol{a}_{i+1}^L$ in the lower trapezoidal membership function ($\tilde{\boldsymbol{A}}^L$), where, $\boldsymbol{H}_1(\tilde{\boldsymbol{A}}^U), \boldsymbol{H}_2(\tilde{\boldsymbol{A}}^U), \boldsymbol{H}_1(\tilde{\boldsymbol{A}}^L), \boldsymbol{H}_2(\tilde{\boldsymbol{A}}^L) \in [0, 1]$.

If $\tilde{A}^U = \tilde{A}^l$, the trapezoidal interval type-2 fuzzy number $\tilde{A} = (\tilde{A}^U, \tilde{A}^l)$ is converted to the trapezoidal type-1 fuzzy number $\tilde{\boldsymbol{A}} = (\boldsymbol{a}_1, \boldsymbol{a}_2, \boldsymbol{a}_3, \boldsymbol{a}_4; \boldsymbol{H}_1(\tilde{\mathbf{A}}), \boldsymbol{H}_2(\tilde{\mathbf{A}}))$.

As trapezoidal interval type-2 fuzzy numbers have piece-wise linear upper and lower membership functions, such numbers can be useful to represent equivalent quantitative values of linguistic terms used in evaluation procedures. As an instance, Zhang and Zhang [24] has proposed the conversion logics of Table 1.1.

Definition 1.45 [5] Assume the trapezoidal interval type-2 fuzzy numbers $\tilde{A} = (\tilde{A}^U, \tilde{A}^l) = ((a_1^U, a_2^U, a_3^U, a_4^U; H_1(\tilde{A}^U), H_2(\tilde{A}^U)), (a_1^L, a_2^L, a_3^L, a_4^L; H_1(\tilde{A}^L), H_2(\tilde{A}^L)))$ and $\tilde{B} = (\tilde{B}^U, \tilde{B}^l) = ((b_1^U, b_2^U, b_3^U, b_4^U; H_1(\tilde{B}^U), H_2(\tilde{B}^U)), (b_1^L, b_2^L, b_3^L, b_4^L; H_1(\tilde{B}^L), H_2(\tilde{B}^L)))$. The following operations are proposed for them.

- $\tilde{A} = (\tilde{A}^U, \tilde{A}^l) \leq \tilde{B} = (\tilde{B}^U, \tilde{B}^l) \Leftrightarrow \begin{cases} a_i^U \leq b_i^U, & i = 1, 2, 3, 4, \\ a_i^L \leq b_i^L, & i = 1, 2, 3, 4, \\ H_i(\tilde{A}^U) \leq H_i(\tilde{B}^U), & i = 1, 2, \\ H_i(\tilde{A}^L) \leq H_i(\tilde{B}^L), & i = 1, 2, \end{cases}$

Table 1.1 A set of linguistic variables and their equivalent trapezoidal interval type-2 fuzzy numbers

Linguistic terms	Trapezoidal interval type-2 fuzzy numbers
Very low (VL)	((0,0,0.1,0.2;1,1),(0,0,0.05,0.15;1,1))
Low (L)	((0.05,0.15,0.25,0.35;1,1),(0.1,0.2,0.2,0.3;0.8,0.8))
Medium low (ML)	((0.2,0.3,0.4,0.5;1,1),(0.25,0.35,0.35,0.45; 0.8,0.8))
Medium (M)	((0.35,0.45,0.55,0.65;1,1),(0.4,0.5,0.5,0.6;0.8,0.8))
Medium high (MH)	((0.5,0.6,0.7,0.8;1,1),(0.55,0.65,0.65,0.75; 0.8,0.8))
High (H)	((0.65,0.75,0.85,0.95;1,1),(0.7,0.8,0.8,0.9;0.8,0.8))
Very high (VH)	((0.8,0.9,1,1; 1,1),(0.85,0.95,1,1; 1,1))

- $$\tilde{A} \cup \tilde{B} = \begin{pmatrix} \begin{pmatrix} a_1^U \vee b_1^U, a_2^U \vee b_2^U, a_3^U \vee b_3^U, a_4^U \vee b_4^U; \\ H_1(\tilde{A}^U) \wedge H_1(\tilde{B}^U), H_2(\tilde{A}^U) \wedge H_2(\tilde{B}^U) \end{pmatrix}, \\ \begin{pmatrix} a_1^L \vee b_1^L, a_2^L \vee b_2^L, a_3^L \vee b_3^L, a_4^L \vee b_4^L; \\ H_1(\tilde{A}^L) \wedge H_1(\tilde{B}^L), H_2(\tilde{A}^L) \wedge H_2(\tilde{B}^L) \end{pmatrix} \end{pmatrix}$$
- $$\tilde{A} \cap \tilde{B} = \begin{pmatrix} \begin{pmatrix} a_1^U \wedge b_1^U, a_2^U \wedge b_2^U, a_3^U \wedge b_3^U, a_4^U \wedge b_4^U; \\ H_1(\tilde{A}^U) \wedge H_1(\tilde{B}^U), H_2(\tilde{A}^U) \wedge H_2(\tilde{B}^U) \end{pmatrix}, \\ \begin{pmatrix} a_1^L \wedge b_1^L, a_2^L \wedge b_2^L, a_3^L \wedge b_3^L, a_4^L \wedge b_4^L; \\ H_1(\tilde{A}^L) \wedge H_1(\tilde{B}^L), H_2(\tilde{A}^L) \wedge H_2(\tilde{B}^L) \end{pmatrix} \end{pmatrix}$$
- $$\tilde{A} \oplus \tilde{B} = \begin{pmatrix} \begin{pmatrix} a_1^U + b_1^U, a_2^U + b_2^U, a_3^U + b_3^U, a_4^U + b_4^U; \\ H_1(\tilde{A}^U) + H_1(\tilde{B}^U) - H_1(\tilde{A}^U) \times H_1(\tilde{B}^U), \\ H_2(\tilde{A}^U) + H_2(\tilde{B}^U) - H_2(\tilde{A}^U) \times H_2(\tilde{B}^U) \end{pmatrix}, \\ \begin{pmatrix} a_1^L + b_1^L, a_2^L + b_2^L, a_3^L + b_3^L, a_4^L + b_4^L; \\ H_1(\tilde{A}^L) + H_1(\tilde{B}^L) - H_1(\tilde{A}^L) \times H_1(\tilde{B}^L), \\ H_2(\tilde{A}^L) + H_2(\tilde{B}^L) - H_2(\tilde{A}^L) \times H_2(\tilde{B}^L) \end{pmatrix} \end{pmatrix}$$
- $$\tilde{A} - \tilde{B} = \begin{pmatrix} \begin{pmatrix} a_1^U - b_4^U, a_2^U - b_3^U, a_3^U - b_2^U, a_4^U - b_1^U; \\ H_1(\tilde{A}^U) + H_1(\tilde{B}^U) - H_1(\tilde{A}^U) \times H_1(\tilde{B}^U), \\ H_2(\tilde{A}^U) + H_2(\tilde{B}^U) - H_2(\tilde{A}^U) \times H_2(\tilde{B}^U) \end{pmatrix}, \\ \begin{pmatrix} a_1^L - b_4^L, a_2^L - b_3^L, a_3^L - b_2^L, a_4^L - b_1^L; \\ H_1(\tilde{A}^L) + H_1(\tilde{B}^L) - H_1(\tilde{A}^L) \times H_1(\tilde{B}^L), \\ H_2(\tilde{A}^L) + H_2(\tilde{B}^L) - H_2(\tilde{A}^L) \times H_2(\tilde{B}^L) \end{pmatrix}, \end{pmatrix}$$
- $$\tilde{A} \otimes \tilde{B} = \begin{pmatrix} (x_{11}^U, x_{12}^U, x_{13}^U, x_{14}^U; H_1(\tilde{A}^U) \times H_1(\tilde{B}^U), H_2(\tilde{A}^U) \times H_2(\tilde{B}^U)), \\ (x_{11}^L, x_{12}^L, x_{13}^L, x_{14}^L; H_1(\tilde{A}^L) \times H_1(\tilde{B}^L), H_2(\tilde{A}^L) \times H_2(\tilde{B}^L)) \end{pmatrix}$$

where, $x_{1i}^T = \min\left\{a_i^T b_i^T, a_i^T b_{(5-i)}^T, a_{(5-i)}^T b_i^T, a_{(5-i)}^T b_{(5-i)}^T\right\}, T \in \{U, L\}, i \in \{1, 2\}$ and $x_{1j}^T = \min\left\{a_{(5-j)}^T b_{(5-j)}^T, a_{(5-j)}^T b_j^T, a_j^T b_{(5-j)}^T, a_j^T b_j^T\right\}, T \in \{U, L\}, i \in \{3, 4\}$.

- For $\lambda > 0$, $$\lambda \times \tilde{A} = \begin{pmatrix} \begin{pmatrix} \lambda a_1^U, \lambda a_2^U, \lambda a_3^U, \lambda a_4^U; \\ 1 - (1 - H_1(\tilde{A}^U))^\lambda, 1 - (1 - H_2(\tilde{A}^U))^\lambda \end{pmatrix}, \\ \begin{pmatrix} \lambda a_1^L, \lambda a_2^L, \lambda a_3^L, \lambda a_4^L; \\ 1 - (1 - H_1(\tilde{A}^L))^\lambda, 1 - (1 - H_2(\tilde{A}^L))^\lambda \end{pmatrix} \end{pmatrix}$$
- For $\lambda > 0$, $$\tilde{A}^\lambda = \begin{pmatrix} \begin{pmatrix} (a_1^U)^\lambda, (a_2^U)^\lambda, (a_3^U)^\lambda, (a_4^U)^\lambda; \\ (H_1(\tilde{A}^U))^\lambda, (H_2(\tilde{A}^U))^\lambda \end{pmatrix}, \\ \begin{pmatrix} (a_1^L)^\lambda, (a_2^L)^\lambda, (a_3^L)^\lambda, (a_4^L)^\lambda; \\ (H_1(\tilde{A}^L))^\lambda, (H_2(\tilde{A}^L))^\lambda \end{pmatrix} \end{pmatrix}$$
- Distance of $\tilde{A}$ and $\tilde{B}$: $d(\tilde{A}, \tilde{B}) =$

$$\sqrt{\sum_{i=1}^{4}\left(a_i^U - b_i^U\right)^2 + \sum_{i=1}^{4}\left(a_i^L - b_i^L\right)^2 + \sum_{i=1}^{2}(H_i(\tilde{A}^U) - H_i(\tilde{B}^U))^2 + \sum_{i=1}^{2}(H_i(\tilde{A}^L) - H_i(\tilde{B}^L))^2}$$

Theorem 1.20 [24] *Assume the trapezoidal interval type-2 fuzzy numbers* $\tilde{A}$ and $\tilde{B}$, *then,* $\left(\tilde{A} \cup \tilde{B}\right)^c = \tilde{A}^c \cap \tilde{B}^c$ *and* $\left(\tilde{A} \cap \tilde{B}\right)^c = \tilde{A}^c \cup \tilde{B}^c$.

Theorem 1.21 [24] *Assume the trapezoidal interval type-2 fuzzy numbers* $\tilde{A}$, $\tilde{B}$, *and* $\tilde{C}$, *then,* $\tilde{A} \cup \left(\tilde{B} \cap \tilde{C}\right) = \left(\tilde{A} \cup \tilde{B}\right) \cap \left(\tilde{A} \cup \tilde{C}\right)$ *and* $\tilde{A} \cap \left(\tilde{B} \cup \tilde{C}\right) = \left(\tilde{A} \cap \tilde{B}\right) \cup \left(\tilde{A} \cap \tilde{C}\right)$.

Definition 1.46 [5] Assume the trapezoidal interval type-2 fuzzy number $\tilde{A} = (\tilde{A}^U, \tilde{A}^l) = ((a_1^U, a_2^U, a_3^U, a_4^U; H_1(\tilde{A}^U), H_2(\tilde{A}^U)), (a_1^L, a_2^L, a_3^L, a_4^L; H_1(\tilde{A}^L), H_2(\tilde{A}^L)))$. Its expected value is defined as $E(\tilde{A}) = \frac{1}{2}\left(\frac{1}{4}\sum_{i=1}^{4}\left(a_i^L + a_i^U\right)\right) \times \frac{1}{4}\left(\sum_{i=1}^{2}(H_i(\tilde{A}^L) + H_i(\tilde{A}^U))\right)$.

For the trapezoidal interval type-2 fuzzy numbers $\tilde{A}$ and $\tilde{B}$, then, $\tilde{A} > \tilde{B}$ if and only if $E(\tilde{A}) > E(\tilde{B})$.

1.8 Distance Between Fuzzy Sets

Definition 1.47 Distance in a set such as X (for example $x_1, x_2, x_3 \in X$) is a function such as d which assigns a non-negative real value to any pair of entities of X with the below properties.

- $d(x_1, x_1) = 0$
- $d(x_1, x_2) = 0$ if and only if $x_1 = x_2$
- $d(x_1, x_2) = d(x_2, x_1)$
- $d(x_1, x_3) \le d(x_1, x_2) + d(x_2, x_3)$

Definition 1.48 Assume the reference set $X = \{x_1, x_2, \ldots, x_n\}$, and $\tilde{A}$ and $\tilde{B}$ as two fuzzy sets with membership functions of $\mu_{\tilde{A}}(x)$ and $\mu_{\tilde{B}}(x)$ respectively. The following distances are defined for these sets.

- $d^1(\tilde{A}, \tilde{B}) = \sum_{i=1}^{n}\left|\mu_{\tilde{A}}(x_i) - \mu_{\tilde{B}}(x_i)\right|$
- $d^2(\tilde{A}, \tilde{B}) = \frac{1}{n}\sum_{i=1}^{n}\left|\mu_{\tilde{A}}(x_i) - \mu_{\tilde{B}}(x_i)\right|$
- $d^3(\tilde{A}, \tilde{B}) = \sqrt{\sum_{i=1}^{n}\left(\mu_{\tilde{A}}(x_i) - \mu_{\tilde{B}}(x_i)\right)^2}$
- $d^4(\tilde{A}, \tilde{B}) = \sqrt{\frac{1}{n}\sum_{i=1}^{n}\left(\mu_{\tilde{A}}(x_i) - \mu_{\tilde{B}}(x_i)\right)^2}$

1.9 Application: Type-2 Fuzzy Transportation Problem

A transportation problem with type-2 fuzzy parameters is considered in this section. The problem is converted to a crisp form using the credibility measure described in this chapter. The formulation of this problem is as below.

$$\min Z = \sum_{i=1}^{M}\sum_{j=1}^{N}\tilde{c}_{ij}x_{ij} \tag{1.50}$$

s.t.

$$\sum_{j=1}^{N} x_{ij} \leq \tilde{a}_i, \tag{1.51}$$

$$\sum_{i=1}^{M} x_{ij} \geq \tilde{b}_j, \tag{1.52}$$

$$x_{ij} \geq 0, \tag{1.53}$$

In this formulation, the sources are indexed by $i = 1, 2, \ldots, M$, the destinations are indexed by $j = 1, \ldots, N$, $\tilde{c}_{ij} = \left(c_{ij}^1, c_{ij}^2, c_{ij}^3; \theta_{l,ij}, \theta_{r,ij}\right)$ is the type-2 fuzzy cost for transporting a unit of product from source i to destination j, $\tilde{a}_i = \left(a_i^1, a_i^2, a_i^3; \theta_{l,i}, \theta_{r,i}\right)$ is the type-2 availability at source i, $\tilde{b}_j = \left(b_j^1, b_j^2, b_j^3; \theta_{l,j}, \theta_{r,j}\right)$ is the fuzzy type-2 demand of destination j, and x_{ij} is a non-negative continuous variable for the amount of product transported from source i to destination j. The objective function minimizes total transportation costs, and the constraints satisfy the availabilities and demands at sources and destinations.

In order to solve this problem, first the type-2 fuzzy parameters are reduced to type-1 fuzzy type by two methods of critical value and gravity center separately. Then, the type-1 fuzzy problem is solved.

Assume $\tilde{\boldsymbol{c}}_{ij}^1$, $\tilde{\boldsymbol{a}}_i^1$, and $\tilde{\boldsymbol{b}}_j^1$ as type-1 fuzzy values which are obtained from independent type-2 fuzzy values $\tilde{c}_{ij} = \left(c_{ij}^1, c_{ij}^2, c_{ij}^3; \theta_{l,ij}, \theta_{r,ij}\right)$, $\tilde{a}_i = \left(a_i^1, a_i^2, a_i^3; \theta_{l,i}, \theta_{r,i}\right)$, and $\tilde{b}_j = \left(b_j^1, b_j^2, b_j^3; \theta_{l,j}, \theta_{r,j}\right)$ respectively using the critical value method. Therefore, the following generalized credibility-based chance-constrained programming formulation is obtained.

$$\min \overline{Z} \tag{1.54}$$

Subject to

$$\tilde{C}r\left\{\sum_{i=1}^{M}\sum_{j=1}^{N}\tilde{c}_{ij}^{1}x_{ij} \le \overline{Z}\right\} \ge \gamma, \tag{1.55}$$

$$\tilde{C}r\left\{\sum_{j=1}^{N}x_{ij} - \tilde{a}_i^1 \le 0\right\} \ge \alpha_i, \quad i = 1, 2, \ldots, M, \tag{1.56}$$

$$\tilde{C}r\left\{\sum_{i=1}^{M}x_{ij} - \tilde{b}_j^1 \ge 0\right\} \ge \beta_j, \quad j = 1, 2, \ldots, N, \tag{1.57}$$

$$x_{ij} \ge 0, \quad i = 1, 2, \ldots, M, j = 1, 2, \ldots, N. \tag{1.58}$$

In this formulation, the main objective function is converted to a constraint. Also, the confidence level values $\gamma, \alpha_i, \beta_j \in 0, 1$ are determined by decision makers in advance.

According to Theorem 1.19 and Corollary 1.4, for the case of $\alpha \in 0, 0.25$, the below crisp formulation is obtained,

$$\min \sum_{i=1}^{M}\sum_{j=1}^{N}\left[\frac{\left(1 - 2\alpha + (1 - 4\alpha)\theta_{r,ij}\right)c_{ij}^{1}x_{ij} + 2\alpha c_{ij}^{2}x_{ij}}{1 + (1 - 4\alpha)\theta_{r,ij}}\right] \tag{1.59}$$

s.t.

$$\sum_{j=1}^{N}x_{ij} \le F_{a_i}, \quad i = 1, 2, \ldots, M, \tag{1.60}$$

$$\sum_{i=1}^{M}x_{ij} \ge F_{b_j}, \quad j = 1, 2, \ldots, N, \tag{1.61}$$

$$x_{ij} \ge 0 \quad i = 1, 2, \ldots, M, j = 1, 2, \ldots, N. \tag{1.62}$$

where,

$$F_{a_i} = \begin{cases} \frac{\left(1-2\alpha_i+(1-4\alpha_i)\theta_{l,i}\right)a_i^3+2\alpha_i a_i^2}{1+(1-4\alpha_i)\theta_{l,i}}, & \alpha_i \in 0, 0.25, \\ \frac{(1-2\alpha_i)a_i^3+\left(2\alpha_i+(4\alpha_i-1)\theta_{r,i}\right)a_i^2}{1+(4\alpha_i-1)\theta_{r,i}}, & \alpha_i \in 0.25, 0.5, \\ \frac{(2\alpha_i-1)a_i^1+\left(2(1-\alpha_i)+(3-4\alpha_i)\theta_{r,i}\right)a_i^2}{1+(3-4\alpha_i)\theta_{r,i}}, & \alpha_i \in 0.5, 0.75, \\ \frac{\left(2\alpha_i-1+(4\alpha_i-3)\theta_{l,i}\right)a_i^1+2(1-\alpha_i)a_i^2}{1+(4\alpha_i-3)\theta_{l,i}}, & \alpha_i \in 0.75, 1, \end{cases} \tag{1.63}$$

$$F_{b_j} = \begin{cases} \frac{(1-2\beta_j+(1-4\beta_j)\theta_{r,j})b_j^1+2\beta_j b_j^2}{1+(1-4\beta_j)\theta_{r,j}}, & \beta_j \in 0, 0.25 \\ \frac{(1-2\beta_j)b_j^1+(2\beta_j+(4\beta_j-1)\theta_{l,j})b_j^2}{1+(4\beta_j-1)\theta_{l,j}}, & \beta_j \in 0.25, 0.5 \\ \frac{(2\beta_j-1)b_j^3+(2(1-\beta_j)+(3-4\beta_j)\theta_{l,j})b_j^2}{1+(3-4\beta_j)\theta_{l,j}}, & \beta_j \in 0.5, 0.75 \\ \frac{(2\beta_j-1+(4\beta_j-3)\theta_{r,j})b_j^3+2(1-\beta_j)b_j^2}{1+(4\beta_j-3)\theta_{r,j}}, & \beta_j \in 0.75, 1 \end{cases} \quad (1.64)$$

According to Theorem 1.19 and Corollary 1.3, for the case of $\alpha \in 0.25, 0.5$, the below crisp formulation is obtained,

$$\min \sum_{i=1}^{M} \sum_{j=1}^{N} \left[\frac{(1-2\alpha)c_{ij}^1 x_{ij} + \left(2\alpha + (4\alpha - 1)\theta_{l,ij}\right)c_{ij}^2 x_{ij}}{1 + (4\alpha - 1)\theta_{l,ij}} \right] \quad (1.65)$$

s.t.

$$\sum_{j=1}^{N} x_{ij} \le F_{a_i}, \quad i = 1, 2, \ldots, M, \quad (1.66)$$

$$\sum_{i=1}^{M} x_{ij} \ge F_{b_j}, \quad j = 1, 2, \ldots, N, \quad (1.67)$$

$$x_{ij} \ge 0 \quad i = 1, 2, \ldots, M, j = 1, 2, \ldots, N. \quad (1.68)$$

where, F_{a_i} and F_{b_j} are the same as above.

According to Theorem 1.19 and Corollary 1.3, for the case of $\alpha \in 0.5, 0.75$, the below crisp formulation is obtained,

$$\min \sum_{i=1}^{M} \sum_{j=1}^{N} \left[\frac{\left(2(1-\alpha) + (3-4\alpha)\theta_{l,ij}\right)c_{ij}^2 x_{ij} + (2\alpha - 1)c_{ij}^3 x_{ij}}{1 + (3 - 4\alpha)\theta_{l,ij}} \right] \quad (1.69)$$

Subject to

$$\sum_{j=1}^{N} x_{ij} \le F_{a_i}, \quad i = 1, 2, \ldots, M, \quad (1.70)$$

$$\sum_{i=1}^{M} x_{ij} \ge F_{b_j}, \quad j = 1, 2, \ldots, N, \quad (1.71)$$

$$x_{ij} \ge 0 \quad i = 1, 2, \ldots, M, j = 1, 2, \ldots, N. \quad (1.72)$$

where, F_{a_i} and F_{b_j} are the same as above.

According to Theorem 1.19 and Corollary 1.3, for the case of $\alpha \in 0.75, 1$, the below crisp formulation is obtained,

$$\min \sum_{i=1}^{M} \sum_{j=1}^{N} \left[\frac{2(1-\alpha)c_{ij}^{2}x_{ij} + \left(2\alpha - 1 + (4\alpha - 3)\theta_{r,ij}\right)c_{ij}^{3}x_{ij}}{1 + (4\alpha - 3)\theta_{r,ij}} \right] \tag{1.73}$$

s.t.

$$\sum_{j=1}^{N} x_{ij} \le F_{a_i}, \quad i = 1, 2, \ldots, M, \tag{1.74}$$

$$\sum_{i=1}^{M} x_{ij} \ge F_{b_j}, \quad j = 1, 2, \ldots, N, \tag{1.75}$$

$$x_{ij} \ge 0 \quad i = 1, 2, \ldots, M, j = 1, 2, \ldots, N. \tag{1.76}$$

where, F_{a_i} and F_{b_j} are the same as above.

The above-mentioned crisp formulations are considered instead of the formulation (1.50)–(1.53) and can be easily solved by any optimization solver.

Example 1.16 A transportation problem with 3 sources and 2 destinations and the data of Table 1.1 is considered.

Considering the confidence level values $\gamma = 0.9$, $\alpha_i = 0.9$ (for $i = 1, 2, 3$), and $\beta_j = 0.9$ (for $\boldsymbol{j} = 1, 2$), the below crisp model is obtained,

Table 1.2 The type-2 fuzzy data of Example 1.16

		Destinations		$\tilde{a}_i$
		1	2	
Sources	1	(2,3,4;0.4,0.7)	(3.5,4,4.2;0.5,0.8)	(16,18,20;0.5,0.8)
	2	(4,5,6;0.7,0.8)	(4,6,7;0.6,0.4)	(15,16,18;0.4,0.6)
	3	(4.4,4.6,5;0.5,0.5)	(5,5.5,6;0.5,0.7)	(14,15,16;0.6,0.9)
$\tilde{b}_j$		(20,22,24;0.5,0.5)	(18,19,20;0.6,0.8)	

$$\min \sum_{i=1}^{3}\sum_{j=1}^{2}\left[\frac{(0.8+0.6\theta_{r,ij})c_{ij}^{3}x_{ij}+0.2c_{ij}^{2}x_{ij}}{1+0.6\theta_{r,ij}}\right]$$

s.t.

$$\sum_{j=1}^{2} x_{ij} \le F_{a_i}, \quad i = 1, 2, 3,$$

$$\sum_{i=1}^{3} x_{ij} \ge F_{b_j}, \quad j = 1, 2,$$

$$x_{ij} \ge 0 \quad i = 1, 2, 3, j = 1, 2.$$

where, the values of $F_{a_1} = 16.30$, $F_{a_2} = 15.16$, $F_{a_3} = 14.14$, $F_{b_1} = 23.69$, and $F_{b_2} = 19.86$ are calculated. Finally, by solving the crisp model, the optimal solution of $x_{11} = 0$, $x_{12} = 16.3$, $x_{21} = 9.55$, $x_{22} = 3.56$, $x_{31} = 14.14$, and $x_{32} = 0$, and the optimal objective function value of $z^* = 218.205$ is obtained.

References

1. Bellman, R., Zadeh, L.A.: Decision making in a fuzzy environment. Manage. Sci. **17**(4), 141–164 (1970)
2. Dubois, D., Prade, H.: Fuzzy Sets and Systems Theory and Applications. Academic Press, New York (1980)
3. Dombi, J., Jonas, T.: Preference implication-based approach to ranking fuzzy numbers. Iran. J. Fuzzy Syst. **18**, 51–68 (2021)
4. Hosseinzadeh Lotfi, F., Sanei, M., Hosseinzadeh, A., Niroomand, S., Mahmoodirad, A.: Uncertainty in data envelopment analysis. In: Fuzzy and Belief Degree-Based Uncertainties, 1st Edition, 9780323994453 (2023)
5. Hu, J., Zhang, Y., Chen, X., Liu, Y.: Multi-criteria decision making method based on possibility degree of interval type-2 fuzzy number. Knowl.-Based Syst. **43**, 21–29 (2013)
6. Jamshidi, M., Saneie, M., Mahmoodirad, A., Lotfi, F.H., Tohidi, G.: Uncertain RUSSEL data envelopment analysis model: a case study in Iranian banks. J. Intell. Fuzzy Syst. **37**(2), 2937–2951 (2019)
7. Kundu, P., Kar, S., Maiti, M.: Fixed charge transportation problem with type-2 fuzzy variables. Inf. Sci. **255**, 170–186 (2014)
8. Lai, Y.J., Hwang, C.L.: Fuzzy Mathematical Programming. Springer, Berlin (1992)
9. Liu, B.: Toward fuzzy optimization without mathematical ambiguity. Fuzzy Optim. Decis. Making **1**(1), 43–63 (2002)
10. Liu, B., Liu, Y.K.: Expected value of fuzzy variable and fuzzy expected value models. IEEE Trans. Fuzzy Syst. **10**(4), 445–450 (2002)
11. Mahmoodirad, A., Sanei, M.: Solving a multi-stage multi-product solid supply chain network design problem by meta-heuristics. Sci. Iranica **23**(3), 1428–1440 (2016)
12. Mahmoodirad, A., Niroomand, S., Shafiee, M.: A closed loop supply chain network design problem with multi-mode demand satisfaction in fuzzy environment. J. Intell. Fuzzy Syst. **39**(1), 503–524 (2020)
13. Mirzaei, N., Mahmoodirad, A., Niroomand, S.: An uncertain multi-objective assembly line balancing problem: a credibility-based fuzzy modeling approach. Int. J. Fuzzy Syst. **21**(8), 2392–2404 (2019)

14. Mahmoodirad, A., Pamucar, D., Niroomand, S.: A new intuitionistic fuzzy scheme of data envelopment analysis for evaluating rural comprehensive health service centers. Socioecon. Plann. Sci. **95**, 102004 (2024)
15. Molla-Alizadeh-Zavardehi, S., Mahmoodirad, A., Rahimian, M.: Step fixed charge transportation problems via genetic algorithm. Indian J. Sci. Technol. **7**(7), 949 (2014)
16. Qin, R., Liu, Y.K., Liu, Z.Q.: Methods of critical value reduction for type-2 fuzzy variables and their applications. J. Comput. Appl. Math. **235**, 1454–1481 (2011)
17. Xu, J., Zhou, X.: Approximation based fuzzy multi-objective models with expected objectives and chance constraints: application to earth-rock work allocation. Inf. Sci. **238**, 75–95 (2013)
18. Yager, R.R.: Ranking fuzzy subsets over the unit interval. In: Decision and Control including the 17th Symposium on Adaptive Processes, 1978 IEEE Conference, pp. 1435–1437 (1978)
19. Yager, R.R.: A procedure for ordering fuzzy subsets of the unit interval. Inf. Sci. **24**, 143–161 (1981)
20. Zadeh, L.A.: Fuzzy sets. Inform. Control **8**(3), 338–353 (1965)
21. Zadeh, L.A.: The concept of a linguistic variable and its application to approximate reasoning—I. Inf. Sci. **8**(3), 199–249 (1975)
22. Zadeh, L.A.: Fuzzy sets as a basis for a theory of possibility. Fuzzy Sets Syst. **1**, 3–28 (1978)
23. Zimmermann, H.J.: Fuzzy Sets, Decision Making and Expert Systems. Kluwer Academic Publishers, Boston (1987)
24. Zhang, Z., Zhang, S.: A novel approach to multi attribute group decision making based on trapezoidal interval type-2 fuzzy soft sets. Appl. Math. Model. **37**, 4948–4971 (2013)

Chapter 2
Intuitionistic Fuzzy Set

2.1 Introduction

Despite the extensive application of fuzzy set theory in modeling real life cases and problems, it has some weaknesses in representing all aspects of the uncertain events of real-world problems. The classical fuzzy logics have two main shortages such as (1) for each property of a system a certain membership function should be defined, and (2) for the choices that there is not enough information available or the choices with equal agree and disagree opinions the classical fuzzy set theory cannot be useful. Therefore, the classical fuzzy set theory is not useful for cases with missing or unrelated data and cases with complex membership functions.

For example, in image processing for classification of an image, we can consider two cases such as (1) an image belongs to a class with membership function of 0.5 and not belong to that class with membership function of 0.5, and (2) the image belongs to the given class with membership function of 0.5 but there is not enough information to determine the membership function value for not belonging to the class (or the information is inexact, not clear, and vague). For the first case, the classical fuzzy sets and numbers can be used while the classical fuzzy sets and numbers cannot be used for the second case. To model such inexact and vague cases, the intuitionistic fuzzy logic can be used. For this aim, intuitionistic fuzzy set theory was introduced by Atanassov [1] for the first time.

The intuitionistic fuzzy set $\tilde{A}^I$, is a set that each of its members takes membership function and non-membership function values from the interval [0, 1]. The membership function and non-membership function of an intuitionistic fuzzy set are shown by $\mu_{\tilde{A}^I}(x)$ and $\nu_{\tilde{A}^I}(x)$ respectively. Considering the intuitionistic fuzzy set $\tilde{A}^I$ and its membership and non-membership functions $\mu_{\tilde{A}^I}(x)$ and $\nu_{\tilde{A}^I}(x)$, for $x \in X$, if $\mu_{\tilde{A}^I}(x)$ takes a value close to 1, means that x is highly belonging to the set $\tilde{A}^I$ and if $\mu_{\tilde{A}^I}(x)$ takes a value close to 0, means that x is slightly belonging to the set $\tilde{A}^I$. On the other hand, if $\nu_{\tilde{A}^I}(x)$ takes a value close to 1, means that x is highly not belonging to the set $\tilde{A}^I$ and if $\nu_{\tilde{A}^I}(x)$ take a value close to 0, means that x is slightly belonging to the

A. Mahmoodirad et al., *Recent Advances on Fuzzy Sets*, Studies in Fuzziness and Soft Computing 440, https://doi.org/10.1007/978-3-032-08776-8_2

set $\tilde{A}^I$. Therefore, the value of $\mu_{\tilde{A}^I}(x)$ is the degree of accepting x as a member of $\tilde{A}^I$ and the value of $\nu_{\tilde{A}^I}(x)$ is the degree of not accepting x as a member of $\tilde{A}^I$.

In intuitionistic fuzzy theory, the uncertainty is described by membership and non-membership function values that may not be complement of each other. This is the difference between intuitionistic fuzzy sets and classical fuzzy sets where the sum of membership and non-membership function values in classical fuzzy sets is equal to 1, but this sum in intuitionistic fuzzy sets may not be equal to 1. Therefore, classical fuzzy sets are a special case of intuitionistic fuzzy sets.

In the rest of this chapter, intuitionistic fuzzy sets are defined, their ranking functions are introduced, intuitionistic fuzzy optimization is described, the possibility theory of intuitionistic fuzzy numbers is introduced, interval-valued and type-2 intuitionistic fuzzy numbers are described, and application of intuitionistic fuzzy sets and numbers in data envelopment analysis is studied.

2.2 Intuitionistic Fuzzy Sets

Definition 2.1 Intuitionistic fuzzy set $\tilde{A}^I$ for the reference set X is defined by the triplet set $\tilde{A}^I = \{(x, \mu_{\tilde{A}^I}(x), \nu_{\tilde{A}^I}(x)) : x \in X\}$, where, $\mu_{\tilde{A}^I}(x)$ and $\nu_{\tilde{A}^I}(x)$ are respectively the membership and non-membership function values of $x \in X$ in $\tilde{A}^I$, with the condition $0 \leq \mu_{\tilde{A}^I}(x) + \nu_{\tilde{A}^I}(x) \leq 1$.

Hesitation degree of $\tilde{A}^I$ in X is obtained as $h(x) = 1 - \mu_{\tilde{A}^I}(x) - \nu_{\tilde{A}^I}(x)$ where $0 \leq h(x) \leq 1$.

Definition 2.2 Intuitionistic fuzzy set $\tilde{A}^I = \{(x, \mu_{\tilde{A}^I}(x), \nu_{\tilde{A}^I}(x)) : x \in X\}$ is called intuitionistic fuzzy normal if there exist two points such as $x_0, x_1 \in X$ where $\mu_A(x_0) = 1$ and $\nu_A(x_1) = 1$. Therefore, an intuitionistic fuzzy set is normal if at least one point is certainly belonging to the set and at least one point is certainly not belonging to the set.

Definition 2.3 For $\alpha, \beta \in [0, 1]$ and $\alpha + \beta \leq 1$, (α, β)-cut of the intuitionistic fuzzy set $\tilde{A}^I$ is defined as below,

$$\begin{aligned} A_{\alpha,\beta}(x) &= \{(x, \mu_{\tilde{A}^I}(x), \nu_{\tilde{A}^I}(x)), \mu_{\tilde{A}^I}(x) \geq \alpha, \nu_{\tilde{A}^I}(x) \leq \beta, x \in \mathbb{R}\} \\ &= \{[A_1(\alpha), A_2(\alpha)], [B_1(\beta), B_2(\beta)]\} \end{aligned} \tag{2.1}$$

where,

- $A_1(\alpha)$ and $B_2(\beta)$ are continuous, monotonic increasing functions of α, respective, β,
- $A_2(\alpha)$ and $B_1(\beta)$ are continuous, monotonic increasing functions of α, respective, β,
- $A_1(1) = A_2(1)$, and $B_1(0) = B_2(0)$.

Definition 2.4 The intuitionistic fuzzy set $\tilde{A}^I = \{(x, \mu_{\tilde{A}^I}(x), \nu_{\tilde{A}^I}(x)) : x \in X\}$ is intuitionistic fuzzy convex, if for any $x_1, x_2 \in X$ and $\lambda \in [0, 1]$,

$$\mu_A(\lambda x_1 + (1-\lambda)x_2) \geq \min(\mu_A(x_1), \mu_A(x_2)) \tag{2.2}$$

$$\nu_A(\lambda x_1 + (1-\lambda)x_2) \leq \min(\nu_A(x_1), \nu_A(x_2)) \tag{2.3}$$

2.3 Intuitionistic Fuzzy Numbers

Definition 2.5 The intuitionistic fuzzy set $\tilde{A}^I = \{(x, \mu_{\tilde{A}^I}(x), \nu_{\tilde{A}^I}(x)) : x \in X\}$ is called an intuitionistic fuzzy number if:

- There is an $r \in \mathbb{R}$ such that $\mu_{\tilde{A}^I}(r) = 1$ and $\nu_{\tilde{A}^I}(r) = 0$, which means $\tilde{A}^I$ be normal.
- The functions $\mu_{\tilde{A}^I}(x)$ and $\nu_{\tilde{A}^I}(x)$ be piecewise continuous on $\mathbb{R}$ and the condition $0 \leq \mu_{\tilde{A}^I}(x) + \nu_{\tilde{A}^I}(x) \leq 1$ for $\forall x \in X$ is held.

The membership and non-membership functions of the intuitionistic fuzzy number $\tilde{A}^I$ are as below.

$$\mu_{\tilde{A}^I}(x) = \begin{cases} g_1(x), & r-\alpha \leq x \leq r, \\ 1, & x = r, \\ h_1(x) & r \leq x \leq r+\beta, \\ 0, & 0.w. \end{cases} \tag{2.4}$$

$$\nu_{\tilde{A}^I}(x) = \begin{cases} g_2(x), & r-\alpha' \leq x \leq r, 0 \leq g_1(x)+g_2(x) \leq 1 \\ 0, & x = r, \\ h_2(x), & r \leq x \leq r+\beta', 0 \leq h_1(x)+h_2(x) \leq 1 \\ 1, & 0.w. \end{cases} \tag{2.5}$$

In these functions, $g_1(x)$ and $h_1(x)$ are respectively strictly increasing and strictly decreasing on $[r-\alpha, r]$ and $[r, r+\beta]$. The functions $g_2(x)$ and $h_2(x)$ are respectively strictly decreasing and strictly increasing on $[r-\alpha', r]$ and $[r, r+\beta']$. The notation r is the mean value of $\tilde{A}^I$. The values α and β are the left and right extensions of the membership function $\mu_{\tilde{A}^I}(x)$, and the values α' and β' are the left and right extensions of the non-membership function $\nu_{\tilde{A}^I}(x)$.

Definition 2.6 Based on the previous definition, trapezoidal intuitionistic fuzzy number (TrIFN) $\tilde{A}^I$ is defined by the below membership and non-membership functions [17].

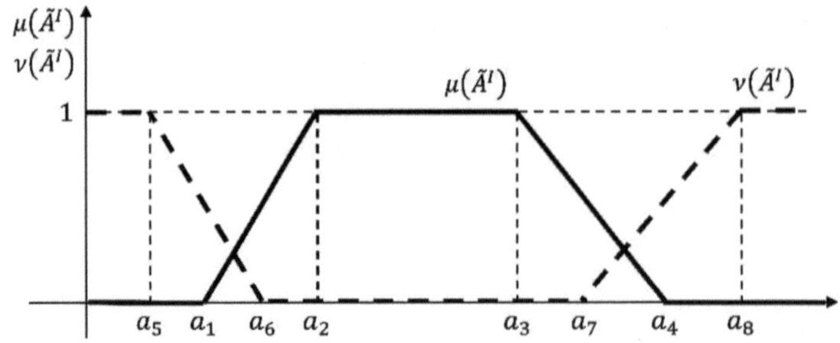

Fig. 2.1 The schematic representation of a trapezoidal intuitionistic fuzzy number

$$\mu_{\tilde{A}^I}(x) = \begin{cases} \frac{x-a_1}{a_2-a_1}, & a_1 \le x \le a_2 \\ 1, & a_2 \le x \le a_3 \\ \frac{x-a_4}{a_3-a_4}, & a_3 \le x \le a_4 \\ 0, & o.w \end{cases} \tag{2.6}$$

$$\nu_{\tilde{A}^I}(x) = \begin{cases} \frac{x-a_6}{a_5-a_6}, & a_5 \le x \le a_6 \\ 0, & a_6 \le x \le a_7 \\ \frac{x-a_7}{a_8-a_7}, & a_7 \le x \le a_8 \\ 1, & 0.w \end{cases} \tag{2.7}$$

This number is shown as $\tilde{A}^I = (a_1, a_2, a_3, a_4; a_5, a_6, a_7, a_8)$ where $a_5 \le a_1 \le a_6 \le a_2 \le a_3 \le a_7 \le a_4 \le a_8$. The schematic representation of this TrIFN is depicted by Fig. 2.1.

Remark 2.1 The trapezoidal intuitionistic fuzzy number $\tilde{A}^I = (a_1, a_2, a_3, a_4; a_5, a_6, a_7, a_8)$ is reduced to a triangular intuitionistic fuzzy number (TIFN) if $a_2 = a_3$ and $a_6 = a_7$.

Definition 2.7 The trapezoidal intuitionistic fuzzy number $\tilde{A}^I = (a_1, a_2, a_3, a_4; a_5, a_6, a_7, a_8)$ is non-negative if and only if $a_5 \ge 0$.

Definition 2.8 The trapezoidal intuitionistic fuzzy numbers $\tilde{A}^I = (a_1, a_2, a_3, a_4; a_5, a_6, a_7, a_8)$ and $\tilde{B}^I = (b_1, b_2, b_3, b_4; b_5, b_6, b_7, b_8)$ are equal $\left(\tilde{A}^I \approx \tilde{B}^I\right)$ if and only if $a_i = b_i$ for $i = 1, \ldots, 8$.

2.3.1 Arithmetic Operations in Intuitionistic Fuzzy Numbers

For two trapezoidal intuitionistic fuzzy numbers $\tilde{A}^I = (a_1, a_2, a_3, a_4; a_5, a_6, a_7, a_8)$ and $\tilde{B}^I = (b_1, b_2, b_3, b_4; b_5, b_6, b_7, b_8)$ the below arithmetic operations are defined.

$$\tilde{A}^I \oplus t\tilde{B}^I = \begin{matrix}(a_1 + tb_1, a_2 + tb_2, a_3 + tb_3, a_4 + tb_4; \\ a_5 + tb_5, a_6 + tb_6, a_7 + tb_7, a_8 + tb_8)\end{matrix} \quad t > 0, \tag{2.8}$$

$$\tilde{A}^I \oplus t\tilde{B}^I = \begin{matrix}(a_1 + tb_4, a_2 + tb_3, a_3 + tb_2, a_4 + tb_1; \\ a_5 + tb_8, a_6 + tb_7, a_7 + tb_6, a_8 + tb_5)\end{matrix} \quad t < 0, \tag{2.9}$$

$$\tilde{A}^I \otimes \tilde{B}^I \approx (k_1, k_2, k_3, k_4; k_5, k_6, k_7, k_8) \tag{2.10}$$

where

$$\begin{aligned}
&k_1 = \min\{a_1b_1, a_1b_4, a_4b_1, a_4b_4\}, k_2 = \min\{a_2b_2, a_2b_3, a_3b_2, a_3b_3\},\\
&k_3 = \max\{a_1b_1, a_1b_4, a_4b_1, a_4b_4\}, k_4 = \max\{a_1b_1, a_1b_4, a_4b_1, a_4b_4\},\\
&k_5 = \min\{a_5b_5, a_5b_8, a_8b_5, a_8b_8\}, k_6 = \min\{a_6b_6, a_6b_7, a_7b_6, a_7b_7\},\\
&k_7 = \max\{a_6b_6, a_6b_7, a_7b_6, a_7b_7\}, k_8 = \max\{a_5b_5, a_5b_8, a_8b_5, a_8b_8\}.
\end{aligned}$$

$$\tilde{A}^I \otimes \tilde{B}^I = (a_1b_1, a_2b_2, a_3b_3, a_4b_4; a_5b_5, a_6b_6, a_7b_7, a_8b_8), \quad a_1 \geq 0, b_1 \geq 0,$$

Definition 2.9 An intuitionistic fuzzy number can be shown as (μ, ν) where μ is its membership function and ν is its non-membership function [21]. The below arithmetic operations are defined for the numbers $\tilde{A}^I = (\mu_1, \nu_1)$ and $\tilde{B}^I = (\mu_2, \nu_2)$.

$$\tilde{A}^I \oplus \tilde{B}^I = (\mu_1 + \mu_2 - \mu_1\mu_2, \nu_1\nu_2) \tag{2.11}$$

$$\tilde{A}^I \otimes \tilde{B}^I = (\mu_1\mu_2, \nu_1 + \nu_2 - \nu_1\nu_2) \tag{2.12}$$

$$\lambda\tilde{A}^I = \left(1 - (1 - \mu_1)^\lambda, \nu_1^\lambda\right) \tag{2.13}$$

$$\tilde{A}^{I^\lambda} = \left(\mu_1^\lambda, 1 - (1 - \nu_1)^\lambda\right), \lambda > 0. \tag{2.14}$$

Definition 2.10 For $\alpha, \beta \in [0, 1]$ and $\alpha + \beta \leq 1$, (α, β)-cut of the trapezoidal intuitionistic fuzzy number $\tilde{A}^I = (a_1, a_2, a_3, a_4; b_1, b_2, b_3, b_4)$ is a sub-set of real numbers which is defined as below,

$$\begin{aligned}
A_{\alpha,\beta}(x) &= \left\{\left(x, \mu_{\tilde{A}^I}(x), \nu_{\tilde{A}^I}(x)\right), \mu_{\tilde{A}^I}(x) \geq \alpha, \nu_{\tilde{A}^I}(x) \leq \beta, x \in \mathbb{R}\right\}\\
&= \{[A_1(\alpha), A_2(\alpha)], [B_1(\beta), B_2(\beta)]\},
\end{aligned}$$

where,$A_1(\alpha) = a_1 + \alpha(a_2 - a_1)$, $A_2(\alpha) = a_4 - \alpha(a_4 - a_3)$,$B_1(\beta) = b_2 - \beta(b_2 - b_1)$, and $B_2(\beta) = b_3 + \beta(b_4 - b_3)$.

2.3.2 Generalized Intuitionistic Fuzzy Numbers

Definition 2.11 The intuitionistic fuzzy number $\tilde{A}^I$ is a generalized intuitionistic fuzzy number if it has the below properties.

- There should be an $m \in \mathbb{R}$ such that $\mu_{\tilde{A}^I}(m) = w$ and $\nu_{\tilde{A}^I}(m) = 0$ where $0 < w \leq 1$.
- $\mu_{\tilde{A}^I}$ should be a continuous function from $\mathbb{R}$ to the interval $(0, w]$ and for $x \in \mathbb{R}$ there should be $0 \leq \mu_{\tilde{A}^I}(x) + \nu_{\tilde{A}^I}(x) \leq w$.

The membership and non-membership functions of the generalized intuitionistic fuzzy number $\tilde{A}^I$ are defined as below,

$$\mu_{\tilde{A}^I}(x) = \begin{cases} wg_1(x), & r - \alpha \leq x \leq r, \\ w, & x = r, \\ wh_1(x) & r \leq x \leq r + \beta, \\ 0, & 0.w. \end{cases} \tag{2.15}$$

$$\nu_{\tilde{A}^I}(x) = \begin{cases} wg_2(x), & r - \alpha' \leq x \leq r, 0 \leq w(g_1(x) + g_2(x)) \leq 1, \\ 0, & x = r, \\ wh_2(x) & r \leq x \leq r + \beta', 0 \leq w(h_1(x) + h_2(x)) \leq 1, \\ w, & 0.w. \end{cases} \tag{2.16}$$

In these functions, $g_1(x)$ and $h_1(x)$ are respectively strictly increasing and strictly decreasing on $[r - \alpha, r]$ and $[r, r + \beta]$. The functions $g_2(x)$ and $h_2(x)$ are respectively strictly decreasing and strictly increasing on $\left[r - \alpha', r\right]$ and $\left[r, r + \beta'\right]$. The notation r is the mean value of $\tilde{A}^I$. The values α and β are the left and right extensions of the membership function $\mu_{\tilde{A}^I}(x)$, and the values α' and β' are the left and right extensions of the non-membership function $\nu_{\tilde{A}^I}(x)$.

Definition 2.12 Based on the previous definition, generalized trapezoidal intuitionistic fuzzy number (GTrIFN) $\tilde{A}^I = ((a_1, a_2, a_3, a_4; w), (a_5, a_6, a_7, a_8; w))$ where $a_5 \leq a_1 \leq a_6 \leq a_2 \leq a_3 \leq a_7 \leq a_4 \leq a_8$, is defined by the below membership and non-membership functions.

$$\mu_{\tilde{A}^I}(x) = \begin{cases} w\frac{x-a_1}{a_2-a_1}, & a_1 \leq x \leq a_2 \\ w, & a_2 \leq x \leq a_3 \\ w\frac{x-a_4}{a_3-a_4}, & a_3 \leq x \leq a_4 \\ 0, & o.w \end{cases} \tag{2.17}$$

$$\nu_{\tilde{A}^I}(x) = \begin{cases} w\frac{x-a_6}{a_5-a_6}, & a_5 \leq x \leq a_6 \\ 0, & a_6 \leq x \leq a_7 \\ w\frac{x-a_7}{a_8-a_7}, & a_7 \leq x \leq a_8 \\ w, & 0.w \end{cases} \tag{2.18}$$

The generalized trapezoidal intuitionistic fuzzy number $\tilde{A}^I = ((a_1, a_2, a_3, a_4; w), (a_5, a_6, a_7, a_8; w))$ is reduced to a generalized triangular intuitionistic fuzzy number (GTIFN) if $a_2 = a_3$ and $a_6 = a_7$.

The generalized trapezoidal intuitionistic fuzzy number $\tilde{A}^I((a_1, a_2, a_3, a_4; w), (a_5, a_6, a_7, a_8; w))$ is non-negative if and only if $a_5 \geq 0$.

Definition 2.13 The GTrIFNs $\tilde{A}^I = ((a_1, a_2, a_3, a_4; w_1), (a_5, a_6, a_7, a_8; w_1))$ and $\tilde{B}^I = ((b_1, b_2, b_3, b_4; w_2), (b_5, b_6, b_7, b_8; w_2))$ are equal $\left(\tilde{A}^I \approx \tilde{B}^I\right)$ if and only if $a_i = b_i$ for $i = 1, \ldots, 8$, and $w_1 = w_2$.

Definition 2.14 For $\alpha, \beta \in [0, 1]$ and $\alpha + \beta \leq 1$, (α, β)-cut of the trapezoidal intuitionistic fuzzy number $\tilde{A}^I = ((a_1, a_2, a_3, a_4; w), (a_5, a_6, a_7, a_8; w))$ is a sub-set of real numbers which is defined as below,

$$\begin{aligned} A_{\alpha,\beta}(x) &= \left\{\left(x, \mu_{\tilde{A}^I}(x), \nu_{\tilde{A}^I}(x)\right), \mu_{\tilde{A}^I}(x) \geq \alpha, \nu_{\tilde{A}^I}(x) \leq \beta, x \in \mathbb{R}\right\} \\ &= \{[A_1(\alpha), A_2(\alpha)], [B_1(\beta), B_2(\beta)]\} \end{aligned}$$

where, $A_1(\alpha) = a_1 + \frac{\alpha}{w}(a_2 - a_1)$, $A_2(\alpha) = a_4 - \frac{\alpha}{w}(a_4 - a_3)$, $B_1(\beta) = b_2 - \frac{\beta}{w}(b_2 - b_1)$, and $B_2(\beta) = b_3 + \frac{\beta}{w}(b_4 - b_3)$.

2.3.3 *Arithmetic Operations in Generalized Intuitionistic Fuzzy Numbers*

Supposing two generalized TrIFNs $\tilde{A}^I = ((a_1, a_2, a_3, a_4; w_1), (a_5, a_6, a_7, a_8; w_1))$ and $\tilde{B}^I = ((b_1, b_2, b_3, b_4; w_2), (b_5, b_6, b_7, b_8; w_2))$, then the below operations are defined where $0 < w = \min\{w_1, w_2\} \leq 1$:

for $t > 0$,

$$\begin{aligned} \tilde{A}^I \oplus t\tilde{B}^I = &((a_1 + tb_1, a_2 + tb_2, a_3 + tb_3, a_4 + tb_4; w), \\ &(a_5 + tb_5, a_6 + tb_6, a_7 + tb_7, a_8 + tb_8; w)), \end{aligned} \tag{2.19}$$

for $t < 0$,

$$\begin{aligned} \tilde{A}^I \oplus t\tilde{B}^I = &((a_1 + tb_4, a_2 + tb_3, a_3 + tb_2, a_4 + tb_1; w), \\ &(a_5 + tb_8, a_6 + tb_7, a_7 + tb_6, a_8 + tb_5; w). \end{aligned} \tag{2.20}$$

2.3.4 The Extension Principle in Intuitionistic Fuzzy Sets

Like the fuzzy sets and numbers, the extension principle is a basic concept in intuitionistic fuzzy sets. Assume the sets X and Y, the function $f : X \to Y$, and the intuitionistic fuzzy set $\tilde{A}^I$ from X. The output of f on $\tilde{A}^I$ is defined as below [19],

$$\tilde{B}^I = \left\{ \left(y, \mu_{\tilde{B}^I}(y), \nu_{\tilde{B}^I}(y)\right) : y = f(x), x \in X \right\}$$

where,

$$\mu_{\tilde{B}^I}(y) = \begin{cases} \sup\limits_{y=f(x)} \mu_{\tilde{A}^I}(x), & y \in f(X) \\ 0 & y \notin f(X) \end{cases}$$

$$\mu_{\tilde{B}^I}(y) = \begin{cases} \inf\limits_{y=f(x)} \nu_{\tilde{A}^I}(x), & y \in f(X) \\ 1, & y \notin f(X) \end{cases}$$

and $f^{-1}(y) \neq \phi$ is the inverse image of Y.

2.4 Ranking Function of Intuitionistic Fuzzy Numbers

Definition 2.15 The accuracy function of the TrIFN $\tilde{A}^I = (a_1, a_2, a_3, a_4; a_5, a_6, a_7, a_8)$ is defined as.

$$H\left(\tilde{A}^I\right) = \frac{a_1 + a_2 + a_3 + a_4 + a_5 + a_6 + a_7 + a_8}{8}.$$

Theorem 2.1 *For the TrIFNs $\tilde{A}^I$ and $\tilde{B}^I$, the relationship $H\left(\tilde{A}^I + t\tilde{B}^I\right) = H\left(\tilde{A}^I\right) + tH\left(\tilde{B}^I\right)$ is held for any $t \in \mathbb{R}$ [13].*

Definition 2.16 According to Mahmoodirad et al. [13], for the TrIFNs $\tilde{A}^I$ and $\tilde{B}^I$ which are not equal to zero, $\tilde{A}^I \leq (\geq)\tilde{B}^I$ if and only if $H\left(\tilde{A}^I\right) \leq (\geq)H\left(\tilde{B}^I\right)$.

Definition 2.17 For the intuitionistic funny number $\tilde{A}^I = (\mu, \nu)$, the score function is defied as $S\left(\tilde{A}^I\right) = \mu - \nu$ [4], and the accuracy function is defined as $H\left(\tilde{A}^I\right) = \mu + \nu$ [9]. So, for two intuitionistic funny number $\tilde{A}^I = (\mu, \nu)$ and $\tilde{B}^I = (\mu_2, \nu_2)$, we have:

$$\tilde{A}^I < \tilde{B}^I \text{ if and only if } S\left(\tilde{A}^I\right) < S\left(\tilde{B}^I\right), \tag{2.21}$$

$$\tilde{A}^I < \tilde{B}^I \text{ if and only if } S\left(\tilde{A}^I\right) = S\left(\tilde{B}^I\right), H\left(\tilde{A}^I\right) < H\left(\tilde{B}^I\right). \tag{2.22}$$

2.5 Intuitionistic Fuzzy Optimization

An optimization problem consists of one or some objective functions which is to be optimized subject to some constraints. Such problem with fuzzy objective functions and constraints where the parameters are of intuitionistic fuzzy numbers are first converted to a crisp form. The crisp form can be obtained either by using the ranking function for the fuzzy parameters or using the possibility theory measures for the fuzzy objective functions and fuzzy constraints. Then as a solution approach, the fuzzy optimization approach can be used to obtain the Pareto-optimal solutions of the crisp form. Here, the procedure of the fuzzy optimization approach is extended to an intuitionistic fuzzy environment. Therefore, first a crisp (deterministic) form obtained for a fuzzy multi-objective problem is considered as below.

$$\min f(x) = (f_1(x), f_2(x), \ldots, f_b(x)) \tag{2.23}$$

Subject to

$$x \in S = \{x | g_j(x) \leq b_j, j = 1, 2, \ldots, m\},$$

In this problem, $x \in S \subset \mathbb{R}$ are the decision variables, S is the feasible space, $f(x)$ is the set of b objective functions, and m is the number of constraints.

Definition 2.18 The solution vector $x^* \in S$ is called a Pareto-optimal solution, if there is no other solution such as $x \in S$ with the condition $f_t(x) \leq f_t(x^*)$ for $t = 1, 2, \ldots, b$. Therefore, in a Pareto-optimal solution, for at least one of the objective functions such as s, the condition $f_s(x) < f_s(x^*)$ should be held.

In order to obtain a Pareto-optimal solution from the possible set of Pareto-optimal solutions for a multi-objective optimization problem such as the above-mentioned problem, the fuzzy optimization approach [8] has been introduced which considers the problem structure and decision maker preferences simultaneously.

Classically, the fuzzy optimization approach considers the membership functions of the objective functions to introduce a Pareto-optimal solution. But, in the cases that both of the membership and non-membership functions exist and are no complement to each other, the intuitionistic fuzzy optimization approach can be used for finding a Pareto-optimal solution of the problem with multiple objective functions and a set of constraints. Therefore, objective function f_t (where $t = 1, 2, \ldots, b$) and its membership and non-membership functions μ_{f_t} and ν_{f_t} which are calculated based on the positive and negative ideal solutions of the objective function are considered and the below steps are defined for representing the intuitionistic fuzzy optimization approach.

Step 1. Each objective function is considered separately and solved subject to all of the constraints. Therefore, the ideal solution x_t^* is obtained for objective function t. Then, the minimum and maximum possible values for objective function t are obtained as $m_t = min_{1 \leq q \leq b} f_t\left(x_q^*\right)$ and $M_t = max_{1 \leq q \leq b} f_t\left(x_q^*\right)$.

Step 2. The membership and non-membership functions of objective function t are obtained as below where $\mu_{f_t}(x)$ and $\nu_{f_t}(x)$ are respectively strictly increasing and strictly decreasing functions of $f_t(x)$.

$$\mu_{f_t}(x) = \begin{cases} 1, & f_t(x) \le m_t \\ \frac{M_t - f_t(x)}{M_t - m_t}, & m_t \le f_t(x) \le M_t \\ 0, & f_t(x) \ge M_t \end{cases} \tag{2.24}$$

$$\nu_{f_t}(x) = \begin{cases} 0, & f_t(x) \le m_t \\ \frac{f_t(x) - m_t}{M_t - m_t}, & m_t \le f_t(x) \le M_t \\ 1, & f_t(x) \ge M_t \end{cases} \tag{2.25}$$

Step 3. The equivalent deterministic optimization problem of the problem (2.23) is defined as below.

$$\begin{gathered} \max \alpha, \\ \min \beta \\ \text{Subject to} \\ \mu_{f_t}(x) \ge \alpha, \\ \nu_{f_t}(x) \le \beta, \\ \alpha \ge \beta, \\ \alpha + \beta \le 1, \\ \alpha, \beta \ge 0. \end{gathered} \tag{2.26}$$

In this problem, α and β are the minimum satisfaction degree and maximum rejection degree of the objective functions respectively. This model can be converted to the below form easily.

$$\begin{gathered} \max \alpha - \beta \\ \text{Subject to} \\ \mu_{f_t}(x) \ge \alpha, \\ \nu_{f_t}(x) \le \beta, \\ \alpha \ge \beta, \\ \alpha + \beta \le 1 \\ \alpha, \beta \ge 0. \end{gathered} \tag{2.27}$$

The model (2.27) has an optimal solution, and its optimal solution is the Pareto-optimal solution of the multi-objective problem (2.23).

2.6 Possibility Theory and Expected Value in Intuitionistic Fuzzy Environment

2.6.1 Possibility Theory

Here, the classical possibility theory is extended to an intuitionistic fuzzy environment. The required definitions are stated in continue.

Definition 2.19 For the IFNs $\tilde{A}$ and $\tilde{B}$ with membership functions $\mu_{\tilde{A}}$ and μ_B, and non-membership functions $\upsilon_{\tilde{A}}$ and $\upsilon_{\tilde{B}}$, the following possibility and necessity measures (*PS* and *NC*) are defined.

$$PS_\mu\{\tilde{A} * \tilde{B}\} = \sup\{\min(\mu_{\tilde{A}}, \mu_{\tilde{B}}), x, y \in X, x * y\}, \tag{2.28}$$

$$PS_\upsilon\{\tilde{A}*\tilde{B}\} = \sup\{\min(\upsilon_{\tilde{A}}, \upsilon_{\tilde{B}}), x, y \in X, x*y\}, \tag{2.29}$$

$$NC_\mu\{\tilde{A}*\tilde{B}\} = \inf\{\max(\mu_{\tilde{A}}, \mu_{\tilde{B}}), x, y \in X, x*y\}, \tag{2.30}$$

$$NC_\upsilon\{\tilde{A}*\tilde{B}\} = \inf\{\max(\upsilon_{\tilde{A}}, \upsilon_{\tilde{B}}), x, y \in X, x*y\}. \tag{2.31}$$

In Eqs. (2.28)–(2.31), instead of the sign $*$, any of the signs $<$, $>$, $\leq$, $\geq$, and $=$ can be considered.

Definition 2.20 For the IFN $\tilde{A}$ membership function $\mu_{\tilde{A}}$ and non-membership functions $\upsilon_{\tilde{A}}$, the following credibility measures (*CR*) are defined.

$$CR_\mu\{\tilde{A}\} = \frac{1}{2}(PS_\mu\{\tilde{A}\} + NC_\mu\{\tilde{A}\}), \tag{2.32}$$

$$CR_\upsilon\{\tilde{A}\} = \frac{1}{2}(PS_\upsilon\{\tilde{A}\} + NC_\upsilon\{\tilde{A}\}). \tag{2.33}$$

The main disadvantage of the credibility measure is that it considers only the middle of optimistic and pessimistic view points of an event. In order to be more flexible, Xu and Zhou [22] introduced another measure called *ME*. This measure is defined in continue.

Definition 2.21 For the IFN $\tilde{A}$, its *ME* measure for membership and non-membership functions are defined as below,

$$ME_\mu\{\tilde{A}\} = \lambda PS_\mu\{\tilde{A}\} + (1-\lambda) NC_\mu\{\tilde{A}\}, \tag{2.34}$$

$$ME_\upsilon\{\tilde{A}\} = \lambda PS_\upsilon\{\tilde{A}\} + (1-\lambda) NC_\upsilon\{\tilde{A}\}. \tag{2.35}$$

where λ is optimistic-pessimistic parameter of the *ME* measure and $0 \leq \lambda \leq 1$. For this measure, the following issues can be considered.

- If $\lambda = 1$, then $ME_\mu\{\tilde{A}\} = PS_\mu\{\tilde{A}\}$, and $ME_\upsilon\{\tilde{A}\} = PS_\upsilon\{\tilde{A}\}$.
- If $\lambda = 0$, then $ME_\mu\{\tilde{A}\} = NC_\mu\{\tilde{A}\}$, and $ME_\upsilon\{\tilde{A}\} = NC_\upsilon\{\tilde{A}\}$.
- If $\lambda = 0.5$, then $ME_\mu\{\tilde{A}\} = CR_\mu\{\tilde{A}\}$, and $ME_\upsilon\{\tilde{A}\} = CR_\upsilon\{\tilde{A}\}$.

Definition 2.22 For the TrIFN $\tilde{A} = (a_1, a_2, a_3, a_4; a_5, a_2, a_3, a_6)$, the *ME* measures are represented as below,

$$ME_\mu\{\tilde{A} \leq r\} = \begin{cases} 0, & r \leq a_1, \\ \lambda \frac{r-a_1}{a_2-a_1}, & a_1 \leq r \leq a_2, \\ \lambda, & a_2 \leq r \leq a_3, \\ \lambda + (1-\lambda)\frac{r-a_3}{a_4-a_3}, & a_3 \leq r \leq a_4, \\ 1, & \text{otherwise}, \end{cases} \tag{2.36}$$

$$ME_v\{\tilde{A} \leq r\} = \begin{cases} 1, & r \leq a_5, \\ (1-\lambda) + \lambda \frac{a_2-r}{a_2-a_5}, & a_5 \leq r \leq a_2, \\ 1-\lambda, & a_2 \leq r \leq a_3, \\ (1-\lambda)\frac{a_6-r}{a_6-a_3}, & a_3 \leq r \leq a_6, \\ 0, & \text{otherwise}, \end{cases} \tag{2.37}$$

$$ME_\mu\{\tilde{A} \geq r\} = \begin{cases} 1, & r \leq a_1, \\ \lambda + (1-\lambda)\frac{a_2-r}{a_2-a_1}, & a_1 \leq r \leq a_2, \\ \lambda, & a_2 \leq r \leq a_3, \\ \lambda \frac{a_4-r}{a_4-a_3}, & a_3 \leq r \leq a_4, \\ 0, & \text{otherwise}, \end{cases} \tag{2.38}$$

$$ME_v\{\tilde{A} \geq r\} = \begin{cases} 0, & r \leq a_5, \\ (1-\lambda)\frac{r-a_5}{a_2-a_5}, & a_5 \leq r \leq a_2, \\ 1-\lambda, & a_2 \leq r \leq a_3, \\ (1-\lambda) + \lambda \frac{r-a_3}{a_6-a_3}, & a_3 \leq r \leq a_6, \\ 1 & \text{otherwise}, \end{cases} \tag{2.39}$$

According to the *ME* measures (2.36)–(2.39), the following chance constraints can be defined. The confidence levels α and β should follow the condition, $0 \leq \alpha + \beta \leq 1$.

$$ME_\mu\{\tilde{A} \leq r\} \geq \alpha \Leftrightarrow \begin{cases} \lambda \frac{r-a_1}{a_2-a_1} \geq \alpha, & \alpha \in [0, \lambda], \\ \lambda + (1-\lambda)\frac{r-a_3}{a_4-a_3} \geq \alpha, & \alpha \in (\lambda,\ 1], \end{cases} \tag{2.40}$$

$$ME_\upsilon\{\tilde{A} \leq r\} \leq \beta \Leftrightarrow \begin{cases} (1-\lambda) + \lambda \frac{a_2-r}{a_2-a_5} \leq \beta, & \beta \in (1-\lambda,\ 1], \\ (1-\lambda)\frac{a_6-r}{a_6-a_3} \leq \beta, & \beta \in [0, 1-\lambda], \end{cases} \tag{2.41}$$

$$ME_{\mu}\{\tilde{A} \geq r\} \geq \alpha \Leftrightarrow \begin{cases} \lambda + (1-\lambda)\frac{a_2-r}{a_2-a_1} \geq \alpha, & \alpha \in (\lambda, 1], \\ \lambda \frac{a_4-r}{a_4-a_3} \geq \alpha, & \alpha \in [0, \lambda], \end{cases} \tag{2.42}$$

$$ME_{\upsilon}\{\tilde{A} \geq r\} \leq \beta \Leftrightarrow \begin{cases} (1-\lambda)\frac{r-a_5}{a_2-a_5} \leq \beta, & \beta \in [0, 1-\lambda], \\ (1-\lambda) + \lambda \frac{r-a_3}{a_6-a_3} \leq \beta, & \beta \in (1-\lambda, 1]. \end{cases} \tag{2.43}$$

Considering the below facts,

- as $0 \leq \alpha + \beta \leq 1$, if $\alpha \in [0, \lambda]$, then either $\beta \in [0, 1-\lambda]$ or $\beta \in [1-\lambda, 1]$,
- as $0 \leq \alpha + \beta \leq 1$, if $\alpha \in [\lambda, 1]$ then $\beta \in [0, 1-\lambda]$,

then, the following conversions are proposed for the claims (2.40)–(2.44).

$$ME_{\mu}\{\tilde{A} \leq r\} \geq \alpha \Leftrightarrow \lambda \frac{r-a_1}{a_2-a_1} \geq \alpha$$
$$\Leftrightarrow r \geq \left(1-\frac{\alpha}{\lambda}\right)a_1 + \frac{\alpha}{\lambda}a_2, \quad \alpha \leq \lambda, \tag{2.44}$$

$$ME_{\mu}\{\tilde{A} \leq r\} \geq \alpha \Leftrightarrow \lambda + (1-\lambda)\frac{r-a_3}{a_4-a_3} \geq \alpha$$
$$\Leftrightarrow r \geq \frac{1-\alpha}{1-\lambda}a_3 + \frac{\alpha-\lambda}{1-\lambda}a_4, \quad \alpha > \lambda, \tag{2.45}$$

$$ME_{\upsilon}\{\tilde{A} \leq r\} \leq \beta \Leftrightarrow (1-\lambda) + \lambda \frac{a_2-r}{a_2-a_5} \leq \beta$$
$$\Leftrightarrow r \geq \frac{1-\beta}{\lambda}a_2 + \frac{\beta+\lambda-1}{\lambda}a_5, \quad \beta > 1-\lambda, \tag{2.46}$$

$$ME_{\upsilon}\{\tilde{A} \leq r\} \leq \beta \Leftrightarrow (1-\lambda)\frac{a_6-r}{a_6-a_3} \leq \beta$$
$$\Leftrightarrow r \geq \frac{\beta}{1-\lambda}a_3 + \left(1-\frac{\beta}{1-\lambda}\right)a_6, \quad \beta \leq 1-\lambda, \tag{2.47}$$

$$ME_{\mu}\{\tilde{A} \geq r\} \geq \alpha \Leftrightarrow \lambda + (1-\lambda)\frac{a_2-r}{a_2-a_1} \geq \alpha$$
$$\Leftrightarrow r \leq \frac{\alpha-\lambda}{1-\lambda}a_1 + \frac{1-\alpha}{1-\lambda}a_2, \quad \alpha > \lambda, \tag{2.48}$$

$$ME_{\mu}\{\tilde{A} \geq r\} \geq \alpha \Leftrightarrow \lambda \frac{a_4-r}{a_4-a_3} \geq \alpha$$
$$\Leftrightarrow r \leq \frac{\alpha}{\lambda}a_3 + \frac{\lambda-\alpha}{\lambda}a_4, \quad \alpha \leq \lambda, \tag{2.49}$$

$$ME_{\upsilon}\{\tilde{A} \geq r\} \leq \beta \Leftrightarrow \frac{r-a_5}{2(a_2-a_5)} \leq \beta$$

$$\Leftrightarrow r \le \frac{\beta}{1-\lambda} a_2 + \frac{1-\lambda-\beta}{1-\lambda} a_5, \quad \beta \le 1-\lambda, \tag{2.50}$$

$$ME_\upsilon\{\tilde{A} \ge r\} \le \beta \Leftrightarrow (1-\lambda) + \lambda \frac{r-a_3}{a_6-a_3} \le \beta$$
$$\Leftrightarrow r \le \left(1 - \frac{1}{\lambda} + \frac{\beta}{\lambda}\right) a_6 + \left(\frac{1-\beta}{\lambda}\right) a_3, \quad \beta > 1-\lambda, \tag{2.51}$$

2.6.2 *Expected Value*

For the trapezoidal intuitionistic fuzzy number $\tilde{A}^I = (a_1, a_2, a_3, a_4; a_1', a_2, a_3, a_4')$, the expected value of its membership and non-membership functions are obtained as below.

$$E_\mu^{ME}[\tilde{A}^I] = \int_0^\infty ME\{\tilde{A}^I \ge x\} dx - \int_{-\infty}^0 ME\{\tilde{A}^I \le x\} dx \tag{2.52}$$

$$E_\nu^{ME}[\tilde{A}^I] = \int_0^\infty ME\{\tilde{A}^I \le x\} dx - \int_{-\infty}^0 ME\{\tilde{A}^I \ge x\} dx \tag{2.53}$$

These relationships can be converted to the below relationships.

$$E_\mu^{ME}[\tilde{A}^I] = \begin{cases} \frac{\lambda}{2}(a_1+a_2) + \frac{1-\lambda}{2}(a_4' + a_3), & a_4 \le 0, \\ \frac{\lambda}{2}(a_1+a_2) + \frac{\lambda a_4^2 - (1-\lambda)a_3^2}{2(a_4-a_3)}, & a_3 \le 0 \le a_4, \\ \frac{\lambda}{2}(a_1+a_2+a_3+a_4), & a_2 \le 0 \le a_3, \\ \frac{(1-\lambda)a_2^2 - \lambda a_1^2}{2(a_2-a_1)} + \frac{\lambda}{2}(a_3+a_4), & a_1 \le 0 \le a_2, \\ \frac{1-\lambda}{2}(a_1+a_2) + \frac{\lambda}{2}(a_3+a_4), & a_1 \ge 0, \end{cases} \tag{2.54}$$

$$E_\nu^{ME}[\tilde{A}^I] = \begin{cases} \frac{1-\lambda}{2}(2a_4' - a_1' - a_2) + \frac{\lambda}{2}(a_4' - a_3) - a_4', & a_4' \le 0, \\ \frac{1-\lambda}{2}\left(\frac{a_4'^2}{a_4' - a_3 + a_1' + a_2}\right) - \lambda \frac{a_3^2}{2(a_4'-a_3)}, & a_3 \le 0 \le a_4', \\ \frac{1-\lambda}{2}(a_1' + a_2 + a_3 + a_4'), & a_2 \le 0 \le a_3, \\ \frac{\lambda a_2^2}{2(a_2 - a_1')} + \frac{1-\lambda}{2}\left(\frac{a_2a_3 - a_3a_1' + a_4'a_1' - a_1'^2}{a_2 - a_1'}\right), & a_1' \le 0 \le a_2, \\ a_1' + \frac{\lambda}{2}(a_2 - a_1') + \frac{1-\lambda}{2}(a_3 + a_4' - 2a_1'), & a_1' \ge 0, \end{cases} \tag{2.55}$$

Now, for the number $\tilde{A}^I = (a_1, a_2, a_3, a_4; a_1', a_2, a_3, a_4')$ the expected value of $E(\tilde{A}^I)$ is obtained as below.

$$E(\tilde{A}^I) = \frac{E_\mu^{CR}[\tilde{A}^I] + E_\nu^{CR}[\tilde{A}^I]}{2} = \frac{a_1 + 2a_2 + 2a_3 + a_4 + a_{1'} + a_{4'}}{8} \tag{2.56}$$

2.7 Interval-Valued Intuitionistic Fuzzy Sets

Interval-valued intuitionistic fuzzy set [2] is an extension of the intuitionistic fuzzy set where the membership and non-membership function values are interval type values.

Definition 2.23 Assume X as a reference set and $D[0, 1]$ as the set of all closed subintervals of the interval $[0, 1]$. An interval-values intuitionistic fuzzy set ($\tilde{A}^{IVI}$) in X is defined as $\tilde{A}^{IVI} = \{(x, \mu_{\tilde{A}^{IIV}}(x), \nu_{\tilde{A}^{IIV}}(x)) : x \in X\}$, where, $\mu_{\tilde{A}^{IIV}}(x) : X \to D[0, 1]$, $\nu_{\tilde{A}^{IIV}}(x) : X \to D[0, 1]$, and $0 \leq \sup \mu_{\tilde{A}^{IIV}}(x) + \sup \nu_{\tilde{A}^{IIV}}(x) \leq 1$ for any $x \in X$.

Assume $\mu_{\tilde{A}^{IIV}}(x) = [a, b]$ and $\nu_{\tilde{A}^{IIV}}(x) = [c, d]$ where the intervals $[a, b]$ and $[c, d]$ are from $D[0, 1]$ with the condition $b + d \leq 1$. An interval-valued intuitionistic set can be shown as $\tilde{A}^{IVI} = \{(x, [a, b], [c, d]) : x \in X\}$. Therefore, the notation $([a, b], [c, d])$ is an interval-valued intuitionistic fuzzy value. Clearly, $\tilde{A}^+ = ([1, 1], [0, 0])$ and $\tilde{A}^- = ([0, 0], [1, 1])$ are the largest and smallest interval-valued intuitionistic fuzzy values respectively.

Definition 2.24 Assume X as a reference set, and $\tilde{A}_1^{IIV} = ([a_i, b_i], [c_i, d_i])$ and $\tilde{A}_2^{IIV} = ([e_i, f_i], [g_i, h_i])$ (where $i = 1, \ldots, n$) as two intuitionistic fuzzy values from X. The normalized Hamming distance and weighted Hamming distance of these values are defined as below respectively.

$$d(\tilde{A}_1^{IIV}, \tilde{A}_2^{IIV}) = \frac{1}{4}\sum_{i=1}^{n}(|a_i - e_i| + |b_i - f|_i + |c_i - g_i| + |d_i - h_i|) \tag{2.57}$$

$$d(\tilde{A}_1^{IIV}, \tilde{A}_2^{IIV}) = \frac{1}{4}\sum_{i=1}^{n} w_i(|a_i - e_i| + |b_i - f_i| + |c_i - g_i| + |d_i - h_i|) \tag{2.58}$$

The values of $w = (w_1, \ldots, w_n)$ are the vector of weight values, where $w_i \in [0, 1]$ and $\sum_{i=1}^n w_i = 1$.

Definition 2.25 Consider the intuitionistic fuzzy values $\tilde{A}_1^{IIV} = ([a_1, b_1], [c_1, d_1])$ and $\tilde{A}_2^{IIV} = ([a_2, b_2], [c_2, d_2])$. Then,

$$\tilde{A}_1^{IIV} \leq \tilde{A}_2^{IIV} \text{iff} a_1 \leq a_2, b_1 \leq b_2, c_1 \geq c_2, d_1 \geq d_2, \tag{2.59}$$

$$\tilde{A}_1^{IIV} = \tilde{A}_2^{IIV} \text{iff} a_1 = a_2, b_1 = b_2, c_1 = c_2, d_1 = d_2. \tag{2.60}$$

where, the inequality (2.59) may not be held in all situations (see the next definition).

Definition 2.26 [20] Consider the intuitionistic fuzzy values $\tilde{A}_1^{IIV} = ([a_1, b_1], [c_1, d_1])$ and $\tilde{A}_2^{IIV} = ([a_2, b_2], [c_2, d_2])$ with $S(\tilde{A}_1^{IIV}) = \frac{a_1+b_1-c_1-d_1}{2}$ and $S(\tilde{A}_2^{IIV}) = \frac{a_2+b_2-c_2-d_2}{2}$ score functions and $H(\tilde{A}_1^{IIV}) = \frac{a_1+b_1+c_1+d_1}{2}$ and $H(\tilde{A}_2^{IIV}) = \frac{a_2+b_2+c_2+d_2}{2}$ accuracy functions respectively. Then, the below logics are considered.

- If $S(\tilde{A}_1^{IIV}) < S(\tilde{A}_2^{IIV})$ then $\tilde{A}_1^{IIV} \prec \tilde{A}_2^{IIV}$,
- If $S(\tilde{A}_1^{IIV}) > S(\tilde{A}_2^{IIV})$ then $\tilde{A}_1^{IIV} \succ \tilde{A}_2^{IIV}$,
- If $S(\tilde{A}_1^{IIV}) = S(\tilde{A}_2^{IIV})$, then,
  - if $H(\tilde{A}_1^{IIV}) < H(\tilde{A}_2^{IIV})$ then $\tilde{A}_1^{IIV} \prec \tilde{A}_2^{IIV}$,
  - if $H(\tilde{A}_1^{IIV}) > H(\tilde{A}_2^{IIV})$ then $\tilde{A}_1^{IIV} \succ \tilde{A}_2^{IIV}$,
  - if $H(\tilde{A}_1^{IIV}) = H(\tilde{A}_2^{IIV})$ then $\tilde{A}_1^{IIV} \approx \tilde{A}_2^{IIV}$..

Definition 2.27 Consider the intuitionistic fuzzy values $\tilde{A}_1^{IIV} = ([a_1, b_1], [c_1, d_1])$ and $\tilde{A}_2^{IIV} = ([a_2, b_2], [c_2, d_2])$. Then, $\tilde{A}_1^{IIV} + \tilde{A}_2^{IIV}$ and $\lambda\tilde{A}_1^{IIV}$ are intuitionistic fuzzy values and obtained as below.

$$\tilde{A}_1^{IIV} + \tilde{A}_2^{IIV} = ([a_1 + a_2 - a_1a_2, b_1 + b_2 - b_1b_2], [c_1c_2, d_1d_2]), \tag{2.61}$$

$$\lambda\tilde{A}_1^{IIV} = \left(\left[1 - (1 - a_1)^\lambda, 1 - (1 - b_1)^\lambda\right], \left[c_1^\lambda, d_1^\lambda\right]\right), \lambda > 0, \tag{2.62}$$

Definition 2.28 Consider the intuitionistic fuzzy values $\tilde{A}_1^{IIV} = ([a_1, b_1], [c_1, d_1])$ and $\tilde{A}_2^{IIV} = ([a_2, b_2], [c_2, d_2])$, and $\lambda, \lambda_1, \lambda_2 > 0$. Then,

$$\tilde{A}_1^{IIV} + \tilde{A}_2^{IIV} = \tilde{A}_2^{IIV} + \tilde{A}_1^{IIV}, \tag{2.63}$$

$$\lambda(\tilde{A}_1^{IIV} + \tilde{A}_2^{IIV}) = \lambda\tilde{A}_2^{IIV} + \lambda\tilde{A}_1^{IIV}, \tag{2.64}$$

$$\lambda_1\tilde{A}_1^{IIV} + \lambda_2\tilde{A}_1^{IIV} = (\lambda_1 + \lambda_2)\tilde{A}_1^{IIV}. \tag{2.65}$$

2.8 Type-2 Intuitionistic Fuzzy Sets

Type-2 intuitionistic fuzzy set (T2IFS) was introduced by Singh and Garg [18] and is useful for the cases that the decision makers want to state their fuzzy priorities of their choices when facing with different limitations. An T2IFS include primary and secondary membership functions.

Definition 2.29 The IT2FS $\tilde{A}$ in X is defined as $\tilde{A} = \left\{\left(x, u, \mu_{\tilde{A}}(x, u), \nu_{\tilde{A}}(x, u)\right) : x \in X, u \in J_x \subseteq [0, 1]\right\}$. The functions $\mu_{\tilde{A}} : X \times J_x \to [0, 1]$ and $\nu_{\tilde{A}} : X \times J_x \to [0, 1]$ are the membership and non-membership functions of $u \in J_x$. For any $x \in X$ and $u \in J_x$ the condition $0 \leq \mu_{\tilde{A}}(x,u) + \nu_{\tilde{A}}(x,u) \leq 1$ should be held.

For a discrete element of the reference set $\tilde{A} = \sum_{x\in X}\left(\sum_{u\in J_x}\left(\mu_{\tilde{A}}(x,u), \nu_{\tilde{A}}(x,u)\right)/u\right)/x, J_x \subseteq [0,1]$ and for a continuous element of the reference set $\tilde{A} = \int_{x\in X}\left(\int_{u\in J_x}\left(\mu_{\tilde{A}}(x,u), \nu_{\tilde{A}}(x,u)\right)/u\right)/x, J_x \subseteq [0,1]$ are defined.

Example 2.1 Suppose the youngness set is shown by the T2IFS $\tilde{A}$. Youngness is the primary membership function, and the degree of youngness and the degree of growth are secondary membership and non-membership functions respectively. Assume $X = \{8, 10, 14\}$ and $J_8 = \{0.8, 0.9, 1\}$,

$J_{10} = \{0.6, 0.7, 0.8\}$, and $J_{14} = \{0.4, 0.5, 0.6\}$. Then, a discrete IT2FS of $\tilde{A}$ is obtained as below.

$$\begin{aligned}\tilde{A} = &((0.9,0)/0.8 + (0.7,0.1)/0.9 + (0.6,0.3)/1)/8\\ &+ ((0.8,0.1)/0.6 + (0.7,0.2)/0.7 + (0.6,0.4)/0.8)/10\\ &+ ((0.9,0.1)/0.4 + (0.8,0.2)/0.5 + (0.5,0.4)/0.6)/14\end{aligned}$$

2.8.1 Operations on Type-2 Intuitionistic Fuzzy Sets

Let $\tilde{A} = \int_{x\in X}\left(\int_{u\in J_x^u}\left(\mu_{\tilde{A}}(x,u), \nu_{\tilde{A}}(x,u)\right)/u\right)/x$, and $\tilde{B} = \int_{x\in X}\left(\int_{u\in J_x^v}\left(\mu_{\tilde{B}}(x,v), \nu_{\tilde{B}}(x,v)\right)/v\right)/x$, are two T2IFSs on X, some operations on these numbers such as union, intersection, and complement are described as continue.

- **Union**: $\tilde{A} \cup \tilde{B} = \int_{x\in X}\left(\int_{w\in J_x^w}\left(\mu_{\tilde{A}\cup\tilde{B}}(x,w), \nu_{\tilde{A}\cup\tilde{B}}(x,w)\right)/w\right)/x, J_x^u \cup J_x^v = J_x^w \subseteq [0,1]$ where $\mu_{\tilde{A}\cup\tilde{B}}(x,w) = \int_{u\in J_x^u}\int_{v\in J_x^v}\left(\mu_{\tilde{A}}(x,u) \wedge \mu_{\tilde{B}}(x,v)\right)/(u \vee v)$ and $\nu_{\tilde{A}\cup\tilde{B}}(x,w) = \int_{u\in J_x^u}\int_{v\in J_x^v}\left(\nu_{\tilde{A}}(x,u) \vee \nu_{\tilde{B}}(x,v)\right)/(u \vee v)$,
- **Intersection**: $\tilde{A} \cap \tilde{B} = \int_{x\in X}\left(\int_{w\in J_x^w}\left(\mu_{\tilde{A}\cap\tilde{B}}(x,w), \nu_{\tilde{A}\cap\tilde{B}}(x,w)\right)/w\right)/x, J_x^u \cap J_x^v = J_x^w \subseteq [0,1]$ where $\mu_{\tilde{A}\cap\tilde{B}}(x,w) = \int_{u\in J_x^u}\int_{v\in J_x^v}\left(\mu_{\tilde{A}}(x,u) \wedge \mu_{\tilde{B}}(x,v)\right)/(u \wedge v)$ and $\nu_{\tilde{A}\cup\tilde{B}}(x,w) = \int_{u\in J_x^u}\int_{v\in J_x^v}\left(\nu_{\tilde{A}}(x,u) \vee \nu_{\tilde{B}}(x,v)\right)/(u \wedge v)$,
- **Complement**: $\tilde{A}^c = \int_{x\in X}\left(\int_{u\in J_x^u}\left(\mu_{\tilde{A}}(x,u), \nu_{\tilde{A}}(x,u)\right)/(1-u)\right)/x$.

Example 2.2 [6] Considering Example 2.1, the below T2IFSs are considered to represent the youngness set.

$$\begin{aligned}\tilde{A} = &((0.9,0)/0.8 + (0.7,0.1)/0.9 + (0.6,0.3)/1)/8\\ &+ ((0.8,0.1)/0.6 + (0.7,0.2)/0.7 + (0.6,0.4)/0.8)/10\end{aligned}$$

$$+((0.9, 0.1)/0.4 + (0.8, 0.2)/0.5 + (0.5, 0.4)/0.6)/14,$$

$$\begin{aligned}\tilde{B} = {} & ((0.8, 0.1)/0.7 + (0.6, 0.3)/0.8 + (0.5, 0.5)/0.9)/8 \\ & + ((0.9, 0.1)/0.4 + (0.7, 0.2)/0.5 + (0.5, 0.3)/0.6)/10 \\ & + ((0.8, 0.2)/0.5 + (0.7, 0.2)/0.6 + (0.6, 0.3)/0.7)/14\end{aligned}$$

Now, for element 8 from the youngness set the secondary membership of $((0.9, 0)/0.8 + (0.7, 0.1)/0.9 + (0.6, 0.3)/1)/8$, and the secondary non-membership of $((0.8, 0.1)/0.7 + (0.6, 0.3)/0.8 + (0.5, 0.5)/0.9)/8$, are obtained.

Then for $x = 8$, the union of $\tilde{A}$ and $\tilde{B}\big(\mu_{\tilde{A}\cup\tilde{B}}(8), \nu_{\tilde{A}\cup\tilde{B}}(8)\big)$ is obtained as below.

$$\begin{aligned}& ((0.9, 0)/0.8 + (0.7, 0.1)/0.9 + (0.6, 0.3)/1)/8 \\ & \quad \vee ((0.8, 0.1)/0.7 + (0.6, 0.3)/0.8 + (0.5, 0.5)/0.9)/8 \\ & = ((0.9 \wedge 0.8), (0 \vee 0.1))/(0.8 \vee 0.7) \\ & \quad + ((0.9 \wedge 0.6), (0 \vee 0.3))/(0.8 \vee 0.5) \\ & \quad + ((0.9 \wedge 0.5), (0 \vee 0.5))/(0.8 \vee 0.9) \\ & \quad + ((0.7 \wedge 0.8), (0.1 \vee 0.1))/(0.9 \vee 0.7) \\ & \quad + ((0.7 \wedge 0.6), (0.1 \vee 0.3))/(0.9 \vee 0.8) \\ & \quad + ((0.7 \wedge 0.5), (0.1 \vee 0.5))/(0.9 \vee 0.9) \\ & \quad + ((0.6 \wedge 0.8), (0.3 \vee 0.1))/(1 \vee 0.7) \\ & \quad + ((0.6 \wedge 0.6), (0.3 \vee 0.3))/(1 \vee 0.8) \\ & \quad + ((0.6 \wedge 0.5), (0.3 \vee 0.5))/(1 \vee 0.9))/8 \\ & = ((0.8, 0.1)/0.8 + (0.6, 0.3)/0.8 + (0.5, 0.5)/0.9 \\ & \quad + (0.7, 0.1)/0.9 + (0.6, 0.3)/0.9 + (0.5, 0.5)/0.9 \\ & \quad +(0.6, 0.3)/1 + (0.6, 0.3)/1 + (0.5, 0.5)/1)/8 \\ & = (\max\{0.8, 0.6\}, \min\{0.1, 0.3\})/0.8 \\ & \quad + (\max\{0.5, 0.7, 0.6, 0.5\}, \min\{0.5, 0.1, 0.3, 0.5\})/0.9 \\ & \quad + (\max\{0.6, 0.6, 0.5\}, \min\{0.3, 0.3, 0.5\})/1)/8 \\ & = ((0.8, 0.1)/0.8 + (0.7, 0.1)/0.9 + (0.6, 0.3)/1)/8\end{aligned}$$

For $x = 8$, the intersection of $\tilde{A}$ and $\tilde{B}\big(\mu_{\tilde{A}\cap\tilde{B}}(x, w), \nu_{\tilde{A}\cap\tilde{B}}(x, w)\big)$ is obtained as below.

$$\begin{aligned}& ((0.9, 0)/0.8 + (0.7, 0.1)/0.9 + (0.6, 0.3)/1)/8 \\ & \quad \wedge ((0.8, 0.1)/0.7 + (0.6, 0.3)/0.8 + (0.5, 0.5)/0.9)/8 \\ & = ((0.9 \wedge 0.8), (0 \vee 0.1))/(0.8 \wedge 0.7) \\ & \quad + ((0.9 \wedge 0.6), (0 \vee 0.3))/(0.8 \wedge 0.8) \\ & \quad + ((0.9 \wedge 0.5), (0 \vee 0.5))/(0.8 \wedge 0.9) \\ & \quad + ((0.7 \wedge 0.8), (0.1 \vee 0.1))/(0.9 \wedge 0.7)\end{aligned}$$

$$
\begin{aligned}
&+ ((0.7 \wedge 0.6), (0.1 \vee 0.3))/(0.9 \wedge 0.8) \\
&+ ((0.7 \wedge 0.5), (0.1 \vee 0.5))/(0.9 \wedge 0.9) \\
&+ ((0.6 \wedge 0.8), (0.3 \vee 0.1))/(1 \wedge 0.7) \\
&+ ((0.6 \wedge 0.6), (0.3 \vee 0.3))/(1 \wedge 0.8) \\
&+ ((0.6 \wedge 0.5), (0.3 \vee 0.5))/(1 \wedge 0.9))/8 \\
&= ((0.8, 0.1)/0.7 + (0.6, 0.3)/0.8 + (0.5, 0.5)/0.8 \\
&+ (0.7, 0.1)/0.7 + (0.6, 0.3)/0.8 + (0.5, 0.5)/0.9 \\
&+ (0.6, 0.3)/0.7 + (0.6, 0.3)/0.8 + (0.5, 0.5)/0.9)/8 \\
&= (\max\{0.8, 0.7, 0.6\}, \min\{0.1, 0.1, 0.3\})/0.7 \\
&+ (\max\{0.6, 0.5, 0.6, 0.6\}, \min\{0.3, 0.5, 0.3, 0.3\})/0.8 \\
&+ (\max\{0.5, 0.5\}, \min\{0.5, 0.5\})/0.9)/8 \\
&= ((0.8, 0.1)/0.7 + (0.6, 0.3)/0.8 + (0.5, 0.5)/0.9)/8
\end{aligned}
$$

For $x = 8$, the complement of $\tilde{A}$ is obtained as below.

$$
\begin{aligned}
\tilde{A}^c = &((0.9, 0)/0.2 + (0.7, 0.1)/0.1 + (0.6, 0.3)/0)/8 \\
&+ ((0.8, 0.1)/0.4 + (0.7, 0.2)/0.3 + (0.6, 0.4)/0.2)/10 \\
&+ ((0.9, 0.1)/0.6 + (0.8, 0.2)/0.5 + (0.5, 0.4)/0.4)/14
\end{aligned}
$$

Theorem 2.2 *Assume the T2IFSs $\tilde{A}$, $\tilde{B}$, and $\tilde{C}$. Then,*

$$\tilde{A} \cup \tilde{A} = \tilde{A}, \tag{2.66}$$

$$\tilde{A} \cup \tilde{B} = \tilde{B} \cup \tilde{A}, \tilde{A} \cap \tilde{B} = \tilde{B} \cap \tilde{A}, \tag{2.67}$$

$$(\tilde{A} \cup \tilde{B}) \cup \tilde{C} = \tilde{A} \cup (\tilde{B} \cup \tilde{C}), (\tilde{A} \cap \tilde{B}) \cap \tilde{C} = \tilde{A} \cap (\tilde{B} \cap \tilde{C}), \tag{2.68}$$

$$\tilde{A} \cup (\tilde{B} \cap \tilde{C}), (\tilde{A} \cup \tilde{B}) \cap (\tilde{A} \cup \tilde{C}), \tilde{A} \cap (\tilde{B} \cup \tilde{C}), (\tilde{A} \cap \tilde{B}) \cup (\tilde{A} \cap \tilde{C}), \tag{2.69}$$

$$(\tilde{A}^c)^c = \tilde{A}, \tag{2.70}$$

$$(\tilde{A} \cup \tilde{B})^c = \tilde{A}^c \cap \tilde{B}^c, (\tilde{A} \cap \tilde{B})^c = \tilde{A}^c \cup \tilde{B}^c. \tag{2.71}$$

Proof See Dan et al. [6].

Definition 2.30 Assume $F_2^I(x)$ as the collection of T2IFSs on X. The real function $d : F_2^I(x) \times F_2^I(x) \to [0, 1]$ is a distance function according to the below conditions.

- $0 \le d(\tilde{A}_1, \tilde{A}_2) \le 1, \forall \tilde{A}_1, \tilde{A}_2 \in F_2^I(x)$

- $d(\tilde{A}_1, \tilde{A}_2) = 0$, if $A_1 \approx A_2$,
- $d(\tilde{A}_1, \tilde{A}_2) = d(\tilde{A}_2, \tilde{A}_1)$
- If $d(\tilde{A}_1, \tilde{A}_2) = 0, d(\tilde{A}_1, \tilde{A}_3) = 0, \tilde{A}_3 \in F_{\tilde{I}}^2(X) \Rightarrow d(\tilde{A}_2, \tilde{A}_3) = 0$.

Definition 2.31 For the T2IFSs $\tilde{A} = \{(x, u, \mu_{\tilde{A}}(x, u), \nu_{\tilde{A}}(x, u)) : x \in X\}$ and $\tilde{B} = \{(x, u, \mu_{\tilde{B}}(x, u), \nu_{\tilde{B}}(x, u)) : x \in X\}$, the below distances are defined [6].

Hamming distance:

$$d_H(\tilde{A}, \tilde{B}) = \sum_{x \in X} \sum_{u,v \in J_x} \sum_{j=1}^{L(x,u,u';\tilde{A},\tilde{B})} \left(\left| \mu_{\tilde{A}}^j(x, u) - \mu_{\tilde{B}}^j(x, u') \right| + \left| \nu_{\tilde{A}}^j(x, u) - \nu_{\tilde{B}}^j(x, u') \right| \right) \tag{2.72}$$

Euclidean distance:

$$d_E(\tilde{A}, \tilde{B}) = \sum_{x \in X} \sum_{u,v \in J_x} \sum_{j=1}^{L(x,u,u';\tilde{A},\tilde{B})} \left(\left| \mu_{\tilde{A}}^j(x, u) - \mu_{\tilde{B}}^j(x, u') \right|^2 + \left| \nu_{\tilde{A}}^j(x, u) - \nu_{\tilde{B}}^j(x, u') \right|^2 \right)^{\frac{1}{2}} \tag{2.73}$$

In the above formulas u and u' are the primary membership functions of $\tilde{A}$ and $\tilde{B}$ respectively. Moreover, $\mu_{\tilde{A}}$ and $\nu_{\tilde{A}}$ are the secondary membership and non-membership functions of $\tilde{A}$ which is defined similarly for $\tilde{B}$.

Example 2.3 Assume the reference set $X \neq \emptyset$. The IT2FSs $\tilde{A}$ and $\tilde{B}$, and their distances are defined as below.

$$\tilde{A} = (0.7, 0.2)/0.8/x + (0.6, 0.1)/0.6/y + (0.5, 0.3)/0.9/z$$

$$\tilde{B} = (0.6, 0.4)/0.6/x + (0.5, 0.3)/0.3/y + (0.4, 0.2)/0.4/z$$

$$\begin{aligned} d_H(\tilde{A}, \tilde{B}) &= |0.7 - 0.6| + |0.2 - 0.4| + |0.6 - 0.5| \\ &\quad + |0.1 - 0.3| + |0.5 - 0.4| + |0.3 - 0.2| \\ &= 0.8 \end{aligned}$$

$$\begin{aligned} d_E(\tilde{A}, \tilde{B}) &= \left(|0.7 - 0.6|^2 + |0.2 - 0.4|^2 + |0.6 - 0.5|^2 \right. \\ &\quad \left. + |0.1 - 0.3|^2 + |0.5 - 0.4|^2 + |0.3 - 0.2|^2\right)^{\frac{1}{2}} \\ &= 0.35 \end{aligned}$$

2.8.2 Generalized Trapezoidal Type-2 Intuitionistic Fuzzy Numbers

The concepts and operations of generalized trapezoidal type-2 intuitionistic fuzzy numbers (GTrT2IFNs) and their symmetric forms are presented.

Definition 2.32 A GTrT2IFN is shown by $\tilde{A} = \left(\tilde{A}_1, \tilde{A}_2, \tilde{A}_3; \mu_{\tilde{A}}, \nu_{\tilde{A}}\right)$ where $\tilde{A}_1$, $\tilde{A}_2$, and $\tilde{A}_3$ are generalized trapezoidal type-1 intuitionistic fuzzy numbers as $\tilde{A}_1 = \left(a_{11}, a_{12}, a_{13}, a_{14}; \mu_{\tilde{A}_1}, \nu_{\tilde{A}_1}\right)$, $\tilde{A}_2 = \left(a_{21}, a_{22}, a_{23}, a_{24}; \mu_{\tilde{A}_2}, \nu_{\tilde{A}_2}\right)$, and $\tilde{A}_3 = \left(a_{31}, a_{32}, a_{33}, a_{34}; \mu_{\tilde{A}_3}, \nu_{\tilde{A}_3}\right)$ [12]. Therefore, $\tilde{A} = ((a_{11}, a_{12}, a_{13}, a_{14}; \mu_{\tilde{A}_1}, \nu_{\tilde{A}_1}), (a_{21}, a_{22}, a_{23}, a_{24}; \mu_{\tilde{A}_2}, \nu_{\tilde{A}_2}), (a_{31}, a_{32}, a_{33}, a_{34}; \mu_{\tilde{A}_3}, \nu_{\tilde{A}_3}); \mu_{\tilde{A}}, \nu_{\tilde{A}})$ is a GTrT2IFN with the membership function of $\mu_{\tilde{A}} = \min\left\{\mu_{\tilde{A}_1}, \mu_{\tilde{A}_2}, \mu_{\tilde{A}_3}\right\}$ and the non-membership function of $\nu_{\tilde{A}} = \min\left\{\nu_{\tilde{A}_1}, \nu_{\tilde{A}_2}, \nu_{\tilde{A}_3}\right\}$, where $\mu_{\tilde{A}_1} \geq \mu_{\tilde{A}_2} \geq \mu_{\tilde{A}_3} \geq \nu_{\tilde{A}_3} \geq \nu_{\tilde{A}_2} \geq \nu_{\tilde{A}_1}$.

If $\boldsymbol{a}_{12} = \boldsymbol{a}_{13}$, $\boldsymbol{a}_{22} = \boldsymbol{a}_{23}$, and $\boldsymbol{a}_{32} = \boldsymbol{a}_{33}$, then the GTrT2IFN is converted to a generalized triangular type-2 intuitionistic fuzzy number (GTT2IFN).

Definition 2.33 In the previous definition, if $a_{12} = a_{22} = a_{32} = m_1$ and $a_{13} = a_{23} = a_{33} = m_2$, then the GTrT2IFN is reduced to a generalized symmetric trapezoidal type-2 intuitionistic fuzzy number (GSTrT2IFN) and shown as $\tilde{A} = ((a_{11}, m_1, m_2, a_{14}; \mu_{\tilde{A}_1}, \nu_{\tilde{A}_1}), (a_{21}, m_1, m_2, a_{24}; \mu_{\tilde{A}_2}, \nu_{\tilde{A}_2}), (a_{31}, m_1, m_2, a_{34}; \mu_{\tilde{A}_3}, \nu_{\tilde{A}_3}); \mu_{\tilde{A}}, \nu_{\tilde{A}})$ where $\tilde{A}_1 = \left(a_{11}, m_1, m_2, a_{14}; \mu_{\tilde{A}_1}, \nu_{\tilde{A}_1}\right)$, $\tilde{A}_2 = \left(a_{21}, m_1, m_2, a_{24}; \mu_{\tilde{A}_2}, \nu_{\tilde{A}_2}\right)$, and $\tilde{A}_3 = \left(a_{31}, m_1, m_2, a_{34}; \mu_{\tilde{A}_3}, \nu_{\tilde{A}_3}\right)$ are generalized symmetric trapezoidal type-1 intuitionistic fuzzy numbers (GSTrT1IFNs).

If $\boldsymbol{m}_1 = \boldsymbol{m}_2$, then the GSTrT2IFN is converted to a generalized symmetric triangular type-2 intuitionistic fuzzy number (GSTT2IFN).

2.8.2.1 Arithmetic Operations on GSTrT2IFNs

Definition 2.34 Consider the GSTrT2IFNs $\tilde{A} = ((a_{11}, m_1, m_2, a_{14}; \mu_{\tilde{A}_1}, \nu_{\tilde{A}_1}), (a_{21}, m_1, m_2, a_{24}; \mu_{\tilde{A}_2}, \nu_{\tilde{A}_2}),, (a_{31}, m_1, m_2, a_{34}; \mu_{\tilde{A}_3}, \nu_{\tilde{A}_3}); \mu_{\tilde{A}}, \nu_{\tilde{A}})$ and $\tilde{B} = ((b_{11}, m_3, m_4, b_{14}; \mu_{\tilde{B}_1}, \nu_{\tilde{B}_1}),, (b_{21}, m_3, m_4, b_{24}; \mu_{\tilde{B}_2}, \nu_{\tilde{B}_2}), (b_{31}, m_3, m_4, b_{34}; \mu_{\tilde{B}_3}, \nu_{\tilde{B}_3}); \mu_{\tilde{B}}, \nu_{B})$. The arithmetic operations below are defined.

- **Addition**

$$
\begin{aligned}
&\tilde{A} + \tilde{B} \\
&= \left(\left(a_{11} + b_{11}, m_1 + m_3, m_2 + m_4, a_{14} + b_{14}; \min\left\{\mu_{\tilde{A}_1}, \mu_{\tilde{B}_1}\right\}, \max\left\{\nu_{\tilde{A}_1}, \nu_{\tilde{B}_1}\right\}\right),\right. \\
&\quad \left(a_{21} + b_{21}, m_1 + m_3, m_2 + m_4, a_{24} + b_{24}; \min\left\{\mu_{\tilde{A}_2}, \mu_{\tilde{B}_2}\right\}, \max\left\{\nu_{\tilde{A}_2}, \nu_{\tilde{B}_2}\right\}\right), \\
&\quad \left(a_{31} + b_{31}, m_1 + m_3, m_2 + m_4, a_{34} + b_{34}; \min\left\{\mu_{\tilde{A}_3}, \mu_{\tilde{B}_3}\right\}, \max\left\{\nu_{\tilde{A}_3}, \nu_{\tilde{B}_3}\right\}\right) \\
&\quad \left.; \mu_{\tilde{A}+\tilde{B}}, \nu_{\tilde{A}+\tilde{B}}\right)
\end{aligned}
$$

where, $\mu_{\tilde{A}+\tilde{B}} = \min\{\min\{\mu_{\tilde{A}_1}, \mu_{\tilde{B}_1}\}, \min\{\mu_{\tilde{A}_2}, \mu_{\tilde{B}_2}\}, \min\{\mu_{\tilde{A}_3}, \mu_{\tilde{B}_3}\}\}$ and $\nu_{\tilde{A}+\tilde{B}} = \max\{\max\{\nu_{\tilde{A}_1}, \nu_{\tilde{B}_1}\}, \max\{\nu_{\tilde{A}_2}, \nu_{\tilde{B}_2}\}, \max\{\nu_{\tilde{A}_3}, \nu_{\tilde{B}_3}\}\}$.

- **Subtraction**

$$\begin{aligned}
&\tilde{A} - \tilde{B} \\
&= ((a_{11} - b_{34}, m_1 - m_4, m_2 - m_3, a_{14} - b_{31}; \min\{\mu_{\tilde{A}_1}, \mu_{\tilde{B}_3}\}, \max\{\nu_{\tilde{A}_1}, \nu_{\tilde{B}_3}\}), \\
&\quad (a_{21} - b_{24}, m_1 - m_4, m_2 - m_3, a_{24} - b_{21}; \min\{\mu_{\tilde{A}_2}, \mu_{\tilde{B}_2}\}, \max\{\nu_{\tilde{A}_2}, \nu_{\tilde{B}_2}\}), \\
&\quad (a_{31} + b_{31}, m_1 - m_4, m_2 - m_3, a_{34} - b_{11}; \min\{\mu_{\tilde{A}_3}, \mu_{\tilde{B}_1}\}, \max\{\nu_{\tilde{A}_3}, \nu_{\tilde{B}_1}\}); \\
&\quad \mu_{\tilde{A}-\tilde{B}}, \nu_{\tilde{A}-\tilde{B}})
\end{aligned}$$

where, $\mu_{\tilde{A}-\tilde{B}} = \min\{\min\{\mu_{\tilde{A}_1}, \mu_{\tilde{B}_3}\}, \min\{\mu_{\tilde{A}_2}, \mu_{\tilde{B}_2}\}, \min\{\mu_{\tilde{A}_3}, \mu_{\tilde{B}_1}\}\}$ and $\nu_{\tilde{A}-\tilde{B}} = \max\{\max\{\nu_{\tilde{A}_1}, \nu_{\tilde{B}_3}\}, \max\{\nu_{\tilde{A}_2}, \nu_{\tilde{B}_2}\}, \max\{\nu_{\tilde{A}_3}, \nu_{\tilde{B}_1}\}\}$.

- **Scalar multiplication** if $k \geq 0$:

$$k\tilde{A} = \begin{pmatrix} (ka_{11}, km_1, km_2, ka_{14}; \mu_{\tilde{A}_1}, \nu_{\tilde{A}_1}), (ka_{21}, km_1, km_2, ka_{24}; \mu_{\tilde{A}_2}, \nu_{\tilde{A}_2}), \\ (ka_{31}, km_1, km_2, ka_{34}; \mu_{\tilde{A}_3}, \nu_{\tilde{A}_3}); \mu_{\tilde{A}}, \nu_{\tilde{A}} \end{pmatrix},$$

- **Scalar multiplication** if $k < 0$:

$$k\tilde{A} = \begin{pmatrix} (ka_{14}, km_2, km_1, ka_{11}; \mu_{\tilde{A}_1}, \nu_{\tilde{A}_1}), (ka_{24}, km_2, km_1, ka_{21}; \mu_{\tilde{A}_2}, \nu_{\tilde{A}_2}), \\ (ka_{34}, km_2, km_1, ka_{31}; \mu_{\tilde{A}_3}, \nu_{\tilde{A}_3}); \mu_{\tilde{A}}, \nu_{\tilde{A}} \end{pmatrix},$$

2.8.2.2 Ranking Function for GSTrT2IFNs

For the GSTrT2IFN

$$\begin{aligned}
\tilde{A} = ((&a_{11}, m_1, m_2, a_{14}; \mu_{\tilde{A}_1}, \nu_{\tilde{A}_1}), (a_{21}, m_1, m_2, a_{24}; \mu_{\tilde{A}_2}, \nu_{\tilde{A}_2}), \\
&(a_{31}, m_1, m_2, a_{34}; \mu_{\tilde{A}_3}, \nu_{\tilde{A}_3}); \mu_{\tilde{A}}, \nu_{\tilde{A}})
\end{aligned}$$

where $\tilde{A}_1 = (a_{11}, m_1, m_2, a_{14}; \mu_{\tilde{A}_1}, \nu_{\tilde{A}_1})$, $\tilde{A}_2 = (a_{21}, m_1, m_2, a_{24}; \mu_{\tilde{A}_2}, \nu_{\tilde{A}_2})$, and $\tilde{A}_3 = (a_{31}, m_1, m_2, a_{34}; \mu_{\tilde{A}_3}, \nu_{\tilde{A}_3})$, the below ranking function is defined [12]. This ranking function has the property of $\Re(k\tilde{A}) = k\Re(\tilde{A})$.

$$\begin{aligned}
\Re(\tilde{A}) = \frac{1}{24}&[(a_{11} + m_1 + m_2 + a_{14})(\mu_{\tilde{A}_1} - \nu_{\tilde{A}_1} + 1) \\
&+ (a_{21} + m_1 + m_2 + a_{24})(\mu_{\tilde{A}_2} - \nu_{\tilde{A}_2} + 1) \\
&\times (a_{31} + m_1 + m_2 + a_{34})(\mu_{\tilde{A}_3} - \nu_{\tilde{A}_3} + 1)]
\end{aligned} \tag{2.74}$$

Definition 2.35 For the GSTrT2IFNs $\tilde{A}$ and $\tilde{B}$, the below relationships are defined [12].

$$\tilde{A} \preceq \tilde{B} \Leftrightarrow \Re(\tilde{A}) \leq \Re(\tilde{B}), \tag{2.75}$$

$$\tilde{A} \approx \tilde{B} \Leftrightarrow \Re(\tilde{A}) = \Re(\tilde{B}), \tag{2.76}$$

$$\tilde{A} \succeq \tilde{B} \Leftrightarrow \Re(\tilde{A}) \geq \Re(\tilde{B}), \tag{2.77}$$

$$\min(\tilde{A}, \tilde{B}) = \tilde{A} \Leftrightarrow \tilde{A} \preceq \tilde{B}. \tag{2.78}$$

2.9 Distance of Intuitionistic Fuzzy Sets

As any fuzzy set in X such as $\tilde{A}$ can be represented to an intuitionistic fuzzy set in X such as $\tilde{A}^I = \{(x, \mu_{\tilde{A}}(x), 1 - \mu_{\tilde{A}}(x)) : x \in X\}$, the distances of fuzzy sets also can be extended to intuitionistic fuzzy sets.

Definition 2.36 Assume $\tilde{A}^I = \{(x, \mu_{\tilde{A}^I}(x), \nu_{\tilde{A}^I}(x)) : x \in X\}$ and $\tilde{B}^I = \{(x, \mu_{\tilde{B}^I}(x), \nu_{\tilde{B}^I}(x)) : x \in X\}$ in the reference set $X = \{x_1, x_2, \ldots, x_n\}$. Then, some distances of a pair of intuitionistic fuzzy sets are as below,

- $d_1^I(\tilde{A}^I, \tilde{B}^I) = \sum_{i=1}^{n} (|\mu_{\tilde{A}^I}(x_i) - \mu_{\tilde{B}^I}(x_i)| + |\nu_{\tilde{A}^I}(x_i) - \nu_{\tilde{B}^I}(x_i)| + |\pi_{\tilde{A}^I}(x_i) - \pi_{\tilde{B}^I}(x_i)|)$,
- $d_2^I(\tilde{A}^I, \tilde{B}^I) = \frac{1}{2} \sum_{i=1}^{n} (|\mu_{\tilde{A}^I}(x_i) - \mu_{\tilde{B}^I}(x_i)| + |\nu_{\tilde{A}^I}(x_i) - \nu_{\tilde{B}^I}(x_i)| + |\pi_{\tilde{A}^I}(x_i) - \pi_{\tilde{B}^I}(x_i)|)$,
- $d_3^I(\tilde{A}^I, \tilde{B}^I) = \sqrt{\sum_{i=1}^{n} (\mu_{\tilde{A}}(x_i) - \mu_{\tilde{B}}(x_i))^2 + (\nu_{\tilde{A}^I}(x_i) - \nu_{\tilde{B}^I}(x_i))^2 + (\pi_{\tilde{A}^I}(x_i) - \pi_{\tilde{B}^I}(x_i))^2}$,
- $d_4^I(\tilde{A}^I, \tilde{B}^I) = \sqrt{\frac{1}{2} \sum_{i=1}^{n} (\mu_{\tilde{A}}(x_i) - \mu_{\tilde{B}}(x_i))^2 + (\nu_{\tilde{A}^I}(x_i) - \nu_{\tilde{B}^I}(x_i))^2 + (\pi_{\tilde{A}^I}(x_i) - \pi_{\tilde{B}^I}(x_i))^2}$
- $d_5^I(\tilde{A}^I, \tilde{B}^I) = \frac{1}{2n} \sum_{i=1}^{n} (|\mu_{\tilde{A}^I}(x_i) - \mu_{\tilde{B}^I}(x_i)| + |\nu_{\tilde{A}^I}(x_i) - \nu_{\tilde{B}^I}(x_i)| + |\pi_{\tilde{A}^I}(x_i) - \pi_{\tilde{B}^I}(x_i)|)$
- $d_6^I(\tilde{A}^I, \tilde{B}^I) = \sqrt{\frac{1}{2n} \sum_{i=1}^{n} (\mu_{\tilde{A}}(x_i) - \mu_{\tilde{B}}(x_i))^2 + (\nu_{\tilde{A}^I}(x_i) - \nu_{\tilde{B}^I}(x_i))^2 + (\pi_{\tilde{A}^I}(x_i) - \pi_{\tilde{B}^I}(x_i))^2}$

2.10 Application: Data Envelopment Analysis with Intuitionistic Fuzzy Data

Data envelopment analysis (DEA) is a non-parametric approach for measuring performance of a system which takes inputs and outputs of the system in order to measure its efficiency [15]. Here, the system is denoted as decision making unit (DMU). In the case of multiple DMUs, further than measuring their efficiencies, they may be ranked according to the measured performances. DEA for the first time was proposed by Charnes et al. [5] based on the idea of Farrell [7]. The DEA model proposed by Charnes et al. [5] which is called CCR model is represented by below formulation.

$$
\begin{aligned}
&\max \ \frac{\sum_{r=1}^{s} y_{rp} U_r}{\sum_{i=1}^{m} x_{ip} V_i} \\
&\text{subject to} \\
&\frac{\sum_{r=1}^{s} y_{rj} U_r}{\sum_{i=1}^{m} x_{ij} V_i} \le 1, \quad j = 1, \ldots, n, \\
&U_r \ge 0, \qquad r = 1, \ldots, s, \\
&V_i \ge 0. \qquad i = 1, \ldots, m,
\end{aligned}
\tag{2.79}
$$

The parameters and variables of the model (2.80) and the intuitionistic fuzzy DEA model that will be presented are summarized by Table 2.1.

Table 2.1 Notations used in the DEA models of the paper

Notation	Type	Description
n	Parameter	The number of DMUs,
m	Parameter	The number of inputs,
s	Parameter	The number of outputs,
j	Index	The index used for DMUs,
p	Index	The index used for the DMU being evaluated,
i	Index	The index used for inputs
r	Index	The index used for outputs,
x_{ij}	Parameter	The value of j-th DMU in i-th input,
y_{rj}	Parameter	The value of j-th DMU in r-th output,
$\tilde{x}_{ij}$	Parameter	The intuitionistic fuzzy value of j-th DMU in i-th inpu $\left(\tilde{x}_{ij} = \left(x_{ij}^1, x_{ij}^2, x_{ij}^3, x_{ij}^4; x_{ij}^5, x_{ij}^2, x_{ij}^3, x_{ij}^6\right)\right)$,
$\tilde{y}_{rj}$	Parameter	The intuitionistic fuzzy value of j-th DMU in r-th output $\left(\tilde{y}_{rj} = \left(y_{rj}^1, y_{rj}^2, y_{rj}^3, y_{rj}^4; y_{rj}^5, y_{rj}^2, y_{rj}^3, y_{rj}^6\right)\right)$,
V_i	Variable	The weight of i-th input,
U_r	Variable	The weight of r-th output

As the CCR model (2.80) is a non-linear model, it can be easily linearized according to the Charnes-Cooper transformation [3] as below,

$$\begin{aligned}
&\max \sum_{r=1}^{s} y_{rp} U_r \\
&\text{subject to} \\
&\sum_{i=1}^{m} x_{ip} V_i = 1, \\
&\sum_{r=1}^{s} y_{rj} U_r - \sum_{i=1}^{m} x_{ij} V_i \leq 0, \; j = 1, \ldots, n, \\
&U_r \geq 0, \qquad r = 1, \ldots, s, \\
&V_i \geq 0. \qquad i = 1, \ldots, m,
\end{aligned} \tag{2.80}$$

In model (2.80), the input and output values are positive. This model is called the input-oriented multiplier form CCR model (CCR-IO model) for measuring the efficiency value of p-th DMU. The intuitionistic fuzzy form of the CCR-IO model (2.80) for DMU p is defined as below,

$$\begin{aligned}
&\max \sum_{r=1}^{s} \tilde{y}_{rp} U_r \\
&\text{subject to} \\
&\sum_{i=1}^{m} \tilde{x}_{ip} V_i \leq 1, \\
&\sum_{r=1}^{s} \tilde{y}_{rj} U_r - \sum_{i=1}^{m} \tilde{x}_{ij} V_i \leq 0, \; j = 1, \ldots, n, \\
&U_r \geq 0, \qquad r = 1, \ldots, s, \\
&V_i \geq 0. \qquad i = 1, \ldots, m,
\end{aligned} \tag{2.81}$$

where, the intuitionistic fuzzy coefficients of this model are shown as below,

$$\begin{aligned}
\tilde{x}_{ij} &= \left(x_{ij}^{1}, x_{ij}^{2}, x_{ij}^{3}, x_{ij}^{4}; x_{ij}^{5}, x_{ij}^{2}, x_{ij}^{3}, x_{ij}^{6}\right), \quad i = 1, \ldots, m, j = 1, \ldots, n, \\
\tilde{y}_{rj} &= \left(y_{rj}^{1}, y_{rj}^{2}, y_{rj}^{3}, y_{rj}^{4}; y_{rj}^{5}, y_{rj}^{2}, y_{rj}^{3}, y_{rj}^{6}\right), \quad r = 1, \ldots, s, j = 1, \ldots, n.
\end{aligned} \tag{2.82}$$

It is notable to mention that, the constraint $\sum_{i=1}^{m} \tilde{x}_{ip} V_i = 1$, in deterministic model (2.80) is converted to the intuitionistic fuzzy constraint $\sum_{i=1}^{m} \tilde{x}_{ip} V_i \leq 1$, in intuitionistic fuzzy model (2.81) [14].

In continue, a possibility-theory-based approach is proposed to handle the fuzziness of the CCR-IO model. Possibility theory is the simplest uncertainty theory devoted to the modelling of incomplete information [10, 11]. This approach is an efficient method that relies on high mathematical fuzzy concepts and can support different kinds of fuzzy numbers such as triangular and trapezoidal fuzzy numbers. This method enables decision maker to satisfy some chance constraints at some given confidence levels. In this method, decision maker sets possibility distributions for the parameters of a fuzzy model.

As mentioned, the *ME* measure defined in Sect. 2.5 is an effective measure of the possibility theory which provides more flexibility to decision makers comparing to the classical measures such as credibility, necessity, and possibility measures. Based on the literature, the *ME* measure has not been applied to the intuitionistic fuzzy DEA models yet. In next, the *ME* measure is applied to intuitionistic fuzzy CCR-IO model. For this aim, first the new continuous variable f_p is considered and the below equivalent chance-constrained form is obtained for model (2.81) using the *ME* measure.

$$
\begin{aligned}
& \max f_p \\
& \text{subject to} \\
& ME_\mu\left\{\sum_{r=1}^{s} \tilde{y}_{rp} U_r \geq f_p\right\} \geq \theta_1, \\
& ME_\upsilon\left\{\sum_{r=1}^{s} \tilde{y}_{rp} U_r \geq f_p\right\} \leq \theta_2, \\
& ME_\mu\left\{\sum_{i=1}^{m} \tilde{x}_{ip} V_i \leq 1\right\} \geq \theta_3, \\
& ME_\upsilon\left\{\sum_{i=1}^{m} \tilde{x}_{ip} V_i \leq 1\right\} \leq \theta_4, \\
& ME_\mu\left\{\sum_{r=1}^{s} \tilde{y}_{rj} U_r - \sum_{i=1}^{m} \tilde{x}_{ij} V_i \leq 0\right\} \geq \beta_j,\ j = 1, \ldots, n, \\
& ME_\upsilon\left\{\sum_{r=1}^{s} \tilde{y}_{rj} U_r - \sum_{i=1}^{m} \tilde{x}_{ij} V_i \leq 0\right\} \leq \gamma_j,\ j = 1, \ldots, n, \\
& U_r \geq 0, \qquad r = 1, \ldots, s, \\
& V_i \geq 0. \qquad i = 1, \ldots, m,
\end{aligned}
\tag{2.83}
$$

In this model, $\theta_1, \theta_2, \theta_3, \theta_4, \beta_j$, and γ_j, are confidence levels for occurrence of the chance constraints. The following conditions should be held for these confidence levels,

- $0 \leq \theta_1 + \theta_2 \leq 1$,
- $0 \leq \theta_3 + \theta_4 \leq 1$,
- $0 \leq \beta_j + \gamma_j \leq 1 \quad (j = 1, \ldots, n)$.

Because of the chance constraints, model (2.83) is a complicated model to be solved by traditional methods of optimization software. Therefore, first the model should be converted to an equivalent linear form. For this aim, the below proposition is proposed. This proposition is used to convert the previously mentioned claims (2.40)–(2.43) to their equivalent crisp form. Now, model (2.83) can be converted to a solvable form using the above-mentioned definition. Therefore, the equivalent crisp form of model (2.83) is easily obtained as below model,

$$
\begin{aligned}
&\max f_p \\
&\text{subject to} \\
&\left(\tfrac{\theta_1-\lambda}{1-\lambda}\right)\sum_{r=1}^{s} y_{rp}^{1}U_r + \left(\tfrac{1-\theta_1}{1-\lambda}\right)\sum_{r=1}^{s} y_{rp}^{2}U_r \geq f_p, && \theta_1 > \lambda, \\
&\tfrac{\theta_1}{\lambda}\sum_{r=1}^{s} y_{rp}^{3}U_r + \left(\tfrac{\lambda-\theta_1}{\lambda}\right)\sum_{r=1}^{s} y_{rp}^{4}U_r \geq f_p, && \theta_1 \leq \lambda, \\
&\left(\tfrac{\theta_2}{1-\lambda}\right)\sum_{r=1}^{s} y_{rp}^{2}U_r + \left(\tfrac{1-\lambda-\theta_2}{1-\lambda}\right)\sum_{r=1}^{s} y_{rp}^{5}U_r \geq f_p, && \theta_2 \leq 1-\lambda, \\
&\left(1-\tfrac{1}{\lambda}+\tfrac{\theta_2}{\lambda}\right)\sum_{r=1}^{s} y_{rp}^{6}U_r + \left(\tfrac{1-\theta_2}{\lambda}\right)\sum_{r=1}^{s} y_{rp}^{3}U_r \geq f_p, && \theta_2 > 1-\lambda, \\
&\left(1-\tfrac{\theta_3}{\lambda}\right)\sum_{i=1}^{m} x_{ip}^{1}V_i + \tfrac{\theta_3}{\lambda}\sum_{i=1}^{m} x_{ip}^{2}V_i \leq 1, && \theta_3 \leq \lambda, \\
&\left(\tfrac{1-\theta_3}{1-\lambda}\right)\sum_{i=1}^{m} x_{ip}^{3}V_i + \left(\tfrac{\theta_3-\lambda}{1-\lambda}\right)\sum_{i=1}^{m} x_{ip}^{4}V_i \leq 1, && \theta_3 > \lambda, \\
&\left(\tfrac{\theta_4}{1-\lambda}\right)\sum_{i=1}^{m} x_{ip}^{3}V_i + \left(1-\tfrac{\theta_4}{1-\lambda}\right)\sum_{i=1}^{m} x_{ip}^{6}V_i \leq 1, && \theta_4 \leq 1-\lambda, \\
&\left(\tfrac{1-\theta_4}{\lambda}\right)\sum_{i=1}^{m} x_{ip}^{3}V_i + \left(\tfrac{\theta_4+\lambda-1}{\lambda}\right)\sum_{i=1}^{m} x_{ip}^{5}V_i \leq 1, && \theta_4 > 1-\lambda \\
&\left(\frac{1-\beta_j}{1-\lambda}\right)\sum_{r=1}^{s} y_{rj}^{3}U_r + \left(\frac{\beta_j-\lambda}{1-\lambda}\right)\sum_{r=1}^{s} y_{rj}^{4}U_r \\
&\quad -\left(\frac{\beta_j-\lambda}{1-\lambda}\right)\sum_{i=1}^{m} x_{ij}^{1}V_i - \left(\frac{1-\beta_j}{1-\lambda}\right)\sum_{i=1}^{m} x_{ij}^{2}V_i \leq 0, && \beta_j > \lambda,\ j=1,2,\ldots,n, \\
&\left(1-\frac{\beta_j}{\lambda}\right)\sum_{r=1}^{s} y_{rj}^{1}U_r + \frac{\beta_j}{\lambda}\sum_{r=1}^{s} y_{rj}^{2}U_r \\
&\quad -\left(1-\frac{\beta_j}{\lambda}\right)\sum_{i=1}^{m} x_{ij}^{4}V_i - \frac{\beta_j}{\lambda}\sum_{i=1}^{m} x_{ij}^{3}V_i \leq 0, && \beta_j \leq \lambda,\ j=1,2,\ldots,n, \\
&\left(\frac{1-\gamma_j}{\lambda}\right)\sum_{r=1}^{s} y_{rj}^{2}U_r + \left(\frac{\gamma_j+\lambda-1}{\lambda}\right)\sum_{r=1}^{s} y_{rj}^{5}U_r \\
&\quad -\left(\frac{\gamma_j+\lambda-1}{\lambda}\right)\sum_{i=1}^{m} x_{ij}^{6}V_i - \left(\frac{1-\gamma_j}{\lambda}\right)\sum_{i=1}^{m} x_{ij}^{3}V_i \leq 0, && \gamma_j > 1-\lambda,\ j=1,2,\ldots,n, \\
&\left(\frac{\gamma_j}{1-\lambda}\right)\sum_{r=1}^{s} y_{rj}^{3}U_r + \left(1-\frac{\gamma_j}{1-\lambda}\right)\sum_{r=1}^{s} y_{rj}^{6}U_r \\
&\quad -\left(1-\frac{\gamma_j}{1-\lambda}\right)\sum_{i=1}^{m} x_{ij}^{5}V_i - \left(\frac{\gamma_j}{1-\lambda}\right)\sum_{i=1}^{m} x_{ij}^{2}V_i \leq 0, && \gamma_j \leq 1-\lambda,\ j=1,2,\ldots,n, \\
&U_r \geq 0, && r=1,2,\ldots,s, \\
&V_i \geq 0. && i=1,2,\ldots,m,
\end{aligned}
\tag{2.84}
$$

Model (2.84) is an equivalent crisp form of the chance-constrained model (2.83). As can be seen, it involves all possible values of the confidence levels and λ. As, optimistic-pessimistic parameter λ and also the confidence levels can take different values from 0 to 1, their values should determined by decision maker. Actually, decision maker set these values and solves model (2.84) for each DMU in order to determine efficiency or inefficiency situation of the DMU. When the values of these confidence levels and λ are set, according to the values the extra constrains of model (2.84) are removed and a sub-model is obtained.

According to what mentioned above, the below definitions can be mentioned in order to decide about situation of DMUs [14, 16].

Definition 2.37 For a certain set of values considered for optimistic-pessimistic parameter λ and confidence levels $\theta_1, \theta_2, \theta_3, \theta_4, \beta_j$, and γ_j, the associated sub-model of this set of values is determined and solved for each DMU. A DMU is efficient (E) if and only if the optimal value of f_p be greater than or equal to 1.

Definition 2.38 A DMU is marginal-efficient (E+) if its f_p values for at least one set of values of optimistic-pessimistic parameter λ and confidence levels $\theta_1, \theta_2, \theta_3, \theta_4, \beta_j$, and γ_j, be greater than or equal to 1.

Definition 2.39 A DMU is ultra-efficient (E++) if its f_p values for all sets of values of optimistic-pessimistic parameter λ and confidence levels $\theta_1, \theta_2, \theta_3, \theta_4, \beta_j$, and γ_j, be greater than or equal to 1.

Definition 2.40 A DMU is inefficient (E-) if its f_p values for all sets of values of optimistic-pessimistic parameter λ and confidence levels $\theta_1, \theta_2, \theta_3, \theta_4, \beta_j$, and γ_j, be less than 1.

References

1. Atanassov, K.T.: Intuitionistic fuzzy sets. Fuzzy Sets Syst. **20**, 87–96 (1986)
2. Atanassov, K., Gargov, G.: Interval valued intuitionistic fuzzy sets. Fuzzy Sets Syst. **31**(3), 343–349 (1989)
3. Charnes, A., Cooper, W.W.: Chance-constrained programming. Manage. Sci. **6**(1), 73–79 (1959)
4. Chen, S.M., Tan, J.M.: Handling multi-criteria fuzzy decision making problems based on vague set theory. Fuzzy Sets Syst. **67**(2), 163–172 (1994)
5. Charnes, A., Cooper, W.W., Rhodes, E.: Measuring the efficiency of decision making units. Eur. J. Oper. Res. **2**(6), 429–444 (1978)
6. Dan, S., Kar, M.B., Majumder, S., Roy, B., Kar, S., Pamucar, D.: Intuitionistic type-2 fuzzy set and its properties. Symmetry **11**(6), 808 (2019)
7. Farrell, M.J.: The measurement of productive efficiency. J. R. Stat. Soc. Ser. A (General) **120**(3), 253–290 (1957)
8. Garg, H., Rani, M., Sharma, S.P., Vishwakarma, Y.: Intuitionistic fuzzy optimization technique for solving multi-objective reliability optimization problems in interval environment. Expert Syst. Appl. **41**(7), 3157–3167 (2014)
9. Hong, D.H., Choi, C.H.: Multi-criteria fuzzy decision-making problems based on vague set theory. Fuzzy Sets Syst. **114**(1), 103–113 (2000)
10. Jamshidi, M., Sanei, M., Mahmoodirad, A., Lotfi, F.H., Tohidi, G.: Uncertain SBM data envelopment analysis model: a case study in Iranian banks. Int. J. Financ. Econ. **26**(2), 2674–2689 (2021)

11. Jamshidi, M., Saneie, M., Mahmoodirad, A., Lotfi, F.H., Tohidi, G.: Uncertain RUSSEL data envelopment analysis model: a case study in Iranian banks. J. Intell. Fuzzy Syst. **37**(2), 2937–2951 (2019)
12. Jansi Rani, J., Manivannan, A.: An application of generalized symmetric type-2 intuitionistic fuzzy variables to a transportation problem with the effect of a new ranking function. Expert Syst. Appl. **237**, 121384 (2024)
13. Mahmoodirad, A., Allahviranloo, T., Niroomand, S.: A new effective solution method for fully intuitionistic fuzzy transportation problem. Soft. Comput. **23**, 4521–4530 (2019)
14. Mahmoodirad, A., Pamucar, D., Niroomand, S.: A new intuitionistic fuzzy scheme of data envelopment analysis for evaluating rural comprehensive health service centers. Socioecon. Plann. Sci. **95**, 102004 (2024)
15. Modhej, D., Sanei, M., Shoja, N., HosseinzadehLotfi, F.: Integrating inverse data envelopment analysis and neural network to preserve relative efficiency values. J. Intell. Fuzzy Syst. **32**(6), 4047–4058 (2017)
16. Peykani, P., Mahmoodirad, A., Amirteimoori, A.: A novel adjustable intuitionistic fuzzy framework for two-stage data envelopment analysis: an application in the banking sector. Inform. Sci. 122372 (2025)
17. Singh, S.K., Yadav, S.P.: A new approach for solving intuitionistic fuzzy transportation problem of type-2. Ann. Oper. Res. **243**(1), 349–363 (2014)
18. Singh, S., Garg, H.: Distance measures between type-2 intuitionistic fuzzy sets and their application to multicriteria decision-making process. Appl. Intell. **46**, 788–799 (2017)
19. Stoeva, M.: Intuitionistic fuzzy extension principle. NIFS **5**, 29–30 (1999)
20. Xu, Z.S., Chen, J.: An approach to group decision making based on interval-valued intuitionistic judgment matrices. Syst. Eng. Theory Pract. **27**(4), 126–133 (2007)
21. Xu, Z.S.: Intuitionistic preference relations and their application in group decision making. Inf. Sci. **177**(11), 2363–2379 (2007)
22. Xu, J., Zhou, X.: Approximation based fuzzy multi-objective models with expected objectives and chance constraints: application to earth-rock work allocation. Inf. Sci. **238**, 75–95 (2013)

Chapter 3
Pythagorean Fuzzy Sets

3.1 Introduction

After introducing the fuzzy sets by [1], this was used by many researchers to represent the uncertain nature of the optimization problems [2]. Therefore, some extensions of the classical fuzzy sets and numbers have been introduced in the literature. As of important extensions of fuzzy sets, Pythagorean fuzzy sets can be mentioned which was introduced by [3].

Pythagorean fuzzy sets are one step after intuitionistic fuzzy sets. Generally, the aim of Pythagorean fuzzy sets is to consider more vagueness in representing uncertain events. Therefore, despite the intuitionistic fuzzy sets, in the Pythagorean fuzzy sets, the sum of membership and non-membership of an element of the reference set is allowed to be greater than 1 (therefore, $\mu(x) + \nu(x) \geq 0$) with the limitation of $0 \leq \mu(x)^2 + \nu(x)^2 \leq 1$. The below example can explain this idea.

Example 3.1 If an expert wants to obtain the satisfaction degree of a service with membership function degree of $\frac{\sqrt{7}}{3}$ and to remove the dissatisfaction degree of the service with non-membership function degree of $\frac{1}{3}$, as $\frac{\sqrt{7}}{3} + \frac{1}{3} \geq 1$, therefore, the intuitionistic fuzzy sets are not suitable to represent such uncertain situation. This uncertain situation can be represented by Pythagorean fuzzy sets as $\left(\frac{\sqrt{7}}{3}\right)^2 + \left(\frac{1}{3}\right)^2 \leq 1$.

3.2 Pythagorean Fuzzy Sets

The concepts and definitions related to Pythagorean fuzzy sets are introduced in this section.

A. Mahmoodirad et al., *Recent Advances on Fuzzy Sets*, Studies in Fuzziness and Soft Computing 440, https://doi.org/10.1007/978-3-032-08776-8_3

Definition 3.1 [3] Consider the reference set X. The Pythagorean fuzzy set $\tilde{P}$ on this reference set is defined as below.

$$\tilde{P} = \left\{ \left(x, \mu_{\tilde{P}}(x), \nu_{\tilde{P}}(x)\right) : x \in X \right\} \tag{3.1}$$

In this fuzzy set, $\mu_{\tilde{P}}(x) : X \to [0, 1]$ and $\nu_{\tilde{P}}(x) : X \to [0, 1]$ are the membership and non-membership function of $x \in X$ where $0 \leq \left(\mu_{\tilde{P}}(x)\right)^2 + \left(\nu_{\tilde{P}}(x)\right)^2 \leq 1$.

Definition 3.2 [3] Hesitancy degree of $x \in X$ in the Pythagorean fuzzy set $\tilde{P}$ is shown as $\pi_{\tilde{P}}(x)$ and is calculated by the below formula.

$$\pi_{\tilde{P}}(x) = \sqrt{1 - \left(\left(\mu_{\tilde{P}}(x)\right)^2 + \left(\nu_{\tilde{P}}(x)\right)^2\right)} \tag{3.2}$$

Definition 3.3 [4] A Pythagorean fuzzy number is shown as $\tilde{P}\left(\mu_{\tilde{P}}(x), \nu_{\tilde{P}}(x)\right)$. By assuming $\mu_{\tilde{P}}(x) = \mu_\beta$ and $\nu_{\tilde{P}}(x) = \nu_\beta$, the number is shown as $\beta = \tilde{P}\left(\mu_\beta, \nu_\beta\right)$ where $\mu_\beta, \nu_\beta \in [0, 1]$, $\pi_\beta = \sqrt{1 - \left(\mu_\beta\right)^2 - \left(\nu_\beta\right)^2}$, and $\left(\mu_\beta\right)^2 + \left(\nu_\beta\right)^2 \leq 1$.

Therefore, $\tilde{P}(1, 0)$ is the largest Pythagorean fuzzy number (called the positive ideal Pythagorean fuzzy number) and $\tilde{P}(0, 1)$ is the smallest Pythagorean fuzzy number (called the negative ideal Pythagorean fuzzy number).

Definition 3.4 [4] For the Pythagorean fuzzy numbers $\beta_1 = \tilde{P}\left(\mu_{\beta_1}, \nu_{\beta_1}\right)$, $\beta_2 = \tilde{P}\left(\mu_{\beta_2}, \nu_{\beta_2}\right)$, and $\beta_3 = \tilde{P}\left(\mu_{\beta_3}, \nu_{\beta_3}\right)$, the below operations are defined.

$$\beta_1 \cup \beta_2 = \tilde{P}\left(\max\left\{\mu_{\beta_1}(x), \mu_{\beta_2}(x)\right\}, \min\left\{\nu_{\beta_1}(x), \nu_{\beta_2}(x)\right\}\right) \tag{3.3}$$

$$\beta_1 \cap \beta_2 = \tilde{P}\left(\min\left\{\mu_{\beta_1}(x), \mu_{\beta_2}(x)\right\}, \max\left\{\nu_{\beta_1}(x), \nu_{\beta_2}(x)\right\}\right) \tag{3.4}$$

$$\beta_1^c = \tilde{P}\left(\nu_{\beta_1}, \mu_{\beta_1}\right) \tag{3.5}$$

$$\beta_1 \oplus \beta_2 = \tilde{P}\left(\sqrt{\mu_{\beta_1}^2 + \mu_{\beta_2}^2 - \mu_{\beta_1}^2 \mu_{\beta_2}^2}, \nu_{\beta_1} \nu_{\beta_2}\right) \tag{3.6}$$

$$\beta_1 \ominus \beta_2 = \tilde{P}\left(\sqrt{\frac{\mu_{\beta_1}^2 - \mu_{\beta_2}^2}{1 - \mu_{\beta_2}^2}}, \frac{\nu_{\beta_1}}{\nu_{\beta_2}}\right) \text{if} \quad \mu_{\beta_1} \geq \mu_{\beta_2};$$

$$\nu_{\beta_1} \leq \min\left\{\nu_{\beta_2}, \frac{\nu_{\beta_2} \pi_{\beta_2}}{\pi_{\beta_2}}\right\} \tag{3.7}$$

$$\beta_1 \otimes \beta_2 = \tilde{P}\left(\mu_{\beta_1} \mu_{\beta_2}, \sqrt{\nu_{\beta_1}^2 + \nu_{\beta_2}^2 - \nu_{\beta_1}^2 \nu_{\beta_2}^2}\right) \tag{3.8}$$

$$\beta_1 \oslash \beta_2 = \tilde{P}\left(\frac{\mu_{\beta_1}}{\mu_{\beta_2}}, \sqrt{\frac{\nu_{\beta_1}^2 - \nu_{\beta_2}^2}{1 - \nu_{\beta_2}^2}}\right); \text{if } \nu_{\beta_1} \geq \nu_{\beta_2};$$

$$\nu_{\beta_1} \leq \min\left\{\mu_{\beta_2}, \frac{\mu_{\beta_2}\pi_{\beta_2}}{\pi_{\beta_2}}\right\} \tag{3.9}$$

$$\lambda\beta_1 = \tilde{P}\left(\sqrt{1 - \left(1 - \mu_{\beta_1}^2\right)^{\lambda}}, \left(\nu_{\beta_1}\right)^{\lambda}\right), \lambda > 0, \tag{3.10}$$

$$(\beta_1)^{\lambda} = \tilde{P}\left(\mu_{\beta_1}^{\lambda}, \sqrt{1 - \left(1 - \nu_{\beta_1}^2\right)^{\lambda}}\right), \lambda > 0, \tag{3.11}$$

Theorem 3.1 [4] *For the Pythagorean fuzzy numbers* $\beta_1 = \tilde{P}\left(\mu_{\beta_1}, \nu_{\beta_1}\right)$, $\beta_2 = \tilde{P}\left(\mu_{\beta_2}, \nu_{\beta_2}\right)$, *and* $\beta_3 = \tilde{P}\left(\mu_{\beta_3}, \nu_{\beta_3}\right)$, *the below relations are true.*

$$\beta_1 \oplus \beta_2 = \beta_2 \oplus \beta_1 \tag{3.12}$$

$$\beta_1 \otimes \beta_2 = \beta_2 \otimes \beta_1 \tag{3.13}$$

$$\lambda(\beta_1 \oplus \beta_2) = \lambda\beta_1 \oplus \lambda\beta_2, \lambda > 0, \tag{3.14}$$

$$\lambda_1\beta_1 \oplus \lambda_2\beta_1 = (\lambda_1 + \lambda_2)\beta_1, \lambda_1, \lambda_2 > 0, \tag{3.15}$$

$$(\beta_1 \otimes \beta_2)^{\lambda} = \beta_1^{\lambda} \otimes \beta_2^{\lambda}, \lambda > 0, \tag{3.16}$$

$$\beta_1^{\lambda_1} \otimes \beta_1^{\lambda_2} = \beta_1^{\lambda_1 + \lambda_2}, \lambda_1, \lambda_2 > 0, \tag{3.17}$$

$$(\beta_1)^{\lambda} = \tilde{P}\left(\mu_{\beta_1}^{\lambda}, \sqrt{1 - \left(1 - \nu_{\beta_1}^2\right)^{\lambda}}\right), \lambda > 0, \tag{3.18}$$

Example 3.2 Consider the Pythagorean fuzzy numbers $\beta_1 = \tilde{P}\left(\frac{\sqrt{3}}{2}, \frac{1}{2}\right)$ and $\beta_2 = \tilde{P}\left(\frac{\sqrt{2}}{2}, \frac{\sqrt{2}}{2}\right)$. Then,

$$\beta_1 \oplus \beta_2 = \tilde{P}\left(\sqrt{\mu_{\beta_1}^2 + \mu_{\beta_2}^2 - \mu_{\beta_1}^2\mu_{\beta_2}^2}, \nu_{\beta_1}\nu_{\beta_2}\right) = \tilde{P}\left(\frac{7}{8}, \frac{\sqrt{2}}{4}\right),$$

$$\beta_1 \otimes \beta_2 = \tilde{P}\left(\mu_{\beta_1}\mu_{\beta_2}, \sqrt{\nu_{\beta_1}^2 + \nu_{\beta_2}^2 - \nu_{\beta_1}^2\nu_{\beta_2}^2}\right) = \tilde{P}\left(\frac{\sqrt{6}}{4}, \frac{5}{8}\right),$$

$$3\beta_1 = \tilde{P}\left(\sqrt{1-\left(1-\mu_{\beta_1}^2\right)^3}, \left(\nu_{\beta_1}\right)^3\right) = \tilde{P}\left(\frac{\sqrt{63}}{8}, \frac{1}{8}\right),$$

$$(\beta_1)^2 = \tilde{P}\left(\mu_{\beta_1}^2, \sqrt{1-\left(1-\nu_{\beta_1}^2\right)^2}\right) = \tilde{P}\left(\frac{3}{4}, \frac{\sqrt{7}}{4}\right),$$

3.3 Ranking Function for Pythagorean Fuzzy Numbers

In the definitions of this section, some ranking functions of Pythagorean fuzzy numbers are introduced and detailed.

Definition 3.5 [4] For the Pythagorean fuzzy number $\beta_1 = \tilde{P}\left(\mu_{\beta_1}, \nu_{\beta_1}\right)$, the score function of is $S(\beta_1) = \left(\mu_{\beta_1}\right)^2 - \left(\nu_{\beta_1}\right)^2$ defined. Based on this function, for the Pythagorean fuzzy numbers $\beta_1 = \tilde{P}\left(\mu_{\beta_1}, \nu_{\beta_1}\right)$ and $\beta_2 = \tilde{P}\left(\mu_{\beta_2}, \nu_{\beta_2}\right)$, the relationships below are defined.

$$S(\beta_1) < S(\beta_2) \Rightarrow \beta_1 < \beta_2,$$

$$S(\beta_1) > S(\beta_2) \Rightarrow \beta_1 > \beta_2,$$

$$S(\beta_1) = S(\beta_2) \Rightarrow \beta_1 = \beta_2$$

Example 3.3 For the Pythagorean fuzzy numbers $\beta_1 = \tilde{P}\left(\frac{\sqrt{3}}{2}, \frac{1}{2}\right)$ and $\beta_2 = \tilde{P}\left(\frac{\sqrt{2}}{2}, \frac{\sqrt{2}}{2}\right)$ the below results are obtained.

$$S(\beta_1) = \left(\mu_{\beta_1}\right)^2 - \left(\nu_{\beta_1}\right)^2 = \frac{1}{2}, S(\beta_2) = \left(\mu_{\beta_2}\right)^2 - \left(\nu_{\beta_2}\right)^2$$
$$= 0, S(\beta_1) > S(\beta_2) \Rightarrow \beta_1 \succ \beta_2,$$

Definition 3.6 [5] Assume the Pythagorean fuzzy number $\beta = \tilde{P}\left(\mu_\beta, \nu_\beta\right)$ and the set of Pythagorean fuzzy numbers PFS. A ranking function such as $\Re : PFS \to \mathbb{R}$ for β is defined as below.

$$\Re(\boldsymbol{\beta}) = \frac{\left(\boldsymbol{\mu_\beta}\right)^2 + \left(\boldsymbol{\nu_\beta}\right)^2 - \left(\boldsymbol{\mu_\beta} - \boldsymbol{\nu_\beta}\right)^2}{2}. \tag{3.19}$$

Based on this function, for the Pythagorean fuzzy numbers $\beta_1 = \tilde{P}\left(\mu_{\beta_1}, \nu_{\beta_1}\right)$ and $\beta_2 = \tilde{P}\left(\mu_{\beta_2}, \nu_{\beta_2}\right)$, the relationships below are defined.

$$\Re(\beta_1) < \Re(\beta_2) \Rightarrow \beta_1 < \beta_2,$$

$$\Re(\beta_1) > \Re(\beta_2) \Rightarrow \beta_1 > \beta_2,$$

$$\Re(\beta_1) = \Re(\beta_2) \Rightarrow \beta_1 = \beta_2$$

Definition 3.7 [3] For the Pythagorean fuzzy number $\beta = \tilde{P}\big(\mu_\beta, \nu_\beta\big)$, the below score function ($S(\beta)$) and accuracy function ($A(\beta)$) are defined.

$$S(\beta) = \frac{1}{2}\Big(1 - (\mu_\beta)^2 - (\nu_\beta)^2\Big), \tag{3.20}$$

$$A(\beta) = (\mu_\beta)^2 + (\nu_\beta)^2. \tag{3.21}$$

Based on the functions (3.20) and (3.21), for the Pythagorean fuzzy numbers $\beta_1 = \tilde{P}\big(\mu_{\beta_1}, \nu_{\beta_1}\big)$ and $\beta_2 = \tilde{P}\big(\mu_{\beta_2}, \nu_{\beta_2}\big)$, the relationships below are defined.

$$S(\beta_1) < S(\beta_2) \Rightarrow \beta_1 \prec \beta_2,$$

$$S(\beta_1) > S(\beta_2) \Rightarrow \beta_1 \succ \beta_2,$$

$$S(\beta_1) = S(\beta_2), A(\beta_1) > A(\beta_2), \Rightarrow \beta_1 \succ \beta_2,$$

$$S(\beta_1) = S(\beta_2), A(\beta_1) < A(\beta_2), \Rightarrow \beta_1 \prec \beta_2,$$

$$S(\beta_1) = S(\beta_2), A(\beta_1) = A(\beta_2), \Rightarrow \beta_1 \approx \beta_2.$$

This is notable to mention that the functions presented by the definitions of this section are of non-linear type. Therefore, as a difficulty, these functions are not useful for being employed by a linear optimization model.

Definition 3.8 [6] A triangular Pythagorean fuzzy number is shown as $\tilde{A}^p = \big\{(a, b, c), \mu_{\tilde{A}^p}(x), \nu_{\tilde{A}^p}(x) : x \in X\big\}$ where $\mu_{\tilde{A}^p}(x) \geq 0$, $\nu_{\tilde{A}^p}(x) \geq 0$, $\big(\mu_{\tilde{A}^p}(x)\big)^2 + \big(\nu_{\tilde{A}^p}(x)\big)^2 \leq 1$, and $0 \leq a \leq b \leq c$ ((a, b, c) is a triangular fuzzy number). The membership and non-membership functions of this number are defined as below.

$$\mu_{\tilde{A}^p}(x) = \begin{cases} \frac{x-a}{b-a}; & a \leq x \leq b \\ \frac{c-x}{c-b}; & b \leq x \leq c \\ 0; & otherwise \end{cases} \tag{3.22}$$

$$\nu_{\tilde{A}^p}(x) = \begin{cases} \sqrt{1 - \left(\frac{x-a}{b-a}\right)^2}; & a \le x \le b \\ \sqrt{1 - \left(\frac{c-x}{c-b}\right)^2}; & b \le x \le c \\ 1; & otherwise, \end{cases} \tag{3.23}$$

Definition 3.9 [6] The below arithmetic operations are defined for the triangular Pythagorean fuzzy numbers $\tilde{A}^p = \left\{(a_1, b_1, c_1), \mu_{\tilde{A}^p}(x), \nu_{\tilde{A}^p}(x) : x \in X\right\}$ and $\tilde{B}^p = \left\{(a_2, b_2, c_2), \mu_{\tilde{B}^p}(x), \nu_{\tilde{B}^p}(x) : x \in X\right\}$ where $k \ge 0$.

$$\begin{aligned} \tilde{A}^p \oplus \tilde{B}^p = \{&(a_1 + a_2, b_1 + b_2, c_1 + c_2), \\ &\sqrt{\mu^2_{\tilde{A}^p}(x) + \mu^2_{\tilde{B}^p}(x) - \mu^2_{\tilde{A}^p}(x)\mu^2_{\tilde{B}^p}(x)}, \\ &\nu_{\tilde{A}^p}(x)\nu_{\tilde{B}^p}(x) : x \in X\big\}, \end{aligned} \tag{3.24}$$

$$\begin{aligned} \tilde{A}^p \odot \tilde{B}^p = \big\{&(a_1a_2, b_1b_2, c_1c_2), \mu_{\tilde{A}^p}(x)\mu_{\tilde{B}^p}(x), \\ &\sqrt{\nu^2_{\tilde{A}^p}(x) + \nu^2_{\tilde{B}^p}(x) - \nu^2_{\tilde{A}^p}(x)\nu^2_{\tilde{B}^p}(x)} : x \in X\big\}, \end{aligned} \tag{3.25}$$

$$k\tilde{A}^p = \left\{(ka_1, kb_1, kc_1), \sqrt{1 - \left(1 - \mu^2_{\tilde{A}^p}(x)\right)^k}, \nu^k_{\tilde{A}^p}(x) : x \in X\right\}, \tag{3.26}$$

$$\left(\tilde{A}^p\right)^k = \left\{\left(a_1^k, b_1^k, c_1^k\right), \mu^k_{\tilde{A}^p}(x), \sqrt{1 - \left(1 - \nu^2_{\tilde{A}^p}(x)\right)^k} : x \in X\right\}. \tag{3.27}$$

Definition 3.10 [6] The below score and accuracy functions are defined for the triangular Pythagorean fuzzy numbers $\tilde{A}^p = \left\{(a_1, b_1, c_1), \mu_{\tilde{A}^p}(x), \nu_{\tilde{A}^p}(x) : x \in X\right\}$.

$$S\left(\tilde{A}^p\right) = \frac{a+b+c}{3}\left(\mu^2_{\tilde{A}^p} - \nu^2_{\tilde{A}^p}\right), S\left(\tilde{A}^p\right) \in [-1, 1], \tag{3.28}$$

$$A\left(\tilde{A}^p\right) = \frac{a+b+c}{3}\left(\mu^2_{\tilde{A}^p} + \nu^2_{\tilde{A}^p}\right), M\left(\tilde{A}^p\right) \in [0, 1], \tag{3.29}$$

Definition 3.11 [6] The below ranking function is for the triangular Pythagorean fuzzy numbers $\tilde{A}^p = \left\{(a_1, b_1, c_1), \mu_{\tilde{A}^p}(x), \nu_{\tilde{A}^p}(x) : x \in X\right\}$ where $\lambda \in (0, 1)$.

$$\begin{aligned} \Re\left(\tilde{A}^p\right) &= \frac{\lambda \int_a^c x\mu^2_{\tilde{A}^p}(x)dx + (1-\lambda)\int_a^c x\nu^2_{\tilde{A}^p.}(x)dx}{\lambda \int_a^c \mu^2_{\tilde{A}^p}(x)dx + (1-\lambda)\int_a^c \nu^2_{\tilde{A}^p}(x)dx} \\ &= \frac{\lambda(a+2b+c) + (1-\lambda)(5a-2b+5c)}{4(2-\lambda)}. \end{aligned} \tag{3.30}$$

Theorem 3.2 [6] *Assume the triangular Pythagorean fuzzy numbers* $\tilde{A}^p = \left\{(a_1, b_1, c_1), \mu_{\tilde{A}^p}(x), \nu_{\tilde{A}^p}(x) : x \in X\right\}$ *and* $\tilde{B}^p =$

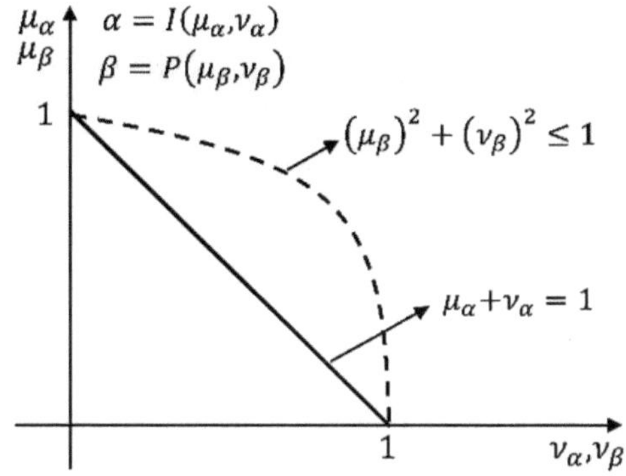

Fig. 3.1 The comparison of intuitionistic and Pythagorean fuzzy numbers

$\{(a_2, b_2, c_2), \mu_{\tilde{B}^p}(x), \nu_{\tilde{B}^p}(x) : x \in X\}$. *Then* $\Re(\alpha\tilde{A}^p + \beta\tilde{B}^p) = \alpha\Re(\tilde{A}^p) + \beta\Re(\tilde{B}^p)$, *where* $\alpha, \beta \geq 0$. *This means that the ranking function* (3.30) *is linear.*

3.4 A Comparison Between Intuitionistic and Pythagorean Fuzzy Numbers

The condition for the intuitionistic fuzzy number $\tilde{\alpha} = \tilde{I}(\mu_{\tilde{\alpha}}(x), \nu_{\tilde{\alpha}}(x))$ is that $0 \leq \mu_{\tilde{\alpha}}(x) + \nu_{\tilde{\alpha}}(x) \leq 1$. Furthermore, the condition for the Pythagorean fuzzy number $\tilde{\beta} = \tilde{P}(\mu_{\tilde{\beta}}(x), \nu_{\tilde{\beta}}(x))$ is that $0 \leq \mu_{\tilde{\alpha}}(x) + \nu_{\tilde{\alpha}}(x) \leq 1$. As for any point such as $a, b \in [0, 1]$, if $a + b \leq 1$ then $a^2 + b^2 \leq 1$, [3] showed that the Pythagorean membership degree is greater than the intuitionistic membership degree. Therefore, an intuitionistic fuzzy number is a Pythagorean fuzzy number (see Fig. 3.1).

3.5 Distance and Similarity Criteria in Pythagorean Fuzzy Sets

Theorem 3.3 [7] *Assume the Pythagorean fuzzy sets* $\tilde{M}$ *and* $\tilde{N}$ *from the reference set* X. *The below distance criteria are defined for these sets.*

$$d^1_{PFS}(\tilde{M}, \tilde{N}) = \frac{1}{2|X|}\sum_{x\in X}\left(\left|\mu^2_{\tilde{M}}(x) - \mu^2_{\tilde{N}}(x)\right| + \left|\nu^2_{\tilde{M}}(x) - \nu^2_{\tilde{N}}(x)\right| + \left|\pi^2_{\tilde{M}}(x) - \pi^2_{\tilde{N}}(x)\right|\right), \tag{3.31}$$

$$d^2_{PFS}(\tilde{M}, \tilde{N}) = \frac{1}{2|X|}\sum_{x\in X}\left(\left|\mu^2_{\tilde{M}}(x) - \mu^2_{\tilde{N}}(x) - (\nu^2_{\tilde{M}}(x) - \nu^2_{\tilde{N}}(x)\right|\right), \tag{3.32}$$

$$d_{PFS}^3(\tilde{M},\tilde{N}) = \frac{1}{4|X|}\left(\sum_{x\in X}\left(\left|\mu_{\tilde{M}}^2(x) - \mu_{\tilde{N}}^2(x)\right| + \left|\nu_{\tilde{M}}^2(x) - \nu_{\tilde{N}}^2(x)\right| + \left|\pi_{\tilde{M}}^2(x) - \pi_{\tilde{N}}^2(x)\right| + \sum_{x\in X}\left|\mu_{\tilde{M}}^2(x) - \nu_{\tilde{M}}^2(x)\right| + \left|\mu_{\tilde{N}}^2(x) - \nu_{\tilde{N}}^2(x)\right|\right)\right) \tag{3.33}$$

$$d_{PFS}^4(\tilde{M},\tilde{N}) = \frac{1}{|X|}\sum_{x\in X}\left(\left|\mu_{\tilde{M}}^2(x) - \mu_{\tilde{N}}^2(x)\right| \vee \left|\nu_{\tilde{M}}^2(x) - \nu_{\tilde{N}}^2(x)\right|\right), \tag{3.34}$$

$$d_{PFS}^5(\tilde{M},\tilde{N}) = \frac{2}{|X|}\sum_{x\in X}\frac{\left|\mu_{\tilde{M}}^2(x) - \mu_{\tilde{N}}^2(x)\right| \vee \left|\nu_{\tilde{M}}^2(x) - \nu_{\tilde{N}}^2(x)\right|}{1 + \left|\mu_{\tilde{M}}^2(x) - \mu_{\tilde{N}}^2(x)\right| \vee \left|\nu_{\tilde{M}}^2(x) - \nu_{\tilde{N}}^2(x)\right|}, \tag{3.35}$$

$$d_{PFS}^6(\tilde{M},\tilde{N}) = 1 - \alpha\frac{\sum_{x\in X}\left(\mu_{\tilde{M}}^2(x) \wedge \mu_{\tilde{N}}^2(x)\right)}{\sum_{x\in X}\left(\mu_{\tilde{M}}^2(x) \vee \mu_{\tilde{N}}^2(x)\right)} - \beta\frac{\sum_{x\in X}\left(\nu_{\tilde{M}}^2(x) \wedge \nu_{\tilde{N}}^2(x)\right)}{\sum_{x\in X}\left(\nu_{\tilde{M}}^2(x) \vee \nu_{\tilde{N}}^2(x)\right)}, \quad \alpha + \beta = 1, \alpha, \beta \in [0, 1], \tag{3.36}$$

$$d_{PFS}^7(\tilde{M},\tilde{N}) = 1 - \frac{\alpha}{|X|}\sum_{x\in X}\frac{\left(\mu_{\tilde{M}}^2(x) \wedge \mu_{\tilde{N}}^2(x)\right)}{\left(\mu_{\tilde{M}}^2(x) \vee \mu_{\tilde{N}}^2(x)\right)} - \frac{\beta}{|X|}\sum_{x\in X}\frac{\left(\nu_{\tilde{M}}^2(x) \wedge \nu_{\tilde{N}}^2(x)\right)}{\left(\nu_{\tilde{M}}^2(x) \vee \nu_{\tilde{N}}^2(x)\right)}, \alpha + \beta = 1, \alpha, \beta \in [0, 1], \tag{3.37}$$

$$d_{PFS}^8(\tilde{M},\tilde{N}) = 1 - \frac{1}{|X|}\sum_{x\in X}\frac{\left(\mu_{\tilde{M}}^2(x) \wedge \mu_{\tilde{N}}^2(x)\right) + \left(\nu_{\tilde{M}}^2(x) \wedge \nu_{\tilde{N}}^2(x)\right)}{\left(\mu_{\tilde{M}}^2(x) \vee \mu_{\tilde{N}}^2(x)\right) + \left(\nu_{\tilde{M}}^2(x) \vee \nu_{\tilde{N}}^2(x)\right)}, \tag{3.38}$$

$$d_{PFS}^9(\tilde{M},\tilde{N}) = 1 - \frac{\sum_{x\in X}\left(\left(\mu_{\tilde{M}}^2(x) \wedge \mu_{\tilde{N}}^2(x)\right) + \left(\nu_{\tilde{M}}^2(x) \wedge \nu_{\tilde{N}}^2(x)\right)\right)}{\sum_{x\in X}\left(\left(\mu_{\tilde{M}}^2(x) \vee \mu_{\tilde{N}}^2(x)\right) + \left(\nu_{\tilde{M}}^2(x) \vee \nu_{\tilde{N}}^2(x)\right)\right)}, \tag{3.39}$$

$$d_{PFS}^{10}(\tilde{M},\tilde{N}) = 1 - \frac{1}{|X|}\sum_{x\in X}\frac{\left(\mu_{\tilde{M}}^2(x) \wedge \mu_{\tilde{N}}^2(x)\right) + \left(1 - \nu_{\tilde{M}}^2(x)\right) \wedge \left(1 - \nu_{\tilde{N}}^2(x)\right)}{\left(\mu_{\tilde{M}}^2(x) \vee \mu_{\tilde{N}}^2(x)\right) + \left(1 - \nu_{\tilde{M}}^2(x)\right) \vee \left(1 - \nu_{\tilde{N}}^2(x)\right)}, \tag{3.40}$$

$$d_{PFS}^{11}(\tilde{M},\tilde{N}) = 1 - \frac{\sum_{x\in X}\left(\left(\mu_{\tilde{M}}^2(x) \wedge \mu_{\tilde{N}}^2(x)\right) + \left(1 - \nu_{\tilde{M}}^2(x)\right) \wedge \left(1 - \nu_{\tilde{N}}^2(x)\right)\right)}{\sum_{x\in X}\left(\left(\mu_{\tilde{M}}^2(x) \vee \mu_{\tilde{N}}^2(x)\right) + \left(1 - \nu_{\tilde{M}}^2(x)\right) \vee \left(1 - \nu_{\tilde{N}}^2(x)\right)\right)}, \tag{3.41}$$

$$d_{PFS}^{12}(\tilde{M},\tilde{N}) = \frac{2\sum_{x\in X}\left(\left|\mu_{\tilde{M}}^2(x) - \mu_{\tilde{N}}^2(x)\right| \vee \left|\nu_{\tilde{M}}^2(x) - \nu_{\tilde{N}}^2(x)\right|\right)}{\sum_{x\in X}\left(1 + \left|\mu_{\tilde{M}}^2(x) - \mu_{\tilde{N}}^2(x)\right| \vee \left|\nu_{\tilde{M}}^2(x) - \nu_{\tilde{N}}^2(x)\right|\right)} \tag{3.42}$$

Definition 3.12 [7] Assume the Pythagorean fuzzy sets $\tilde{M}$, $\tilde{N}$, and $\tilde{O}$ from the reference set X. A similarity index such as $s_{PFS} : PFS(X) \times PFS(X) \rightarrow [0, 1]$ with the below properties is defined for these sets.

- $0 \leq s_{PFS}(\tilde{M}, \tilde{N}) \leq 1$,
- $S_{PFS}(\tilde{M}, \tilde{N}) = s_{PFS}(\tilde{N}, \tilde{M})$,
- $s_{PFS}(\tilde{M}, \tilde{N}) = 1$, if and only if $\tilde{M} = \tilde{N}$,
- $s_{PFS}(\tilde{M}, \tilde{M}^c) = 0$, if and only if $\tilde{M}$ is a crisp set,
- If $\tilde{M} \subseteq \tilde{N} \subseteq \tilde{O}$, then, $s_{PFS}(\tilde{M}, \tilde{O}) \leq s_{PFS}(\tilde{M}, \tilde{N})$, $s_{PFS}(\tilde{M}, \tilde{O}) \leq s_{PFS}(\tilde{N}, \tilde{O})$.

Theorem 3.4 [7] *Assume the Pythagorean fuzzy sets* $\tilde{\boldsymbol{M}}$ *and* $\tilde{N}$ *from the reference set* X. *The below similarity criteria are defined for these sets.*

$$s^1_{PFS}(\tilde{M}, \tilde{N}) = 1 - \frac{1}{2|X|}\sum_{x \in X}\left(\left|\mu^2_{\tilde{M}}(x) - \mu^2_{\tilde{N}}(x)\right| + \left|\nu^2_{\tilde{M}}(x) - \nu^2_{\tilde{N}}(x)\right| + \left|\pi^2_{\tilde{M}}(x) - \pi^2_{\tilde{N}}(x)\right|\right), \tag{3.43}$$

$$s^2_{PFS}(\tilde{M}, \tilde{N}) = 1 - \frac{1}{2|X|}\sum_{x \in X}\left(\left|\mu^2_{\tilde{M}}(x) - \mu^2_{\tilde{N}}(x) - (\nu^2_{\tilde{M}}(x) - \nu^2_{\tilde{N}}(x)\right|\right), \tag{3.44}$$

$$\begin{aligned} s^3_{PFS}(\tilde{M}, \tilde{N}) = 1 &- \frac{1}{4|X|}\left(\sum_{x \in X}\left(\left|\mu^2_{\tilde{M}}(x) - \mu^2_{\tilde{N}}(x)\right| + \left|\nu^2_{\tilde{M}}(x) - \nu^2_{\tilde{N}}(x)\right|\right.\right. \\ &+ \left.\left|\pi^2_{\tilde{M}}(x) - \pi^2_{\tilde{N}}(x)\right|\right) + \sum_{x \in X}\left|\left(\mu^2_{\tilde{M}}(x) - \nu^2_{\tilde{M}}(x)\right)\right. \\ &\left.\left. - \left(\mu^2_{\tilde{N}}(x) - \nu^2_{\tilde{N}}(x)\right)\right|\right) \end{aligned} \tag{3.45}$$

$$s^4_{PFS}(\tilde{M}, \tilde{N}) = 1 - \frac{1}{|X|}\sum_{x \in X}\left(\left|\mu^2_{\tilde{M}}(x) - \mu^2_{\tilde{N}}(x)\right| \vee \left|\nu^2_{\tilde{M}}(x) - \nu^2_{\tilde{N}}(x)\right|\right), \tag{3.46}$$

$$s^5_{PFS}(\tilde{M}, \tilde{N}) = \frac{1}{|X|}\sum_{x \in X}\frac{1 - \left|\mu^2_{\tilde{M}}(x) - \mu^2_{\tilde{N}}(x)\right| \vee \left|\nu^2_{\tilde{M}}(x) - \nu^2_{\tilde{N}}(x)\right|}{1 + \left|\mu^2_{\tilde{M}}(x) - \mu^2_{\tilde{N}}(x)\right| \vee \left|\nu^2_{\tilde{M}}(x) - \nu^2_{\tilde{N}}(x)\right|}, \tag{3.47}$$

$$\begin{aligned} s^6_{PFS}(\tilde{M}, \tilde{N}) &= \alpha\frac{\sum_{x \in X}\left(\mu^2_{\tilde{M}}(x) \wedge \mu^2_{\tilde{N}}(x)\right)}{\sum_{x \in X}\left(\mu^2_{\tilde{M}}(x) \vee \mu^2_{\tilde{N}}(x)\right)} \\ &+ \beta\frac{\sum_{x \in X}\left(\nu^2_{\tilde{M}}(x) \wedge \nu^2_{\tilde{N}}(x)\right)}{\sum_{x \in X}\left(\nu^2_{\tilde{M}}(x) \vee \nu^2_{\tilde{N}}(x)\right)}, \\ \alpha + \beta &= 1, \alpha, \beta \in [0, 1], \end{aligned} \tag{3.48}$$

$$s^7_{PFS}(\tilde{M},\tilde{N}) = \frac{\alpha}{|X|}\sum_{x\in X}\frac{\left(\mu^2_{\tilde{M}}(x)\wedge\mu^2_{\tilde{N}}(x)\right)}{\left(\mu^2_{\tilde{M}}(x)\vee\mu^2_{\tilde{N}}(x)\right)} + \frac{\beta}{|X|}\sum_{x\in X}\frac{\left(\nu^2_{\tilde{M}}(x)\wedge\nu^2_{\tilde{N}}(x)\right)}{\left(\nu^2_{\tilde{M}}(x)\vee\nu^2_{\tilde{N}}(x)\right)}, \alpha+\beta=1, \alpha,\beta\in[0,1], \quad (3.49)$$

$$s^8_{PFS}(\tilde{M},\tilde{N}) = \frac{1}{|X|}\sum_{x\in X}\frac{\left(\mu^2_{\tilde{M}}(x)\wedge\mu^2_{\tilde{N}}(x)\right)+\left(\nu^2_{\tilde{M}}(x)\wedge\nu^2_{\tilde{N}}(x)\right)}{\left(\mu^2_{\tilde{M}}(x)\vee\mu^2_{\tilde{N}}(x)\right)+\left(\nu^2_{\tilde{M}}(x)\vee\nu^2_{\tilde{N}}(x)\right)}, \quad (3.50)$$

$$s^9_{PFS}(\tilde{M},\tilde{N}) = \frac{\sum_{x\in X}\left(\left(\mu^2_{\tilde{M}}(x)\wedge\mu^2_{\tilde{N}}(x)\right)+\left(\nu^2_{\tilde{M}}(x)\wedge\nu^2_{\tilde{N}}(x)\right)\right)}{\sum_{x\in X}\left(\left(\mu^2_{\tilde{M}}(x)\vee\mu^2_{\tilde{N}}(x)\right)+\left(\nu^2_{\tilde{M}}(x)\vee\nu^2_{\tilde{N}}(x)\right)\right)}, \quad (3.51)$$

$$s^{10}_{PFS}(\tilde{M},\tilde{N}) = \frac{1}{|X|}\sum_{x\in X}\frac{\left(\mu^2_{\tilde{M}}(x)\wedge\mu^2_{\tilde{N}}(x)\right)+\left(1-\nu^2_{\tilde{M}}(x)\right)\wedge\left(1-\nu^2_{\tilde{N}}(x)\right)}{\left(\mu^2_{\tilde{M}}(x)\vee\mu^2_{\tilde{N}}(x)\right)+\left(1-\nu^2_{\tilde{M}}(x)\right)\vee\left(1-\nu^2_{\tilde{N}}(x)\right)}, \quad (3.52)$$

$$s^{11}_{PFS}(\tilde{M},\tilde{N}) = \frac{\sum_{x\in X}\left(\left(\mu^2_{\tilde{M}}(x)\wedge\mu^2_{\tilde{N}}(x)\right)+\left(1-\nu^2_{\tilde{M}}(x)\right)\wedge\left(1-\nu^2_{\tilde{N}}(x)\right)\right)}{\sum_{x\in X}\left(\left(\mu^2_{\tilde{M}}(x)\vee\mu^2_{\tilde{N}}(x)\right)+\left(1-\nu^2_{\tilde{M}}(x)\right)\vee\left(1-\nu^2_{\tilde{N}}(x)\right)\right)}, \quad (3.53)$$

$$s^{12}_{PFS}(\tilde{M},\tilde{N}) = \frac{\sum_{x\in X}\left(1-\left|\mu^2_{\tilde{M}}(x)-\mu^2_{\tilde{N}}(x)\right|\vee\left|\nu^2_{\tilde{M}}(x)-\nu^2_{\tilde{N}}(x)\right|\right)}{\sum_{x\in X}\left(1+\left|\mu^2_{\tilde{M}}(x)-\mu^2_{\tilde{N}}(x)\right|\vee\left|\nu^2_{\tilde{M}}(x)-\nu^2_{\tilde{N}}(x)\right|\right)} \quad (3.54)$$

Example 3.4 [7] Assume the set of patients $P=\{p_1,p_2,p_3,p_4\}$, the set of diseases $D=\{d_1,d_2,d_3,d_4,d_5\}$, and the set of symptoms $S=\{s_1,s_2,s_3,s_4,s_5\}$. The Pythagorean fuzzy values for each symptom in each disease and each symptom in each patient are represented by Tables 3.1 and 3.2.

Now, based on the similarity criterion s^1_{PFS}, the similarity value of each disease and each patient is obtained and represented by Table 3.3.

According to Table 3.3 the illness of each patient can be distinguished where patient p_1 has disease d_2, etc.

Definition 3.13 [4] For the set of Pythagorean fuzzy numbers $\xi_i=\left(\mu_{\xi_i},\nu_{\xi_i}\right)$ where $i=1,\ldots,n$, and their associated set of weight values $w=(w_1,\ldots,w_n)$ where $\sum_{i=1}^{n}w_i=1$, the Pythagorean fuzzy weighted arithmetic mean (PFWA) and the

Table 3.1 The Pythagorean fuzzy number of each symptom in each disease

		Symptom				
		s_1	s_2	s_3	s_4	s_5
Disease	d_1	(0.4, 0)	(0.3, 0.5)	(0.1, 0.7)	(0.4, 0.3)	(0.1, 0.7)
	d_2	(0.7, 0)	(0.2, 0.6)	(0, 0.9)	(0.7, 0)	(0.1, 0.8)
	d_3	(0.3, 0.3)	(0.6, 0.1)	(0.2, 0.7)	(0.2, 0.6)	(0.1, 0.9)
	d_4	(0.1, 0.7)	(0.2, 0.4)	(0.8, 0)	(0.2, 0.7)	(0.2, 0.7)
	d_5	(0.1, 0.8)	(0, 0.8)	(0.2, 0.8)	(0.2, 0.8)	(0.8, 0.1)

Table 3.2 The Pythagorean fuzzy number of each symptom in each patient

		Symptom				
		s_1	s_2	s_3	s_4	s_5
Patient	p_1	(0.8, 0.1)	(0.6, 0.1)	(0.2, 0.8)	(0.6, 0.1)	(0.1, 0.6)
	p_2	(0, 0.8)	(0.4, 0.4)	(0.6, 0.1)	(0.1, 0.7)	(0.1, 0.8)
	p_3	(0.8, 0.1)	(0.8, 0.1)	(0, 0.6)	(0.2, 0.7)	(01, 0.5)
	p_4	(0.6, 0.1)	(0.5, 0.4)	(0.3, 0.4)	(0.7, 0.2)	(0.3, 0.4)

Table 3.3 The similarity values calculated for the diseases and the patients

		Symptom				
		d_1	d_2	d_3	d_4	d_5
Patient	p_1	0.746	**0.782**	0.7	0.554	0.496
	p_2	0.648	0.574	0.694	**0.854**	0.614
	p_3	0.634	0.58	**0.66**	0.570	0.524
	p_4	**0.728**	0.698	0.63	0.596	0.452

The bold values show the highest similarity for each patient

Pythagorean fuzzy weighted geometric mean (PFWG) values are obtained by the below equations.

$$PFWA_w(\tilde{\xi}_1, \tilde{\xi}_2, \ldots, \tilde{\xi}_n) = w_1\tilde{\xi}_1 \oplus w_2\tilde{\xi}_2 \oplus \ldots \oplus w_n\tilde{\xi}_n = \left(\sqrt{1 - \prod_{i=1}^{n}\left(1 - \mu_{\tilde{\xi}_i}^2\right)^{w_i}}, \prod_{i=1}^{n} \nu_{\tilde{\xi}_i}^{w_i}\right), \tag{3.55}$$

$$PFWG_w(\tilde{\xi}_1, \tilde{\xi}_2, \ldots, \tilde{\xi}_n) = w_1\tilde{\xi}_1 \otimes w_2\tilde{\xi}_2 \otimes \ldots \otimes w_n\tilde{\xi}_n = \left(\prod_{i=1}^{n} \mu_{\tilde{\xi}_i}^{w_i}, \sqrt{1 - \prod_{i=1}^{n}\left(1 - \nu_{\tilde{\xi}_i}^2\right)^{w_i}}\right). \tag{3.56}$$

Definition 3.14 [8] For the set of Pythagorean fuzzy numbers $\tilde{\xi}_i = \left(\mu_{\tilde{\xi}_i}, \nu_{\tilde{\xi}_i}\right)$ where $i = 1, \ldots, n$, and their associated set of weight values $w = (w_1, \ldots, w_n)$ where $\sum_{i=1}^{n} w_i = 1$, the Pythagorean fuzzy weighted powered arithmetic (PFWPA) and the Pythagorean fuzzy weighted powered geometric (PFWPG) operators are obtained by the below equations.

$$PFWPA_w(\tilde{\xi}_1, \tilde{\xi}_2, \ldots, \tilde{\xi}_n) = \left(\sqrt{w_i \mu_{w_i}^2}, \sqrt{w_i \nu_{w_i}^2}\right), \tag{3.57}$$

$$PFWPG_w(\tilde{\xi}_1, \tilde{\xi}_2, \ldots, \tilde{\xi}_n) = \left(\sqrt{1 - \prod_{i=1}^{n} \left(1 - \mu_{\tilde{\xi}_i}^2\right)^{w_i}}, \sqrt{1 - \prod_{i=1}^{n} \left(1 - \nu_{\tilde{\xi}_i}^2\right)^{w_i}} \right). \tag{3.58}$$

3.6 Interval-Valued Pythagorean Set

In order to show the uncertain situations with higher degree of uncertainty, interval-values Pythagorean sets (IVPFSs) and numbers (IFPFNs) are defined.

Definition 3.15 [9] Considering the reference set X, an interval-valued Pythagorean fuzzy set (IVPFS) is defined as below,

$$\tilde{A}_I = \left\{ \left(x, \mu_{\tilde{A}_I}(x), \nu_{\tilde{A}_I}(x)\right) : x \in X \right\} \tag{3.59}$$

where, $\mu_{\tilde{A}_I} : X \to [0, 1]$ and $\nu_{\tilde{A}_I} : X \to [0, 1]$ are respectively the membership and non-membership function of $x \in X$ for the fuzzy set $\tilde{A}_I$. These functions are represented as interval values such as $\mu_{\tilde{A}_I}(x) = \left[\mu_{\tilde{A}_I}^-(x), \mu_{\tilde{A}_I}^+(x)\right]$ and $\nu_{\tilde{A}_I}(x) = \left[\nu_{\tilde{A}_I}^-(x), \nu_{\tilde{A}_I}^+(x)\right]$ where $0 \le \mu_{\tilde{A}_I}^-(x) \le \mu_{\tilde{A}_I}^+(x) \le 1$ and $0 \le \nu_{\tilde{A}_I}^-(x) \le \nu_{\tilde{A}_I}^+(x) \le 1$. Also, for $x \in X$ the condition $\left(\mu_{\tilde{A}_I}^+(x)\right)^2 + \left(\nu_{\tilde{A}_I}^+(x)\right)^2 \le 1$ should be held. The interval-valued Pythagorean fuzzy set (3.59) also can be shown as below.

$$\tilde{A}_I = \left\{ \left(x, \left[\mu_{\tilde{A}_I}^-(x), \mu_{\tilde{A}_I}^+(x)\right], \left[\nu_{\tilde{A}_I}^-(x), \nu_{\tilde{A}_I}^+(x)\right]\right) : x \in X \right\} \tag{3.60}$$

Similar to the Pythagorean fuzzy sets, the hesitancy degree of $x \in X$ for $\tilde{A}_I$ is obtained as $\pi_{\tilde{A}_I} = \left[\pi_{\tilde{A}_I}^-, \pi_{\tilde{A}_I}^+\right] = \left[\sqrt{1 - \left(\mu_{\tilde{A}_I}^+(x)\right)^2 - \left(\nu_{\tilde{A}_I}^+(x)\right)^2}, \sqrt{1 - \left(\mu_{\tilde{A}_I}^-(x)\right)^2 - \left(\nu_{\tilde{A}_I}^-(x)\right)^2}.\right.$

For simplicity, the interval $\left(\left[\mu_{\tilde{A}_I}^-(x), \mu_{\tilde{A}_I}^+(x)\right], \left[\nu_{\tilde{A}_I}^-(x), \nu_{\tilde{A}_I}^+(x)\right]\right)$ can be shown as $\left(\left[\mu_{\tilde{A}_I}^-, \mu_{\tilde{A}_I}^+\right], \left[\nu_{\tilde{A}_I}^-, \nu_{\tilde{A}_I}^+\right]\right)$. For the case of $\mu_{\tilde{A}_I}^- = \mu_{\tilde{A}_I}^+$ and $\nu_{\tilde{A}_I}^- = \nu_{\tilde{A}_I}^+$, the interval-valued Pythagoran fuzzy set is converted to a Pythagoran fuzzy set.

Clearly, the largest interval-valued Pythagorean fuzzy number (IVPFN) is $\tilde{P}^+ = ([1, 1], [0, 0])$, and the lowest interval-valued Pythagorean fuzzy number is $\tilde{P}^- = ([0, 0], [1, 1])$.

Definition 3.16 [10] Consider the IVPFNs $\tilde{P}_1 = \left(\left[\mu_1^-, \mu_1^+\right], \left[\nu_1^-, \nu_1^+\right]\right)$ and $\tilde{P}_2 = \left(\left[\mu_2^-, \mu_2^+\right], \left[\nu_2^-, \nu_2^+\right]\right)$. Then,

$$\tilde{P}_1 = \tilde{P}_2 \text{ if and only if } \mu_1^- = \mu_2^-, \mu_1^+ = \mu_2^+, \nu_1^- = \nu_2^-, \nu_1^+ = \nu_2^+, \tag{3.61}$$

$$\tilde{P}_1 \prec \tilde{P}_2 \text{ if and only if } \mu_1^- \leq \mu_2^-, \mu_1^+ \leq \mu_2^+, \nu_1^- \geq \nu_2^-, \nu_1^+ \geq \nu_2^+ \tag{3.62}$$

Definition 3.17 [9] Consider the IVPFNs $\tilde{P}_1 = \left(\left[\mu_1^-, \mu_1^+\right], \left[\nu_1^-, \nu_1^+\right]\right)$ and $\tilde{P}_2 = \left(\left[\mu_2^-, \mu_2^+\right], \left[\nu_2^-, \nu_2^+\right]\right)$. Then, the below arithmetic operations are defined.

$$\tilde{P}_1 \cup \tilde{P}_2 = \left(\left[\max\left\{\mu_1^-, \mu_2^-\right\}, \max\left\{\mu_1^+, \mu_2^+\right\}\right], \left[\min\left\{\nu_1^-, \nu_2^-\right\}, \min\left\{\nu_1^+, \nu_2^+\right\}\right]\right) \tag{3.63}$$

$$\tilde{P}_1 \cap \tilde{P}_2 = \left(\left[\min\left\{\mu_1^-, \mu_2^-\right\}, \min\left\{\mu_1^+, \mu_2^+\right\}\right], \left[\max\left\{\nu_1^-, \nu_2^-\right\}, \max\left\{\nu_1^+, \nu_2^+\right\}\right]\right) \tag{3.64}$$

$$\tilde{P}_1 \oplus \tilde{P}_2 = \left(\left[\sqrt{\left(\mu_1^-\right)^2 + \left(\mu_2^-\right)^2 - \left(\mu_1^-\right)^2\left(\mu_2^-\right)^2}\right.\right.$$
$$\left.\sqrt{\left(\mu_1^+\right)^2 + \left(\mu_2^+\right)^2 - \left(\mu_1^+\right)^2\left(\mu_2^+\right)^2}\right], \left[\nu_1^-\nu_2^-, \nu_1^+\nu_2^+\right] \tag{3.65}$$

$$\tilde{P}_1 \ominus \tilde{P}_2 = \left(\left[\sqrt{\frac{\mu_1^{-2} - \mu_2^{-2}}{1 - \mu_2^{-2}}}, \sqrt{\frac{\mu_1^{+2} - \mu_2^{+2}}{1 - \mu_2^{+2}}}\right], \left[\frac{\nu_1^-}{\nu_2^-}, \frac{\nu_1^+}{\nu_2^+}\right]\right) \text{ if } \mu_1^- \geq \mu_2^-, \mu_1^+ \geq \mu_2^+$$
$$\nu_1^- \leq \min\left\{\nu_2^-, \frac{\nu_2^-\pi_2^-}{\pi_2^-}\right\}, \nu_1^+ \leq \min\left\{\nu_2^+, \frac{\nu_2^+\pi_2^+}{\pi_2^+}\right\} \tag{3.66}$$

$$\tilde{P}_1 \otimes \tilde{P}_2 = \left(\left[\mu_1^-\mu_2^-, \mu_1^+\mu_2^+\right], \left[\sqrt{\left(\nu_1^-\right)^2 + \left(\nu_2^-\right)^2 - \left(\nu_1^-\right)^2\left(\nu_2^-\right)^2},\right.\right.$$
$$\left.\left.\sqrt{\left(\nu_1^+\right)^2 + \left(\nu_2^+\right)^2 - \left(\nu_1^+\right)^2\left(\nu_2^+\right)^2}\right]\right) \tag{3.67}$$

$$\lambda\tilde{P}_1 = \left(\left[\sqrt{1 - \left(1 - \left(\mu_1^-\right)^2\right)^\lambda}, \sqrt{1 - \left(1 - \left(\mu_1^+\right)^2\right)^\lambda}\right], \left[\left(\nu_1^-\right)^\lambda, \left(\nu_1^+\right)^\lambda\right]\right) \tag{3.68}$$

$$\tilde{P}_1^\lambda = \left(\left[\left(\mu_1^-\right)^\lambda, \left(\mu_1^+\right)^\lambda\right], \left[\sqrt{1 - \left(1 - \left(\nu_1^-\right)^2\right)^\lambda}, \sqrt{1 - \left(1 - \left(\nu_1^+\right)^2\right)^\lambda}\right]\right) \tag{3.69}$$

$$\tilde{P}_1 \subseteq \tilde{P}_2 \text{if and only if } \mu_1^- \leq \mu_2^-, \mu_1^+ \leq \mu_2^+,$$
$$\nu_1^- \geq \nu_2^-, \nu_1^+ \geq \nu_2^+, \text{ for } \forall x \in X, \tag{3.70}$$

$$\tilde{P}_1^c = \left(\left[\nu_1^-, \nu_1^+\right], \left[\mu_1^-, \mu_1^+\right]\right). \tag{3.71}$$

Example 3.5 Consider the IVPFNs $\tilde{P}_1 = ([0.5, 0.7], [0.2, 0.3])$ and $\tilde{P}_2 = ([0.4, 0.6], [0.1, 0.4])$. Then, the below calculations are done.

$$\begin{aligned}\tilde{P}_1 \cup \tilde{P}_2 &= ([\max\{0.5, 0.4\}, \max\{0.7, 0.6\}], \\ &\quad [\min\{0.2, 0.1\}], \min\{0.3, 0.4\}) \\ &= ([0.5, 0.7], [0.1, 0.3])\end{aligned}$$

$$\begin{aligned}\tilde{P}_1 \cap \tilde{P}_2 &= ([\min\{0.5, 0.4\}, \min\{0.7, 0.6\}], \\ &\quad [\max\{0.2, 0.1\}], \max\{0.3, 0.4\}) \\ &= ([0.4, 0.6], [0.2, 0.4])\end{aligned}$$

$$\begin{aligned}\tilde{P}_1 \oplus \tilde{P}_2 &= \left(\left[\sqrt{0.5^2 + 0.4^2 - 0.5^2 0.4^2}, \sqrt{0.7^2 + 0.6^2 - 0.7^2 0.6^2}\right],\right. \\ &\quad [0.2 \times 0.1, 0.3 \times 0.4]) = \left(\left[\sqrt{0.37}, \sqrt{0.6736}\right], [0.02, 0.12]\right) \\ &= ([0.608, 0.821], [0.02, 0.12])\end{aligned}$$

$$\begin{aligned}\tilde{P}_1 \otimes \tilde{P}_2 &= \left([0.5 \times 0.4, 0.7 \times 0.6], \left[\sqrt{0.2^2 + 0.1^2 - 0.2^2 \times 0.1^2},\right.\right. \\ &\quad \left.\sqrt{0.3^2 + 0.4^2 - 0.3^2 \times 0.4^2}\right] = \left([0.2, 0.42], \left[\sqrt{0.0496}, \sqrt{0.2356}\right]\right) \\ &= ([0.2, 0.42], [0.223, 0.485])\end{aligned}$$

$$\begin{aligned}2\tilde{P}_1 &= \left(\left[\sqrt{1 - \left(1 - (0.5)^2\right)^2}, \sqrt{1 - \left(1 - (0.7)^2\right)^2}\right], \left[(0.2)^2, (0.3)^2\right]\right) \\ &= ([0.661, 0.860], [0.04, 0.09])\end{aligned}$$

$$\begin{aligned}\tilde{P}_1^2 &= \left(\left[(0.5)^2, (0.7)^2\right], \left[\sqrt{1 - \left(1 - (0.2)^2\right)^2}, \sqrt{1 - \left(1 - (0.3)^2\right)^2}\right]\right) \\ &= ([0.25, 0.49], [0.28, 0.414])\end{aligned}$$

Definition 3.18 [10] The score and accuracy functions of the IVPFN $\tilde{P}_I = \left(\left[\mu^-, \mu^+\right], \left[\nu^-, \nu^+\right]\right) = ([a, b], [c, d])$ can be defined as below.

$$S(\tilde{P}_I) = \frac{a^2 + b^2 - c^2 - d^2}{2}, \quad S(\tilde{P}_I) \in [-1, 1] \tag{3.72}$$

$$A(\tilde{P}_I) = \frac{a^2 + b^2 + c^2 + d^2}{2}, \quad A(\tilde{P}_I) \in [0, 1] \tag{3.73}$$

According to the Eqs. (3.72) and (3.73), the IVPFNs can be ranked by the below rules.

$$S(\tilde{P}_1) < S(\tilde{P}_2) \Rightarrow \tilde{P}_1 < \tilde{P}_2,$$

$$S(\tilde{P}_1) > S(\tilde{P}_2) \Rightarrow \tilde{P}_1 > \tilde{P}_2,$$

$$S(\tilde{P}_1) = S(\tilde{P}_2), A(\tilde{P}_1) > A(\tilde{P}_2) \Rightarrow \tilde{P}_1 \succ \tilde{P}_2,$$

$$S(\tilde{P}_1) = S(\tilde{P}_2), A(\tilde{P}_1) < A(\tilde{P}_2) \Rightarrow \tilde{P}_1 \prec \tilde{P}_2,$$

$$S(\tilde{P}_1) = S(\tilde{P}_2), A(\tilde{P}_1) = A(\tilde{P}_2) \Rightarrow \tilde{P}_1 \approx \tilde{P}_2.$$

Example 3.6 Consider the IVPFNs $\tilde{P}_1 = ([0.2, 0.4], [0.2, 0.4])$ and $\tilde{P}_2 = ([0.5, 0.6], [0.5, 0.6])$. According to (3.72) and (3.73), as $S(\tilde{P}_1) = S(\tilde{P}_2) = 0$ and $A(\tilde{P}_1) = 0.2 < A(\tilde{P}_2) = 0.61$, then $\tilde{P}_1 \prec \tilde{P}_2$.

Example 3.7 Consider the IVPFNs $\tilde{P}_1 = ([0.2, 0.4], [0.2, 0.4])$ and $\tilde{P}_2 = ([0.5, 0.6], [0.5, 0.6])$. According to Definition 3.18, for these two numbers the values of $S(\tilde{P}_1) = S(\tilde{P}_2) = -0.125$ and $A(\tilde{P}_1) = A(\tilde{P}_2) = 0.375$ are obtained resulting that $\tilde{P}_1 \approx \tilde{P}_2$. But the result is not correct as clearly $\tilde{P}_1 \neq \tilde{P}_2$. Therefore, the functions (3.72) and (3.73) may not result in a correct ranking. For this aim the below definition has been introduced by [9].

Definition 3.19 [9] The ranking function $\Re(\tilde{P}_I)$ for the IVPFN $\tilde{P}_I = ([\mu^-, \mu^+], [\nu^-, \nu^+]) = ([a, b], [c, d])$ is defined as below.

$$\Re(\tilde{P}_I) = \frac{(a^2 - c^2)\left(1 + \sqrt{1 - a^2 - c^2}\right) + (b^2 - d^2)\left(1 + \sqrt{1 - b^2 - d^2}\right)}{2},$$
$$\Re(\tilde{P}_I) \in [-1, 1] \tag{3.74}$$

According to the function (3.74), the IVPFNs can be ranked by the below rules.

$$\Re(\tilde{\mathrm{P}}_1) < \Re(\tilde{\mathrm{P}}_2) \Rightarrow \tilde{\mathrm{P}}_1 \prec \tilde{\mathrm{P}}_2,$$

$$\Re(\tilde{\mathrm{P}}_1) > \Re(\tilde{\mathrm{P}}_2) \Rightarrow \tilde{\mathrm{P}}_1 \succ \tilde{\mathrm{P}}_2,$$

$$\Re(\tilde{\mathrm{P}}_1) = \Re(\tilde{\mathrm{P}}_2), \Rightarrow \tilde{\mathrm{P}}_1 \approx \tilde{\mathrm{P}}_2.$$

Example 3.8 Consider the IVPFNs $\tilde{P}_1 = ([0.2, 0.4], [0.2, 0.4])$ and $\tilde{P}_2 = ([0.5, 0.6], [0.5, 0.6])$. Based on the ranking function (3.74), $\Re(\tilde{P}_1) = -0.1912$ and $\Re(\tilde{P}_2) = -0.2246$. Then, $\tilde{P}_1 \succ \tilde{P}_2$.

Definition 3.20 [11] The score and accuracy functions of the IVPFN $\tilde{P}_I = ([\mu^-, \mu^+], [\nu^-, \nu^+]) = ([a, b], [c, d])$ can be defined as below.

$$S(\tilde{P}_I) = \begin{cases} \frac{2a^2+2b^2-c^2-d^2-1}{3}, & \tilde{P}_I > 0, \\ 0, & \tilde{P}_I \leq 0, \end{cases} S(\tilde{P}_I) \in \left[-\frac{1}{3}, \frac{1}{3}\right], \tag{3.75}$$

$$A(\tilde{P}_I) = \begin{cases} \frac{2a^2+2b^2+c^2+d^2-1}{3}, & \tilde{P}_I > 0 \\ 0, & \tilde{P}_I \leq 0, \end{cases} \tag{3.76}$$

According to the functions (3.75) and (3.76), the IVPFNs can be ranked by the below rules.

$$S(\tilde{P}_1) < S(\tilde{P}_2) \Rightarrow \tilde{P}_1 \prec \tilde{P}_2,$$

$$S(\tilde{P}_1) > S(\tilde{P}_2) \Rightarrow \tilde{P}_1 \succ \tilde{P}_2,$$

$$S(\tilde{P}_1) = S(\tilde{P}_2), A(\tilde{P}_1) > A(\tilde{P}_2), \Rightarrow \tilde{P}_1 \succ \tilde{P}_2,$$

$$S(\tilde{P}_1) = S(\tilde{P}_2), A(\tilde{P}_1) < A(\tilde{P}_2), \Rightarrow \tilde{P}_1 \prec \tilde{P}_2,$$

$$S(\tilde{P}_1) = S(\tilde{P}_2), A(\tilde{P}_1) = A(\tilde{P}_2), \Rightarrow \tilde{P}_1 \approx \tilde{P}_2.$$

Example 3.9 Consider the IVPFNs $\tilde{P}_1 = ([0.5, 0.9], [0.2, 0.8])$ and $\tilde{P}_2 = ([0.2, 0.8], [0.1, 0.4])$. Based on the score functions defined above, the values of Table 3.4 are obtained.

Definition 3.21 [12] For the IVPFN $\tilde{z} = ([\mu^-, \mu^+], [\nu^-, \nu^+])$, the below ranking function is defined.

$$I(\tilde{z}) = \frac{\sqrt{(\mu^-)^2 + (\mu^+)^2} + \sqrt{(1-\nu^-)^2 + (1-\nu^+)^2} + (\mu^- \times \mu^+) - (1-\nu^-) \times (1-\nu^+)}{3} \tag{3.77}$$

Definition 3.22 [13] Consider the IVPFNs $\tilde{\boldsymbol{P}}_1 = ([\boldsymbol{\mu}_1^-, \boldsymbol{\mu}_1^+], [\boldsymbol{\nu}_1^-, \boldsymbol{\nu}_1^+])$ and $\tilde{\boldsymbol{P}}_2 = ([\boldsymbol{\mu}_2^-, \boldsymbol{\mu}_2^+], [\boldsymbol{\nu}_2^-, \boldsymbol{\nu}_2^+])$ with the hesitancy degrees $\pi_{\tilde{P}_1} = \left[\pi_{\tilde{P}_1}^-, \pi_{\tilde{P}_1}^+\right]$ and $\pi_{\tilde{P}_2} = \left[\pi_{\tilde{P}_2}^-, \pi_{\tilde{P}_2}^+\right]$ respectively. The below distance value for these numbers are defined.

Table 3.4 Different score values obtained for the numbers of Example 3.8

The score function	$\tilde{P}_1 = ([0.5, 0.9], [0.2, 0.8])$	$\tilde{P}_2 = ([0.2, 0.8], [0.1, 0.4])$
(3.72)	0.465	0.255
(3.74)	0.66717	0.37689
(3.75)	0.33	0.063

$$d(\tilde{P}_1, \tilde{P}_2) = \frac{\sqrt{(\mu_1^2 - \mu_2^2)\left(1 - \frac{\pi_{\tilde{P}_1}\pi_{\tilde{P}_2}}{2}\right)} + \sqrt{(\mu_1^{+2} - \mu_2^{+2})\left(1 - \frac{\pi_{\tilde{P}_1}^{+} - \pi_{\tilde{P}_2}^{+}}{2}\right)}}{2}, \quad (3.78)$$

Definition 3.23 [12] For a set of IVPFNs such as $\tilde{P}_i = \left(\left[\mu_i^-, \mu_i^+\right], \left[\nu_i^-, \nu_i^+\right]\right)$ where $i = 1, \ldots, n$, the below Pythagorean fuzzy average aggregation operator is defined.

$$PFAA = \left(\left[\prod_{i=1}^{n}(\mu_i^-)^{\frac{1}{n}}, \prod_{i=1}^{n}(\mu_i^+)^{\frac{1}{n}},\right], \left[\sqrt{1 - \left(\prod_{i=1}^{n}\left(1 - (\nu_i^-)^2\right)\right)^{\frac{1}{n}}}, \sqrt{1 - \left(\prod_{i=1}^{n}\left(1 - (\nu_i^+)^2\right)\right)^{\frac{1}{n}}}\right]\right). \quad (3.79)$$

3.7 Type-2 Pythagorean Fuzzy Numbers

Type-2 Pythagorean fuzzy numbers (T2PFNs) are a complex extension of Pythagorean fuzzy numbers. T2PFNs are used to show more complex uncertainty in rea-world events. A T2PFN has one more uncertain layer comparing to Pythagorean fuzzy numbers. Therefore, membership degree and non-membership degree of a T2PFN are represented by the concept of hesitancy degree. Use of this type of fuzzy numbers allow decision makers to make more robust decisions and consider more complex scenarios with contradictory factors and criteria. In the rest of this section the concepts and operations of type-2 Pythagorean fuzzy numbers are introduced.

Definition 3.24 [14] A triangular type-2 Pythagorean fuzzy number (TT2PFN) on X is shown by $\tilde{\tilde{A}} = \left((\tilde{A}_1, \tilde{A}_2, \tilde{A}_3); \mu_{\tilde{\tilde{A}}}, \nu_{\tilde{\tilde{A}}}\right)$. In this number $\tilde{A}_1$, $\tilde{A}_2$, and $\tilde{A}_3$ are Pythagorean fuzzy numbers. Also, $\mu_{\tilde{\tilde{A}}}$ and $\nu_{\tilde{\tilde{A}}}$ are the membership and non-membership functions of $\tilde{\tilde{A}}$ respectively. Therefore, $\tilde{\tilde{A}}$ can be rewritten as below,

$$\begin{aligned}\tilde{\tilde{A}} &= \langle \tilde{a}_{ij}, \tilde{b}_{ij}, \tilde{c}_{ij}; \omega_1, \omega_2 \rangle \\ &= \left\langle \left(a_{ij}^1, a_{ij}^2, a_{ij}^3; \mu_{\tilde{a}_{ij}}, \nu_{\tilde{a}_{ij}}\right), \left(b_{ij}^1, b_{ij}^2, b_{ij}^3; \mu_{\tilde{b}_{ij}}, \nu_{\tilde{b}_{ij}}\right), \right. \\ &\quad \left. \left(c_{ij}^1, c_{ij}^2, c_{ij}^3; \mu_{\tilde{c}_{ij}}, \nu_{\tilde{c}_{ij}}\right); \omega_1, \omega_2 \right\rangle \end{aligned} \quad (3.80)$$

where, $\omega_1 = \min\left\{\mu_{\tilde{a}_{ij}}, \mu_{\tilde{b}_{ij}}, \mu_{\tilde{c}_{ij}}\right\}$ and $\omega_2 = \max\left\{\mu_{\tilde{a}_{ij}}, \mu_{\tilde{b}_{ij}}, \mu_{\tilde{c}_{ij}}\right\}$ are the membership and non-membership functions of $\tilde{\tilde{A}}$ respectively such that $0 \leq \omega_1, \omega_2 \leq 1$ and $\omega_1^2 + \omega_2^2 \leq 1$.

Example 3.10 [14] Consider a transportation firm with three delivery seasons. In the first season, the firm aims to deliver 1000 containers of goods with the lowest estimation of 800 and highest estimation of 1200 containers. But because of environmental uncertainties such as weather, safe delivery, etc. the exact number of delivered containers cannot be estimated. In this season, the acceptance rate for delivery is estimated as 70% and the unacceptance rate for delivery is estimated as just 10%. In the second season, the firm aims to deliver 900 containers of goods with the lowest estimation of 700 and highest estimation of 1100 containers. In this season, the acceptance rate for delivery is estimated as 60% and the unacceptance rate for delivery is estimated as just 20%. In the third season, the firm aims to deliver 1200 containers of goods with the lowest estimation of 1000 and highest estimation of 1400 containers. In this season, the acceptance rate for delivery is estimated as 70% and the unacceptance rate for delivery is estimated as just 10%. The given information can be represented as the below TT2PFN which shows the uncertainty of the delivery plan of this firm.

$$\tilde{\tilde{A}} = \langle (800, 1000, 1200; 0.7, 0.1), (700, 900, 1100; 0.6, 0.2),$$
$$(1000, 1200, 1400; 0.6, 0.2) \rangle$$

Definition 3.25 [14] For the triangular type-2 Pythagorean fuzzy numbers $\tilde{\tilde{A}} = \left\langle \left(\left(a_1^1, a_1^2, a_1^3; \mu_{\tilde{a}_{11}}, \nu_{\tilde{a}_{11}} \right), \left(a_2^1, a_2^2, a_2^3; \mu_{\tilde{a}_{22}}, \nu_{\tilde{a}_{22}} \right), \left(a_3^1, a_3^2, a_3^3; \mu_{\tilde{a}_{33}}, \nu_{\tilde{a}_{33}} \right); \theta_1, \theta_2 \right) \right\rangle$ and $\tilde{\tilde{B}} = \left(\left(b_1^1, b_1^2, b_1^3; \mu_{\tilde{b}_{11}}, \nu_{\tilde{b}_{11}} \right), \left(b_2^1, b_2^2, b_2^3; \mu_{\tilde{b}_{22}}, \nu_{\tilde{b}_{22}} \right), \left(b_3^1, b_3^2, b_3^3; \mu_{\tilde{b}_{33}}, \nu_{\tilde{b}_{33}} \right); \theta_3, \theta_4 \right)$, where $\theta_1 = \min\left\{ \mu_{\tilde{a}_{11}}, \mu_{\tilde{a}_{22}}, \mu_{\tilde{a}_{33}} \right\}$, $\theta_2 = \max\left\{ \nu_{\tilde{a}_{11}}, \nu_{\tilde{a}_{22}}, \nu_{\tilde{a}_{33}} \right\}$, $\theta_3 = \min\left\{ \mu_{\tilde{b}_{11}}, \mu_{\tilde{b}_{22}}, \mu_{\tilde{b}_{33}} \right\}$, and $\theta_4 = \max\left\{ \nu_{\tilde{b}_{11}}, \nu_{\tilde{b}_{22}}, \nu_{\tilde{b}_{33}} \right\}$, the below operators are defined,

$$\tilde{\tilde{A}} + \tilde{\tilde{B}} = \left\langle \left(\left(a_1^1 + b_1^1, a_1^2 + b_1^2, a_1^3 + b_1^3; \mu_{\tilde{a}_{11}} \wedge \mu_{\tilde{b}_{11}}, \nu_{\tilde{a}_{11}} \vee \nu_{\tilde{b}_{11}} \right), \left(a_2^1 + b_2^1, a_2^2 + b_2^2, a_2^3 \right. \right. \right.$$
$$\left. + b_2^3; \mu_{\tilde{a}_{22}} \wedge \mu_{\tilde{b}_{22}}, \nu_{\tilde{a}_{22}} \vee \nu_{\tilde{b}_{22}} \right),$$
$$\left. \left. \left(a_3^1 + b_3^1, a_3^2 + b_3^2, a_3^3 + b_3^3; \mu_{\tilde{a}_{33}} \wedge \mu_{\tilde{b}_{33}}, \nu_{\tilde{a}_{33}} \vee \nu_{\tilde{b}_{33}} \right) \right); \gamma_1, \gamma_2 \right\rangle$$

$$\tilde{\tilde{A}} - \tilde{\tilde{B}} = \langle \left(\left(a_1^1 - b_3^3, a_1^2 - b_3^2, a_1^3 - b_3^1; \mu_{\tilde{a}_{11}} \wedge \mu_{\tilde{b}_{11}}, \nu_{\tilde{a}_{11}} \vee \nu_{\tilde{b}_{11}} \right), \left(a_2^1 - b_2^3, a_2^2 - b_2^2, a_2^3 \right. \right.$$
$$\left. - b_2^1; \mu_{\tilde{a}_{22}} \wedge \mu_{\tilde{b}_{22}}, \nu_{\tilde{a}_{22}} \vee \nu_{\tilde{b}_{22}} \right),$$
$$\left(a_3^1 - b_1^3, a_3^2 - b_1^2, a_3^3 - b_1^1; \mu_{\tilde{a}_{33}} \wedge \mu_{\tilde{b}_{33}}, \nu_{\tilde{a}_{33}} \vee \nu_{\tilde{b}_{33}} \right); \gamma_3, \gamma_4$$

$$k\tilde{\tilde{A}} = \left(\left(ka_1^1, ka_1^2, ka_1^3; \mu_{\tilde{a}_{11}}, \nu_{\tilde{a}_{11}} \right), \left(ka_2^1, ka_2^2, ka_2^3; \mu_{\tilde{a}_{22}}, \nu_{\tilde{a}_{22}} \right), \right.$$
$$\left. \left(ka_3^1, ka_3^2, ka_3^3; \mu_{\tilde{a}_{33}}, \nu_{\tilde{a}_{33}} \right); \theta_1, \theta_2 \right)$$

where, $\gamma_1 = \min\left\{ \mu_{\tilde{a}_{11}} \wedge \mu_{\tilde{b}_{11}}, \mu_{\tilde{a}_{22}} \wedge \mu_{\tilde{b}_{22}}, \mu_{\tilde{a}_{33}} \wedge \mu_{\tilde{b}_{33}} \right\}$, $\gamma_2 = \max\left\{ \nu_{\tilde{a}_{11}} \vee \nu_{\tilde{b}_{11}}, \nu_{\tilde{a}_{22}} \vee \nu_{\tilde{b}_{22}}, \nu_{\tilde{a}_{33}} \vee \nu_{\tilde{b}_{33}} \right\}$, $\gamma_3 = \min\left\{ \mu_{\tilde{a}_{11}} \wedge \mu_{\tilde{b}_{11}}, \mu_{\tilde{a}_{22}} \wedge \mu_{\tilde{b}_{22}}, \mu_{\tilde{a}_{33}} \wedge \mu_{\tilde{b}_{33}} \right\}$, and $\gamma_4 = \max\left\{ \nu_{\tilde{a}_{11}} \vee \nu_{\tilde{b}_{11}}, \nu_{\tilde{a}_{22}} \vee \nu_{\tilde{b}_{22}}, \nu_{\tilde{a}_{33}} \vee \nu_{\tilde{b}_{33}} \right\}$.

Definition 3.26 [14] A ranking function such as $\Re : F\left(\tilde{\tilde{A}}\right) \to \mathbb{R}$ where $F\left(\tilde{\tilde{A}}\right)$ is the set of triangular type-2 Pythagorean fuzzy numbers and $\mathbb{R}$ is the set of real numbers is defined as below (where $\tilde{\tilde{A}} = \left\langle\left(\left(a_1^1, a_1^2, a_1^3; \mu_{\tilde{a}_{11}}, \nu_{\tilde{a}_{11}}\right), \left(a_2^1, a_2^2, a_2^3; \mu_{\tilde{a}_{22}}, \nu_{\tilde{a}_{22}}\right), \left(a_3^1, a_3^2, a_3^3; \mu_{\tilde{a}_{33}}, \nu_{\tilde{a}_{33}}\right); \theta_1, \theta_2\right)\right\rangle$).

$$\Re\left(\tilde{\tilde{A}}\right) = \left(\frac{\theta_1 + \theta_2}{2}\right)\left(\frac{1}{3}\right)\left(\frac{a_1^1 + a_1^2 + a_1^3 + a_2^1 + a_2^2 + a_2^3 + a_3^1 + a_3^2 + a_3^3}{3}\right) \quad (3.81)$$

Theorem 3.5 [14] *The ranking function* (3.81) *is a total order relation for a given set of triangular type-2 Pythagorean fuzzy numbers.*

Theorem 3.6 [14] *The ranking function* (3.81) *is linear. Therefore, for the triangular type-2 Pythagorean fuzzy numbers $\tilde{\tilde{A}}$ and $\tilde{\tilde{B}}$, and the real numbers k_1 and k_2, there is* $\Re\left(k_1\tilde{\tilde{A}} + k_2\tilde{\tilde{B}}\right) = k_1\Re\left(\tilde{\tilde{A}}\right) + k_2\Re\left(\tilde{\tilde{B}}\right)$. *The below rules also are true.*

$$\Re\left(\tilde{\tilde{A}}\right) > \Re\left(\tilde{\tilde{B}}\right) \text{ iff } \tilde{\tilde{A}} \succ \tilde{\tilde{B}},$$

$$\Re\left(\tilde{\tilde{A}}\right) < \Re\left(\tilde{\tilde{B}}\right) \text{ iff } \tilde{\tilde{A}} \prec \tilde{\tilde{B}},$$

$$\Re\left(\tilde{\tilde{A}}\right) = \Re\left(\tilde{\tilde{B}}\right) \text{ iff } \tilde{\tilde{A}} \approx \tilde{\tilde{B}}$$

3.8 Application: Pythagorean Fuzzy Multi-criteria Decision Making

An extension of the EDAS approach [15] to an interval-valued Pythagorean fuzzy environment [12] is presented here. Therefore, the below steps are introduced for the interval-valued Pythagorean fuzzy EDAS approach. In general, these steps follow the steps of the classical EDAS approach.

Step 1. Determine the alternatives (indexed by *i* and totally *m* alternatives) and criteria (indexed by *j* and totally *n* criteria) which are classified as benefit criteria (*B*) and cost criteria (*NB*).

Step 2. Determine an interval-valued Pythagorean fuzzy decision matrix and a set of criteria weight values shown as the below relations respectively.

$$\tilde{X} = \begin{pmatrix} \tilde{x}_{11} & \dots & \tilde{x}_{1n} \\ \vdots & \ddots & \vdots \\ \tilde{x}_{m1} & \cdots & \tilde{x}_{mn} \end{pmatrix} = \begin{pmatrix} \left(\left[\mu_{11}^-, \mu_{11}^+\right], \left[\nu_{11}^-, \nu_{11}^+\right]\right) & \dots & \left(\left[\mu_{1n}^-, \mu_{1n}^+\right], \left[\nu_{1n}^-, \nu_{1n}^+\right]\right) \\ \vdots & \ddots & \vdots \\ \left(\left[\mu_{n1}^-, \mu_{n1}^+\right], \left[\nu_{n1}^-, \nu_{n1}^+\right]\right) & \cdots & \left(\left[\mu_{m1}^-, \mu_{m1}^+\right], \left[\nu_{mn}^-, \nu_{mn}^+\right]\right) \end{pmatrix} \tag{3.82}$$

$$\tilde{W} = (\tilde{w}_1, \dots, \tilde{w}_n) = \left(\left[\mu_{\tilde{w}_1}^-, \mu_{\tilde{w}_1}^+\right], \left[\nu_{\tilde{w}_1}^-, \nu_{\tilde{w}_1}^+\right], \dots, \left[\mu_{\tilde{w}_n}^-, \mu_{\tilde{w}_n}^+\right], \left[\nu_{\tilde{w}_n}^-, \nu_{\tilde{w}_n}^+\right]\right) \tag{3.83}$$

Step 3. For each criterion calculate the aggregated Pythagorean fuzzy arithmetic mean of its values according to the below formula.

$$\widetilde{AV}_j = \left(\left[\sqrt{1 - \left(\prod_{i=1}^{n}\left(1 - \left(\mu_{ij}^-\right)^2\right)\right)^{\frac{1}{n}}}, \sqrt{1 - \left(\prod_{i=1}^{n}\left(1 - \left(\mu_{ij}^+\right)^2\right)\right)^{\frac{1}{n}}}\right], \left[\prod_{i=1}^{n}\left(\nu_{ij}^-\right)^{\frac{1}{n}}, \prod_{i=1}^{n}\left(\nu_{ij}^+\right)^{\frac{1}{n}}\right]\right). \tag{3.84}$$

Step 4. The Pythagorean fuzzy positive distance from the average ($PDA = \left[\widetilde{PDA}_{ij}\right]$) and the Pythagorean fuzzy negative distance from the average ($NDA = \left[\widetilde{NDA}_{ij}\right]$) are calculated from the below formulas where $\tilde{z}_{ij} = \left(\left[\mu_{ij}^-, \mu_{ij}^+\right], \left[\nu_{ij}^-, \nu_{ij}^+\right]\right)$ is a Pythagorean fuzzy number.

$$\widetilde{PDA}_{ij} = \frac{\max\left(0, \tilde{z}_{ij} \ominus \widetilde{AV}_j\right)}{I\left(\widetilde{AV}_j\right)} \text{if } j \in B,$$
$$\max\left(0, \tilde{z}_{ij} \ominus \widetilde{AV}_j\right) = \begin{cases} \tilde{z}_{ij} \ominus \widetilde{AV}_j \text{if } I\left(\tilde{z}_{ij}\right) > I\left(\widetilde{AV}_j\right) \\ ([0, 0], [1, 1]) O \cdot W \end{cases} \tag{3.85}$$

$$\widetilde{PDA}_{ij} = \frac{\max\left(0, \tilde{z}_{ij} \ominus \widetilde{AV}_j\right)}{I\left(\widetilde{AV}_j\right)} \text{if } j \in NB,$$
$$\max\left(0, \tilde{z}_{ij} \ominus \widetilde{AV}_j\right) = \begin{cases} \tilde{z}_{ij} \ominus \widetilde{AV}_j \text{if } I\left(\tilde{z}_{ij}\right) < I\left(\widetilde{AV}_j\right) \\ ([0, 0], [1, 1]) O.W \end{cases} \tag{3.86}$$

$$\widetilde{NDA}_{ij} = \frac{\max\left(0, \widetilde{AV}_j \ominus \tilde{z}_{ij}\right)}{I\left(\widetilde{AV}_j\right)} \text{if } j \in B,$$
$$\max\left(0, \widetilde{AV}_j \ominus \tilde{z}_{ij}\right) = \begin{cases} \widetilde{AV}_j \ominus \tilde{z}_{ij} \text{if } I\left(\tilde{z}_{ij}\right) < I\left(\widetilde{AV}_j\right) \\ ([0, 0], [1, 1]) O.W \end{cases} \tag{3.87}$$

$$\widetilde{NDA}_{ij} = \frac{\max\left(0, \widetilde{AV}_j \ominus \tilde{z}_{ij}\right)}{I\left(\widetilde{AV}_j\right)} \text{if } j \in NB,$$

$$\max\left(0, \widetilde{AV}_j \ominus \tilde{z}_{ij}\right) = \begin{cases} \widetilde{AV}_j \ominus \tilde{z}_{ij} \text{if } I\left(\tilde{z}_{ij}\right) > I\left(\widetilde{AV}_j\right) \\ ([0, 0], [1, 1]) O \cdot W \end{cases} \tag{3.88}$$

In the above formulas, the Eq. (3.66) is used for subtraction operators. For defuzzification, we can use the Eq. (3.77). The division operator can be expressed as the product of two numbers, thus, $\frac{\tilde{z}_{ij} \ominus \tilde{A}V_j}{I(\tilde{A}V_j)} = \frac{1}{I(\tilde{A}V_j)} \times \tilde{z}_{ij} \ominus \tilde{A}V_j$. Therefore,

$$\frac{\tilde{z}_{ij} \ominus AV_j}{I(AV_j)} = \left(\left[\sqrt{1 - \left(1 - \frac{\mu_{ij}^{-2} - \mu_j^{-2}}{1 - \mu_j^{-2}}\right)^{\frac{1}{I(AV_j)}}}, \sqrt{1 - \left(1 - \frac{\mu_{ij}^{+2} - \mu_j^{+2}}{1 - \mu_j^{+2}}\right)^{\frac{1}{I(AV_j)}}} \right], \right.$$
$$\left. \left[\left(\frac{\nu_{ij}^-}{\nu_j^-}\right)^{\frac{1}{I(AV_j)}}, \left(\frac{\nu_{ij}^+}{\nu_j^+}\right)^{\frac{1}{I(AV_j)}} \right] \right) \tag{3.89}$$

Step 5. The positive ($\widetilde{SP}_i$) and negative ($\widetilde{SN}_i$) weighted sum of each alternative are calculated as below,

$$\widetilde{SP}_i = \sum_{j=1}^{n} \tilde{w}_j \otimes \widetilde{PDA}_{ij} = \left(\tilde{w}_1 \otimes \widetilde{PDA}_{i1}\right) \oplus \ldots \oplus \left(\tilde{w}_n \otimes \widetilde{PDA}_{in}\right) \tag{3.90}$$

$$\widetilde{SN}_i = \sum_{j=1}^{n} \tilde{w}_j \otimes \widetilde{NDA}_{ij} = \left(\tilde{w}_1 \otimes \widetilde{NDA}_{i1}\right) \oplus \ldots \oplus \left(\tilde{w}_n \otimes \widetilde{NDA}_{in}\right). \tag{3.91}$$

Step 6. The normalized values of $\widetilde{SP}_i$ and $\widetilde{SN}_i$ are calculated as below.

$$\widetilde{NSP}_i = \frac{\widetilde{SP}_i}{\max_i\left(I\left(\widetilde{SP}_i\right)\right)} \tag{3.92}$$

$$\widetilde{NSN}_i = 1 - \frac{\widetilde{SN}_i}{\max_i\left(I\left(\widetilde{SN}_i\right)\right)} \tag{3.93}$$

Here, if $\frac{\widetilde{SN}_i}{\max_i(I(\widetilde{SN}_i))} = \left(\left[\mu_{11}^-, \mu_{11}^+\right], \left[\nu_{11}^-, \nu_{11}^+\right]\right)$ then $\widetilde{NSN}_i = 1 - \left(\left[\mu_{11}^-, \mu_{11}^+\right], \left[\nu_{11}^-, \nu_{11}^+\right]\right) = \left(\left[\nu_{11}^-, \nu_{11}^+\right], \left[\mu_{11}^-, \mu_{11}^+\right]\right)$.

Step 7. The score of each alternative is calculated as $\widetilde{AS}_i = \frac{\widetilde{NSP}_i \oplus \widetilde{NSN}_i}{2}$.

Table 3.5 The Pythagorean fuzzy decision matrix of Example 3.11

Alternatives/criterion	Design	Safety	Fuel-economy
A_1	([0.59439, 0.79581], [0.21735, 0.31308])	([0.49324, 0.69521], [0.31308, 0.41142])	([0.59439, 0.79581], [0.21735, 0.31308])
A_2	([0.49324, 0.69521], [0.31308, 0.41142])	([0.60000, 0.80000], [0.20000, 0.30000])	([0.43089, 0.63164], [0.37052, 0.47024])
A_3	([0.60000, 0.80000], [0.20000, 0.30000])	([0.46416, 0.66494], [0.33758, 0.43720])	([0.50000, 0.70000], [0.30000, 0.40000])

Table 3.6 The cumulative Pythagorean fuzzy arithmetic mean values of the criteria

Criteria	$\widetilde{AV}_j$
Design	([0.56648, 0.76904], [0.23875, 0.33808])
Safety	([0.52508, 0.72807], [0.27649, 0.37789])
Fuel-economy	([0.51604, 0.71930], [0.28909, 0.38906])

Table 3.7 The defuzzified values for $\widetilde{AV}_j$

Alternatives	$I(\widetilde{AV}_j)$
A_1	0.6319
A_2	0.5937
A_3	0.5865

Step 8. The score of each alternative is ranked by $I(\widetilde{AS}_i)$. Then, the alternatives are ranked by the decreasing order of $I(\widetilde{AS}_i)$ values.

Example 3.11 [12] Consider three types of cars such as A_1, A_2, and A_3, to be evaluated and ranked according to the criteria such as design, safety, and fuel-economy by the proposed EDAS approach. The decision matrix of Table 3.5 has been obtained for the problem.

The calculations of this example according to the steps of the proposed Pythagorean fuzzy EDAS approach are completed and the results of all steps are represented by Tables 3.6, 3.7, 3.8, 3.9, 3.10, 3.11, 3.12, 3.13, 3.14 and 3.15. According to the final results presented by Table 3.15 the ranking of A_1, A_3, and A_2 is obtained for the cars.

Table 3.8 The obtained *PDA* matrix

Alternatives/criterion	Design	Safety	Fuel-economy
A_1	([0.27288, 0.39667], [0.86190, 0.88552])	([0, 0], [1, 1])	([0.44405, 0.61153], [0.61489, 0.69043])
A_2	([0, 0], [1, 1])	([0.43339, 0.60094], [0.58007, 0.67833])	([0, 0], [1, 1])
A_3	([0.29931, 0.42606], [0.75556, 0.82769])	([0, 0], [1, 1])	([0, 0], [1, 1])

Table 3.9 The obtained *NDA* matrix

Alternatives/criterion	Design	Safety	Fuel-economy
A_1	([0, 0], [1, 1])	([0.26645, 0.38404], [0.81142, 0.86677])	([0, 0], [1, 1])
A_2	([0.39671, 0.55706], [0.65122, 0.73293])	([0, 0], [1, 1])	([0.40357, 0.55871], [0.65498, 0.72387])
A_3	([0, 0], [1, 1])	([0.35463, 0.50059], [0.71489, 0.78259])	([0.19172, 0.29976], [0.93879, 0.95380])

Table 3.10 The positive distance value of each alternative

Alternatives	$\widetilde{SP}_i$
A_1	([0.23173, 0.39413], [0.63298, 0.89783])
A_2	([0.19503, 0.33051], [0.68618, 0.78958])
A_3	([0.13469, 0.23433], [0.81103, 0.88336])

Table 3.11 The negative distance value of each alternative

Alternatives	$\widetilde{SN}_i$
A_1	([0.11190, 0.21122], [0.85298, 0.90914])
A_2	([0.25259, 0.42360], [0.54266, 0.80359])
A_3	([0.18089, 0.31769], [0.7418, 0.85045])

Table 3.12 The normalized positive distance value of each alternative

Alternatives	$\widetilde{NSP}_i$
A_1	([0.41161, 0.65819], [0.21481, 0.69596])
A_2	([0.34967, 0.56772], [0.28178, 0.45178])
A_3	([0.24437, 0.41592], [0.49440, 0.65896])

Table 3.13 The normalized negative distance value of each alternative

Alternatives	$\widetilde{NSN}_i$
A_1	([0.62299, 0.75315], [0.20537, 0.35638])
A_2	([0.16217, 0.521166], [0.42207, 0.66697])
A_3	([0.41342, 0.61749], [0.30702, 0.52095])

Table 3.14 The Pythagorean fuzzy score obtained for each alternative

Alternatives	$\tilde{AS}_i$
A_1	([0.53582, 0.71045], [0.21004, 0.49802])
A_2	([0.27483, 0.54559], [0.34486, 0.54893])
A_3	([0.34215, 0.53359], [0.38960, 0.58590])

Table 3.15 The defuzzified score and ranking of each alternative

Alternatives	Defuzzication of $\tilde{AS}_i$	Ranking
A_1	0.60332	1
A_2	0.42024	3
A_3	0.43375	2

References

1. Zadeh, L.A.: Fuzzy sets. Inf. Control. **8**, 338–353 (1965)
2. Garg, H., Mahmoodirad, A., Niroomand, S.: Fractional two-stage transshipment problem under uncertainty: application of the extension principle approach. Complex Intel. Syst. **7**(2), 807–822 (2021)
3. Yager, R.R.: Pythagorean fuzzy subsets. In: 2013 Joint IFSA World Congress and NAFIPS Annual Meeting (IFSA/NAFIPS), pp. 57–61 (2013)
4. Zhang, X., Xu, Z.: Extension of TOPSIS to multiple criteria decision making with Pythagorean fuzzy sets. Int. J. Intell. Syst. **29**, 1061–1078 (2014)
5. Hemalatha, K., Venkateswarlu, B.: Pythagorean fuzzy transportation problem: new way of ranking for Pythagorean fuzzy sets and mean square approach. Heliyon **9**(10), e20775 (2023)
6. Ghosh, S., Roy, S.K., Fügenschuh, A.: The multi-objective solid transportation problem with preservation technology using Pythagorean fuzzy sets. Int. J. Fuzzy Syst. **24**, 2687–2704 (2022)
7. Peng, X., Yuan, H., Yang, Y.: Pythagorean fuzzy information measures and their applications. Int. J. Intell. Syst. **32**, 991–1029 (2017)
8. Peng, X., Yang, Y.: Some results for Pythagorean fuzzy sets. Int. J. Intell. Syst. **30**(11), 1133–1160 (2015)
9. Garg, H.: A new improved score function of an interval-valued Pythagorean fuzzy set based TOPSIS method. Int. J. Uncertainty Quantification **7**(5), 2017020197 (2017)
10. Peng, X.D., Yang, Y.: Fundamental properties of interval-valued Pythagorean fuzzy aggregation operators. Int. J. Intell. Syst. **31**(5), 444–487 (2016)
11. Ritu, Kumar, T., Jahnvi, Kumar, K., Dhiman, N., Sharma, M.K.: Novel Pythagorean fuzzy score function to optimize fuzzy transportation models. Results Eng. **25**, 104048 (2025)
12. Yanmaz, O., Turgut, Y., Can, E.N., Kahraman, C.: Interval-valued Pythagorean fuzzy EDAS method: an application to car selection problem J. . Intel. Fuzzy Syst. **38**, 4061–4077 (2020)
13. Karasan, A., Ilbahar, E., Kahraman, C.: A novel pythagorean fuzzy AHP and its application to landfill site selection problem. Soft. Comput. **23**, 10953–10968 (2019)
14. Yu, V.F., Bera, A., Das, S.K., Manna, S., Jhulki, P.K., Dey, B., Ali, S.K.A.: Optimizing green solid transportation with carbon cap and trade: a multi-objective two-stage approach in a type-2 Pythagorean fuzzy context. Soft. Comput. **28**, 11015–11039 (2024)
15. Keshavarz Ghorabaee, M., Zavadskas, E.K., Olfat, L., Turskis, Z.: Multi-criteria inventory classification using a new method of evaluation based on distance from average solution (EDAS). Informatica **26**(3), 435–451 (2015)

Chapter 4
Fermatean Fuzzy Sets

4.1 Introduction

In this chapter another extension of fuzzy sets and numbers is focused. For this aim the Fermatean fuzzy sets, proposed by [1], are considered and detailed here. This type of fuzzy sets is an extension of Pythagorean fuzzy sets where there is a condition that the sum of cubic values of the membership and non-membership values could not exceed 1 (say, $0 \leq \mu(x)^3 + \nu(x)^3 \leq 1$). Like other types of fuzzy sets, the Fermatean fuzzy sets and numbers can be used to reflect the uncertain nature of the real-world problems.

4.2 Fermatean Fuzzy Set

The concepts and definitions related to Fermatean fuzzy sets are introduced in this section. These definitions are taken from Senapati and Yager [1–3].

Definition 4.1 Consider the reference set X. The Fermatean fuzzy set $\tilde{F}$ on this reference set is defined as below.

$$\tilde{F} = \left\{ \left(x, \mu_{\tilde{F}}(x), \nu_{\tilde{F}}(x) \right) : x \in X \right\} \tag{4.1}$$

In this fuzzy set, $\mu_{\tilde{F}}(x) : X \to [0, 1]$ and $\nu_{\tilde{F}}(x) : X \to [0, 1]$ are the membership and non-membership function of $x \in X$ where $0 \leq \left(\mu_{\tilde{F}}(x)\right)^3 + \left(\nu_{\tilde{F}}(x)\right)^3 \leq 1$.

Definition 4.2 Hesitancy degree of $x \in X$ in the Fermatean fuzzy set $\tilde{F}$ is shown as $\pi_{\tilde{F}}(x)$ and is calculated by the below formula.

$$\pi_{\tilde{F}}(x) = \sqrt[3]{1 - \left(\left(\mu_{\tilde{F}}(x)\right)^3 + \left(\nu_{\tilde{P}}(x)\right)^3 \right)}, \tag{4.2}$$

A. Mahmoodirad et al., *Recent Advances on Fuzzy Sets*, Studies in Fuzziness and Soft Computing 440, https://doi.org/10.1007/978-3-032-08776-8_4

For more simplicity, in continue, the Fermatean fuzzy set $\tilde{F} = \{(x, \mu_{\tilde{F}}(x), \nu_{\tilde{F}}(x)) : x \in X\}$ is shown as $\tilde{F} = (\mu_{\tilde{F}}, \nu_{\tilde{F}})$.

Example 4.1 Consider a decision making problem where the performance of alternative i in criterion j has membership degree of 0.9 and non-membership degree of 0.6. The conditions $0.9+0.6 > 1$, $(0.9)^2+(0.6)^2 = 1.17 > 1$, and $(0.9)^3+(0.6)^3 = 0.945 < 1$ are considered. Therefore, the performance of alternative i in criterion j can be only represented by the Fermatean fuzzy sets ($\tilde{F} = (\mu_{\tilde{F}}, \nu_{\tilde{F}}) = (0.9, 0.6)$) and it cannot be defined as an intuitionistic fuzzy set or a Pythagorean fuzzy set.

Definition 4.3 For the Fermatean fuzzy sets $\tilde{F} = (\mu_{\tilde{F}}, \nu_{\tilde{F}})$, $\tilde{F}_1 = (\mu_{\tilde{F}_1}, \nu_{\tilde{F}_1})$, and $\tilde{F}_2 = (\mu_{\tilde{F}_2}, \nu_{\tilde{F}_2})$, the below operations are defined.

$$\tilde{F}_1 \cap \tilde{F}_2 = \left(\min\{\mu_{\tilde{F}_1}, \mu_{\tilde{F}2}\}, \max\{\nu_{\tilde{F}_1}, \nu_{\tilde{F}2}\}\right), \tag{4.3}$$

$$\tilde{F}_1 \cup \tilde{F}_2 = \left(\max\{\mu_{\tilde{F}_1}, \mu_{\tilde{F}2}\}, \min\{\nu_{\tilde{F}_1}, \nu_{\tilde{F}2}\}\right), \tag{4.4}$$

$$\tilde{F}^c = (\nu_{\tilde{F}}, \mu_{\tilde{F}}), \tag{4.5}$$

$$\tilde{F}_1 \oplus \tilde{F}_2 = \left(\sqrt[3]{\mu^3_{\tilde{F}_1} + \mu^3_{\tilde{F}_2} - \mu^3_{\tilde{F}_1}\mu^3_{\tilde{F}_2}}, \nu_{\tilde{F}_1}\nu_{\tilde{F}_2}\right), \tag{4.6}$$

$$\tilde{F}_1 \otimes \tilde{F}_2 = \left(\nu_{\tilde{F}_1}\nu_{\tilde{F}_2}, \sqrt[3]{\nu^3_{\tilde{F}_1} + \nu^3_{\tilde{F}_2} - \nu^3_{\tilde{F}_1}\nu^3_{\tilde{F}_2}}\right), \tag{4.7}$$

$$\lambda\tilde{F} = \left(\sqrt[3]{1 - (1 - \mu^3_{\tilde{F}})^\lambda}, (\nu_{\tilde{F}})^\lambda\right), \lambda > 0, \tag{4.8}$$

$$(\tilde{F})^\lambda = \left(\mu^\lambda_{\tilde{F}}, \sqrt[3]{1 - (1 - \nu^3_{\tilde{F}})^\lambda}\right), \lambda > 0, \tag{4.9}$$

Theorem 4.1 *For the Fermatean fuzzy sets* $\tilde{F} = (\mu_{\tilde{F}}, \nu_{\tilde{F}})$, $\tilde{F}_1 = (\mu_{\tilde{F}_1}, \nu_{\tilde{F}_1})$, *and* $\tilde{F}_2 = (\mu_{\tilde{F}_2}, \nu_{\tilde{F}_2})$, *the below relations are true.*

$$\tilde{F}_1 \cap \tilde{F}_2 = \tilde{F}_2 \cap \tilde{F}_1 \tag{4.10}$$

$$\tilde{F}_1 \cup \tilde{F}_2 = \tilde{F}_2 \cup \tilde{F}_1 \tag{4.11}$$

$$\tilde{F}_1(\tilde{F}_2 \cap \tilde{F}_3) = (\tilde{F}_1 \cap \tilde{F}_2) \cap \tilde{F}_3 \tag{4.12}$$

$$\tilde{F}_1(\tilde{F}_2 \cup \tilde{F}_3) = (\tilde{F}_1 \cup \tilde{F}_2) \cap \tilde{F}_3 \tag{4.13}$$

$$\lambda(\tilde{F}_1 \cup \tilde{F}_2) = \lambda\tilde{F}_1 \cup \lambda\tilde{F}_2 \tag{4.14}$$

$$\left(\tilde{F}_1 \cup \tilde{F}_2\right)^{\lambda} = \tilde{F}_1^{\lambda} \cup \tilde{F}_2^{\lambda} \tag{4.15}$$

$$\left(\tilde{F}_1 \cap \tilde{F}_2\right)^{c} = \tilde{F}_1^{c} \cup \tilde{F}_2^{c} \tag{4.16}$$

$$\left(\tilde{F}_1 \cup \tilde{F}_2\right)^{c} = \tilde{F}_1^{c} \cap \tilde{F}_2^{c} \tag{4.17}$$

$$\left(\tilde{F}_1^{c}\right)^{\lambda} = \left(\lambda \tilde{F}_1\right)^{c} \tag{4.18}$$

$$\lambda \tilde{F}_1^{c} = \left(\tilde{F}_1^{\lambda}\right)^{c}. \tag{4.19}$$

Theorem 4.2 *For the Fermatean fuzzy sets* $\tilde{F} = \left(\mu_{\tilde{F}}, \nu_{\tilde{F}}\right)$, $\tilde{F}_1 = \left(\mu_{\tilde{F}_1}, \nu_{\tilde{F}_1}\right)$, *and* $\tilde{F}_2 = \left(\mu_{\tilde{F}_2}, \nu_{\tilde{F}_2}\right)$, *the below relations are true.*

$$\tilde{F}_1 \oplus \tilde{F}_2 = \tilde{F}_2 \oplus \tilde{F}_1 \tag{4.20}$$

$$\tilde{F}_1 \otimes \tilde{F}_2 = \tilde{F}_2 \otimes \tilde{F}_1 \tag{4.21}$$

$$\lambda\left(\tilde{F}_1 \oplus \tilde{F}_2\right) = \lambda \tilde{F}_1 \oplus \lambda \tilde{F}_2, \lambda > 0, \tag{4.22}$$

$$(\lambda_1 + \lambda_2)\tilde{F}_1 = \lambda_1 \tilde{F}_1 \oplus \lambda_2 \tilde{F}_1, \lambda_1, \lambda_2 > 0, \tag{4.23}$$

$$\left(\tilde{F}_1 \otimes \tilde{F}_2\right)^{\lambda} = \tilde{F}_1^{\lambda} \otimes \tilde{F}_2^{\lambda}, \lambda > 0, \tag{4.24}$$

$$\left(\tilde{F}_1\right)^{\lambda_1+\lambda_2} = \tilde{F}_1^{\lambda_1} \otimes \tilde{F}_1^{\lambda_2}, \lambda_1, \lambda_2 > 0, \tag{4.25}$$

$$\left(\tilde{F}_1 \oplus \tilde{F}_2\right)^{c} = \tilde{F}_1^{c} \otimes \tilde{F}_2^{c} \tag{4.26}$$

$$\left(\tilde{F}_1 \otimes \tilde{F}_2\right)^{c} = \tilde{F}_1^{c} \oplus \tilde{F}_2^{c} \tag{4.27}$$

$$\left(\tilde{F}_1 \cap \tilde{F}_2\right) \oplus \tilde{F}_3 = \left(\tilde{F}_1 \oplus \tilde{F}_3\right) \cap \left(\tilde{F}_2 \oplus \tilde{F}_3\right) \tag{4.28}$$

$$\left(\tilde{F}_1 \cup \tilde{F}_2\right) \oplus \tilde{F}_3 = \left(\tilde{F}_1 \oplus \tilde{F}_3\right) \cup \left(\tilde{F}_2 \oplus \tilde{F}_3\right) \tag{4.29}$$

$$\left(\tilde{F}_1 \cap \tilde{F}_2\right) \otimes \tilde{F}_3 = \left(\tilde{F}_1 \otimes \tilde{F}_3\right) \cap \left(\tilde{F}_2 \otimes \tilde{F}_3\right) \tag{4.30}$$

$$\left(\tilde{F}_1 \cup \tilde{F}_2\right) \otimes \tilde{F}_3 = \left(\tilde{F}_1 \otimes \tilde{F}_3\right) \cup \left(\tilde{F}_2 \otimes \tilde{F}_3\right). \tag{4.31}$$

Definition 4.4 A Fermatean fuzzy number is a Fermatean fuzzy set on $\mathbb{R}$ with the below properties and conditions.

- It should be normal such as there should exist a $x \in \mathbb{R}$ where $\mu_{\tilde{F}}(x) = 1$ and $\nu_{\tilde{F}}(x) = 0$.
- Its membership function should be convex such that $\mu_{\tilde{F}}(\delta x + (1-\delta)y) \geq \min\{\mu_{\tilde{F}}(x), \mu_{\tilde{F}}(y)\}, \forall x, y \in \mathbb{R}, \forall \delta \in [0, 1]$.
- Its non-membership function should be concave such that $\nu_{\tilde{F}}(\delta x + (1-\delta)y) \leq \max\{\nu_{\tilde{F}}(x), \nu_{\tilde{F}}(y)\}, \forall x, y \in \mathbb{R}, \forall \delta \in [0, 1]$.

Example 4.2 Consider the Fermatean fuzzy numbers $\tilde{F} = (0.8, 0.3)$, $\tilde{F}_1 = (0.4, 0.7)$, and $\tilde{F}_2 = (0.5, 0.4)$, and the scalar $\lambda = 2$. Then,

$$\tilde{F}_1 \oplus \tilde{F}_2 = (0.4, 0.7) \oplus (0.5, 0.4) = (0.0026, 0.28),$$

$$\tilde{F}_1 \otimes \tilde{F}_2 = (0.4, 0.7) \otimes (0.5, 0.4) = (0.2, 0.5723),$$

$$\lambda\tilde{F} = (0.07938, 0.09),$$

$$\left(\tilde{F}\right)^{\lambda} = (0.8, 0.3)^2 = (0.64, 0.4481),$$

$$\tilde{F}_1 \cup \tilde{F}_2 = (0.5, 0.4),$$

$$\tilde{F}_1 \cap \tilde{F}_2 = (0.4, 0.7),$$

$$\tilde{F}^c = (0.8, 0.3)^c = (0.3, 0.8).$$

Definition 4.5 The below natural quasi-ordering is defined for the Fermatean fuzzy sets $\tilde{F}_1 = \left(\mu_{\tilde{F}_1}, \nu_{\tilde{F}_1}\right)$ and $\tilde{F}_2 = \left(\mu_{\tilde{F}_2}, \nu_{\tilde{F}_2}\right)$.

Definition 4.6 For the Fermatean fuzzy set $\tilde{F} = \left(\mu_{\tilde{F}}, \nu_{\tilde{F}}\right)$ the score function of $S\left(\tilde{F}\right) = \left(\mu_{\tilde{F}}\right)^3 - \left(\nu_{\tilde{F}}\right)^3$ is defined where $S\left(\tilde{F}\right) \in [-1, 1]$.

Corollary 4.1 *According to the last definition, for* $\tilde{F} = (0, 1)$, *the score of* $S\left(\tilde{F}\right) = -1$, *and for* $\tilde{F} = (1, 0)$, *the score of* $S\left(\tilde{F}\right) = 1$, *are obtained.*

Some other score functions for the number $\tilde{F} = \left(\mu_{\tilde{F}}, \nu_{\tilde{F}}\right)$ *are as follows* [4],

$$S\left(\tilde{F}\right) = \frac{1}{2}\left(1 + \mu_{\tilde{F}}^3 - \nu_{\tilde{F}}^3\right) \tag{4.32}$$

$$S\left(\tilde{F}\right) = \frac{1}{3}\left(1 + 2\mu_{\tilde{F}}^3 - \nu_{\tilde{F}}^3\right) \tag{4.33}$$

$$S(\tilde{F}) = \frac{1}{2}\left(1 + \mu_{\tilde{F}}^2 - \nu_{\tilde{F}}^2\right)\left|\mu_{\tilde{F}} - \nu_{\tilde{F}}\right|, \tag{4.34}$$

where, clearly $S(\tilde{F}) \in [0, 1]$.

Definition 4.7 For the Fermatean fuzzy set $\tilde{F} = (\mu_{\tilde{F}}, \nu_{\tilde{F}})$ the accuracy function of $A(\tilde{F}) = (\mu_{\tilde{F}})^3 + (\nu_{\tilde{F}})^3$ is defined where $A(\tilde{F}) \in [0, 1]$.

Definition 4.8 Consider the Fermatean fuzzy sets $\tilde{F}_1 = (\mu_{\tilde{F}_1}, \nu_{\tilde{F}_1})$ and $\tilde{F}_2 = (\mu_{\tilde{F}_2}, \nu_{\tilde{F}_2})$ with the score functions $S(\tilde{F}_1)$ and $S(\tilde{F}_2)$, and the accuracy function $A(\tilde{F}_1)$ and $A(\tilde{F}_2)$ respectively. Then,

$$S(\tilde{F}_1) > S(\tilde{F}_2), \text{ then } \tilde{F}_1 \succ \tilde{F}_2 \tag{4.35}$$

$$S(\tilde{F}_1) < S(\tilde{F}_2), \text{ then } \tilde{F}_1 \prec \tilde{F}_2 \tag{4.36}$$

$$S(\tilde{F}_1) = S(\tilde{F}_2), A(\tilde{F}_1) > A(\tilde{F}_2), \text{ then } \tilde{F}_1 \succ \tilde{F}_2, \tag{4.37}$$

$$S(\tilde{F}_1) = S(\tilde{F}_2), A(\tilde{F}_1) < A(\tilde{F}_2), \text{ then } \tilde{F}_1 \prec \tilde{F}_2 \tag{4.38}$$

$$S(\tilde{F}_1) = S(\tilde{F}_2), A(\tilde{F}_1) = A(\tilde{F}_2), \text{ then } \tilde{F}_1 \approx \tilde{F}_2. \tag{4.39}$$

Example 4.3 For the Fermatean fuzzy numbers $\tilde{F}_2 = \left(\frac{\sqrt[3]{4}}{2}, \frac{\sqrt[3]{4}}{2}\right)$ and $\tilde{F}_2 = \left(\frac{1}{2}, \frac{1}{2}\right)$, according to Definition 4.6 the scores of $S(\tilde{F}_1) = S(\tilde{F}_2) = 0$ are obtained but clearly $\tilde{F}_1 \neq \tilde{F}_2$. For these numbers, according to Definition 4.8 the accuracy values of $A(\tilde{F}_1) = 1$ and $A(\tilde{F}_2) = 0.25$ are obtained and the order of $\tilde{F}_1 > \tilde{F}_2$ is concluded.

Definition 4.9 Suppose X as a non-empty set and the Fermatean fuzzy sets $\tilde{F}_1, \tilde{F}_2, \tilde{F}_3 \in FFSs(X)$ where $FFSs(X)$ is the set of Fermatean fuzzy sets on X. The distance of $\tilde{F}_1$ and $\tilde{F}_2$ is a function of $FFSs(X) \times FFSs(X) \to [0, 1]$ with the below conditions,

$$d(\tilde{F}_1, \tilde{F}_2) = d(\tilde{F}_2, \tilde{F}_1),$$

$$d(\tilde{F}_1, \tilde{F}_2) = 0, \text{ if and if } \tilde{F}_1 = \tilde{F}_2,$$

$$0 \leq d(\tilde{F}_1, \tilde{F}_2) \leq \sqrt{2},$$

$$d(\tilde{F}_1, \tilde{F}_2) \leq d(\tilde{F}_1, \tilde{F}_3) + d(\tilde{F}_2, \tilde{F}_3).$$

Definition 4.10 Consider the Fermatean fuzzy sets $\tilde{F}_1 = \left(\mu_{\tilde{F}_1}, \nu_{\tilde{F}_1}\right)$ and $\tilde{F}_2 = \left(\mu_{\tilde{F}_2}, \nu_{\tilde{F}_2}\right)$. The Euclidean, Hamming, and Mikowski distances of these numbers are defined as below respectively.

$$d_E\left(\tilde{F}_1, \tilde{F}_2\right) = \sqrt{\frac{1}{2}\left[\left(\mu^3_{\tilde{F}_1} - \mu^3_{\tilde{F}_2}\right)^2 + \left(\nu^3_{\tilde{F}_1} - \nu^3_{\tilde{F}_2}\right)^2 + \left(\pi^3_{\tilde{F}_1} - \pi^3_{\tilde{F}_2}\right)^2\right]} \tag{4.40}$$

$$d_H\left(\tilde{F}_1, \tilde{F}_2\right) = \frac{1}{2}\left(\left|\mu^3_{\tilde{F}_1} - \mu^3_{\tilde{F}_2}\right| + \left|\nu^3_{\tilde{F}_1} - \nu^3_{\tilde{F}_2}\right| + \left|\pi^3_{\tilde{F}_1} - \pi^3_{\tilde{F}_2}\right|\right) \tag{4.41}$$

$$d_M\left(\tilde{F}_1, \tilde{F}_2\right) = \left(\frac{1}{2}\left[\left(\mu^3_{\tilde{F}_1} - \mu^3_{\tilde{F}_2}\right)^p + \left(\nu^3_{\tilde{F}_1} - \nu^3_{\tilde{F}_2}\right)^p + \left(\pi^3_{\tilde{F}_1} - \pi^3_{\tilde{F}_2}\right)^p\right]\right)^{\frac{1}{p}} \tag{4.42}$$

The Minkowski distance $d_M\left(\tilde{F}_1, \tilde{F}_2\right)$, is converted to Euclidean and Hamming distances where $p = 1$ and $p = 2$ respectively.

Example 4.4 Consider the Fermatean fuzzy sets $\tilde{F}_1 = (0.7, 0.5)$ and $\tilde{F}_2 = (0.6, 0.4)$, with $\pi_{\tilde{F}_1} = \sqrt[3]{1 - 0.7^3 - 0.5^3} = \sqrt[3]{0.532} = 0.81$ and $\pi_{\tilde{F}_2} = \sqrt[3]{1 - 0.6^3 - 0.4^3} = \sqrt[3]{0.72} = 0.896$. Now, the below distances are calculated.

$$\begin{aligned} d_E\left(\tilde{F}_1, \tilde{F}_2\right) &= \sqrt{\frac{1}{2}\left[\left(0.7^3 - 0.6^3\right)^2 + \left(0.5^3 - 0.4^3\right)^2 + (0.532 - 0.72)^2\right]} \\ &= 0.166 \end{aligned}$$

$$\begin{aligned} d_H\left(\tilde{F}_1, \tilde{F}_2\right) &= \frac{1}{2}\left[\left|0.7^3 - 0.6^3\right| + \left|0.5^3 - 0.4^3\right| + |0.532 - 0.72|\right] \\ &= 0.188. \end{aligned}$$

Theorem 4.3 [1] *Consider the Fermatean fuzzy sets* $\tilde{F}_1 = \left(\mu_{\tilde{F}_1}, \nu_{\tilde{F}_1}\right)$ *and* $\tilde{F}_2 = \left(\mu_{\tilde{F}_2}, \nu_{\tilde{F}_2}\right)$, *and their Euclidean distance* $d_E\left(\tilde{F}_1, \tilde{F}_2\right)$. *Then,*

$$d_E\left(\tilde{F}_1, \tilde{F}_2\right) = d_E\left(\tilde{F}_2, \tilde{F}_1\right), \tag{4.43}$$

$$d_E\left(\tilde{F}_1, \tilde{F}_2\right) = 0, \text{ if and only if } \tilde{F}_1 = \tilde{F}_2, \tag{4.44}$$

$$0 \leq d_E\left(\tilde{F}_1, \tilde{F}_2\right) \leq \sqrt{2}. \tag{4.45}$$

Theorem 4.4 [1] *Consider the Fermatean fuzzy sets* $\tilde{F}_1 = \left(\mu_{\tilde{F}_1}, \nu_{\tilde{F}_1}\right)$, $\tilde{F}_2 = \left(\mu_{\tilde{F}_2}, \nu_{\tilde{F}_2}\right)$, *and* $\tilde{F}_3 = \left(\mu_{\tilde{F}_3}, \nu_{\tilde{F}_3}\right)$, *where* $\tilde{F}_1 \leq \tilde{F}_2 \leq \tilde{F}_3$. *Then,* $d_E\left(\tilde{F}_1, \tilde{F}_2\right) \leq d_E\left(\tilde{F}_1, \tilde{F}_3\right)$ *and* $d_E\left(\tilde{F}_2, \tilde{F}_3\right) \leq d_E\left(\tilde{F}_1, \tilde{F}_3\right)$.

4.3 Trapezoidal Fermatean Fuzzy Numbers

Definition 4.11 [5] A trapezoidal Fermatean fuzzy number (TrFFN) is a Fermatean fuzzy set with the below membership ($\mu_{\tilde{F}}(x)$) and non-membership ($\nu_{\tilde{F}}(x)$) functions.

$$\mu_{\tilde{F}}(x) = \begin{cases} p\frac{x-a}{b-a}; & a \le x \le b \\ p; & b \le x \le c \\ p\frac{d-x}{d-c}; & c \le x \le d \\ 0; & otherwise \end{cases}$$

$$\nu_{\tilde{F}}(x) = \begin{cases} \frac{b-x+q(x-a)}{b-a}; & a \le x \le b \\ q; & b \le x \le c \\ \frac{x-c+q(d-x)}{d-c}; & c \le x \le d \\ 1; & otherwise \end{cases} \tag{4.46}$$

Here, p is the maximum value of $\mu_{\tilde{F}}$, and q is the minimum value of $\nu_{\tilde{F}}$, where, $p, q \in [0, 1]$ and $0 \le p^3 + q^3 \le 1$. This TrFFN is represented by $\tilde{F} = (a, b, c, d; p, q)$.

Definition 4.12 For the numbers $\tilde{F}_1 = (a_1, b_1, c_1, d_1; p_1, q_1)$ and $\tilde{F}_2 = (a_2, b_2, c_2, d_2; p_2, q_2)$, the below operations are defined.

$$\tilde{F}_1 \oplus \tilde{F}_2 = \left(a_1 + a_2, b_1 + b_2, c_1 + c_2, d_1 + d_2; \sqrt[3]{p_1^3 + p_2^3 - p_1^3 p_2^3}, q_{1_1} q_2\right) \tag{4.47}$$

$$\tilde{F}_1 \otimes \tilde{F}_2 = \left(a_1 a_2, b_1 b_2, c_1 c_2, d_1 d_2; q_{1_1} q_2, \sqrt[3]{p_1^3 + p_2^3 - p_1^3 p_2^3}\right) \tag{4.48}$$

$$\lambda \tilde{F}_1 = \left(\lambda a_1, \lambda b_1, \lambda c_1, \lambda d_1; \sqrt[3]{1 - \left(1 - p_1^3\right)^\lambda}, (q_1)^\lambda\right), \lambda > 0, \tag{4.49}$$

$$\left(\tilde{F}_1\right)^\lambda = \left(a_1^\lambda, b_1^\lambda, c_1^\lambda, d_1^\lambda; p_1^\lambda, \sqrt[3]{1 - \left(1 - q_1^3\right)^\lambda}\right), \lambda > 0, \tag{4.50}$$

Theorem 4.5 [6] *For the numbers* $\tilde{F}_1 = (a_1, b_1, c_1, d_1; p_1, q_1)$, $\tilde{F}_2 = (a_2, b_2, c_2, d_2; p_2, q_2)$, *and* $\tilde{F}_2 = (a_2, b_2, c_2, d_2; p_2, q_2)$, *the below relations are true.*

$$\tilde{F}_1 \oplus \tilde{F}_2 = \tilde{F}_2 \oplus \tilde{F}_1, \tag{4.51}$$

$$\tilde{F}_1 \otimes \tilde{F}_2 = \tilde{F}_2 \otimes \tilde{F}_1 \tag{4.52}$$

$$\lambda\left(\tilde{F}_1 \oplus \tilde{F}_2\right) = \lambda \tilde{F}_1 \oplus \lambda \tilde{F}_2, \lambda > 0, \tag{4.53}$$

$$(\lambda_1 + \lambda_2)\tilde{F} = \lambda_1 \tilde{F} + \lambda_2 \tilde{F}, \lambda_1, \lambda_2 > 0, \tag{4.54}$$

$$\left(\tilde{F}_1 \otimes \tilde{F}_2\right)^{\lambda} = \tilde{F}_1^{\lambda} \otimes \tilde{F}_2^{\lambda}, \lambda > 0, \tag{4.55}$$

$$\tilde{F}^{\lambda_1} \otimes \tilde{F}^{\lambda_2} = \tilde{F}^{\lambda_1+\lambda_2}, \lambda_1, \lambda_2 > 0. \tag{4.56}$$

Definition 4.13 [6] For the TrFFN $\tilde{F} = (a, b, c, d; p, q)$, the below score and accuracy functions are defined.

$$S(\tilde{F}) = \frac{a+b+c+d}{4}(p^3 - q^3), S(\tilde{F}) \in [-1, 1], \tag{4.57}$$

$$A(\tilde{F}) = \frac{a+b+c+d}{4}(p^3 + q^3), A(\tilde{F}) \in [0, 1], \tag{4.58}$$

Based on this definition, for two numbers $\tilde{F}_1 = (a_1, b_1, c_1, d_1; p_1, q_1)$ and $\tilde{F}_2 = (a_2, b_2, c_2, d_2; p_2, q_2)$,

- If $S(\tilde{F}_1) > S(\tilde{F}_2)$, then $\tilde{F}_1 > \tilde{F}_2$,
- If $S(\tilde{F}_1) < S(\tilde{F}_2)$, then $\tilde{F}_1 < \tilde{F}_2$,
- If $S(\tilde{F}_1) = S(\tilde{F}_2)$:
  - If $A(\tilde{F}_1) > A(\tilde{F}_2)$, then $\tilde{F}_1 \succ \tilde{F}_2$,
  - If $A(\tilde{F}_1) < A(\tilde{F}_2)$, then $\tilde{F}_1 \prec \tilde{F}_2$,
  - If $A(\tilde{F}_1) = A(\tilde{F}_2)$, then $\tilde{F}_1 \approx \tilde{F}_2$.

Example 4.5 [7] consider the triangular Fermatean fuzzy transportation cost of $(25, 35, 55; 0.9, 0.1)$ between an origin and destination. This cost value is described as below.

- The acceptance degree of this cost value for decision maker is increased if the cost is increased from \$25 to \$35.
- The acceptance degree of this cost value for decision maker is decreased if the cost is increased from \$35 to \$55.
- The unacceptance degree of this cost value for decision maker is decreased if the cost is increased from \$25 to \$35.
- The unacceptance degree of this cost value for decision maker is increased if the cost is increased from \$35 to \$55.
- In the cost of \$35, decision maker has acceptance degree of 0.9 and unacceptance degree of 0.1.

Now, assume that the decision maker prefers the minimum acceptance degree of 0.6 and the maximum unacceptance degree of 0.2. Therefore, by using the concept of α cut for $\alpha = 0.6$ and $\beta = 0.2$, the interval of [33.8, 37.2] is obtained. This interval contains all cost values around \$35 with an acceptance degree from 0.6 to 0.9, and an unacceptance degree from 0.1 to 0.2.

Definition 4.14 The (α, β)-cut of the TrFFN $\tilde{F} = (a, b, c, d; p, q)$ is a sub-set of real numbers shown as $\tilde{F}^C = \{x | \mu_{\tilde{F}}(x) \geq \alpha, \nu_{\tilde{F}}(x) \leq \beta, 0 \leq \alpha \leq p, q \leq \beta \leq 1, \alpha + \beta \leq 1\}$, where,

$$\mu_{\tilde{F}}(x) \geq \alpha \Rightarrow \begin{cases} p\frac{x-a}{b-a} \geq \alpha \\ p\frac{d-x}{d-c} \geq \alpha \end{cases} \Rightarrow \begin{cases} x \geq a + \frac{\alpha}{p}(b-a) \\ x \leq d - \frac{\alpha}{p}(d-c) \end{cases} \tag{4.59}$$

$$\nu_{\tilde{F}}(x) \leq \beta \Rightarrow \begin{cases} \frac{b-x+q(x-a)}{b-a} \leq \beta \\ \frac{x-c+q(d-x)}{d-c} \leq \beta \end{cases} \Rightarrow \begin{cases} x \geq \frac{1-\beta}{1-q}b - \frac{q-\beta}{1-q}a \\ x \leq \frac{\beta-q}{1-q}b + \frac{1-\beta}{1-q}c \end{cases} \tag{4.60}$$

Assuming $\tilde{F}_L^c = \max\left\{a + \frac{\alpha}{p}(b-a), \frac{1-\beta}{1-q}b - \frac{q-\beta}{1-q}a\right\}$ and $\tilde{F}_U^c = \min\left\{d - \frac{\alpha}{p}(d-c), \frac{\beta-q}{1-q}b + \frac{1-\beta}{1-q}c\right\}$, then $q \in \left[\tilde{F}_L^c, \tilde{F}_U^c\right]$. Therefore, the (α, β)-cut converts the TrFFN $\tilde{F} = (a, b, c, d; p, q)$ to an interval with lower bound of $\tilde{F}_L^c$ and upper bound of $\tilde{F}_U^c$.

Considering $p = 1$ and $q = 0$ for Definition 4.11, the TrFFN $\tilde{F} = (a, b, c, d; a', b', c', d')$ is obtained, where,

$$\mu_{\tilde{F}}(x) = \begin{cases} \frac{x-a}{b-a}; & a \leq x \leq b \\ 1; & b \leq x \leq c \\ \frac{d-x}{d-c}; & c \leq x \leq d \\ 0; & otherwise \end{cases} \tag{4.61}$$

$$\nu_{\tilde{F}}(x) = \begin{cases} \frac{b'-x}{b'-a'}; & a' \leq x \leq b' \\ 0; & b' \leq x \leq c' \\ \frac{x-c'}{d'-c'}; & c' \leq x \leq d' \\ 1; & otherwise \end{cases} \tag{4.62}$$

$\tilde{F} = (a, b, c, d; a', b', c', d')$ is a TrFFN where $a' \leq a \leq b' \leq b \leq c \leq c' \leq d \leq d'$. Its schematic representation can be found in Fig. 4.1.

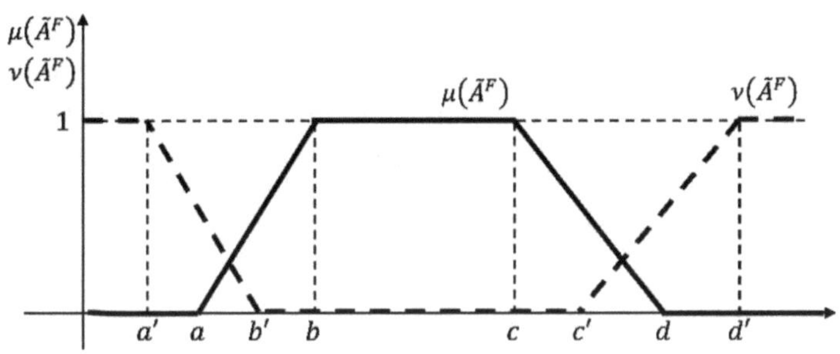

Fig. 4.1 Schematic representation of the TrFFN $\tilde{F} = (a, b, c, d; a', b', c', d')$

Remark 4.1 For the case of $b' = b = c = c'$, the triangular Fermatean fuzzy number (TFFN) $\tilde{F} = (a, b, c; a', b', c')$ is obtained.

Definition 4.15 Consider the TrFFNs $\tilde{A}_1 = (a_1, b_1, c_1, d_1; a_1', b_1', c_1', d_1')$ and $\tilde{A}_2 = (a_2, b_2, c_2, d_2; a_2', b_2', c_2', d_2')$. The relation $\tilde{A}_1 \leq \tilde{A}_2$ is held if and only if $a_1 \leq a_2$, $b_1 \leq b_2$, $c_1 \leq c_2$, $d_1 \leq d_2$, $a_1' \leq a_2'$, $b_1' \leq b_2'$, $c_1' \leq c_2'$, and $d_1' \leq d_2'$.

Definition 4.16 [5] The TrFFN $\tilde{F} = (a, b, c, d; a', b', c', d')$ is non-negative if and only if $a' \geq 0$.

Definition 4.17 [5] Consider the TrFFNs $\tilde{A}_1 = (a_1, b_1, c_1, d_1; a_1', b_1', c_1', d_1')$ and $\tilde{A}_2 = (a_2, b_2, c_2, d_2; a_2', b_2', c_2', d_2')$. Then,

$$\begin{aligned}\tilde{A}_1 \oplus \tilde{A}_2 = (&a_1 + a_2, b_1 + b_2, c_1 + c_2, d_1 + d_2;\\ &a_1' + a_2', b_1' + b_2', c_1' + c_2', d_1' + d_2')\end{aligned} \tag{4.63}$$

$$\begin{aligned}\tilde{A}_1 \ominus \tilde{A}_2 = (&a_1 - d_2, b_1 - c_2, c_1 - b_2, d_1 - a_2;\\ &a_1 - d_2\prime, b_1 - c_2\prime, c_1 - b_2\prime, d_1 - a_2\prime)\end{aligned} \tag{4.64}$$

$$k\tilde{A}_1 = (ka_1, kb_1, kc_1, kd_1; ka_1', kb_1', kc_1', kd_1'), k > 0, \tag{4.65}$$

$$k\tilde{A}_1 = (kd_1, kc_1, kb_1, ka_1; kd_1', kc_1', kb_1', ka_1'), k < 0, \tag{4.66}$$

Definition 4.18 [5] Consider the non-negative TrFFNs $\tilde{A}_1 = (a_1, b_1, c_1, d_1; a_1', b_1', c_1', d_1')$ and $\tilde{A}_2 = (a_2, b_2, c_2, d_2; a_2', b_2', c_2', d_2')$. Then,

$$\tilde{A}_1 \otimes \tilde{A}_2 = (a_1 a_2, b_1 b_2, c_1 c_2, d_1 d_2; a_1' a_2', b_1' b_2', c_1' c_2', d_1' d_2') \tag{4.67}$$

Definition 4.19 [5] The accuracy function of the TrFFN $\tilde{F} = (a, b, c, d; a', b', c', d')$ is represented as,

$$A(\tilde{F}) = \frac{a + b + c + d + a' + b' + c' + d'}{8}. \tag{4.68}$$

Definition 4.20 [5] Consider the TrFFNs $\tilde{A}_1 = (a_1, b_1, c_1, d_1; a_1', b_1', c_1', d_1')$ and $\tilde{A}_2 = (a_2, b_2, c_2, d_2; a_2', b_2', c_2', d_2')$ with the accuracy functions $A(\tilde{A}_1)$ and $A(\tilde{A}_2)$ respectively. Then,

$$\tilde{\mathbf{A}}_1 \prec \tilde{\mathbf{A}}_2 \text{if } A(\tilde{A}_1) < A(\tilde{A}_2), \tag{4.69}$$

$$\tilde{\mathbf{A}}_1 \succ \tilde{\mathbf{A}}_2 \text{if } A(\tilde{A}_1) > A(\tilde{A}_2), \tag{4.70}$$

$$\tilde{A}_1 \approx \tilde{A}_2 \text{if } A(\tilde{A}_1) = A(\tilde{A}_2). \tag{4.71}$$

Definition 4.21 [6] The hamming distance of the TrFFNs $\tilde{F}_1 = (a_1, b_1, c_1, d_1; p_1, q_1)$ and $\tilde{F}_2 = (a_2, b_2, c_2, d_2; p_2, q_2)$ is obtained by the below formula.

$$d_H(\tilde{F}_1, \tilde{F}_2) = \frac{1}{8}\left\{ \begin{array}{l} \left|\sqrt[3]{(1+p_1^3-p_2^3)a_1} - \sqrt[3]{2}\right| + \left|\sqrt[3]{(1+p_1^3-p_2^3)b_1} - \sqrt[3]{2}\right| \\ + \left|\sqrt[3]{(1+p_1^3-p_2^3)c_1} - \sqrt[3]{2}\right| + \left|\sqrt[3]{(1+p_1^3-p_2^3)d_1} - \sqrt[3]{2}\right| \end{array} \right\} \tag{4.72}$$

Definition 4.22 [6] Assume $D = (d_{ij})_{m\times n} = (a_{ij}, b_{ij}, c_{ij}, d_{ij}; p_{ij}, q_{ij})$ where $0 \leq a_{ij} \leq b_{ij} \leq c_{ij} \leq d_{ij} \leq 1$, $0 \leq p_{ij}, q_{ij} \leq 1$, and $0 \leq p_{ij}^3 + q_{ij}^3 \leq 1$. Then the below positive and negative ideal solutions are obtained.

$$\tilde{d}^+ = (a^+, b^+, c^+, d^+; p^+, q^+) = (1, 1, 1, 1; 1, 0) \tag{4.73}$$

$$\tilde{d}^- = (a^-, b^-, c^-, d^-; p^-, q^-) = (0, 0, 0, 0; 0, 1) \tag{4.74}$$

4.4 Ranking Function for Fermatean Fuzzy Numbers

Definition 4.23 Assume the Fermatean fuzzy set $\tilde{F} = (\mu_{\tilde{F}}, \nu_{\tilde{F}})$. A ranking function for this number is defined as below.

$$\Re(\tilde{F}) = \frac{1}{2}(1 + \mu_{\tilde{F}}^2 - \nu_{\tilde{F}}^2) \times \min\{\mu_{\tilde{F}}, \nu_{\tilde{F}}\}. \tag{4.75}$$

Lemma 4.1 *Assume the Fermatean fuzzy set $\tilde{F} = (\mu_{\tilde{F}}, \nu_{\tilde{F}})$. Then, for the ranking function* (4.76), $\Re(\tilde{F}) \in [0, 1]$.

Proof As $\mu_{\tilde{F}}, \nu_{\tilde{F}} \in [0, 1]$, then $\min\{\mu_{\tilde{F}}, \nu_{\tilde{F}}\} \in [0, 1]$. As $\mu_{\tilde{F}}^2, \nu_{\tilde{F}}^2 \geq 0$ and $\mu_{\tilde{F}}^2, \nu_{\tilde{F}}^2 \leq 1$, therefore, $1 - \nu_{\tilde{F}}^2 \geq 0 \Rightarrow 1 + \mu_{\tilde{F}}^2 - \nu_{\tilde{F}}^2 \geq 0 \Rightarrow \frac{1}{2}(1 + \mu_{\tilde{F}}^2 - \nu_{\tilde{F}}^2)(\min\{\mu_{\tilde{F}}, \nu_{\tilde{F}}\}) \geq 0$. On the other hand, $\mu_{\tilde{F}}^2 - \nu_{\tilde{F}}^2 \leq 1$, then $1 + \mu_{\tilde{F}}^2 - \nu_{\tilde{F}}^2 \leq 2$ and $\frac{1}{2}(1 + \mu_{\tilde{F}}^2 - \nu_{\tilde{F}}^2)(\min\{\mu_{\tilde{F}}, \nu_{\tilde{F}}\}) \leq 1$. Therefore, $0 \leq \frac{1}{2}(1 + \mu_{\tilde{F}}^2 - \nu_{\tilde{F}}^2)(\min\{\mu_{\tilde{F}}, \nu_{\tilde{F}}\}) \leq 1$. ■

Definition 4.24 For the TrFFn $\tilde{F} = (a, b, c, d; a', b', c', d')$, the below ranking function is obtained.

$$\Re(\tilde{F}) = \frac{a + 2b + 2c + d + a\prime + 2b\prime + 2c\prime + d\prime}{12}. \tag{4.76}$$

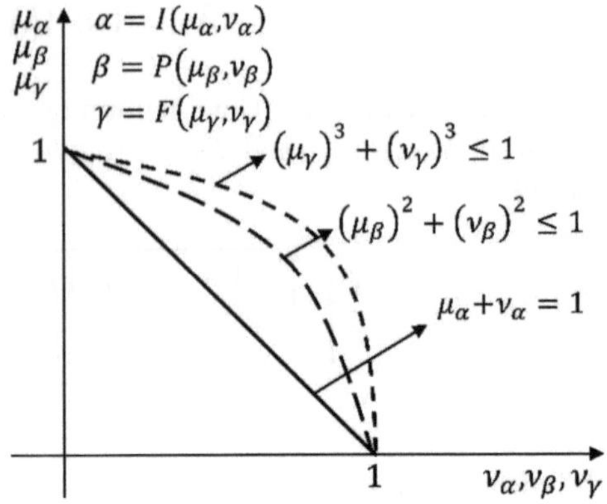

Fig. 4.2 Comparison of Pythagorean and Fermatean fuzzy numbers

4.5 Comparison of Pythagorean and Fermatean Fuzzy Numbers

For the intuitionistic fuzzy number $\tilde{\alpha} = (\mu_{\tilde{\alpha}}(x), \nu_{\tilde{\alpha}}(x))$, the condition of $0 \leq \mu_{\tilde{\alpha}}(x) + \nu_{\tilde{\alpha}}(x) \leq 1$ should be held. On the other hand, for the Pythagorean fuzzy number $\tilde{\beta} = \left(\mu_{\tilde{\beta}}(x), \nu_{\tilde{\beta}}(x)\right)$ and the Fermatean fuzzy number $\tilde{F} = \left(\mu_{\tilde{F}}, \nu_{\tilde{F}}\right)$ the conditions $\left(\mu_{\tilde{\beta}}(x)\right)^2 + \left(\nu_{\tilde{\beta}}(x)\right)^2 \leq 1$ and $0 \leq \left(\mu_{\tilde{F}}\right)^3 + \left(\nu_{\tilde{F}}\right)^3 \leq 1$ should be held respectively. As for a point such as $a, b \in [0, 1]$, the relationship $a + b \leq 1 \Rightarrow a^2 + b^2 \leq 1 \Rightarrow a^3 + b^3 \leq 1$ is held, according to Senapati and Yager [2], the Fermatean membership degree space is greater than the intuitionistic and Pythagorean membership degree spaces. Therefore, any intuitionistic fuzzy number is a Pythagorean fuzzy number, and any Pythagorean fuzzy number is a Fermatean fuzzy number (see Fig. 4.2).

4.6 Interval-Valued Fermatean Fuzzy Sets

The interval-valued Fermatean fuzzy sets (IVFFSs) provides a framework for modeling and presenting the uncertain situations with inexact membership and non-membership degrees where these degrees are shown by interval values. The IVFFSs are more suitable for the systems that the inputs are determined by experts, and they need more flexibility to express their opinions. Therefore, as an advantage, the interval-valued Fermatean fuzzy sets have more flexibility in terms of determining the membership and non-membership function values and a more robust and reliable decision can be made by an expert [8].

Definition 4.25 [9] Assume ***Int***$[0, 1]$ as the set of all sub-intervals of the interval $[0, 1]$. Then, $\tilde{F} = \left\{\left(x, \left[\mu_{\tilde{F}_L}(x), \mu_{\tilde{F}_U}(x)\right], \left[\nu_{\tilde{F}_L}(x), \nu_{\tilde{F}_U}(x)\right]\right) : x \in X\right\}$ is called an interval-valued Fermatean fuzzy set (IVFFS) where for any $x \in X$, the conditions $0 \leq \mu_{\tilde{F}_L}(x) \leq \mu_{\tilde{F}_U}(x) \leq 1$, $0 \leq \nu_{\tilde{F}_L}(x) \leq \nu_{\tilde{F}_U}(x) \leq 1$, and $0 \leq$

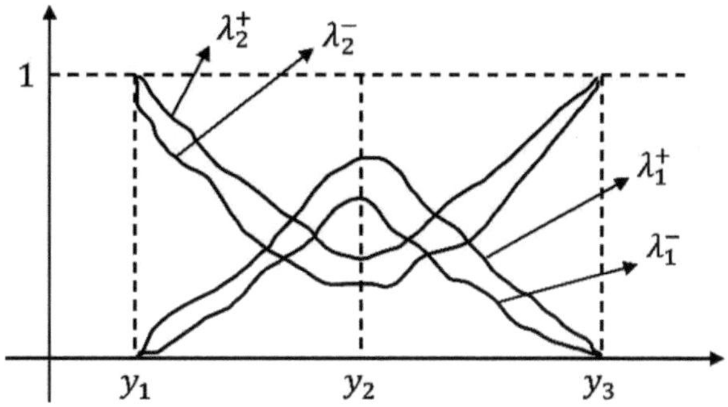

Fig. 4.3 An interval-valued Fermatean fuzzy number

$\mu^3_{\tilde{F}_U}(x) + \nu^3_{\tilde{F}_U}(x) \leq 1$ are held. The values $\left[\mu_{\tilde{F}_L}(x), \mu_{\tilde{F}_U}(x)\right]$ and $\left[\nu_{\tilde{F}_L}(x), \nu_{\tilde{F}_U}(x)\right]$ are the interval-valued membership and non-membership degrees respectively. The hesitancy degree of the IVFFS is obtained by $\pi_{\tilde{F}}(x) = \left[\pi_{\tilde{F}_L}(x), \pi_{\tilde{F}_U}(x)\right] = \left[\sqrt[3]{1 - \mu^3_{\tilde{F}_U}(x) - \nu^3_{\tilde{F}_U}(x)}, \sqrt[3]{1 - \mu^3_{\tilde{F}_L}(x) - \nu^3_{\tilde{F}_L}(x)}\right]$.

Some special cases of the IVFFSs are as below.

- If $\mu_{\tilde{F}_L}(x) = \mu_{\tilde{F}_U}(x)$ and $\nu_{\tilde{F}_L}(x) = \nu_{\tilde{F}_U}(x)$, then the IVFFS is converted to a Fermatean fuzzy set (FFS).
- If $0 \leq \mu_{\tilde{F}_U}(x) + \nu_{\tilde{F}_U}(x) \leq 1$ then the IVFFS is converted to an interval-valued intuitionistic fuzzy set.
- If $0 \leq \mu^2_{\tilde{F}_U}(x) + \nu^2_{\tilde{F}_U}(x) \leq 1$ then the IVFFS is converted to an interval-valued Pythagorean fuzzy set.
- An interval-valued Fermatean fuzzy number (IVFFN) is represented by $\tilde{F} = \left(\left[\mu_{\tilde{F}_L}, \mu_{\tilde{F}_U}\right], \left[\nu_{\tilde{F}_L}, \nu_{\tilde{F}_U}\right]\right)$.

Figure 4.3 represents an interval-valued Fermatean fuzzy number. Also, a comparison among the interval-valued Fermatean fuzzy numbers, interval-valued Pythagoran fuzzy numbers, and interval-valued intuitionistic fuzzy numbers is represented by Fig. 4.4.

Definition 4.26 [9] For the IVFFNs $\tilde{F}_1 = \left(\left[\mu_{\tilde{F}_{1L}}, \mu_{\tilde{F}_{1U}}\right], \left[\nu_{\tilde{F}_{1L}}, \nu_{\tilde{F}_{1U}}\right]\right)$ and $\tilde{F}_2 = \left(\left[\mu_{\tilde{F}_{2L}}, \mu_{\tilde{F}_{2U}}\right], \left[\nu_{\tilde{F}_{2L}}, \nu_{\tilde{F}_{2U}}\right]\right)$, $\tilde{F}_1 \leq \tilde{F}_2$ if and only if $\mu_{\tilde{F}_{1L}} \leq \mu_{\tilde{F}_{2L}}$, $\mu_{\tilde{F}_{1U}} \leq \mu_{\tilde{F}_{1U}}$, $\nu_{\tilde{F}_{1L}} \geq \nu_{\tilde{F}_{2L}}$, and $\nu_{\tilde{F}_{1U}} \geq \nu_{\tilde{F}_{2U}}$.

Theorem 4.6 *Assume the IVFFSs* $\tilde{F} = \left(\left[\mu_{\tilde{F}_L}, \mu_{\tilde{F}_U}\right], \left[\nu_{\tilde{F}_L}, \nu_{\tilde{F}_U}\right]\right)$, $\tilde{F}_1 = \left(\left[\mu_{\tilde{F}_{1L}}, \mu_{\tilde{F}_{1U}}\right], \left[\nu_{\tilde{F}_{1L}}, \nu_{\tilde{F}_{1U}}\right]\right)$, $\tilde{F}_2 = \left(\left[\mu_{\tilde{F}_{2L}}, \mu_{\tilde{F}_{2U}}\right], \left[\nu_{\tilde{F}_{2L}}, \nu_{\tilde{F}_{2U}}\right]\right)$, *and* $\tilde{F}_3 = \left(\left[\mu_{\tilde{F}_{3L}}, \mu_{\tilde{F}_{3U}}\right], \left[\nu_{\tilde{F}_{3L}}, \nu_{\tilde{F}_{3U}}\right]\right)$. *Then, the following rules and logics are held.*

- $\tilde{F}_1 \cup \tilde{F}_2 = \left(\left[\min\left\{\mu_{\tilde{F}_{1L}}, \mu_{\tilde{F}_{2L}}\right\}, \min\left\{\mu_{\tilde{F}_{1U}}, \mu_{\tilde{F}_{2U}}\right\}\right], \left[\max\left\{\mu_{\tilde{F}_{1L}}, \mu_{\tilde{F}_{2L}}\right\}, \max\left\{\mu_{\tilde{F}_{1U}}, \mu_{\tilde{F}_{2U}}\right\}\right]\right)$.
- $\tilde{F}_1 \cup \tilde{F}_2 = \left(\left[\max\left\{\mu_{\tilde{F}_{1L}}, \mu_{\tilde{F}_{2L}}\right\}, \max\left\{\mu_{\tilde{F}_{1U}}, \mu_{\tilde{F}_{2U}}\right\}\right], \left[\min\left\{\mu_{\tilde{F}_{1L}}, \mu_{\tilde{F}_{2L}}\right\}, \min\left\{\mu_{\tilde{F}_{1U}}, \mu_{\tilde{F}_{2U}}\right\}\right]\right)$.
- $\tilde{F}^c = \left(\left[\nu_{\tilde{F}_L}, \nu_{\tilde{F}_U}\right], \left[\mu_{\tilde{F}_L}, \mu_{\tilde{F}_U}\right]\right)$.
- Involution: $\left(\tilde{F}^c\right)^c = \tilde{F}$.

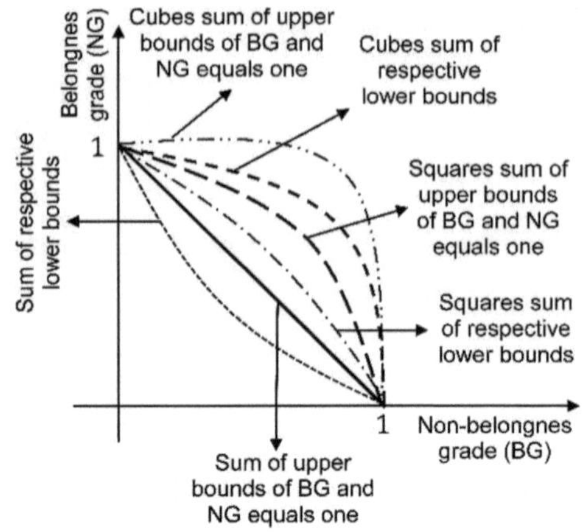

Fig. 4.4 A comparison of the interval-valued versions of the Fermatean, Pythagorean, and intuitionistic fuzzy numbers

- Commutativity: $\tilde{F}_1 \cup \tilde{F}_2 = \tilde{F}_2 \cup \tilde{F}_1$ and $\tilde{F}_1 \cap \tilde{F}_2 = \tilde{F}_2 \cap \tilde{F}_1$.
- De Morgan's law: $(\tilde{F}_1 \cup \tilde{F}_2)^c = \tilde{F}_1^c \cap \tilde{F}_2^c$ and $(\tilde{F}_1 \cap \tilde{F}_2)^c = \tilde{F}_1^c \cup \tilde{F}_2^c$.
- Associativity: $\tilde{F}_1 \cup (\tilde{F}_2 \cup \tilde{F}_3) = (\tilde{F}_1 \cup \tilde{F}_2) \cup \tilde{F}_3$ and $\tilde{F}_1 \cap (\tilde{F}_2 \cap \tilde{F}_3) = (\tilde{F}_1 \cap \tilde{F}_2) \cap \tilde{F}_3$.
- Distributivity: $\tilde{F}_1 \cup (\tilde{F}_2 \cap \tilde{F}_3) = (\tilde{F}_1 \cup \tilde{F}_2) \cap (\tilde{F}_1 \cup \tilde{F}_3)$ and $\tilde{F}_1 \cap (\tilde{F}_2 \cup \tilde{F}_3) = (\tilde{F}_1 \cap \tilde{F}_2) \cup (\tilde{F}_1 \cap \tilde{F}_3)$.
- Idempotence: $\tilde{F}_1 \cup \tilde{F}_1 = \tilde{F}_1$ and $\tilde{F}_1 \cap \tilde{F}_1 = \tilde{F}_1$.
- Absorption: $\tilde{F}_1 \cup (\tilde{F}_1 \cap \tilde{F}_2) = \tilde{F}_1$ and $\tilde{F}_1 \cap (\tilde{F}_1 \cup \tilde{F}_2) = \tilde{F}_1$.

Proof See Jeevaraj [9].

Definition 4.27 [9] For the IVFFNs $\tilde{F} = ([\mu_{\tilde{F}_L}, \mu_{\tilde{F}_U}], [\nu_{\tilde{F}_L}, \nu_{\tilde{F}_U}])$, $\tilde{F}_1 = ([\mu_{\tilde{F}_{1L}}, \mu_{\tilde{F}_{1U}}], [\nu_{\tilde{F}_{1L}}, \nu_{\tilde{F}_{1U}}])$, and $\tilde{F}_2 = ([\mu_{\tilde{F}_{2L}}, \mu_{\tilde{F}_{2U}}], [\nu_{\tilde{F}_{2L}}, \nu_{\tilde{F}_{2U}}])$, the below operators are defined.

- $\tilde{F}_1 \oplus \tilde{F}_2 = \left(\left[\sqrt[3]{\mu^3_{\tilde{F}_{1L}} + \mu^3_{\tilde{F}_{2L}} - \mu^3_{\tilde{F}_{1L}} \times \mu^3_{\tilde{F}_{2L}}}, \sqrt[3]{\mu^3_{\tilde{F}_{1U}} + \mu^3_{\tilde{F}_{2U}} - \mu^3_{\tilde{F}_{1U}} \times \mu^3_{\tilde{F}_{2U}}}\right], \left[\nu^3_{\tilde{F}_{1L}} \nu^3_{\tilde{F}_{2L}}, \nu^3_{\tilde{F}_{1U}} \nu^3_{\tilde{F}_{2U}}\right]\right)$
- $\tilde{F}_1 \otimes \tilde{F}_2 = \left(\left[\mu^3_{\tilde{F}_{1L}} \mu^3_{\tilde{F}_{2L}}, \mu^3_{\tilde{F}_{1U}} \mu^3_{\tilde{F}_{2U}}\right], \left[\sqrt[3]{\nu^3_{\tilde{F}_{1L}} + \nu^3_{\tilde{F}_{2L}} - \nu^3_{\tilde{F}_{1L}} \times \nu^3_{\tilde{F}_{2L}}}, \sqrt[3]{\nu^3_{\tilde{F}_{1U}} + \nu^3_{\tilde{F}_{2U}} - \nu^3_{\tilde{F}_{1U}} \times \nu^3_{\tilde{F}_{2U}}}\right]\right)$
- $\lambda\tilde{F} = \left(\left[\sqrt[3]{1 - \left(1 - \mu^3_{\tilde{F}_L}\right)^{\lambda}}, \sqrt[3]{1 - \left(1 - \mu^3_{\tilde{F}_U}\right)^{\lambda}}\right], \left[\nu^{\lambda}_{\tilde{F}_L}, \nu^{\lambda}_{\tilde{F}_U}\right]\right), \lambda > 0$
- $\tilde{F}^{\lambda} = \left(\left[\mu^{\lambda}_{\tilde{F}_L}, \mu^{\lambda}_{\tilde{F}_U}\right], \left[\sqrt[3]{1 - \left(1 - \nu^3_{\tilde{F}_L}\right)^{\lambda}}, \sqrt[3]{1 - \left(1 - \nu^3_{\tilde{F}_U}\right)^{\lambda}}\right]\right), \lambda > 0.$

Example 4.6 Considering the IVFFN $\tilde{F} = ([1, 1], [0, 0])$ and $\lambda > 0$, then,

$$\lambda\tilde{F} = \left(\left[\sqrt[3]{1 - \left(1 - \mu^3_{\tilde{F}_L}\right)^{\lambda}}, \sqrt[3]{1 - \left(1 - \mu^3_{\tilde{F}_U}\right)^{\lambda}}\right], \left[\nu^{\lambda}_{\tilde{F}_L}, \nu^{\lambda}_{\tilde{F}_U}\right]\right)$$

$$= ([1, 1], [0, 0]) = \tilde{F}$$

and

$$\tilde{F}^{\lambda} = \left(\left[\mu_{\tilde{F}_L}^{\lambda}, \mu_{\tilde{F}_U}^{\lambda}\right], \left[\sqrt[3]{1-\left(1-\nu_{\tilde{F}_L}^{3}\right)^{\lambda}}, \sqrt[3]{1-\left(1-\nu_{\tilde{F}_U}^{3}\right)^{\lambda}}\right]\right)$$
$$= ([1, 1], [0, 0]) = \tilde{F}.$$

Example 4.7 Considering the IVFFN $\tilde{F} = ([0, 0], [1, 1])$ and $\lambda > 0$, then,

$$\lambda\tilde{F} = \left(\left[\sqrt[3]{1-\left(1-\mu_{\tilde{F}_L}^{3}\right)^{\lambda}}, \sqrt[3]{1-\left(1-\mu_{\tilde{F}_U}^{3}\right)^{\lambda}}\right], \left[\nu_{\tilde{F}_L}^{\lambda}, \nu_{\tilde{F}_U}^{\lambda}\right]\right)$$
$$= ([0, 0], [1, 1]) = \tilde{F}$$

and

$$\tilde{F}^{\lambda} = \left(\left[\mu_{\tilde{F}_L}^{\lambda}, \mu_{\tilde{F}_U}^{\lambda}\right], \left[\sqrt[3]{1-\left(1-\nu_{\tilde{F}_L}^{3}\right)^{\lambda}}, \sqrt[3]{1-\left(1-\nu_{\tilde{F}_U}^{3}\right)^{\lambda}}\right]\right)$$
$$= ([0, 0], [1, 1]) = \tilde{F}.$$

Theorem 4.7 *Assume the IVFFSs* $\tilde{F} = \left(\left[\mu_{\tilde{F}_L}, \mu_{\tilde{F}_U}\right], \left[\nu_{\tilde{F}_L}, \nu_{\tilde{F}_U}\right]\right)$, $\tilde{F}_1 = \left(\left[\mu_{\tilde{F}_{1L}}, \mu_{\tilde{F}_{1U}}\right], \left[\nu_{\tilde{F}_{1L}}, \nu_{\tilde{F}_{1U}}\right]\right)$, *and* $\tilde{F}_2 = \left(\left[\mu_{\tilde{F}_{2L}}, \mu_{\tilde{F}_{2U}}\right], \left[\nu_{\tilde{F}_{2L}}, \nu_{\tilde{F}_{2U}}\right]\right)$. *Then, the following relations are held.*

- $\tilde{F}_1 \oplus \tilde{F}_2 = \tilde{F}_2 \oplus \tilde{F}_1$
- $\tilde{F}_1 \otimes \tilde{F}_2 = \tilde{F}_2 \otimes \tilde{F}_1$
- $\lambda\left(\tilde{F}_1 \oplus \tilde{F}_2\right) = \lambda\tilde{F}_2 \oplus \lambda\tilde{F}_1, \forall\lambda > 0$
- $(\lambda_1 + \lambda_2)\tilde{F} = \lambda_1\tilde{F} \oplus \lambda_2\tilde{F}, \forall\lambda_1, \lambda_2 > 0$
- $\left(\tilde{F}_1 \otimes \tilde{F}_2\right)^{\lambda} = \tilde{F}_1^{\lambda} \otimes \tilde{F}_2^{\lambda}, \forall\lambda > 0$
- $\tilde{F}^{\lambda_1} \otimes \tilde{F}^{\lambda_2} = \tilde{F}^{\lambda_1+\lambda_2}, \forall\lambda_1, \lambda_2 > 0$

Proof See Jeevaraj [9].

Theorem 4.8 *Assume the IVFFSs* $\tilde{F} = \left(\left[\mu_{\tilde{F}_L}, \mu_{\tilde{F}_U}\right], \left[\nu_{\tilde{F}_L}, \nu_{\tilde{F}_U}\right]\right)$, $\tilde{F}_1 = \left(\left[\mu_{\tilde{F}_{1L}}, \mu_{\tilde{F}_{1U}}\right], \left[\nu_{\tilde{F}_{1L}}, \nu_{\tilde{F}_{1U}}\right]\right)$, *and* $\tilde{F}_2 = \left(\left[\mu_{\tilde{F}_{2L}}, \mu_{\tilde{F}_{2U}}\right], \left[\nu_{\tilde{F}_{2L}}, \nu_{\tilde{F}_{2U}}\right]\right)$. *Then, the following relations are held.*

- $\tilde{F}_1^c \otimes \tilde{F}_2^c = \left(\tilde{F}_1 \oplus \tilde{F}_2\right)^c$
- $\tilde{F}_1^c \oplus \tilde{F}_2^c = \left(\tilde{F}_1 \otimes \tilde{F}_2\right)^c$
- $\left(\tilde{F}^c\right)^{\lambda} = \left(\lambda\tilde{F}\right)^c$
- $\left(\tilde{F}^{\lambda}\right)^c = \lambda\tilde{F}^c$

Proof See Jeevaraj [9].

Theorem 4.9 *Assume the IVFFSs* $\tilde{F}_1 = ([\mu_{\tilde{F}_{1L}}, \mu_{\tilde{F}_{1U}}], [\nu_{\tilde{F}_{1L}}, \nu_{\tilde{F}_{1U}}])$ and $\tilde{F}_2 = ([\mu_{\tilde{F}_{2L}}, \mu_{\tilde{F}_{2U}}], [\nu_{\tilde{F}_{2L}}, \nu_{\tilde{F}_{2U}}])$. *Then, the following relations are held.*

- $(\tilde{F}_1^c \cup \tilde{F}_2^c) \oplus (\tilde{F}_1^c \cap \tilde{F}_2^c) = \tilde{F}_1 \oplus \tilde{F}_2$
- $(\tilde{F}_1^c \cup \tilde{F}_2^c) \otimes (\tilde{F}_1^c \cap \tilde{F}_2^c) = \tilde{F}_1 \otimes \tilde{F}_2$

Proof See Rani and Mishra [10].

Definition 4.28 [9] For the IVFFS $\tilde{F} = ([\mu_{\tilde{F}_L}, \mu_{\tilde{F}_U}], [\nu_{\tilde{F}_L}, \nu_{\tilde{F}_U}])$, the below functions are defined.

- The hesitancy score is $HS(\tilde{F}) = \frac{1}{2}\left(\mu^3_{\tilde{F}_L} + \mu^3_{\tilde{F}_U} + \nu^3_{\tilde{F}_L} + \nu^3_{\tilde{F}_U}\right) \in [0, 1]$.
- The membership score is $MS(\tilde{F}) = \frac{1}{2}\left(\mu^3_{\tilde{F}_L} + \mu^3_{\tilde{F}_U} - \nu^3_{\tilde{F}_L} - \nu^3_{\tilde{F}_U}\right) \in [-1, 1]$.
- The complete score is $CS(\tilde{F}) = \frac{1}{2}\left(-\mu^3_{\tilde{F}_L} + \mu^3_{\tilde{F}_U} - \nu^3_{\tilde{F}_L} + \nu^3_{\tilde{F}_U}\right) \in \left[-\frac{1}{2}, \frac{1}{2}\right]$.
- The precise score is $PS(\tilde{F}) = \frac{1}{2}\left(-\mu^3_{\tilde{F}_L} + \mu^3_{\tilde{F}_U} + \nu^3_{\tilde{F}_L} - \nu^3_{\tilde{F}_U}\right) \in \left[-\frac{1}{2}, \frac{1}{2}\right]$.

Definition 4.29 [9] For the IVFFSs $\tilde{F}_1 = ([\mu_{\tilde{F}_{1L}}, \mu_{\tilde{F}_{1U}}], [\nu_{\tilde{F}_{1L}}, \nu_{\tilde{F}_{1U}}])$ and $\tilde{F}_2 = ([\mu_{\tilde{F}_{2L}}, \mu_{\tilde{F}_{2U}}], [\nu_{\tilde{F}_{2L}}, \nu_{\tilde{F}_{2U}}])$, then according to Definition 4.23, the below logics are taken.

- If $MS(\tilde{F}_1) > MS(\tilde{F}_2)$, then $\tilde{F}_1 > \tilde{F}_2$, and if $MS(\tilde{F}_1) < MS(\tilde{F}_2)$, then $\tilde{F}_1 < \tilde{F}_2$.
- Holding $MS(\tilde{F}_1) = MS(\tilde{F}_2)$, if $HS(\tilde{F}_1) > HS(\tilde{F}_2)$, then $\tilde{F}_1 > \tilde{F}_2$, and if $HS(\tilde{F}_1) < HS(\tilde{F}_2)$, then $\tilde{F}_1 < \tilde{F}_2$.
- Holding $HS(\tilde{F}_1) = HS(\tilde{F}_2)$, if $PS(\tilde{F}_1) > PS(\tilde{F}_2)$, then $\tilde{F}_1 > \tilde{F}_2$, and if $PS(\tilde{F}_1) < PS(\tilde{F}_2)$, then $\tilde{F}_1 < \tilde{F}_2$.
- Holding $PS(\tilde{F}_1) = PS(\tilde{F}_2)$, if $CS(\tilde{F}_1) > CS(\tilde{F}_2)$, then $\tilde{F}_1 > \tilde{F}_2$, if $CS(\tilde{F}_1) < CS(\tilde{F}_2)$, then $\tilde{F}_1 < \tilde{F}_2$, and if $CS(\tilde{F}_1) = CS(\tilde{F}_2)$, then $\tilde{F}_1 = \tilde{F}_2$.

Theorem 4.10 *Assume the IVFFSs* $\tilde{F}_1 = ([\mu_{\tilde{F}_{1L}}, \mu_{\tilde{F}_{1U}}], [\nu_{\tilde{F}_{1L}}, \nu_{\tilde{F}_{1U}}])$ *and* $\tilde{F}_2 = ([\mu_{\tilde{F}_{2L}}, \mu_{\tilde{F}_{2U}}], [\nu_{\tilde{F}_{2L}}, \nu_{\tilde{F}_{2U}}])$. *If* $MS(\tilde{F}_1) = MS(\tilde{F}_2)$, $HS(\tilde{F}_1) = HS(\tilde{F}_2)$, $PS(\tilde{F}_1) = PS(\tilde{F}_2)$, *and* $CS(\tilde{F}_1) = CS(\tilde{F}_2)$, *then* $\tilde{F}_1 = \tilde{F}_2$. *This means that* $\mu_{\tilde{F}_{1L}} = \mu_{\tilde{F}_{2L}}$, $\mu_{\tilde{F}_{1U}} = \mu_{\tilde{F}_{2U}}$, $\nu_{\tilde{F}_{1L}} = \nu_{\tilde{F}_{2L}}$, *and* $\nu_{\tilde{F}_{1U}} = \nu_{\tilde{F}_{2U}}$.

Proof See Jeevaraj [9].

Example 4.8 Assume the IVFFSs $\tilde{F}_1 = ([0.2, .2], [0.3, 0.4])$ and $\tilde{F}_2 = \left(\left[0.1, \sqrt{0.07}\right], \left[\sqrt{0.05}, \sqrt{0.2}\right]\right)$. As $MS(\tilde{F}_1) = -0.0375 > MS(\tilde{F}_2) = -0.0405$, then $\tilde{F}_1 > \tilde{F}_2$.

Definition 4.30 [9] The Euclidean distance of the IVFFSs $\tilde{F}_1 = ([\mu_{\tilde{F}_{1L}}, \mu_{\tilde{F}_{1U}}], [\nu_{\tilde{F}_{1L}}, \nu_{\tilde{F}_{1U}}])$ and $\tilde{F}_2 = ([\mu_{\tilde{F}_{2L}}, \mu_{\tilde{F}_{2U}}], [\nu_{\tilde{F}_{2L}}, \nu_{\tilde{F}_{2U}}])$ is shown by $D_{GE}(\tilde{F}_1, \tilde{F}_2)$ and calculated as,

$$D_{GE}(\tilde{F}_1, \tilde{F}_2) = \sqrt{\frac{\begin{array}{l}\left(\mu^3_{\tilde{F}_{1L}} - \mu^3_{\tilde{F}_{2L}}\right)^2 + \left(\mu^3_{\tilde{F}_{1U}} - \mu^3_{\tilde{F}_{2U}}\right)^2 \\ \quad + \left(\nu^3_{\tilde{F}_{1L}} - \nu^3_{\tilde{F}_{2L}}\right)^2 + \left(\nu^3_{\tilde{F}_{1U}} - \nu^3_{\tilde{F}_{2U}}\right)^2 \\ \quad + \left(\left(1 - \mu^3_{\tilde{F}_{1U}} - \nu^3_{\tilde{F}_{1U}}\right) - \left(1 - \mu^3_{\tilde{F}_{2U}} - \nu^3_{\tilde{F}_{2U}}\right)\right)^2 \\ \quad + \left(\left(1 - \mu^3_{\tilde{F}_{1L}} - \nu^3_{\tilde{F}_{1L}}\right) - \left(1 - \mu^3_{\tilde{F}_{2L}} - \nu^3_{\tilde{F}_{2L}}\right)\right)^2\end{array}}{6}}.$$

Theorem 4.11 *Assume the IVFFSs* $\tilde{F}_1 = ([\mu_{\tilde{F}_{1L}}, \mu_{\tilde{F}_{1U}}], [\nu_{\tilde{F}_{1L}}, \nu_{\tilde{F}_{1U}}])$ *and* $\tilde{F}_2 = ([\mu_{\tilde{F}_{2L}}, \mu_{\tilde{F}_{2U}}], [\nu_{\tilde{F}_{2L}}, \nu_{\tilde{F}_{2U}}])$. *Then,*

- $0 \le D_{GE}(\tilde{F}_1, \tilde{F}_2) \le 1$,
- $D_{GE}(\tilde{F}_1, \tilde{F}_2) = D_{GE}(\tilde{F}_2, \tilde{F}_1)$,
- $D_{GE}(\tilde{F}_1, \tilde{F}_2) = 0$ *if and only if* $\tilde{F}_1 = \tilde{F}_2$.

Proof See Jeevaraj [9].

Theorem 4.12 *Assume the IVFFSs* $\tilde{F}_1 = ([\mu_{\tilde{F}_{1L}}, \mu_{\tilde{F}_{1U}}], [\nu_{\tilde{F}_{1L}}, \nu_{\tilde{F}_{1U}}])$, $\tilde{F}_2 = ([\mu_{\tilde{F}_{2L}}, \mu_{\tilde{F}_{2U}}], [\nu_{\tilde{F}_{2L}}, \nu_{\tilde{F}_{2U}}])$, *and* $\tilde{F}_3 = ([\mu_{\tilde{F}_{3L}}, \mu_{\tilde{F}_{3U}}], [\nu_{\tilde{F}_{3L}}, \nu_{\tilde{F}_{3U}}])$. *If* $\tilde{F}_1 \le \tilde{F}_2 \le \tilde{F}_3$, *then* $D_{GE}(\tilde{F}_1, \tilde{F}_2) \le D_{GE}(\tilde{F}_1, \tilde{F}_3)$ *and* $D_{GE}(\tilde{F}_2, \tilde{F}_3) \le D_{GE}(\tilde{F}_1, \tilde{F}_3)$.

Proof See Jeevaraj [9].

Definition 4.31 [9] Consider the IVFFSs $\tilde{F}_1 = ([\mu_{\tilde{F}_{1L}}, \mu_{\tilde{F}_{1U}}], [\nu_{\tilde{F}_{1L}}, \nu_{\tilde{F}_{1U}}])$ and $\tilde{F}_2 = ([\mu_{\tilde{F}_{2L}}, \mu_{\tilde{F}_{2U}}], [\nu_{\tilde{F}_{2L}}, \nu_{\tilde{F}_{2U}}])$. The similarity measure of these sets is defined as $S_{GE}(\tilde{F}_1, \tilde{F}_2) = 1 - D_{GE}(\tilde{F}_1, \tilde{F}_2)$.

Theorem 4.13 *Assume the IVFFSs* $\tilde{F}_1 = ([\mu_{\tilde{F}_{1L}}, \mu_{\tilde{F}_{1U}}], [\nu_{\tilde{F}_{1L}}, \nu_{\tilde{F}_{1U}}])$ *and* $\tilde{F}_2 = ([\mu_{\tilde{F}_{2L}}, \mu_{\tilde{F}_{2U}}], [\nu_{\tilde{F}_{2L}}, \nu_{\tilde{F}_{2U}}])$. *Then,*

- $0 \le S_{GE}(\tilde{F}_1, \tilde{F}_2) \le 1$,
- $S_{GE}(\tilde{F}_1, \tilde{F}_2) = S_{GE}(\tilde{F}_2, \tilde{F}_1)$,
- $S_{GE}(\tilde{F}_1, \tilde{F}_2) = 1$ *if and only if* $\tilde{F}_1 = \tilde{F}_2$.

Proof See Jeevaraj [9].

Theorem 4.14 *Assume the IVFFSs* $\tilde{F}_1 = ([\mu_{\tilde{F}_{1L}}, \mu_{\tilde{F}_{1U}}], [\nu_{\tilde{F}_{1L}}, \nu_{\tilde{F}_{1U}}])$, $\tilde{F}_2 = ([\mu_{\tilde{F}_{2L}}, \mu_{\tilde{F}_{2U}}], [\nu_{\tilde{F}_{2L}}, \nu_{\tilde{F}_{2U}}])$, *and* $\tilde{F}_3 = ([\mu_{\tilde{F}_{3L}}, \mu_{\tilde{F}_{3U}}], [\nu_{\tilde{F}_{3L}}, \nu_{\tilde{F}_{3U}}])$. *If* $\tilde{F}_1 \le \tilde{F}_2 \le \tilde{F}_3$, *then* $S_{GE}(\tilde{F}_1, \tilde{F}_2) \ge S_{GE}(\tilde{F}_1, \tilde{F}_3)$ *and* $S_{GE}(\tilde{F}_2, \tilde{F}_3) \ge S_{GE}(\tilde{F}_1, \tilde{F}_3)$.

Proof See Jeevaraj [9].

Definition 4.32 [8] The interval-valued Fermatean fuzzy weighted arithmetic operator (IVFFWA) and the interval-valued Fermatean fuzzy weighted geometric operator (IVFFWG) of a set of IVFFNs $\tilde{F}_1, \ldots, \tilde{F}_n$ with the set of weights $w_1, \ldots, w_n$, where $0 \le w_i \le 1$ and $\sum_{i=1}^{n} w_i = 1$, are defined as below.

$$IVFFWA(\tilde{F}_1, \ldots, \tilde{F}_n) = \left(\left[\sqrt[3]{1 - \prod_{i=1}^{n} \left(1 - \mu_{\tilde{F}_{iL}}^3\right)^{w_i}}, \sqrt[3]{1 - \prod_{i=1}^{n} \left(1 - \mu_{\tilde{F}_{iU}}^3\right)^{w_i}} \right], \left[\prod_{i=1}^{n} \nu_{\tilde{F}_{iL}}^{w_i}, \prod_{i=1}^{n} \nu_{\tilde{F}_{iU}}^{w_i} \right] \right), \tag{4.77}$$

$$IVFFWG(\tilde{F}_1, \ldots, \tilde{F}_n) = \left(\left[\prod_{i=1}^{n} \mu_{\tilde{F}_{iL}}^{w_i}, \prod_{i=1}^{n} \mu_{\tilde{F}_{iU}}^{w_i} \right], \left[\sqrt[3]{1 - \prod_{i=1}^{n} \left(1 - \nu_{\tilde{F}_{iL}}^3\right)^{w_i}}, \sqrt[3]{1 - \prod_{i=1}^{n} \left(1 - \nu_{\tilde{F}_{iU}}^3\right)^{w_i}} \right] \right). \tag{4.78}$$

4.6.1 Trapezoidal Interval-Valued Fermatean Fuzzy Numbers

The trapezoidal interval-valued Fermatean fuzzy number $\tilde{F}$ on the reference set X is shown by $\tilde{F} = \{(x_1, x_2, x_3, x_4), [\mu_{\tilde{F}_L}, \mu_{\tilde{F}_U}], [\nu_{\tilde{F}_L}, \nu_{\tilde{F}_U}]\}$ and Fig. 4.5 where $0 \leq \mu_{\tilde{F}_U}^3(x) + \nu_{\tilde{F}_U}^3(x) \leq 1$. The minimum and maximum values for the membership function are shown by $\mu_{\tilde{F}_L} : X \to [0, 1]$ and $\mu_{\tilde{F}_U} : X \to [0, 1]$ respectively. The minimum and maximum values for the non-membership function are shown by $\nu_{\tilde{F}_L} : X \to [0, 1]$ and $\nu_{\tilde{F}_U} : X \to [0, 1]$ respectively. The following membership and non-membership functions are defined for the trapezoidal interval-valued Fermatean fuzzy number $\tilde{F}$.

$$\mu_{\tilde{F}_L}(x) = \begin{cases} \frac{x - x_1}{x_2 - x_1} \mu_{\tilde{F}_L}; & x_1 \leq x \leq x_2 \\ \mu_{\tilde{F}_L}; & x_2 \leq x \leq x_3 \\ \frac{x_4 - x}{x_4 - x_3} \mu_{\tilde{F}_L}; & x_3 \leq x \leq x_4 \\ 0; & otherwise \end{cases} \tag{4.79}$$

$$\mu_{\tilde{F}_U}(x) = \begin{cases} \frac{x - x_1}{x_2 - x_1} \mu_{\tilde{F}_U}; & x_1 \leq x \leq x_2 \\ \mu_{\tilde{F}_U}; & x_2 \leq x \leq x_3 \\ \frac{x_4 - x}{x_4 - x_3} \mu_{\tilde{F}_U}; & x_3 \leq x \leq x_4 \\ 0; & otherwise \end{cases} \tag{4.80}$$

$$\nu_{\tilde{F}_L}(x) = \begin{cases} 1 - \frac{x - x_1}{x_2 - x_1} \times \left(1 - \nu_{\tilde{F}_L}\right), & x_1 \leq x \leq x_2, \\ \nu_{\tilde{F}_L}, & x_2 \leq x \leq x_3, \\ \nu_{\tilde{F}_L} + \frac{x - x_3}{x_4 - x_3} \times \left(1 - \nu_{\tilde{F}_L}\right), & x_3 \leq x \leq x_4, \\ 1; & otherwise \end{cases} \tag{4.81}$$

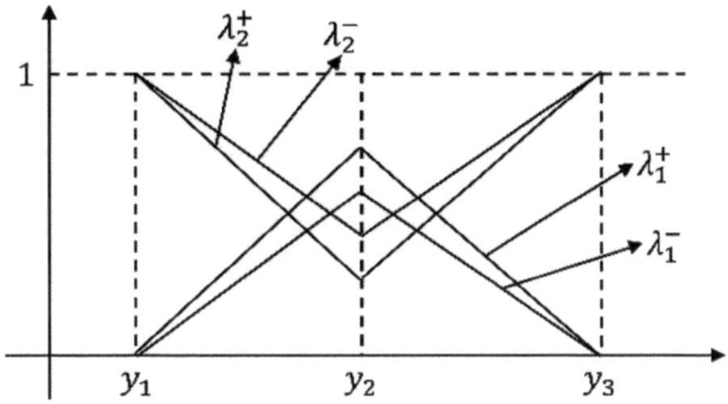

Fig. 4.5 The schematic representation of a trapezoidal interval-valued Fermatean fuzzy number

$$\nu_{\tilde{F}_U}(x) = \begin{cases} 1 - \frac{x-x_1}{x_2-x_1} \times \left(1 - \nu_{\tilde{F}_U}\right), & x_1 \le x \le x_2, \\ \nu_{\tilde{F}_U}, & x_2 \le x \le x_3, \\ \nu_{\tilde{F}_U} + \frac{x-x_3}{x_4-x_3} \times \left(1 - \nu_{\tilde{F}_U}\right), & x_3 \le x \le x_4, \\ 1; & otherwise \end{cases} \tag{4.82}$$

Definition 4.33 Assume the trapezoidal IVFFNs $\tilde{X} = \left\{(x_1, x_2, x_3, x_4), \left[\mu_{\tilde{X}_L}, \mu_{\tilde{X}_U}\right], \left[\nu_{\tilde{X}_L}, \nu_{\tilde{X}_U}\right]\right\}$ and $\tilde{Y} = \{(y_1, y_2, y_3, y_4), \left[\mu_{\tilde{Y}_L}, \mu_{\tilde{Y}_U}\right], \left[\nu_{\tilde{Y}_L}, \nu_{\tilde{Y}_U}\right]\}$. The below operations are defined.

$$\tilde{X} \oplus \tilde{Y} = \left\{ \begin{array}{l} (x_1 + y_1, x_2 + y_2, x_3 + y_3, x_4 + y_4), \\ \left[\min\left\{\mu_{\tilde{X}_L}, \mu_{\tilde{Y}_L}\right\}, \min\left\{\mu_{\tilde{X}_U}, \mu_{\tilde{Y}_U}\right\}\right], \left[\max\left\{\nu_{\tilde{X}_L}, \nu_{\tilde{Y}_L}\right\}, \max\left\{\nu_{\tilde{X}_U}, \nu_{\tilde{Y}_U}\right\}\right] \end{array} \right\} \tag{4.83}$$

$$\tilde{X} \ominus \tilde{Y} = \left\{ \begin{array}{l} (x_1 - y_4, x_2 - y_3, x_3 - y_2, x_4 - y_1), \\ \left[\min\left\{\mu_{\tilde{X}_L}, \mu_{\tilde{Y}_L}\right\}, \min\left\{\mu_{\tilde{X}_U}, \mu_{\tilde{Y}_U}\right\}\right], \left[\max\left\{\nu_{\tilde{X}_L}, \nu_{\tilde{Y}_L}\right\}, \max\left\{\nu_{\tilde{X}_U}, \nu_{\tilde{Y}_U}\right\}\right] \end{array} \right\} \tag{4.84}$$

$$\tilde{X} \otimes \tilde{Y} = \left\{ \begin{array}{l} (x_1 y_1, x_2 y_2, x_3 y_3, x_4 y_4), \\ \left[\min\left\{\mu_{\tilde{X}_L}, \mu_{\tilde{Y}_L}\right\}, \min\left\{\mu_{\tilde{X}_U}, \mu_{\tilde{Y}_U}\right\}\right], \left[\max\left\{\nu_{\tilde{X}_L}, \nu_{\tilde{Y}_L}\right\}, \max\left\{\nu_{\tilde{X}_U}, \nu_{\tilde{Y}_U}\right\}\right] \end{array} \right\}$$
$$\text{if } x_1, y_1 > 0, \tag{4.85}$$

$$\tilde{X} \otimes \tilde{Y} = \left\{ \begin{array}{l} \left(\frac{x_1}{y_4}, \frac{x_2}{y_3}, \frac{x_3}{y_2}, \frac{x_4}{y_1}\right), \\ \left[\min\left\{\mu_{\tilde{X}_L}, \mu_{\tilde{Y}_L}\right\}, \min\left\{\mu_{\tilde{X}_U}, \mu_{\tilde{Y}_U}\right\}\right], \left[\max\left\{\nu_{\tilde{X}_L}, \nu_{\tilde{Y}_L}\right\}, \max\left\{\nu_{\tilde{X}_U}, \nu_{\tilde{Y}_U}\right\}\right] \end{array} \right\}$$
$$\text{if } x_1, y_1 > 0, \tag{4.86}$$

$$k\tilde{X} = \begin{cases} \left\{(kx_1, kx_2, kx_3, kx_4), \left[\mu_{\tilde{X}_L}, \mu_{\tilde{X}_U}\right], \left[\nu_{\tilde{X}_L}, \nu_{\tilde{X}_U}\right]\right\}, k > 0, \\ \left\{(kx_4, kx_3, kx_2, kx_1), \left[\mu_{\tilde{X}_L}, \mu_{\tilde{X}_U}\right], \left[\nu_{\tilde{X}_L}, \nu_{\tilde{X}_U}\right]\right\}, k < 0, \end{cases}, \tag{4.87}$$

Definition 4.34 For the trapezoidal IVFFN $\tilde{A} = \left\{(a_1, a_2, a_3, a_4), \left[\mu_{\tilde{A}_L}, \mu_{\tilde{A}_U}\right], \left[\nu_{\tilde{A}_L}, \nu_{\tilde{A}_U}\right]\right\}$, the below ranking function is defined.

$$\Re(\tilde{A}) = \frac{(a_1 + a_2 + a_3 + a_4)\left(\mu_{\tilde{A}_L} + \mu_{\tilde{A}_U} + 2 - \nu_{\tilde{A}_L} - \nu_{\tilde{A}_U}\right)}{16} \tag{4.88}$$

Based on this ranking function, for the trapezoidal IVFFNs $\tilde{A}$ and $\tilde{B}$, the below logics exist.

$$\Re(\tilde{A}) > \Re(\tilde{B}) \Rightarrow \tilde{A} \succ \tilde{B}, \tag{4.89}$$

$$\Re(\tilde{A}) < \Re(\tilde{B}) \Rightarrow \tilde{A} \prec \tilde{B}, \tag{4.90}$$

$$\Re(\tilde{A}) = \Re(\tilde{B}) \Rightarrow \tilde{A} \approx \tilde{B}. \tag{4.91}$$

Definition 4.35 The trapezoidal IVFFNs $\tilde{A} = \{(a_1, a_2, a_3, a_4), [\mu_{\tilde{A}_L}, \mu_{\tilde{A}_U}], [\nu_{\tilde{A}_L}, \nu_{\tilde{A}_U}]\}$ and $\tilde{B} = \{(b_1, b_2, b_3, b_4), [\mu_{\tilde{B}_L}, \mu_{\tilde{B}_U}], [\nu_{\tilde{B}_L}, \nu_{\tilde{B}_U}]\}$ are equal ($\tilde{A} = \tilde{B}$) if and only if $a_i = b_i$ $(i = 1, 2, 3, 4)$, $\mu_{\tilde{A}_L} = \mu_{\tilde{B}_L}$, $\mu_{\tilde{A}_U} = \mu_{\tilde{B}_U}$, $\nu_{\tilde{A}_L} = \nu_{\tilde{B}_L}$, and $\nu_{\tilde{A}_U} = \nu_{\tilde{B}_U}$.

4.7 Cubic Fermatean Fuzzy Sets

The cubic Fermatean fuzzy sets are constructed based on the cubic sets [11] and provide a more flexible environment for representing uncertain events and help decision makers for more reliable and robust decisions.

Definition 4.36 [12] The cubic Fermatean fuzzy set $\tilde{E}$ on the reference set X is shown by $\tilde{E} = \{(x, \tilde{A}(x), \tilde{B}(x)) | x \in X\}$ where $\tilde{A}(x) = \{(x, [\mu_{\tilde{A}_L}(x), \mu_{\tilde{A}_U}(x)], [\nu_{\tilde{A}_L}(x), \nu_{\tilde{A}_U}(x)]) | x \in X\}$ is an IVFFS and $\tilde{B}(x) = \{(x, \mu_{\tilde{B}}(x), \nu_{\tilde{B}}(x)) | x \in X\}$ is a FFS on X with the conditions $0 \leq \mu_{\tilde{A}_L}(x) \leq \mu_{\tilde{A}_U}(x) \leq 1, 0 \leq \nu_{\tilde{A}_L}(x) \leq \nu_{\tilde{A}_U}(x) \leq 1, 0 \leq \mu^3_{\tilde{A}_U}(x) + \nu^3_{\tilde{A}_U}(x) \leq 1, 0 \leq \mu_{\tilde{B}}(x) \leq \nu_{\tilde{B}}(x) \leq 1$, and $0 \leq \mu^3_{\tilde{B}}(x) + \nu^3_{\tilde{B}}(x) \leq 1$. By combining the IVFFS and the FFS, the cubic Fermatean fuzzy number (CFFN) $\{([\mu_{\tilde{A}_L}(x), \mu_{\tilde{A}_U}(x)], [\nu_{\tilde{A}_L}(x), \nu_{\tilde{A}_U}(x)]), (\mu_{\tilde{B}}(x), \nu_{\tilde{B}}(x))\}$ is obtained and is shown as $\{([\mu_{\tilde{A}_L}, \mu_{\tilde{A}_U}], [\nu_{\tilde{A}_L}, \nu_{\tilde{A}_U}]), (\mu_{\tilde{A}}, \nu_{\tilde{A}})\}$ for more simplicity.

Definition 4.37 [12] For the CFFNs $\tilde{A} = \{([\mu_{\tilde{A}_L}, \mu_{\tilde{A}_U}], [\nu_{\tilde{A}_L}, \nu_{\tilde{A}_U}]), (\mu_{\tilde{A}}, \nu_{\tilde{A}})\}$ and $\tilde{B} = \{([\mu_{\tilde{B}_L}, \mu_{\tilde{B}_U}], [\nu_{\tilde{B}_L}, \nu_{\tilde{B}_U}]), (\mu_{\tilde{B}}, \nu_{\tilde{B}})\}$, the below relations are defined.

$$\tilde{A} \cup \tilde{B} = \left\{ \begin{array}{l} \left(\left[\max\left\{\mu_{\tilde{A}_L}, \mu_{\tilde{B}_L}\right\}, \max\left\{\mu_{\tilde{A}_U}, \mu_{\tilde{B}_U}\right\}\right], \left[\min\left\{\nu_{\tilde{A}_L}, \nu_{\tilde{B}_L}\right\}, \min\left\{\nu_{\tilde{A}_U}, \nu_{\tilde{B}_U}\right\}\right]\right), \\ \left(\max\{\mu_{\tilde{A}}, \mu_{\tilde{B}}\}, \min\{\nu_{\tilde{A}}, \nu_{\tilde{B}}\}\right) \end{array} \right\} \tag{4.92}$$

$$\tilde{A} \cap \tilde{B} = \left\{ \begin{array}{l} \left(\left[\min\left\{\mu_{\tilde{A}_L}, \mu_{\tilde{B}_L}\right\}, \min\left\{\mu_{\tilde{A}_U}, \mu_{\tilde{B}_U}\right\}\right], \left[\max\left\{\nu_{\tilde{A}_L}, \nu_{\tilde{B}_L}\right\}, \max\left\{\nu_{\tilde{A}_U}, \nu_{\tilde{B}_U}\right\}\right]\right), \\ \left(\min\{\mu_{\tilde{A}}, \mu_{\tilde{B}}\}, \max\{\nu_{\tilde{A}}, \nu_{\tilde{B}}\}\right) \end{array} \right\} \tag{4.93}$$

$$\tilde{A}^c = \left\{\left(\left[\nu_{\tilde{A}_L}, \nu_{\tilde{A}_U}\right], \left[\mu_{\tilde{A}_L}, \mu_{\tilde{A}_U}\right]\right), \left(\nu_{\tilde{A}}, \mu_{\tilde{A}}\right)\right\} \tag{4.94}$$

Theorem 4.15 [12] *For the CFFNs* $\tilde{A} = \left\{\left(\left[\mu_{\tilde{A}_L}, \mu_{\tilde{A}_U}\right], \left[\nu_{\tilde{A}_L}, \nu_{\tilde{A}_U}\right]\right), \left(\mu_{\tilde{A}}, \nu_{\tilde{A}}\right)\right\}$ *and* $\tilde{B} = \left\{\left(\left[\mu_{\tilde{B}_L}, \mu_{\tilde{B}_U}\right], \left[\nu_{\tilde{B}_L}, \nu_{\tilde{B}_U}\right]\right), \left(\mu_{\tilde{B}}, \nu_{\tilde{B}}\right)\right\}$, *the below relations are held.*

$$\tilde{A} \cup \tilde{B} = \tilde{B} \cup \tilde{A}, \tag{4.95}$$

$$\tilde{A} \cap \tilde{B} = \tilde{B} \cap \tilde{A}, \tag{4.96}$$

$$\lambda\left(\tilde{A} \cup \tilde{B}\right) = \lambda\tilde{A} \cup \lambda\tilde{B}, \tag{4.97}$$

$$\left(\tilde{A} \cup \tilde{B}\right)^c = \tilde{A}^c \cap \tilde{B}^c, \tag{4.98}$$

$$\left(\tilde{A} \cap \tilde{B}\right)^c = \tilde{A}^c \cup \tilde{B}^c. \tag{4.99}$$

Definition 4.38 [12] For the CFFN $\tilde{A} = \left\{\left(\left[\mu_{\tilde{A}_L}, \mu_{\tilde{A}_U}\right], \left[\nu_{\tilde{A}_L}, \nu_{\tilde{A}_U}\right]\right), \left(\mu_{\tilde{A}}, \nu_{\tilde{A}}\right)\right\}$, the below score and accuracy functions are defined.

$$S\left(\tilde{A}\right) = \frac{1}{2}\left(\frac{1}{4}\left(\mu^3_{\tilde{A}_L} + \mu^3_{\tilde{A}_U} - \nu^3_{\tilde{A}_L} - \nu^3_{\tilde{A}_U}\right) + \frac{1}{2}\left(\mu^3_{\tilde{A}} - \nu^3_{\tilde{A}} + 1\right)\right), S\left(\tilde{A}\right) \in [0, 1], \tag{4.100}$$

$$A\left(\tilde{A}\right) = \frac{1}{2}\left(\frac{1}{2}\left(\mu^3_{\tilde{A}_L} + \mu^3_{\tilde{A}_U} + \nu^3_{\tilde{A}_L} + \nu^3_{\tilde{A}_U}\right) + \mu^3_{\tilde{A}} + \nu^3_{\tilde{A}}\right), A\left(\tilde{A}\right) \in [0, 1]. \tag{4.101}$$

For the CFFNs $\tilde{A} = \left\{\left(\left[\mu_{\tilde{A}_L}, \mu_{\tilde{A}_U}\right], \left[\nu_{\tilde{A}_L}, \nu_{\tilde{A}_U}\right]\right), \left(\mu_{\tilde{A}}, \nu_{\tilde{A}}\right)\right\}$ and $\tilde{B} = \left\{\left(\left[\mu_{\tilde{B}_L}, \mu_{\tilde{B}_U}\right], \left[\nu_{\tilde{B}_L}, \nu_{\tilde{B}_U}\right]\right), \left(\mu_{\tilde{B}}, \nu_{\tilde{B}}\right)\right\}$, if $S\left(\tilde{A}\right) < S\left(\tilde{B}\right)$, then $\tilde{A} < \tilde{B}$, and if $S\left(\tilde{A}\right) > S\left(\tilde{B}\right)$, then $\tilde{A} > \tilde{B}$. For the case of $S\left(\tilde{A}\right) = S\left(\tilde{B}\right)$, if $A\left(\tilde{A}\right) < A\left(\tilde{B}\right)$, then $\tilde{A} < \tilde{B}$, if $A\left(\tilde{A}\right) > A\left(\tilde{B}\right)$, then $\tilde{A} > \tilde{B}$, and if $A\left(\tilde{A}\right) = A\left(\tilde{B}\right)$, then $\tilde{A} = \tilde{B}$.

Definition 4.39 [12] For the CFFNs $\tilde{A} = \left\{\left(\left[\mu_{\tilde{A}_L}, \mu_{\tilde{A}_U}\right], \left[\nu_{\tilde{A}_L}, \nu_{\tilde{A}_U}\right]\right), \left(\mu_{\tilde{A}}, \nu_{\tilde{A}}\right)\right\}$ and $\tilde{B} = \left\{\left(\left[\mu_{\tilde{B}_L}, \mu_{\tilde{B}_U}\right], \left[\nu_{\tilde{B}_L}, \nu_{\tilde{B}_U}\right]\right), \left(\mu_{\tilde{B}}, \nu_{\tilde{B}}\right)\right\}$, the below general distance of $d^\rho\left(\tilde{A}, \tilde{B}\right)$ where $\rho \geq 1$ is defined.

$$d^\rho\left(\tilde{A}, \tilde{B}\right) = \left(\frac{1}{6n}\sum_{j=1}^{n}\left(\begin{array}{r} \left|\mu^3_{\tilde{A}_L}(x_j) - \mu^3_{\tilde{B}_L}(x_j)\right|^\rho + \left|\mu^3_{\tilde{A}_U}(x_j) - \mu^3_{\tilde{B}_U}(x_j)\right|^\rho \\ + \left|\nu^3_{\tilde{A}_L}(x_j) - \nu^3_{\tilde{B}_L}(x_j)\right|^\rho \\ + \left|\nu^3_{\tilde{A}_U}(x_j) - \nu^3_{\tilde{B}_U}(x_j)\right|^\rho + \left|\mu^3_{\tilde{A}}(x_j) - \mu^3_{\tilde{B}}(x_j)\right|^\rho \\ + \left|\nu^3_{\tilde{A}}(x_j) - \nu^3_{\tilde{B}}(x_j)\right|^\rho \end{array}\right)\right)^{\frac{1}{\rho}} \tag{4.102}$$

Using this distance, for $\rho = 1$ the hamming distance and for $\rho = 2$ the Euclidean distance is obtained.

Lemma 4.2 *Assume the arbitrary* CFFNs $\tilde{A}$, $\tilde{B}$, and $\tilde{C}$. *Then the distance* d^{ρ} *is called a cubic Fermatean fuzzy distance, if it follows the below rules.*

$$0 \le d^{\rho}(\tilde{A}, \tilde{B}) \le 1 \tag{4.103}$$

$$d^{\rho}(\tilde{A}, \tilde{B}) = 0 \text{ if and only if } \tilde{A} = \tilde{B} \tag{4.104}$$

$$d^{\rho}(\tilde{A}, \tilde{B}) = d^{\rho}(\tilde{B}, \tilde{A}) \tag{4.105}$$

$$\text{If } \tilde{A} \subseteq \tilde{B} \subseteq \tilde{C}, \text{ then } d^{\rho}(\tilde{A}, \tilde{B}) \le d^{\rho}(\tilde{A}, \tilde{C}), \text{ and } d^{\rho}(\tilde{B}, \tilde{C}) \le d^{\rho}(\tilde{A}, \tilde{C}). \tag{4.106}$$

Proof See Rong et al. [12].

Definition 4.40 [12] Assume the CFFNs $\tilde{F}_1$ and $\tilde{F}_2$. The similarity measure of these CFFNs is calculated as $S(\tilde{F}_1, \tilde{F}_2) = 1 - d^{\rho}(\tilde{F}_1, \tilde{F}_2)$.

Definition 4.41 [12] Based on Definition 4.37 the similarity measure of the CFFNs $\tilde{F}_1$ and $\tilde{F}_2$ has the below properties.

- $0 \le S(\tilde{F}_1, \tilde{F}_2) \le 1$
- $S(\tilde{F}_1, \tilde{F}_2) = 1$ if and only if $\tilde{F}_1 \approx \tilde{F}_2$,
- $S(\tilde{F}_1, \tilde{F}_2) = S(\tilde{F}_2, \tilde{F}_1)$,
- If $\tilde{A} \subseteq \tilde{B} \subseteq \tilde{C}$, then $S(\tilde{A}, \tilde{B}) \le S(\tilde{A}, \tilde{C})$ and $S(\tilde{B}, \tilde{C}) \le S(\tilde{A}, \tilde{C})$.

4.8 Application: Fermatean Fuzzy Technique for Order Preference by Similarity to Ideal Solution (FF TOPSIS)

In this section, a Fermatean fuzzy version of the TOPSIS approach [13] with Fermatean fuzzy decision matrix is represented. For this aim the below steps are proposed.

Step 1. Construct the decision matrix as below.

$$R = \left(C_j(E_i)\right)_{m\times n} = \begin{matrix} \\ E_1 \\ \vdots \\ E_m \end{matrix} \begin{pmatrix} C_1 & \cdots & C_n \\ (\mu_{11}, \nu_{11}) & \ldots & (\mu_{1n}, \nu_{1n}) \\ \vdots & \ddots & \vdots \\ (\mu_{m1}, \nu_{m1}) & \cdots & (\mu_{mn}, \nu_{mn}) \end{pmatrix}, \tag{4.107}$$

The set of alternatives $\{E_i, i = 1, 2, \ldots, m\}$, the set of criteria $\{C_j, j = 1, 2, \ldots, n\}$, and the set of weight values $\{w_j, j = 1, 2, \ldots, n\}$, where $0 \le w_j \le 1$ $(j = 1, 2, \ldots, n)$ and $\sum_{j=1}^{n} w_j = 1$ are defined here. The Fermatean fuzzy number $C_j(E_i) = (\mu_{ij}, \nu_{ij})$ is the performance of alternative i in criterion j.

Step 2. The set of Fermatean fuzzy positive ideal solutions (FFPIS) $E^+ = \{C_1(E^+), C_2(E^+), \ldots, C_n(E^+)\}$ and the set of Fermatean fuzzy negative ideal solutions (FFNIS) $E^- = \{C_1(E^-), C_2(E^-), \ldots, C_n(E^-)\}$ for the criteria are obtained as below.

$$E^+ = \begin{cases} \max\limits_i \{score\{C_j(E_i)\}\}, j = 1, 2, \ldots, n, \text{ if } \mathrm{C}_j \text{ is a benefit criterion} \\ \min\limits_i \{score\{C_j(E_i)\}\}, j = 1, 2, \ldots, n, \text{ if } \mathrm{C}_j \text{ is a cost criterion} \end{cases}$$
$$= \{(\mu_1^+, \nu_1^+), (\mu_2^+, \nu_2^+), \ldots, (\mu_n^+, \nu_n^+)\} \tag{4.108}$$

$$E^- = \begin{cases} \min\limits_i \{score\{C_j(E_i)\}\}, j = 1, 2, \ldots, n, \text{ if } \mathrm{C}_j \text{ is a benefit criterion} \\ \max\limits_i \{score\{C_j(E_i)\}\}, j = 1, 2, \ldots, n, \text{ if } \mathrm{C}_j \text{ is a cost criterion} \end{cases}$$
$$= \{(\mu_1^-, \nu_1^-), (\mu_2^-, \nu_2^-), \ldots, (\mu_n^-, \nu_n^-)\} \tag{4.109}$$

Step 3. The distance of each alternative from the FFPISs and FFNISs is obtained as below.

$$D(E_i, E^+) = \sum_{j=1}^{n} w_j d(C_j(E_i), C_j(E^+))$$

$$D(E_i, E^+) = \sum_{j=1}^{n} w_j \sqrt{\frac{1}{2}\left[\left(\mu_{ij}^3 - \mu_j^{+^3}\right)^2 + \left(\nu_{ij}^3 - \nu_j^{+^3}\right)^2 + \left(\pi_{ij}^3 - \pi_j^{+^3}\right)^2\right]},$$
$$i = 1, 2, \ldots, m, \tag{4.110}$$

$$D(E_i, E^-) = \sum_{j=1}^{n} w_j d(C_j(E_i), C_j(E^-))$$

$$D(E_i, E^-) = \sum_{j=1}^{n} w_j \sqrt{\frac{1}{2}\left[\left(\mu_{ij}^3 - \mu_j^{-^3}\right)^2 + \left(\nu_{ij}^3 - \nu_j^{-^3}\right)^2 + \left(\pi_{ij}^3 - \pi_j^{-^3}\right)^2\right]},$$
$$i = 1, 2, \ldots, m, \tag{4.111}$$

Step 4. The relative closeness index of each alternative is obtained as below,

Table 4.1 The Fermatean fuzzy decision matrix of Example 4.9

	C_1	C_2	C_3	C_4
E_1	(0.6, 0.4)	(0.3, 0.5)	(0.4, 0.4)	(0.4, 0.5)
E_2	(0.4, 0.7)	(0.8, 0.5)	(0.7, 0.3)	(0.5, 0.3)
E_3	(0.4, 0.3)	(0.4, 0.1)	(0.8, 0.2)	(0.7, 0.3)
E_4	(0.7, 0.4)	(0.5, 0.5)	(0.3, 0.2)	(0.4, 0.2)

Table 4.2 The score values of the Fermatean fuzzy decision matrix obtained by Definition 4.6

	C_1	C_2	C_3	C_4
E_1	0.1520	– 0.0980	0	– 0.0610
E_2	– 0.2790	0.3870	0.3160	0.0980
E_3	0.0370	0.0630	0.504	0.3160
E_4	0.2790	0.0610	0.0190	0.0560

Table 4.3 Decisions made using the FFTOPSIS with Euclidean distance (4.41)

	$D(E_i, E^+)$	$D(E_i, E^-)$	$\xi(E_i)$	Rank
E_1	0.2587	0.1218	– 1.4662	4
E_2	0.1313	0.2414	0	1
E_3	0.2664	0.2123	– 1.1499	2
E_4	0.2557	0.1837	– 1.1869	3

$$\xi(E_i) = \frac{D(E_i, E^-)}{D(E_i, E^-)_{\max} \frac{D(E_i, E^+)}{D(E_i, E^+)_{\min}}} \tag{4.112}$$

where $D(E_i, E^-) \max_{1 \le i \le m} (E_i, E^-)_{\max}$ and $D(E_i, E^+) \min_{1 \le i \le m} (E_i, E^+)_{\min}$.

Step 5. The alternatives are ranked based on the decreasing order of the values of $\xi(E_i)$.

Example 4.9 [14] Consider the below Fermatean fuzzy decision matrix with the weigh vector of $W = (0.25, 0.25, 0.25, 0.25)^T$ (Table 4.1).

According to the score function $S(\tilde{F}) = (\mu_{\tilde{F}})^3 - (\nu_{\tilde{F}})^3 \in [-1, 1]$ (Definition 4.6), the score values of Table 4.2 are calculated.

Therefore, the sets of FFPIS and FFNIS are obtained $E^+ = \{(0.7, 0.4), (0.8, 0.5), (0.8, 0.2), (0.4, 0.5)\}$ and $E^- = \{(0.4, 0.7), (0.3, 0.5), (0.4, 0.4), (0.7, 0.3)\}$.

The results of Step 3, Step 4, and Step 5 for Euclidean distance (4.41) and Hamming distance (4.42) are represented by Tables 4.3 and 4.4 respectively, where both of the distances result similar rankings.

Table 4.4 Decisions made using the FFTOPSIS with hamming distance (4.42)

	$D(E_i, E^+)$	$D(E_i, E^-)$	$\xi(E_i)$	Rank
E_1	0.1829	0.0861	− 1.4662	4
E_2	0.0928	0.1707	0	1
E_3	0.1884	0.1501	− 1.1499	2
E_4	0.1808	0.1299	− 1.1869	3

References

1. Senapati, T., Yager, R.R.: Fermatean fuzzy sets. J. Ambient. Intell. Humaniz. Comput.Intell. Humaniz. Comput. **11**(2), 663–674 (2020)
2. Senapati, T., Yager, R.R.: Fermatean fuzzy weighted averaging/geometric operators and its application in multi-criteria decision making methods. Eng. Appl. Artif. Intell.Artif. Intell. **85**, 112–121 (2019)
3. Senapati, T., Yager, R.R.: Some new operations over Fermatean fuzzy numbers and application of Fermatean fuzzy WPM in multiple criteria decision making. Informatica **30**(2), 391–412 (2019)
4. Sahoo, L.: Transportation problem in Fermatean fuzzy environment. RAIRO-Oper. Res. **57**, 145–156 (2023)
5. Akram, M., Umer Shah, S.M., Allahviranloo, T.: A new method to determine the Fermatean fuzzy optimal solution of transportation problems. J. Intell. Fuzzy Syst. **44**(1), 309–328 (2023)
6. Vijayakumar, R., Mahapatra, G.S., Pamucar, D.: Fermatean trapezoidal fuzziness average aggregation scheme for selection of infant clothing by group decision-making. Appl. Soft Comput. **167**, Part A, 112317 (2024)
7. Singh, A., Arora, R., Arora, S.: A new Fermatean fuzzy multi-objective indefinite quadratic transportation problem with an application to sustainable transportation. Int. Trans. Oper. Res.Oper. Res. **32**, 1977–2002 (2025)
8. Kirisci, M., Kuzu, S., Kablan, A., Ongel, V.: The new integrated interval-valued Fermatean fuzzy decision-making approach with the implementation of green supply chain management. Axioms **14**, 187 (2025)
9. Jeevaraj, S.: Ordering of interval-valued Fermatean fuzzy sets and their applications. Expert Syst. Appl. **185**, 115613 (2021)
10. Rani, P., Mishra, A.R.: Interval-valued Fermatean fuzzy sets with multi-criteria weighted aggregated sum product assessment based decision analysis. Neural Comput. Appl.Comput. Appl. **34**, 8051–8067 (2022)
11. Jun, Y.B., Kim, C.S., Yang, K.O.: Cubic sets. Ann. Fuzzy Math. Inform. **4**(1), 83–98 (2012)
12. Rong, Y., Yu, L., Niu, W., Liu, Y., Senapati, T., Mishra, A.R.: MARCOS approach based upon cubic Fermatean fuzzy set and its application in evaluation and selecting cold chain logistics distribution center. Eng. Appl. Artif. Intell.Artif. Intell. **116**, 105401 (2022)
13. Hwang, C.L., Yoon, K.: Multiple Attribute Decision Making: Methods and Applications. Springer, New York (1981)
14. Dharmalingam, M., Mahapatra, G.S., Vijayakumar, R.: Environment-aware multi-criteria decision-making for selecting a site for a mobile tower installation using the Fermatean fuzzy TOPSIS technique. Int. J. Fuzzy Syst. (2025). https://doi.org/10.1007/s40815-024-01892-z

Chapter 5
Picture Fuzzy Sets

5.1 Introduction

The picture fuzzy sets (PFS) for the first time introduced by Cuong [6]. A picture fuzzy set in addition to the membership and non-membership degrees of an event, uses the neutral degree of the event for describing more uncertain and vague situations. It can be widely used for cases where the evaluations are determined by the experts. As an example, for a question, an expert's answer can be "agree", "disagree", and "no opinion". In such cases, the picture fuzzy sets can be applied to model the uncertain event.

Some concepts and applications of the picture fuzzy sets are reviewed in this chapter.

5.2 Picture Fuzzy Sets

Definition 5.1 [5] The picture fuzzy set $\tilde{P}$ on the reference set X is defined as $\tilde{P} = \left\{ \left(x, \mu_{\tilde{P}}(x), \eta_{\tilde{P}}(x), \nu_{\tilde{P}}(x)\right) \middle| x \in X \right\}$, where $\mu_{\tilde{P}}(x) \in [0, 1]$ is the membership degree, $\eta_{\tilde{P}}(x) \in [0, 1]$ is the neutral degree, and $\nu_{\tilde{P}}(x) \in [0, 1]$ is the non-membership degree of $\tilde{P}$. The condition $\mu_{\tilde{P}}(x) + \eta_{\tilde{P}}(x) + \nu_{\tilde{P}}(x) \leq 1$ should be held for the picture fuzzy set $\tilde{P}$.

Based on this definition, $\pi(x) = 1 - \mu_{\tilde{P}}(x) - \eta_{\tilde{P}}(x) - \nu_{\tilde{P}}(x)$ is the refusal degree of the picture fuzzy set $\tilde{P}$ on the reference set X. The collection of all picture fuzzy sets defined in the set X is shown by $PFS(X)$.

In the case of $\eta_{\tilde{P}}(x) = 0$, the picture fuzzy set $\tilde{P}$ is converted to an intuitionistic fuzzy set.

Definition 5.2 [5] For the picture fuzzy sets $\tilde{A} = \left\{ \left(x, \mu_{\tilde{A}}(x), \eta_{\tilde{A}}(x), \nu_{\tilde{A}}(x)\right) \middle| x \in X \right\}$ and $\tilde{B} = \left\{ \left(x, \mu_{\tilde{B}}(x), \eta_{\tilde{B}}(x), \nu_{\tilde{B}}(x)\right) \middle| x \in X \right\}$, the below operators are defined.

A. Mahmoodirad et al., *Recent Advances on Fuzzy Sets*, Studies in Fuzziness and Soft Computing 440, https://doi.org/10.1007/978-3-032-08776-8_5

- $\tilde{A} \subseteq \tilde{B}$ if and only if for $\forall x \in X$, $\mu_{\tilde{A}}(x) \leq \mu_{\tilde{B}}(x)$, $\eta_{\tilde{A}}(x) \leq \eta_{\tilde{B}}(x)$, $\nu_{\tilde{A}}(x) \leq \nu_{\tilde{B}}(x)$,
- $\tilde{A} = \tilde{B}$ if and only if $\tilde{A} \subseteq \tilde{B}$ and $\tilde{B} \subseteq \tilde{A}$,
- $\tilde{A} \cup \tilde{B} = \{(x, \max\{\mu_{\tilde{A}}(x), \mu_{\tilde{B}}(x)\}, \min\{\eta_{\tilde{A}}(x), \eta_{\tilde{B}}(x)\}, \min\{\nu_{\tilde{A}}(x), \nu_{\tilde{B}}(x)\}) | x \in X\}$,
- $\tilde{A} \cap \tilde{B} = \{(x, \min\{\mu_{\tilde{A}}(x), \mu_{\tilde{B}}(x)\}, \min\{\eta_{\tilde{A}}(x), \eta_{\tilde{B}}(x)\}, \max\{\nu_{\tilde{A}}(x), \nu_{\tilde{B}}(x)\}) | x \in X\}$,
- $\tilde{A}^c = \{(x, \nu_{\tilde{A}}(x), \eta_{\tilde{A}}(x), \mu_{\tilde{A}}(x)) | x \in X\}$.

Definition 5.3 [5] Assume the picture fuzzy numbers $\tilde{A} = \{(x, \mu_{\tilde{A}}(x), \eta_{\tilde{A}}(x), \nu_{\tilde{A}}(x)) | x \in X_1\}$ and $\tilde{B} = \{(y, \mu_{\tilde{B}}(y), \eta_{\tilde{B}}(y), \nu_{\tilde{B}}(y)) | y \in X_2\}$, and their associated reference sets X_1 and X_2. Then, the Cartesian product of these sets is defined as below.

- $\tilde{A} \times_1 \tilde{B} = \{((x, y), \mu_{\tilde{A}}(x) \cdot \mu_{\tilde{B}}(y), \eta_{\tilde{A}}(x) \cdot \eta_{\tilde{B}}(y), \nu_{\tilde{A}}(x) \cdot \nu_{\tilde{B}}(y)) | x \in X_1, y \in X_2\}$
- $$\begin{aligned}\tilde{A} \times_2 \tilde{B} = \{&((x, y), \min\{\mu_{\tilde{A}}(x), \mu_{\tilde{B}}(y)\}, \min\{\eta_{\tilde{A}}(x), \eta_{\tilde{B}}(y)\},\\ &\max\{\nu_{\tilde{A}}(x), \nu_{\tilde{B}}(y)\}) | x \in X_1, y \in X_2\}\end{aligned}$$

Definition 5.4 [7] An (α, δ, β)-cut of the picture fuzzy set $\tilde{A}$ on the reference set X, is defined as $C_{\alpha,\delta,\beta}(\tilde{A}) = \{x \in X : \mu_{\tilde{A}}(x) \geq \alpha, \eta_{\tilde{A}}(x) \leq \delta, \nu_{\tilde{A}}(x) \leq \beta\}$. Here, $\alpha, \delta, \beta \in [0, 1]$ and $\alpha + \delta + \beta \leq 1$.

Definition 5.5 [7] Assume the picture fuzzy set $\tilde{A}$ on the reference set X. The height of the membership function, neutral function, and non-membership function are defined as $\sup\{\mu_{\tilde{A}}(x)\}$, $\inf\{\eta_{\tilde{A}}(x)\}$, and $\inf\{\nu_{\tilde{A}}(x)\}$ respectively.

Definition 5.6 [7] Assume the picture fuzzy sets $\tilde{A}$ and $\tilde{B}$ on the reference set X. For the real number $\lambda \in [0, 1]$, the convex combination of $\tilde{C}_\lambda(\tilde{A}, \tilde{B}) = \{(x, \mu_{\tilde{C}_\lambda}(x), \eta_{\tilde{C}_\lambda}(x), \nu_{\tilde{C}_\lambda}(x)) | x \in X\}$ is defined, where $\mu_{\tilde{C}_\lambda}(x) = \lambda\, \mu_{\tilde{A}_\lambda}(x) + (1 - \lambda)\, \mu_{\tilde{B}_\lambda}(x)$, $\eta_{\tilde{C}_\lambda}(x) = \lambda\, \eta_{\tilde{A}_\lambda}(x) + (1 - \lambda)\, \eta_{\tilde{B}_\lambda}(x)$, and $\nu_{\tilde{C}_\lambda}(x) = \lambda\, \nu_{\tilde{A}_\lambda}(x) + (1 - \lambda)\, \nu_{\tilde{B}_\lambda}(x)$ for all $x \in X$.

Theorem 5.1 *For the convex combination of Definition* 5.6, *the below logics are held.*

- If $\lambda = 1$, *then* $\tilde{C}_\lambda(\tilde{A}, \tilde{B}) = \tilde{A}$, *and if* $\lambda = 0$, *then* $\tilde{C}_\lambda(\tilde{A}, \tilde{B}) = \tilde{B}$.
- If $\tilde{A} \subseteq \tilde{B}$, *for any* $\lambda \in [0, 1]$, *there is* $\tilde{A} \subseteq \tilde{C}_\lambda(\tilde{A}, \tilde{B}) \subseteq \tilde{B}$.
- If $\tilde{B} \subseteq \tilde{A}$ *and* $\lambda_1 \geq \lambda_2$, *then* $\tilde{C}_{\lambda_2}(\tilde{A}, \tilde{B}) \subseteq \tilde{C}_{\lambda_1}(\tilde{A}, \tilde{B})$.
- $\tilde{C}_\lambda(\tilde{A} \cap \tilde{B}, \tilde{D}) = \tilde{C}_\lambda(\tilde{A}, \tilde{D}) \cap \tilde{C}_\lambda(\tilde{B}, \tilde{D})$.
- $\tilde{C}_\lambda(\tilde{A} \cup \tilde{B}, \tilde{D}) = \tilde{C}_\lambda(\tilde{A}, \tilde{D}) \cup \tilde{C}_\lambda(\tilde{B}, \tilde{D})$.

5.2.1 Distance of Picture Fuzzy Sets

Definition 5.7 [17] The function $d : PFS(X) \times PFS(X) \to [0, 1]$ is a distance function of the picture fuzzy sets $\tilde{A}$ and $\tilde{B}$, if it holds the below conditions.

- $0 \leq d(\tilde{A}, \tilde{B}) \leq 1$,

- $d(\tilde{A}, \tilde{B}) = d(\tilde{B}, \tilde{A})$,
- $d(\tilde{A}, \tilde{B}) = 0$ if and only if $\tilde{A} = \tilde{B}$,
- If $\tilde{A} \subseteq \tilde{B} \subseteq \tilde{C}$, then $d(\tilde{A}, \tilde{C}) \geq d(\tilde{A}, \tilde{B})$ and $d(\tilde{A}, \tilde{C}) \geq d(\tilde{B}, \tilde{C})$.

Definition 5.8 Consider the picture fuzzy sets $\tilde{A} = \{(x, \mu_{\tilde{A}}(x), \eta_{\tilde{A}}(x), \nu_{\tilde{A}}(x))|x \in X\}$, $\tilde{B} = \{(x, \mu_{\tilde{B}}(x), \eta_{\tilde{B}}(x), \nu_{\tilde{B}}(x))|x \in X\}$ on the reference set $X = \{x_1, x_2, \ldots, x_n\}$. Then, the below distances are defined.

The generalized distance ($p \in [1, \infty)$):

$$d_G(\tilde{A}, \tilde{B}) = \left(\frac{1}{n}\sum_{i=1}^{n}\left(|\mu_{\tilde{A}}(x_i) - \mu_{\tilde{B}}(x_i)|^p + |\eta_{\tilde{A}}(x_i) - \eta_{\tilde{B}}(x_i)|^p \right.\right.$$
$$\left.\left. + |\nu_{\tilde{A}}(x_i) - \nu_{\tilde{B}}(x_i)|^p\right)\right)^{\frac{1}{p}},$$

The Hamming distance ($p = 1$):

$$d_H(\tilde{A}, \tilde{B}) = \frac{1}{n}\sum_{i=1}^{n}\left(|\mu_{\tilde{A}}(x_i) - \mu_{\tilde{B}}(x_i)| + |\eta_{\tilde{A}}(x_i) - \eta_{\tilde{B}}(x_i)| + |\nu_{\tilde{A}}(x_i) - \nu_{\tilde{B}}(x_i)|\right),$$

The normalized Euclidean distance ($p = 2$):

$$d_E(\tilde{A}, \tilde{B}) = \sqrt{\frac{1}{n}\sum_{i=1}^{n}\left((\mu_{\tilde{A}}(x_i) - \mu_{\tilde{B}}(x_i))^2 + (\eta_{\tilde{A}}(x_i) - \eta_{\tilde{B}}(x_i))^2 + (\nu_{\tilde{A}}(x_i) - \nu_{\tilde{B}}(x_i))^2\right)}.$$

Example *5.1* Consider the picture fuzzy sets $\tilde{A} = \{(0.8, 0.1, 0)/x_1, (0.4, 0.2, 0.3)/x_2, (0.5, 0.3, 0)/x_3\}$, $\tilde{B} = \{(0.3, 0.3, 0.2)/x_1, (0.7, 0.1, 0.1)/x_2, (0.4, 0.3, 0.2)/x_3\}$, and $\tilde{C} = \{(0.3, 0.4, 0.1)/x_1, (0.6, 0.2, 0.1)/x_2, (0.4, 0.3, 0.1)/x_3\}$, on the set $X = \{x_1, x_2, x_3\}$. Then the distance values $d_H(\tilde{A}, \tilde{B}) = 0.6$, $d_H(\tilde{B}, \tilde{C}) = 0.17$, $d_H(\tilde{A}, \tilde{C}) = 0.5$, $d_E(\tilde{A}, \tilde{B}) = 0.24$, $d_E(\tilde{B}, \tilde{C}) = 0.075$, and $d_E(\tilde{A}, \tilde{C}) = 0.387$ are obtained.

Definition 5.9 [17] The function $S : PFS(X) \times PFS(X) \rightarrow [0, 1]$ is a similarity measure for the picture fuzzy sets $\tilde{A}$ and $\tilde{B}$, if it follows the below conditions.

- $0 \leq S(\tilde{A}, \tilde{B}) \leq 1$,
- $S(\tilde{A}, \tilde{B}) = S(\tilde{B}, \tilde{A})$,
- $S(\tilde{A}, \tilde{B}) = 1$ if and only if $\tilde{A} = \tilde{B}$,
- If $\tilde{A} \subseteq \tilde{B} \subseteq \tilde{C}$, then $S(\tilde{A}, \tilde{C}) \leq S(\tilde{A}, \tilde{B})$ and $d(\tilde{A}, \tilde{C}) \leq d(\tilde{B}, \tilde{C})$.

Theorem 5.2 *Consider the picture fuzzy sets* $\tilde{A} = \{(x, \mu_{\tilde{A}}(x), \eta_{\tilde{A}}(x), \nu_{\tilde{A}}(x))|x \in X\}$, $\tilde{B} = \{(x, \mu_{\tilde{B}}(x), \eta_{\tilde{B}}(x), \nu_{\tilde{B}}(x))|x \in X\}$ *on the reference set* $X = \{x_1, x_2, \ldots, x_n\}$. *Then, the below functions are a similarity measure of* $PFS(X) \times PFS(X) \rightarrow [0, 1]$

where $f : [0, 1] \to [0, 1]$ *is a strictly decreasing unary function where* $f(0) = 1$, $f(1) = 0$, $p \in [1, \infty)$, $\lambda_i \in [0, 1]$ *(for* $i = 1, \ldots, n$*), and* $\sum_{i=1}^{n} \lambda_i = 1$.

$$S_1(\tilde{A}, \tilde{B}) = \sum_{i=1}^{n} \lambda_i \Big[f\Big(\frac{1}{3} \big(|\mu_{\tilde{A}}(x_i) - \mu_{\tilde{B}}(x_i)|^p + |\eta_{\tilde{A}}(x_i) - \eta_{\tilde{B}}(x_i)|^p + |\nu_{\tilde{A}}(x_i) - \nu_{\tilde{B}}(x_i)|^p \big) \Big) \Big],$$

$$S_2(\tilde{A}, \tilde{B}) = \sum_{i=1}^{n} \lambda_i \Big[f\Big(\frac{1}{4} \big(|\mu_{\tilde{A}}(x_i) - \mu_{\tilde{B}}(x_i)|^p + |\eta_{\tilde{A}}(x_i) - \eta_{\tilde{B}}(x_i)|^p + |\nu_{\tilde{A}}(x_i) - \nu_{\tilde{B}}(x_i)|^p + |\pi_{\tilde{A}}(x_i) - \pi_{\tilde{B}}(x_i)|^p \big) \Big) \Big],$$

$$S_3(\tilde{A}, \tilde{B}) = \sum_{i=1}^{n} \lambda_i \Big[f\big(\max\big(|\mu_{\tilde{A}}(x_i) - \mu_{\tilde{B}}(x_i)|^p, |\eta_{\tilde{A}}(x_i) - \eta_{\tilde{B}}(x_i)|^p, |\nu_{\tilde{A}}(x_i) - \nu_{\tilde{B}}(x_i)|^p \big) \big) \Big],$$

$$S_4(\tilde{A}, \tilde{B}) = \sum_{i=1}^{n} \lambda_i \Big[f\big(\max\big(|\mu_{\tilde{A}}(x_i) - \mu_{\tilde{B}}(x_i)|^p, |\eta_{\tilde{A}}(x_i) - \eta_{\tilde{B}}(x_i)|^p, |\nu_{\tilde{A}}(x_i) - \nu_{\tilde{B}}(x_i)|^p, |\pi_{\tilde{A}}(x_i) - \pi_{\tilde{B}}(x_i)|^p \big) \big) \Big],$$

$$S_5(\tilde{A}, \tilde{B}) = f\Big(\frac{1}{3} \sum_{i=1}^{n} \lambda_i \big(|\mu_{\tilde{A}}(x_i) - \mu_{\tilde{B}}(x_i)|^p + |\eta_{\tilde{A}}(x_i) - \eta_{\tilde{B}}(x_i)|^p + |\nu_{\tilde{A}}(x_i) - \nu_{\tilde{B}}(x_i)|^p \big) \Big),$$

$$S_6(\tilde{A}, \tilde{B}) = f\Big(\frac{1}{4} \sum_{i=1}^{n} \lambda_i \big(|\mu_{\tilde{A}}(x_i) - \mu_{\tilde{B}}(x_i)|^p + |\eta_{\tilde{A}}(x_i) - \eta_{\tilde{B}}(x_i)|^p + |\nu_{\tilde{A}}(x_i) - \nu_{\tilde{B}}(x_i)|^p + |\pi_{\tilde{A}}(x_i) - \pi_{\tilde{B}}(x_i)|^p \big) \Big),$$

$$S_7(\tilde{A}, \tilde{B}) = f\Big(\sum_{i=1}^{n} \lambda_i \max\big(|\mu_{\tilde{A}}(x_i) - \mu_{\tilde{B}}(x_i)|^p, |\eta_{\tilde{A}}(x_i) - \eta_{\tilde{B}}(x_i)|^p, |\nu_{\tilde{A}}(x_i) - \nu_{\tilde{B}}(x_i)|^p \big) \Big),$$

$$S_8(\tilde{A}, \tilde{B}) = f\Big(\sum_{i=1}^{n} \lambda_i \max\big(|\mu_{\tilde{A}}(x_i) - \mu_{\tilde{B}}(x_i)|^p, |\eta_{\tilde{A}}(x_i) - \eta_{\tilde{B}}(x_i)|^p, |\nu_{\tilde{A}}(x_i) - \nu_{\tilde{B}}(x_i)|^p, |\pi_{\tilde{A}}(x_i) - \pi_{\tilde{B}}(x_i)|^p \big) \Big),$$

Proof See Luo et al. [18].

Theorem 5.3 *Consider the picture fuzzy sets* $\tilde{A} = \{(x, \mu_{\tilde{A}}(x), \eta_{\tilde{A}}(x), \nu_{\tilde{A}}(x))|x \in X\}$, $\tilde{B} = \{(x, \mu_{\tilde{B}}(x), \eta_{\tilde{B}}(x), \nu_{\tilde{B}}(x))|x \in X\}$ *on the reference set* $X = \{x_1, x_2, \ldots, x_n\}$. *Then,* $S_9(\tilde{A}, \tilde{B}) = g(d_1(\tilde{A}, \tilde{B}), d_2(\tilde{A}, \tilde{B}))$ *is a similarity measure of* $PFS(X) \times PFS(X) \to [0, 1]$, *where* $g : [0, 1]^2 \to [0, 1]$ *is a strictly decreasing unary function with* $g(0, 0) = 1$, *and* d_1 *and* d_2 *are two distances of picture fuzzy sets.*

Proof See Luo et al. [18].

Definition 5.10 Assume the picture fuzzy sets $\tilde{A}$ and $\tilde{B}$ on the reference set $X = \{x_1, \ldots, x_n\}$. The Hamming-like distance of $\tilde{A}$ and $\tilde{B}$ is defined as below (where $0 < \lambda < \omega \leq 1$).

$$D_{\lambda,\omega}(\tilde{A}, \tilde{B}) = \frac{1}{n}\sum_{j=1}^{n}\begin{pmatrix} \frac{1-\lambda}{1+\omega-\lambda}\left|\mu_{\tilde{A}}(x_j) - \mu_{\tilde{B}}(x_j)\right| \\ +\frac{\omega-\lambda}{1+\omega-\lambda}\left|\eta_{\tilde{A}}(x_j) - \eta_{\tilde{B}}(x_j)\right| \\ +\frac{\lambda}{1+\omega-\lambda}\left|\nu_{\tilde{A}}(x_j) - \nu_{\tilde{B}}(x_j)\right| \end{pmatrix},$$

Here, the associated similarity measure is defined as $S_{\lambda,\omega}(\tilde{A}, \tilde{B}) = 1 - D_{\lambda,\omega}(\tilde{A}, \tilde{B})$.

Definition 5.11 The picture fuzzy weighted arithmetic aggregator (PFWAA) for the picture fuzzy numbers $\tilde{A}_i = (\mu_{\tilde{A}_i}, \eta_{\tilde{A}_i}, \nu_{\tilde{A}_i})$, $i = 1, 2, \ldots, n$, where the weight of $\tilde{A}_i$ is $w_i \in [0, 1]$, and $\sum_{i=1}^{n} w_i = 1$, is as below.

$$\begin{aligned} PFWAA(\tilde{A}_1, \ldots, \tilde{A}_n) &= \sum_{i=1}^{n} w_i\tilde{A}_i \\ &= \Bigg(1 - \prod_{i=1}^{n}\left(1 - \mu_{\tilde{A}_i}\right)^{w_i}, \prod_{i=1}^{n}\left(\eta_{\tilde{A}_i}\right)^{w_i}, \\ &\qquad \prod_{i=1}^{n}\left(\eta_{\tilde{A}_i} + \nu_{\tilde{A}_i}\right)^{w_i} - \prod_{i=1}^{n}\left(\eta_{\tilde{A}_i}\right)^{w_i}\Bigg). \end{aligned} \tag{5.1}$$

Definition 5.12 The picture fuzzy weighted geometric aggregator (PFWGA) for the picture fuzzy numbers $\tilde{A}_i = (\mu_{\tilde{A}_i}, \eta_{\tilde{A}_i}, \nu_{\tilde{A}_i})$, $i = 1, 2, \ldots, n$, where the weight of $\tilde{A}_i$ is $w_i \in [0, 1]$, and $\sum_{i=1}^{n} w_i = 1$, is as below.

$$\begin{aligned} PFWGA(\tilde{A}_1, \ldots, \tilde{A}_n) &= \prod_{i=1}^{n}\left(\tilde{A}_i\right)^{w_i} \\ &= \Bigg(\prod_{i=1}^{n}\left(\eta_{\tilde{A}_i} + \nu_{\tilde{A}_i}\right)^{w_i} - \prod_{i=1}^{n}\left(\eta_{\tilde{A}_i}\right)^{w_i}, \end{aligned}$$

$$\prod_{i=1}^{n}\left(\eta_{\tilde{A}_i}\right)^{w_i}, 1-\prod_{i=1}^{n}\left(1-\mu_{\tilde{A}_i}\right)^{w_i}\Bigg). \tag{5.2}$$

Definition 5.13 The picture fuzzy weighted interaction aggregator (PFWIA) for the picture fuzzy numbers $\tilde{A}_i = \left(\mu_{\tilde{A}_i}, \eta_{\tilde{A}_i}, \nu_{\tilde{A}_i}\right)$, $i = 1, 2, \ldots, n$, where the weight of $\tilde{A}_i$ is $w_i \in [0, 1]$, and $\sum_{i=1}^{n} w_i = 1$, is as below.

$$PFWIG\left(\tilde{A}_1, \ldots, \tilde{A}_n\right)$$
$$= \left(\prod_{i=1}^{n}\left(1-\eta_{\tilde{A}_i}-\nu_{\tilde{A}_i}\right)^{w_i} - \prod_{i=1}^{n}\left(1-\mu_{\tilde{A}_i}-\eta_{\tilde{A}_i}-\nu_{\tilde{A}_i}\right)^{w_i},\right.$$
$$\left.\prod_{i=1}^{n}\left(1-\nu_{\tilde{A}_i}\right)^{w_i} - \prod_{i=1}^{n}\left(1-\eta_{\tilde{A}_i}-\nu_{\tilde{A}_i}\right)^{w_i}, 1-\prod_{i=1}^{n}\left(1-\nu_{\tilde{A}_i}\right)^{w_i}\right). \tag{5.3}$$

***Example** 5.2* Assume the picture fuzzy numbers $\tilde{A}_1 = (0.6, 0, 0.1)$, $\tilde{A}_2 = (0.4, 0.1, 0.2)$, and $\tilde{A}_3 = (0.5, 0.1, 0.3)$, with the weight vector $W = (0.4, 0.3, 0.3)$. Then, $PFWIG\left(\tilde{A}_1, \tilde{A}_2, \tilde{A}_3\right) = (0.523, 0.067, 0.194)$.

5.2.2 *Extension Principle of Picture Fuzzy Sets*

Definition 5.14 [7] Consider the non-empty sets X and Y, and the function $f : X \to Y$. Assume the picture fuzzy sets $\tilde{A}$ and $\tilde{B}$ on the sets X and Y respectively. Then, the projection of Y by the function f is shown as $f\left(\tilde{A}\right)$ and defined as $f\left(\tilde{A}\right)(y) = \left(\mu_{f(\tilde{A})}(y), \eta_{f(\tilde{A})}(y), \nu_{f(\tilde{A})}(y)\right)$ where,

$$f\left(\tilde{A}\right)(y) = \left(\max\left\{\mu_{\tilde{A}}(x) : x \in f^{-1}(y)\right\},\right.$$
$$\left.\min\left\{\eta_{\tilde{A}}(x) : x \in f^{-1}(y)\right\}, \min\left\{\nu_{\tilde{A}}(x) : x \in f^{-1}(y)\right\}\right),$$

Also, the inverse projection of $\tilde{B}$ by the function f is shown as $f^{-1}\left(\tilde{B}\right)$ and defined as $f^{-1}\left(\tilde{B}\right)(x) = \left(\mu_{f^{-1}(\tilde{B})}(x), \eta_{f^{-1}(\tilde{B})}(x), \nu_{f^{-1}(\tilde{B})}(x)\right)$, where $\mu_{f^{-1}(\tilde{B})}(x) = \mu_{\tilde{B}}(f(x))$, $\eta_{f^{-1}(\tilde{B})}(x) = \eta_{\tilde{B}}(f(x))$, $\nu_{f^{-1}(\tilde{B})}(x) = \nu_{\tilde{B}}(f(x))$ and $f^{-1}\left(\tilde{B}\right)(x) = \left(\mu_{\tilde{B}}(f(x)), \eta_{\tilde{B}}(f(x)), \nu_{\tilde{B}}(f(x))\right)$.

According to the extension principle, the picture fuzzy arithmetic operations for the picture fuzzy numbers $\tilde{A}$ and $\tilde{B}$ are defined as below where $* \in \{+, -, \times, \div\}$.

$$\tilde{A} * \tilde{B} = \left\{\left(z, \mu_{\tilde{A}*\tilde{B}}(z), \eta_{\tilde{A}*\tilde{B}}(z), \nu_{\tilde{A}*\tilde{B}}(z)\right)\right\}$$

and

$$\mu_{\tilde{A}*\tilde{B}}(z) = \max\min\{\mu_{\tilde{A}}(x), \mu_{\tilde{B}}(y)\},$$

$$\eta_{\tilde{A}*\tilde{B}}(z) = \min\max\{\eta_{\tilde{A}}(x), \eta_{\tilde{B}}(y)\},$$

$\nu_{\tilde{A}*\tilde{B}}(z) = \min\max\{\nu_{\tilde{A}}(x), \nu_{\tilde{B}}(y)\}$, and $z = x * y$.

Example ***5.3*** [7] Assume the picture fuzzy sets $\tilde{A} = \{(2, 0.4, 0.2, 0.3), (3, 0.7, 0.1, 0.1), (4, 0.6, 0.2, 0.2)\}$ and $\tilde{B} = \{(1, 0.6, 0.1, 0.2), (2, 0.5, 0.2, 0.1)\}$. Then,

$$\tilde{A} + \tilde{B} = \left\{\begin{array}{l} (2+1, \min\{0.4, 0.6\}, \max\{0.2, 0.1\}, \max\{0.3, 0.2\}), \\ (2+2, \min\{0.4, 0.5\}, \max\{0.2, 0.2\}, \max\{0.3, 0.1\}), \\ (3+1, \min\{0.7, 0.6\}, \max\{0.1, 0.1\}, \max\{0.1, 0.2\}), \\ (3+2, \min\{0.7, 0.5\}, \max\{0.1, 0.2\}, \max\{0.1, 0.1\}), \\ (4+1, \min\{0.6, 0.6\}, \max\{0.2, 0.1\}, \max\{0.2, 0.2\}), \\ (4+2, \min\{0.6, 0.5\}, \max\{0.2, 0.2\}, \max\{0.2, 0.1\}) \end{array}\right\}$$

$$= \left\{\begin{array}{l} (3, 0.4, 0.2, 0.3), (4, 0.4, 0.2, 0.3), (4, 0.6, 0.1, 0.2), \\ (5, 0.6, 0.2, 0.1), (5, 0.6, 0.2, 0.2), (6, 0.5, 0.2, 0.2) \end{array}\right\}$$

$$= \left\{\begin{array}{l} (3, 0.4, 0.2, 0.3), (4, \max\{0.4, 0.6\}, \min\{0.2, 0.1\}, \min\{0.3, 0.2\}), \\ (5, \max\{0.5, 0.6\}, \min\{0.2, 0.2\}, \min\{0.1, 0.2\}), (6, 0.5, 0.2, 0.2) \end{array}\right\}$$

$$= \{(3, 0.4, 0.2, 0.3), (4, 0.6, 0.1, 0.2), (5, 0.6, 0.2, 0.1), (6, 0.5, 0.2, 0.2)\}$$

$$\tilde{A} - \tilde{B} = \left\{\begin{array}{l} (2-1, \min\{0.4, 0.6\}, \max\{0.2, 0.1\}, \max\{0.3, 0.2\}), \\ (2-2, \min\{0.4, 0.5\}, \max\{0.2, 0.2\}, \max\{0.3, 0.1\}), \\ (3-1, \min\{0.7, 0.6\}, \max\{0.1, 0.1\}, \max\{0.1, 0.2\}), \\ (3-2, \min\{0.7, 0.5\}, \max\{0.1, 0.2\}, \max\{0.1, 0.1\}), \\ (4-1, \min\{0.6, 0.6\}, \max\{0.2, 0.1\}, \max\{0.2, 0.2\}), \\ (4-2, \min\{0.6, 0.5\}, \max\{0.2, 0.2\}, \max\{0.2, 0.1\}) \end{array}\right\}$$

$$= \left\{\begin{array}{l} (1, 0.4, 0.2, 0.3), (0, 0.4, 0.2, 0.3), (2, 0.6, 0.1, 0.2), (1, 0.5, 0.2, 0.1), \\ (3, 0.6, 0.2, 0.2), (2, 0.5, 0.2, 0.2) \end{array}\right\}$$

$$= \left\{\begin{array}{l} (0, 0.4, 0.2, 0.3), (1, \max\{0.4, 0.5\}, \min\{0.2, 0.2\}, \min\{0.3, 0.1\}), \\ (2, \max\{0.6, 0.5\}, \min\{0.1, 0.2\}, \min\{0.2, 0.2\}), (3, 0.6, 0.2, 0.2) \end{array}\right\}$$

$$\tilde{A} \times \tilde{B} = \left\{ \begin{array}{l} (1 \times 2, \min\{0.2, 0.1\}, \max\{0.3, 0.3\}, \max\{0.1, 0.3\}), \\ (1 \times 3, \min\{0.2, 0.2\}, \max\{0.3, 0.4\}, \max\{0.1, 0.4\}), \\ (2 \times 2, \min\{0.1, 0.1\}, \max\{0.3, 0.3\}, \max\{0.4, 0.3\}), \\ (2 \times 3, \min\{0.1, 0.2\}, \max\{0.3, 0.4\}, \max\{0.4, 0.4\}), \\ (3 \times 2, \min\{0.5, 0.1\}, \max\{0.3, 0.3\}, \max\{0.1, 0.3\}), \\ (3 \times 3, \min\{0.5, 0.2\}, \max\{0.3, 0.4\}, \max\{0.1, 0.4\}) \end{array} \right\}$$

$$= \left\{ \begin{array}{l} (2, 0.1, 0.3, 0.3), (3, 0.2, 0.4, 0.4), (4, 0.1, 0.3, 0.4), \\ (6, 0.1, 0.4, 0.4), (6, 0.1, 0.3, 0.3), (9, 0.2, 0.4, 0.4) \end{array} \right\}$$

$$= \left\{ \begin{array}{l} (2, 0.1, 0.3, 0.3), (3, 0.2, 0.4, 0.4), (4, 0.1, 0.3, 0.4), \\ (6, \max\{0.1, 0.1\}, \min\{0.3, 0.4\}, \min\{0.3, 0.4\}), \\ (9, 0.2, 0.4, 0.4) \end{array} \right\}$$

$$= \left\{ \begin{array}{l} (2, 0.1, 0.3, 0.3), (3, 0.2, 0.4, 0.4), (4, 0.1, 0.3, 0.4), \\ (6, 0.1, 0.3, 0.3), (9, 0.2, 0.4, 0.4) \end{array} \right\}$$

$$\tilde{A} \div \tilde{B} = \left\{ \begin{array}{l} (2 \div 1, \min\{0.1, 0.6\}, \max\{0.3, 0.2\}, \max\{0.2, 0.2\}), \\ (2 \div 2, \min\{0.1, 0.7\}, \max\{0.3, 0.1\}, \max\{0.2, 0.2\}), \\ (4 \div 1, \min\{0.1, 0.6\}, \max\{0.6, 0.2\}, \max\{0.2, 0.2\}), \\ (4 \div 2, \min\{0.1, 0.7\}, \max\{0.6, 0.1\}, \max\{0.2, 0.2\}) \end{array} \right\}$$

$$= \{(2, 0.1, 0.3, 0.2), (1, 0.1, 0.3, 0.2), (4, 0.1, 0.6, 0.2), (2, 0.1, 0.6, 0.2)\}$$

$$= \{(1, 0.1, 0.3, 0.2),$$
$$(2, \max\{0.1, 0.1\}, \min\{0.3, 0.6\}, \min\{0.2, 0.2\}), (4, 0.1, 0.6, 0.2)\}$$

$$= \{(1, 0.1, 0.3, 0.2), (2, 0.1, 0.3, 0.2), (4, 0.1, 0.6, 0.2)\}$$

5.3 Picture Fuzzy Numbers

Definition 5.15 [16] The picture fuzzy number $\tilde{a}$ is shown as $\tilde{a} = (\mu_{\tilde{a}}, \eta_{\tilde{a}}, \nu_{\tilde{a}}; w_1, w_2, w_3)$ with the below membership, neutral, and non-membership functions.

$$\mu_{\tilde{a}}(x) = \begin{cases} f_{\tilde{a}}^{L}(x) & a \leq x \leq b \\ w_1 & x = b \\ f_{\tilde{a}}^{R}(x) & b \leq x \leq c \\ 0 & otherwise, \end{cases} \tag{5.4}$$

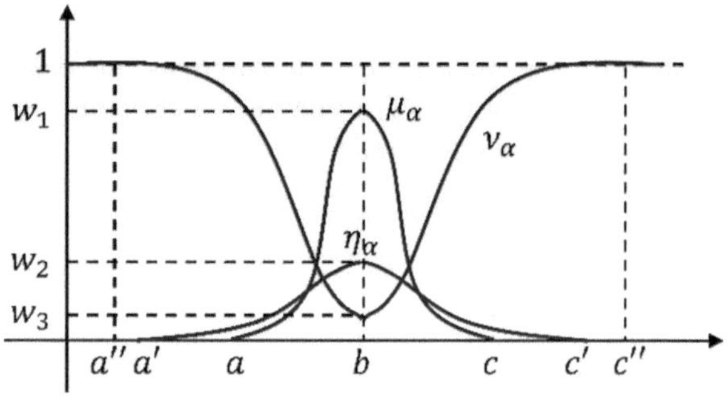

Fig. 5.1 Schematic representation of a picture fuzzy number

$$\eta_{\tilde{a}}(x) = \begin{cases} g_{\tilde{a}}^{L}(x) & a' \le x \le b \\ w_2 & x = b \\ g_{\tilde{a}}^{R}(x) & b \le x \le c' \\ 0 & otherwise, \end{cases} \tag{5.5}$$

$$\nu_{\tilde{a}}(x) = \begin{cases} h_{\tilde{a}}^{L}(x) & a'' \le x \le b \\ w_3 & x = b \\ h_{\tilde{a}}^{R}(x) & b \le x \le c'' \\ 0 & otherwise, \end{cases} \tag{5.6}$$

Here, $f_{\tilde{a}}^{L}(x)$, $g_{\tilde{a}}^{L}(x)$, and $h_{\tilde{a}}^{R}(x)$ are some increasing, and $f_{\tilde{a}}^{R}(x)$, $g_{\tilde{a}}^{R}(x)$, and $h_{\tilde{a}}^{L}(x)$ are some decreasing functions. The values w_1, w_2, and w_3 are respectively the maximum values of the membership, neutral, and non-membership degrees where $w_1, w_2, w_3 \in [0, 1]$ and $0 \le w_1 + w_2 + w_3 \le 1$ (see Fig. 5.1). In the picture fuzzy number $\tilde{a} = (\mu_{\tilde{a}}, \eta_{\tilde{a}}, \nu_{\tilde{a}}; w_1, w_2, w_3)$ if $w_2 = 0$ then an intuitionistic fuzzy number is obtained.

Definition 5.16 [1] The picture fuzzy number $\tilde{A} = \left[(m, n; l_1, r_1; l_2, r_2; l_3, r_3);\ \alpha, \beta, \gamma\right]_{LR}$ is a simple LR flat picture fuzzy number, if it has the below membership, neutral, and non-membership functions.

$$\mu_{\tilde{A}}(x) = \begin{cases} L_1\left(\left(\frac{m-x}{l_1}\right)\alpha\right) & x \le m,\ l_1 > 0, \\ \alpha & m \le x \le n \\ R_1\left(\left(\frac{x-n}{r_1}\right)\alpha\right) & x \ge n,\ r_1 > 0, \end{cases} \tag{5.7}$$

$$\eta_{\tilde{A}}(x) = \begin{cases} L_2\left(\left(\frac{m-x}{l_2}\right)\beta\right) & x \le m, l_2 > 0, \\ \beta & m \le x \le n \\ R_2\left(\left(\frac{x-n}{r_2}\right)\beta\right) & x \ge n,\ r_2 > 0, \end{cases} \tag{5.8}$$

$$\nu_{\tilde{A}}(x) = \begin{cases} L_3\left(\left(\frac{m-x}{l_3}\right)\gamma\right) & x \le m, l_3 > 0, \\ \gamma & m \le x \le n \\ R_3\left(\left(\frac{x-n}{r_3}\right)\gamma\right) & x \ge n,\ r_3 > 0, \end{cases} \tag{5.9}$$

Here, $l_1 \le l_2 \le l_3$, $r_1 \le r_2 \le r_3$, and $0 \le \mu_{\tilde{A}}(x) + \eta_{\tilde{A}}(x) + \nu_{\tilde{A}}(x) \le 1$. The functions L_1 and R_1 are unique, continuous, and decreasing from $[0, +\infty)$ to $[0, \alpha]$. The functions L_2 and R_2 are unique, continuous, and decreasing from $[0, +\infty)$ to $[0, \beta]$. The functions L_3 and R_3 are unique, continuous, and increasing from $[0, +\infty)$ to $\left[\gamma, 1\right]$. Also,

$$L_1(0) = R_1(0) = \alpha, \ \underset{y\to\infty}{Lim}\ R_1(y) = \underset{y\to\infty}{Lim}\ L_1(y) = 0,$$
$$L_2(0) = R_2(0) = \beta, \ \underset{y\to\infty}{Lim}\ R_2(y) = \underset{y\to\infty}{Lim}\ L_2(y) = 0,$$
$$L_3(0) = R_3(0) = \gamma, \ \underset{y\to\infty}{Lim}\ R_3(y) = \underset{y\to\infty}{Lim}\ L_3(y) = 1,$$

Definition 5.17 [1] A picture fuzzy number such as $\tilde{A} = \left[(m, n; l_1, r_1; l_2, r_2; l_3, r_3);\ \alpha, \beta, \gamma\right]_{LR}$ is called a general *LR* flat picture fuzzy number if its membership, neutral, and non-membership functions be as below.

$$\mu_{\tilde{A}}(x) = \begin{cases} L_1\left(\left(\frac{m-x}{l_1}\right)\alpha\right) & x \le m,\ l_1 > 0, \\ \alpha & m \le x \le n \\ R_1\left(\left(\frac{x-n}{r_1}\right)\alpha\right) & x \ge n,\ r_1 > 0, \end{cases} \quad (5.10)$$

$$\eta_{\tilde{A}}(x) = \begin{cases} L_2\left(\left(\frac{m-x}{l_2}\right)\beta\right) & x \le m,\ l_2 > 0, \\ \beta & m \le x \le n \\ R_2\left(\left(\frac{x-n}{r_2}\right)\beta\right) & x \ge n,\ r_2 > 0, \end{cases} \quad (5.11)$$

$$\nu_{\tilde{A}}(x) = \begin{cases} L_3\left(\left(\frac{m-x}{l_3}\right)\gamma\right) & x \le m,\ l_3 > 0, \\ \gamma & m \le x \le n \\ R_3\left(\left(\frac{x-n}{r_3}\right)\gamma\right) & x \ge n,\ r_3 > 0, \end{cases} \quad (5.12)$$

Here, $l_1 \le l_2 \le l_3$, $r_1 \le r_2 \le r_3$, and $0 \le \mu_{\tilde{A}}(x) + \eta_{\tilde{A}}(x) + \nu_{\tilde{A}}(x) \le 1$. The functions L_1 and R_1 are unique, continuous, and decreasing from $[0, +\infty)$ to $[0, \alpha]$. The function L_2 is unique, continuous, and increasing/decreasing from $[0, +\infty)$ to $\left[\beta, \min\left\{1 - (\alpha + \gamma), 1 - \nu_{\tilde{A}}(m - l_2)\right\}\right]/\left[\min\left\{1 - (\alpha + \gamma), 1 - \nu_{\tilde{A}}(m - l_2)\right\}, \beta\right]$. The function R_2 is unique, continuous, and increasing from $[0, +\infty)$ to $\left[\beta, \min\left\{1 - (\alpha + \gamma), 1 - \nu_{\tilde{A}}(n + r_2)\right\}\right]/\left[\min\left\{1 - (\alpha + \gamma), 1 - \nu_{\tilde{A}}(n + r_2)\right\}, \beta\right]$. The functions L_3 and R_3 are unique, continuous, and increasing from $[0, +\infty)$ to $\left[\gamma, 1\right]$. Also,

$$L_1(0) = R_1(0) = \alpha, \ \underset{y\to\infty}{Lim}\ R_1(y) = \underset{y\to\infty}{Lim}\ L_1(y) = 0,$$
$$L_2(0) = R_2(0) = \beta, \ \underset{y\to\infty}{Lim}\ R_2(y) = \underset{y\to\infty}{Lim}\ L_2(y) = 0,$$
$$L_3(0) = R_3(0) = \gamma, \ \underset{y\to\infty}{Lim}\ R_3(y) = \underset{y\to\infty}{Lim}\ L_3(y) = 1,$$

In the rest of this picture, the term "*LR* fuzzy number" refers to the simple form of an *LR* fuzzy number.

Definition 5.18 Consider the *LR* picture fuzzy number $\tilde{A} = \left[(m, n; l_1, r_1; l_2, r_2; l_3, r_3);\ \alpha, \beta, \gamma\right]_{LR}$. Then,

- $\tilde{A}$ is non-negative ($\tilde{A} \geq 0$) if $m - l_3 \geq 0$.
- $\tilde{A}$ is non-positive ($\tilde{A} \leq 0$) if $n + r_3 \leq 0$.
- $\tilde{A}$ is without sign if m or n be of the set of real numbers.
- $\tilde{A}$ is positive if $m - l_3 > 0$.
- $\tilde{A}$ is negative if $n + r_3 < 0$.
- $\tilde{A}$ is zero if $m, n, l_1, r_1, l_2, r_2, l_3, r_3 = 0$.

Definition 5.19 Two picture fuzzy numbers such as $\tilde{A} = \left[\left(m^a, n^a; l_1^a, r_1^a; l_2^a, r_2^a; l_3^a, r_3^a\right);\ \alpha^a, \beta^a, \gamma^a\right]_{LR}$ and $\tilde{B} = \left[\left(m^b, n^b; l_1^b, r_1^b; l_2^b, r_2^b; l_3^b, r_3^b\right);\ \alpha^b, \beta^b, \gamma^b\right]_{LR}$ are equal if $m^a = m^b$, $n^a = n^b$, $l_1^a = l_1^b, r_1^a = r_1^b, l_2^a = l_2^b, r_2^a = r_2^b, l_3^a = l_3^b, r_3^a = r_3^b, \alpha^a = \alpha^b, \beta^a = \beta^b$, and $\gamma^a = \gamma^b$.

Theorem 5.4 *Consider the picture fuzzy* $\tilde{A} = \left[(m, n; l_1, r_1; l_2, r_2; l_3, r_3);\ \alpha, \beta, \gamma\right]_{LR}$. *The* λ, δ, *and* θ *cuts of this number are obtained as* $A_\lambda = \left[m - \frac{l_1}{\alpha} L_1^{-1}(\lambda), n + \frac{r_1}{\alpha} R_1^{-1}(\lambda)\right]$, $A_\delta = \left[m - \frac{l_2}{\beta} L_2^{-1}(\delta), n + \frac{r_2}{\beta} R_2^{-1}(\delta)\right]$, *and* $A_\theta = \left[m - \frac{l_3}{\gamma} L_3^{-1}(\theta), n + \frac{r_3}{\gamma} R_3^{-1}(\theta)\right]$ *respectively (for* $\lambda \in [0, \alpha]$, $\delta \in [0, \beta]$, *and* $\theta \in \left[\gamma, 1\right]$*).*

Theorem 5.5 *Consider the picture fuzzy numbers* $\tilde{A} = \left[\left(m^a, n^a; l_1^a, r_1^a; l_2^a, r_2^a; l_3^a, r_3^a\right);\ \alpha^a, \beta^a, \gamma^a\right]_{LR}$ *and* $\tilde{B} = \left[\left(m^b, n^b; l_1^b, r_1^b; l_2^b, r_2^b; l_3^b, r_3^b\right);\ \alpha^b, \beta^b, \gamma^b\right]_{LR}$. *Then,*

$$\begin{aligned}
\tilde{A} \oplus \tilde{B} = \big[\big(&m^a + m^b, n^a + n^b; l_1^a + l_1^b, r_1^a + r_1^b;\\
&l_2^a + l_2^b, r_2^a + r_2^b; l_3^a + l_3^b, r_3^a + r_3^b\big);\\
&\alpha^a + \alpha^b - \alpha^a\alpha^b, \beta^a\beta^b, \gamma^a\gamma^b\big]_{LR}
\end{aligned}$$

$$k\tilde{A} = \begin{cases} \left[\left(km^a, kn^a; kl_1^a, kr_1^a; kl_2^a, kr_2^a; kl_3^a, kr_3^a\right);\ (\alpha^a)^k, (\beta^a)^k, 1 - (1 - \gamma^a)^k\right]_{LR}, & k \geq 0,\\ \left[\left(kn^a, km^a; -kr_1^a, -kl_1^a; -kr_2^a, -kl_2^a; -kr_3^a, -kl_3^a\right);\ (\alpha^a)^{-k}, (\beta^a)^{-k}, 1 - (1 - \gamma^a)^{-k}\right]_{LR}, & k < 0, \end{cases}$$

$$\begin{aligned}
\tilde{A} \ominus \tilde{B} = \big[\big(&m^a - n^b, n^a - m^b; l_1^a + r_1^b, r_1^a + l_1^b; l_2^a + r_2^b, r_2^a + l_2^b;\\
&l_3^a + r_3^b, r_3^a + l_3^b\big); \alpha^a + \alpha^b - \alpha^a\alpha^b, \beta^a\beta^b, \gamma^a\gamma^b\big]_{LR}
\end{aligned}$$

Proof See Akram et al. [1].

Theorem 5.6 *Assume* $\tilde{A} = \left[\left(m^a, n^a; l_1^a, r_1^a; l_2^a, r_2^a; l_3^a, r_3^a\right);\ \alpha^a, \beta^a, \gamma^a\right]_{LR}$ *as a picture fuzzy number where* $m^a - l_3^a < 0$ *and* $m^a - l_2^a \geq 0$. *Also, assume* $\tilde{B} = \left[\left(m^b, n^b; l_1^b, r_1^b; l_2^b, r_2^b; l_3^b, r_3^b\right);\ \alpha^b, \beta^b, \gamma^b\right]_{LR}$ *as a picture fuzzy number with no sign.*

Then, $\tilde{A} \otimes \tilde{B} = \left[(M, N; L_1, R_1; L_2, R_2; L_3, R_3);\ \alpha^a + \alpha^b - \alpha^a\alpha^b, \beta^a\beta^b, \gamma^a\gamma^b\right]_{LR}$, *where,*

$$\begin{aligned}
M &= \min\{m^a m^b, n^a m^b\},\ N = \min\{m^a n^b, n^a n^b\},\\
L_1 &= \min\{m^a m^b, n^a m^b\} - \min\{m^a m^b - m^a l_1^b - m^b l_1^a + l_1^a l_1^b,\\
&\quad n^a m^b - n^a l_1^b + m^b r_1^a - r_1^a l_1^b\}\\
R_1 &= \max\{m^a n^b + m^a r_1^b - n^b l_1^a - l_1^a r_1^b, n^a n^b + n^a r_1^b + n^b r_1^a + r_1^a r_1^b\}\\
&\quad - \max\{m^a n^b, n^a n^b\}\\
L_2 &= \min\{m^a m^b, n^a m^b\} - \min\{m^a m^b - m^a l_2^b - m^b l_2^a + l_2^a l_2^b,\\
&\quad n^a m^b - n^a l_2^b + m^b r_2^a - r_2^a l_2^b\},\\
R_2 &= \max\{m^a n^b + m^a r_2^b - n^b l_2^a - l_2^a r_2^b, n^a n^b + n^a r_2^b + n^b r_2^a + r_2^a r_2^b\}\\
&\quad - \max\{m^a n^b, n^a n^b\}\\
L_3 &= \min\{m^a m^b, n^a m^b\} - \min\{m^a n^b - m^a r_3^b - n^b l_3^a + l_3^a r_3^b,\\
&\quad n^a m^b - n^a l_3^b + m^b r_3^a - l_3^b r_3^a\},\\
R_3 &= \max\{m^a m^b - m^a l_3^b - m^b l_3^a - l_3^a l_3^b, n^a n^b + n^a r_3^b + n^b r_3^a + r_3^a r_3^b\}\\
&\quad - \max\{m^a n^b, n^a n^b\}.
\end{aligned}$$

Proof See Akram et al. [1].

Definition 5.20 [1] For the picture fuzzy number $\tilde{A} = \left[(m, n; l_1, r_1; l_2, r_2; l_3, r_3);\ \alpha, \beta, \gamma\right]_{LR}$, the below ranking function is defined.

$$\begin{aligned}
\Re(\tilde{A}) = \frac{1}{2}\Bigg[&\int_0^{\alpha}\left(m - \frac{l_1}{\alpha}L_1^{-1}(x)\right)dx + \int_0^{\alpha}\left(n + \frac{r_1}{\alpha}R_1^{-1}(x)\right)dx\\
&+ \int_0^{\beta}\left(m - \frac{l_2}{\beta}L_2^{-1}(x)\right)dx + \int_0^{\beta}\left(n + \frac{r_2}{\beta}R_2^{-1}(x)\right)dx\\
&+ \int_{\gamma}^{1}\left(m - \frac{l_3}{\gamma}L_3^{-1}(x)\right)dx + \int_{\gamma}^{1}\left(n + \frac{r_3}{\gamma}R_3^{-1}(x)\right)dx\Bigg]. \qquad (5.13)
\end{aligned}$$

According to this function, for the *LR* picture fuzzy numbers $\tilde{A}$ and $\tilde{B}$, the below orders can be obtained.

$$\Re(\tilde{A}) > \Re(\tilde{B}) \Rightarrow \tilde{A} > \tilde{B}, \Re(\tilde{A}) < \Re(\tilde{B}) \Rightarrow \tilde{A} < \tilde{B}, \Re(\tilde{A}) = \Re(\tilde{B}) \Rightarrow \tilde{A} = \tilde{B}.$$

Theorem 5.7 *Assume the picture fuzzy numbers* $\tilde{A} = (\mu_{\tilde{A}}, \eta_{\tilde{A}}, \nu_{\tilde{A}})$ *and* $\tilde{B} = (\mu_{\tilde{B}}, \eta_{\tilde{B}}, \nu_{\tilde{B}})$ *on the set* $X = \{x_1, x_2, \ldots, x_n\}$. *Then the similarity measure of these numbers is as follows.*

$$S(\tilde{A},\tilde{B}) = 1 - \frac{\left|3(\mu_{\tilde{A}} - \mu_{\tilde{B}}) - (\eta_{\tilde{A}} - \eta_{\tilde{B}}) - (\nu_{\tilde{A}} - \nu_{\tilde{B}})\right|}{4} \times \left(\frac{\pi_{\tilde{A}} + \pi_{\tilde{B}}}{8}\right)$$
$$- \frac{\left|3(\eta_{\tilde{A}} - \eta_{\tilde{B}}) - (\mu_{\tilde{A}} - \mu_{\tilde{B}}) - (\nu_{\tilde{A}} - \nu_{\tilde{B}})\right|}{4} \times \left(1 - \frac{\pi_{\tilde{A}} + \pi_{\tilde{B}}}{8}\right)$$
$$- \frac{\left|3(\nu_{\tilde{A}} - \nu_{\tilde{B}}) - (\mu_{\tilde{A}} - \mu_{\tilde{B}}) - (\eta_{\tilde{A}} - \eta_{\tilde{B}})\right|}{4} \times \left(1 - \frac{\pi_{\tilde{A}} + \pi_{\tilde{B}}}{8}\right). \quad (5.14)$$

Proof See Cao [4].

5.3.1 *Trapezoidal Picture Fuzzy Numbers*

The picture fuzzy number $\tilde{A} = \left[(m, n; l_1, r_1; l_2, r_2; l_3, r_3);\ \alpha, \beta, \gamma\right]_{LR}$ with the below conditions,

$$L_1(x) = R_1(x) = \begin{cases} \alpha - x, & 0 \le x \le \alpha, \\ 0, & otherwise, \end{cases}$$

$$L_2(x) = R_2(x) = \begin{cases} \beta - x, & 0 \le x \le \beta, \\ 0, & otherwise, \end{cases}$$

$$L_3(x) = R_3(x) = \begin{cases} \frac{x - \gamma x + \gamma^2}{\gamma}, & 0 \le x \le \gamma, \\ 1, & otherwise, \end{cases}$$

is a trapezoidal picture fuzzy number and shown as $\tilde{A} = ((p_1, m, n, q_1; \alpha), (p_2, m, n, q_2; \beta), (p_3, m, n, q_3; \gamma))$ where $p_i = m - l_i$, $q_i = r_i + n$, and $i = 1, 2, 3$.

The trapezoidal picture fuzzy number $\tilde{A} = ((p_1, m, n, q_1; \alpha), (p_2, m, n, q_2; \beta), (p_3, m, n, q_3; \gamma))$ where $(p_1, m, n, q_1) = (p_2, m, n, q_2) = (p_3, m, n, q_3)$, is shown as $\tilde{A} = (p, m, n, q; \alpha, \beta, \gamma)$. In the case of $m = n$, the triangular picture fuzzy number $\tilde{A} = (p, m, q; \alpha, \beta, \gamma)$ is obtained.

Example 5.4 Consider the picture fuzzy number $\tilde{3} = (-1, 2, 4, 6;\ 0.5, 0.1, 0.2)$. When $x = 2$, for the number $\tilde{3}$, membership degree of 0.5, neutral degree of 0.1, non-membership degree of 0.2, and hesitancy degree of 0.2 are obtained.

Definition 5.21 [1] The picture fuzzy number $\tilde{A} = ((a_1, a_2, a_3, a_4; \alpha), (a_5, a_2, a_3, a_6; \beta), (a_7, a_2, a_3, a_8; \gamma))$ is a trapezoidal picture fuzzy number defined on $\mathbb{R}$, where its positive $(\mu_{\tilde{A}})$, neutral $(\eta_{\tilde{A}})$, and negative $(\upsilon_{\tilde{A}})$ membership functions are defined below.

$$\mu_{\tilde{A}}(x) = \begin{cases} \frac{\alpha(x-a_1)}{a_2-a_1} & a_1 \le x \le a_2 \\ \alpha & a_2 \le x \le a_3 \\ \frac{\alpha(a_4-x)}{a_4-a_3} & a_3 \le x \le a_4 \\ 0 & otherwise \end{cases} \tag{5.15}$$

$$\eta_{\tilde{A}}(x) = \begin{cases} \frac{x-a_5}{a_6-a_5}\beta & a_5 \le x \le a_2 \\ \beta & a_2 \le x \le a_3 \\ \frac{a_6-x}{a_6-a_3}\beta & a_3 \le x \le a_6 \\ 0 & otherwise \end{cases} \tag{5.16}$$

$$\upsilon_{\tilde{A}}(x) = \begin{cases} \frac{a_2-x+\gamma(x-a_7)}{a_2-a_7} & a_7 \le x \le a_2 \\ \gamma & a_2 \le x \le a_3 \\ \frac{x-a_3+\gamma(a_8-x)}{a_8-a_3} & a_3 \le x < a_8 \\ 1 & otherwise \end{cases} \tag{5.17}$$

In these equations, the condition $a_7 \le a_5 \le a_1 \le a_2 \le a_3 \le a_4 \le a_6 \le a_8$ must be held. The values α, β, and γ describe the maximum degree of positive membership ($\mu_{\tilde{A}}$), maximum degree of neutral membership ($\eta_{\tilde{A}}$), and minimum degree of negative membership ($\upsilon_{\tilde{A}}$), respectively, such that $\alpha, \beta, \gamma \in [0, 1]$, and $0 \le \alpha + \beta + \gamma \le 1$.

Definition 5.22 The trapezoidal picture fuzzy number $\tilde{A} = ((a_1, a_2, a_3, a_4; \alpha), (a_5, a_2, a_3, a_6; \beta), (a_7, a_2, a_3, a_8; \gamma))$ is non-negative/non-positive, denoted as $\tilde{A} \ge 0 \left(\tilde{A} \le 0\right)$ if $a_9 \ge 0\,(a_{12} \le 0)$.

Definition 5.23 [1] The following operations can be performed on the trapezoidal picture fuzzy numbers $\tilde{A} = ((a_1, a_2, a_3, a_4; \alpha), (a_5, a_2, a_3, a_6; \beta), (a_7, a_2, a_3, a_8; \gamma))$ and $\tilde{B} = \left(\left(b_1, b_2, b_3, b_4; \alpha'\right), \left(b_5, b_2, b_3, b_6; \beta'\right), \left(b_7, b_2, b_3, b_8; \gamma'\right)\right)$, where λ is a real number.

$$\begin{aligned} \tilde{A} \oplus \tilde{B} = \Big(&\left(a_1 + b_1, a_2 + b_2, a_3 + b_3, a_4 + b_4; \alpha + \alpha' - \alpha\alpha'\right), \\ &\left(a_5 + b_5, a_2 + b_2, a_3 + b_3, a_6 + b_6; \beta\beta'\right), \\ &\left(a_7 + b_7, a_2 + b_2, a_3 + b_3, a_8 + b_8; \gamma\gamma'\right)\Big) \end{aligned} \tag{5.18}$$

$$\begin{aligned} -\tilde{A} = (&(-a_4, -a_3, -a_2, -a_1; \alpha), (-a_6, -a_3, -a_2, -a_5; \beta), \\ &(-a_8, -a_3, -a_2, -a_7; \gamma)) \end{aligned} \tag{5.19}$$

$$\begin{aligned} \tilde{A} \Theta \tilde{B} = \Big(&\left(a_1 - b_4, a_2 - b_3, a_3 - b_2, a_4 - b_1; \alpha + \alpha' - \alpha\alpha'\right), \\ &\left(a_5 - b_6, a_2 - b_3, a_3 - b_2, a_6 - b_5; \beta\beta'\right), \\ &\left(a_7 - b_8, a_2 - b_3, a_3 - b_2, a_8 - b_7; \gamma\gamma'\right)\Big) \end{aligned} \tag{5.20}$$

$$k\tilde{A} = \big(\big(ka_1, ka_2, ka_3, ka_4; \alpha^k\big), \big(ka_5, ka_2, ka_3, ka_6; \beta^k\big), \big(ka_7, ka_2, ka_3, ka_8; 1-(1-\gamma)^k\big)\big) \quad k \geq 0 \tag{5.21}$$

$$k\tilde{A} = \big(\big(ka_4, ka_3, ka_2, ka_1; \alpha^{-k}\big), \big(ka_6, ka_3, ka_2, ka_5; \beta^{-k}\big), \big(ka_8, ka_3, ka_2, ka_7; 1-(1-\gamma)^{-k}\big)\big) \quad k < 0 \tag{5.22}$$

$$\tilde{A} \otimes \tilde{B} = \big(\big(l_1, l_2, l_3, l_4; \alpha\alpha'\big), \big(l_5, l_2, l_3, l_6; \beta\beta'\big), \big(l_7, l_2, l_3, l_8; \gamma + \gamma' - \gamma\gamma'\big)\big). \tag{5.23}$$

where,

$$l_1 = \begin{cases} \min\{a_1b_1, a_4b_4\} & a_1 \geq 0 \\ \min\{a_1b_4, a_4b_1\} & a_1 < 0, a_4 \geq 0 \\ \min\{a_1b_4, a_4b_4\} & a_4 < 0 \end{cases},$$

$$l_2 = \begin{cases} \min\{a_2b_2, a_3b_2\} & a_2 \geq 0 \\ \min\{a_2b_3, a_3b_2\} & a_2 < 0, a_3 \geq 0 \\ \min\{a_2b_3, a_3b_3\} & a_3 < 0 \end{cases}$$

$$l_3 = \begin{cases} \max\{a_2b_3, a_3b_3\} & a_2 \geq 0 \\ \max\{a_2b_2, a_3b_3\} & a_2 < 0, b_3 \geq 0 \\ \max\{a_1b_4, a_4b_4\} & b_3 < 0 \end{cases}$$

$$l_4 = \begin{cases} \max\{a_1b_4, a_4b_4\} & a_1 \geq 0 \\ \max\{a_1b_1, a_4b_4\} & a_1 < 0, b_4 \geq 0 \\ \max\{a_1b_1, a_4b_4\} & a_4 < 0 \end{cases}$$

$$l_5 = \begin{cases} \min\{a_5b_5, a_6b_5\} & a_5 \geq 0 \\ \min\{a_5b_6, a_6b_5\} & a_5 < 0, a_6 \geq 0 \\ \min\{a_5b_6, a_6b_6\} & a_6 < 0 \end{cases}$$

$$l_6 = \begin{cases} \max\{a_5b_6, a_6b_6\} & a_5 \geq 0 \\ \max\{a_5b_5, a_6b_6\} & a_5 < 0, a_6 \geq 0 \\ \max\{a_5b_5, a_6b_6\} & a_6 < 0 \end{cases}$$

$$l_7 = \begin{cases} \min\{a_7b_7, a_8b_8\} & a_7 \geq 0 \\ \min\{a_7b_8, a_8b_7\} & a_7 < 0, a_8 \geq 0 \\ \min\{a_7b_8, a_8b_8\} & a_8 < 0 \end{cases}$$

$$l_8 = \begin{cases} \max\{a_7b_8, a_8b_8\} & a_7 \geq 0 \\ \max\{a_7b_8, a_8b_8\} & a_7 < 0, a_8 \geq 0 \\ \max\{a_7b_7, a_8b_8\} & a_8 < 0 \end{cases}$$

Lemma 5.1 *Assume the set of trapezoidal picture fuzzy numbers* $\tilde{A}_i = \left(\left(a_i^1, a_i^2, a_i^3, a_i^4; \alpha_i\right), \left(a_i^5, a_i^2, a_i^3, a_i^6; \beta_i\right), \left(a_i^7, a_i^2, a_i^3, a_i^8; \gamma_i\right)\right)$, *where* $i = 1, 2, \ldots, n$. *The sum of these values is obtained as,*

$$\sum_{i=1}^{n} \tilde{A}_i = \left(\left(\sum_{i=1}^{n} a_i^1, \sum_{i=1}^{n} a_i^2, \sum_{i=1}^{n} a_i^3, \sum_{i=1}^{n} a_i^4; \alpha\right), \left(\sum_{i=1}^{n} a_i^5, \sum_{i=1}^{n} a_i^2, \sum_{i=1}^{n} a_i^3, \sum_{i=1}^{n} a_i^6; \beta\right), \left(\sum_{i=1}^{n} a_i^7, \sum_{i=1}^{n} a_i^2, \sum_{i=1}^{n} a_i^3, \sum_{i=1}^{n} a_i^8; \gamma\right)\right)$$

where, $\alpha = \sum_{i=1}^{n} \alpha_i - \sum_{i<j}^{n} \alpha_i\alpha_j + \sum_{i<j<k}^{n} \alpha_i\alpha_j\alpha_k + \cdots + (-1)^{n-1} \prod_{i=1}^{n} \alpha_i$, $\beta = \prod_{i=1}^{n} \beta_i$, and $\gamma = \prod_{i=1}^{n} \gamma_i$.

Proof It can be done by the mathematical induction method on n.

5.4 Ranking of Picture Fuzzy Numbers

Definition 5.24 For the picture fuzzy number $\tilde{F} = \left(\mu_{\tilde{F}}, \eta_{\tilde{F}}, \nu_{\tilde{F}}\right)$, the below score and accuracy functions are defined.

- $S\left(\tilde{F}\right) = \mu_{\tilde{F}} - \eta_{\tilde{F}} - \nu_{\tilde{F}}, \quad S\left(\tilde{F}\right) \in [-1, 1]$,
- $A\left(\tilde{F}\right) = \mu_{\tilde{F}} + \eta_{\tilde{F}} + \nu_{\tilde{F}}, \quad A\left(\tilde{F}\right) \in [0, 1]$.

The picture fuzzy numbers $\tilde{E} = \left(\mu_{\tilde{E}}, \eta_{\tilde{E}}, \nu_{\tilde{E}}\right)$ and $\tilde{F} = \left(\mu_{\tilde{F}}, \eta_{\tilde{F}}, \nu_{\tilde{F}}\right)$ can be ranked according to the below rules.

If $S\left(\tilde{E}\right) > S\left(\tilde{F}\right) \Rightarrow \tilde{E} \succ \tilde{F}$,
If $S\left(\tilde{E}\right) < S\left(\tilde{F}\right) \Rightarrow \tilde{E} \prec \tilde{F}$,
If $S\left(\tilde{E}\right) = S\left(\tilde{F}\right)$, then.

- If $A\left(\tilde{E}\right) > A\left(\tilde{F}\right) \Rightarrow \tilde{E} \succ \tilde{F}$,
- If $A\left(\tilde{E}\right) < A\left(\tilde{F}\right) \Rightarrow \tilde{E} \prec \tilde{F}$,
- If $A\left(\tilde{E}\right) = A\left(\tilde{F}\right) \Rightarrow \tilde{E} \approx \tilde{F}$.

Definition 5.25 [8] For the picture fuzzy number $\tilde{F} = \left(\mu_{\tilde{F}}, \eta_{\tilde{F}}, \nu_{\tilde{F}}\right)$ the normalized score function $SN\left(\tilde{F}\right) = \frac{1}{3}\left(2 + \mu_{\tilde{F}} - \eta_{\tilde{F}} - \nu_{\tilde{F}}\right)$ is defined where $SN\left(\tilde{F}\right) \in [0, 1]$.

Definition 5.26 [6] For the picture fuzzy numbers $\tilde{E} = \left(\mu_{\tilde{E}}, \eta_{\tilde{E}}, \nu_{\tilde{E}}\right)$ and $\tilde{F} = \left(\mu_{\tilde{F}}, \eta_{\tilde{F}}, \nu_{\tilde{F}}\right)$ the below arithmetic operations are defined.

- $\tilde{E} \oplus \tilde{F} = \left(\mu_{\tilde{E}} + \mu_{\tilde{F}} - \mu_{\tilde{E}}\mu_{\tilde{F}}, \eta_{\tilde{E}}\eta_{\tilde{F}}, \nu_{\tilde{E}}\nu_{\tilde{F}}\right)$,

- $\tilde{E} \otimes \tilde{F} = \left(\mu_{\tilde{E}}\mu_{\tilde{F}}, \eta_{\tilde{E}} + \eta_{\tilde{F}} - \eta_{\tilde{E}}\eta_{\tilde{F}}, \nu_{\tilde{E}} + \nu_{\tilde{F}} - \nu_{\tilde{E}}\nu_{\tilde{F}}\right)$,
- $\tilde{E}^c = \left(\nu_{\tilde{E}}, \eta_{\tilde{E}}, \mu_{\tilde{E}}\right)$.

Definition 5.27 [6] For the picture fuzzy numbers $\tilde{E} = \left(\mu_{\tilde{E}}, \eta_{\tilde{E}}, \nu_{\tilde{E}}\right)$ and $\tilde{F} = \left(\mu_{\tilde{F}}, \eta_{\tilde{F}}, \nu_{\tilde{F}}\right)$, if $\mu_{\tilde{E}} \leq \mu_{\tilde{F}}$, $\eta_{\tilde{E}} \leq \eta_{\tilde{F}}$, and $\nu_{\tilde{E}} \leq \nu_{\tilde{F}}$, then, $\tilde{E} \subseteq \tilde{F}$.

Definition 5.28 [1] For the trapezoidal picture fuzzy number $\tilde{A} = (p_1, m, n, q_1; p_2, m, n, q_2; p_3, m, n, q_3)$, the below ranking function is defined.

$$\Re(\tilde{A}) = \frac{\alpha(p_1 + m + n + q_1) + \beta(p_2 + m + n + q_2) + (1 - \gamma)(p_3 + m + n + q_3)}{4}. \tag{5.24}$$

Definition 5.29 Centroid of the triangular picture fuzzy number $\tilde{A} = ((p_1, m, q_1; \alpha), (p_2, m, q_2; \beta), (p_3, m, q_3; \gamma))$ is obtained as below.

$$X_C(\tilde{A}) = \frac{1}{3}\left(\frac{\alpha k_1 + \beta k_2 + (1 - \gamma)\left(3\left(q_3^2 - p_3^2\right) + k_3\right)}{\alpha(q_1 - p_1) + \beta(q_2 - p_2) + (1 - \gamma)(q_3 - p_3)}\right), \tag{5.25}$$

Here, $k_1 = (q_1 - p_1)(p_1 + m + q_1)$, $k_2 = (q_2 - p_2)(p_2 + m + q_2)$, and $k_3 = (q_3 - p_3)(m - 2q_3 - 2p_3)$.

When $\beta = 0$, the centroid for the generalized intuitionistic fuzzy number $\tilde{A}^I = ((p_1, m, q_1; \alpha), (p_3, m, q_3; \gamma))$ is obtained as below.

$$X_C(\tilde{A}^I) = \frac{1}{3} \times \frac{\alpha(q_1 - p_1)(p_1 + m + q_1) + (1 - \gamma)\left(3\left(q_3^2 - p_3^2\right) + (q_3 - p_3)(m - 2q_3 - 2p_3)\right)}{\alpha(q_1 - p_1) + (1 - \gamma)(q_3 - p_3)}. \tag{5.26}$$

Definition 5.30 [15] For the triangular picture fuzzy number $\tilde{A} = ((p_1, m, q_1; \alpha), (p_2, m, q_2; \beta), (p_3, m, q_3; \gamma))$, the below ranking function is defined.

$$\Re(\tilde{A}) = X_C(\tilde{A})\left(1 + \alpha + \frac{\beta}{2} - \gamma\right). \tag{5.27}$$

The below rules are used to rank the triangular picture fuzzy numbers $\tilde{A}$ and $\tilde{B}$.

- If $\Re(\tilde{A}) > \Re(\tilde{B})$, then $\tilde{A} \succ \tilde{B}$,
- If $\Re(\tilde{A}) < \Re(\tilde{B})$, then $\tilde{A} \prec \tilde{B}$,
- If $\Re(\tilde{A}) = \Re(\tilde{B})$, then $\tilde{A} \approx \tilde{B}$,(in this case, the numbers cannot be ranked).

Example *5.5* [15] Consider the triangular picture fuzzy numbers $\tilde{A} = ((4, 5, 6; 0.512), (3, 5, 7; 0.208), (1, 5, 10; 0.28))$ and $\tilde{B} = ((4, 5, 6; 0.3), (3, 5, 7; 0.42), (1, 5, 10; 0.28))$. As $\Re(\tilde{A}) = 7.026$ and $\Re(\tilde{B}) = 6.453$, then $\tilde{A} > \tilde{B}$.

5.5 Interval-Valued Picture Fuzzy Sets

Definition 5.31 The interval-valued picture fuzzy set $\tilde{A}$ on X is shown as $\tilde{A} = \{(x, \mu_{\tilde{A}}(x), \eta_{\tilde{A}}(x), \nu_{\tilde{A}}(x)) | x \in X\}$, where,

$$\mu_{\tilde{A}} : X \to \text{int}([0,1]), \mu_{\tilde{A}}(x) = [\mu_{\tilde{A}}^L(x), \mu_{\tilde{A}}^U(x)] \in \text{int}([0,1]),$$
$$\eta_{\tilde{A}} : X \to \text{int}([0,1]), \eta_{\tilde{A}}(x) = [\eta_{\tilde{A}}^L(x), \eta_{\tilde{A}}^U(x)] \in \text{int}([0,1]),$$
$$\nu_{\tilde{A}} : X \to \text{int}([0,1]), \nu_{\tilde{A}}(x) = [\nu_{\tilde{A}}^L(x), \nu_{\tilde{A}}^U(x)] \in \text{int}([0,1]),$$

The condition $\sup \mu_{\tilde{A}}(x) + \sup \eta_{\tilde{A}}(x) + \sup \nu_{\tilde{A}}(x) \leq 1$ should be held for any $x \in X$. The set of all interval valued picture fuzzy sets on X is shown by $IVPFS(X)$. Therefore, the interval-valued picture fuzzy set $\tilde{A}$ on X is represented by $\tilde{A} = \left\{\left(x, \left[\mu_{\tilde{A}}^L(x), \mu_{\tilde{A}}^U(x)\right], \left[\eta_{\tilde{A}}^L(x), \eta_{\tilde{A}}^U(x)\right], \left[\nu_{\tilde{A}}^L(x), \nu_{\tilde{A}}^U(x)\right]\right) \middle| x \in X\right\}$, where its hesitancy degree is calculated as below.

$$\tilde{\pi}_{\tilde{A}}(x) = \left[1 - \mu_{\tilde{A}}^U(x) - \eta_{\tilde{A}}^U(x) - \nu_{\tilde{A}}^U(x), 1 - \mu_{\tilde{A}}^L(x) - \eta_{\tilde{A}}^L(x) - \nu_{\tilde{A}}^L(x)\right]$$

Definition 5.32 For the interval-valued picture fuzzy sets $\tilde{A} = \left\{\left(x, \left[\mu_{\tilde{A}}^L(x), \mu_{\tilde{A}}^U(x)\right], \left[\eta_{\tilde{A}}^L(x), \eta_{\tilde{A}}^U(x)\right], \left[\nu_{\tilde{A}}^L(x), \nu_{\tilde{A}}^U(x)\right]\right) \middle| x \in X\right\}$ and $\tilde{B} = \{(x, [\mu_{\tilde{B}}^L(x), \mu_{\tilde{B}}^U(x)], [\eta_{\tilde{B}}^L(x), \eta_{\tilde{B}}^U(x)], [\nu_{\tilde{B}}^L(x), \nu_{\tilde{B}}^U(x)]) | x \in X\}$ on the set X, the below logics are defined.

- $\tilde{A} \subseteq \tilde{B}$, if and only if for all $x \in X$,$\mu_{\tilde{A}}^L(x) \leq \mu_{\tilde{B}}^L(x)$, $\mu_{\tilde{A}}^U(x) \leq \mu_{\tilde{B}}^U(x)$, $\eta_{\tilde{A}}^L(x) \leq \eta_{\tilde{B}}^L(x)$, $\eta_{\tilde{A}}^U(x) \leq \eta_{\tilde{B}}^U(x)$, $\nu_{\tilde{A}}^L(x) \geq \nu_{\tilde{B}}^L(x)$, and $\nu_{\tilde{A}}^U(x) \leq \nu_{\tilde{B}}^U(x)$. Also $\tilde{A} = \tilde{B}$ if and only if $\tilde{A} \subseteq \tilde{B}$ and $\tilde{B} \subseteq \tilde{A}$,
- $\tilde{A} \cup \tilde{B} = \left\{\left(\begin{array}{l} x, \max\{\mu_{\tilde{A}}^L(x), \mu_{\tilde{B}}^L(x)\}, \max\{\mu_{\tilde{A}}^U(x), \mu_{\tilde{B}}^U(x)\}, \\ \min\{\eta_{\tilde{A}}^L(x) \leq \eta_{\tilde{B}}^L(x)\}, \min\{\eta_{\tilde{A}}^U(x) \leq \eta_{\tilde{B}}^U(x)\}, \\ \min\{\nu_{\tilde{A}}^L(x) \leq \nu_{\tilde{B}}^L(x)\}, \min\{\nu_{\tilde{A}}^U(x) \leq \nu_{\tilde{B}}^U(x)\} \end{array}\right) \middle| x \in X\right\}$,
- $\tilde{A} \cap \tilde{B} = \left\{\left(\begin{array}{l} x, \min\{\mu_{\tilde{A}}^L(x), \mu_{\tilde{B}}^L(x)\}, \min\{\mu_{\tilde{A}}^U(x), \mu_{\tilde{B}}^U(x)\}, \\ \min\{\eta_{\tilde{A}}^L(x) \leq \eta_{\tilde{B}}^L(x)\}, \min\{\eta_{\tilde{A}}^U(x) \leq \eta_{\tilde{B}}^U(x)\}, \\ \max\{\nu_{\tilde{A}}^L(x) \leq \nu_{\tilde{B}}^L(x)\}, \max\{\nu_{\tilde{A}}^U(x) \leq \nu_{\tilde{B}}^U(x)\} \end{array}\right) \middle| x \in X\right\}$
- $\tilde{A}^c = \{(x, \nu_{\tilde{A}}(x), \eta_{\tilde{A}}(x), \mu_{\tilde{A}}(x)) | x \in X\}$.

Definition 5.33 [5] Assume the interval-valued picture fuzzy sets $\tilde{A} = \left\{\left(x, \left[\mu_{\tilde{A}}^L(x), \mu_{\tilde{A}}^U(x)\right], \left[\eta_{\tilde{A}}^L(x), \eta_{\tilde{A}}^U(x)\right], \left[\nu_{\tilde{A}}^L(x), \nu_{\tilde{A}}^U(x)\right]\right) \middle| x \in X\right\}$ and $\tilde{B} = \{(x, [\mu_{\tilde{B}}^L(x), \mu_{\tilde{B}}^U(x)], [\eta_{\tilde{B}}^L(x), \eta_{\tilde{B}}^U(x)], [\nu_{\tilde{B}}^L(x), \nu_{\tilde{B}}^U(x)]) | x \in X\}$ on the reference sets X_1 and X_2 respectively. Then, their Cartesian product of these numbers is obtained by the below formulas.

- $\tilde{A} \times_1 \tilde{B} = \left\{ \begin{pmatrix} (x, y), \left(\left[\mu_{\tilde{A}}^L(x).\mu_{\tilde{B}}^L(y), \mu_{\tilde{A}}^U(x).\mu_{\tilde{B}}^L(y) \right] \right), \\ \left(\left[\eta_{\tilde{A}}^L(x).\eta_{\tilde{B}}^L(y), \eta_{\tilde{A}}^U(x).\eta_{\tilde{B}}^L(y) \right] \right) \\ \left(\left[\nu_{\tilde{A}}^L(x).\nu_{\tilde{B}}^L(y), \nu_{\tilde{A}}^U(x).\nu_{\tilde{B}}^L(y) \right] \right) \end{pmatrix} \middle| x \in X_1, y \in X_2 \right\}$
- $\tilde{A} \times_2 \tilde{B} = \left\{ \begin{pmatrix} (x, y), \\ \left(\left[\min\left\{ \mu_{\tilde{A}}^L(x), \mu_{\tilde{B}}^L(y) \right\}, \min\left\{ \mu_{\tilde{A}}^U(x), \mu_{\tilde{B}}^L(y) \right\} \right] \right), \\ \left(\left[\min\left\{ \eta_{\tilde{A}}^L(x), \eta_{\tilde{B}}^L(y) \right\}, \min\left\{ \eta_{\tilde{A}}^U(x), \eta_{\tilde{B}}^L(y) \right\} \right] \right) \\ \left(\left[\max\left\{ \nu_{\tilde{A}}^L(x), \nu_{\tilde{B}}^L(y) \right\}, \max\left\{ \nu_{\tilde{A}}^U(x), \nu_{\tilde{B}}^L(y) \right\} \right] \right) \end{pmatrix} \middle| x \in X_1, y \in X_2 \right\}$

Definition 5.34 For the interval-valued picture fuzzy sets $\tilde{A}$ and $\tilde{B}$ on the reference set X and the real number $\lambda \in [0, 1]$, the convex combination of $\tilde{C}_\lambda(\tilde{A}, \tilde{B}) = \left\{ \left(x, \mu_{\tilde{C}_\lambda}(x), \eta_{\tilde{C}_\lambda}(x), \nu_{\tilde{C}_\lambda}(x) \right) \middle| x \in X \right\}$ is defined, where for any $x \in X$,

$$\begin{aligned} \mu_{\tilde{C}_\lambda}(x) &= \left[\mu_{\tilde{C}_\lambda}^L(x), \mu_{\tilde{C}_\lambda}^U(x) \right] \\ &= \left[\lambda\, \mu_{\tilde{A}}^L(x) + (1-\lambda)\, \mu_{\tilde{B}}^L(x), \lambda\, \mu_{\tilde{A}}^U(x) + (1-\lambda)\, \mu_{\tilde{B}}^U(x) \right] \\ \eta_{\tilde{C}_\lambda}(x) &= \left[\eta_{\tilde{C}_\lambda}^L(x), \eta_{\tilde{C}_\lambda}^U(x) \right] \\ &= \left[\lambda \eta_{\tilde{A}}^L(x) + (1-\lambda)\, \eta_{\tilde{B}}^L(x), \lambda\, \eta_{\tilde{A}}^U(x) + (1-\lambda)\, \eta_{\tilde{B}}^U(x) \right] \\ \nu_{\tilde{C}_\lambda}(x) &= \left[\nu_{\tilde{C}_\lambda}^L(x), \nu_{\tilde{C}_\lambda}^U(x) \right] \\ &= \left[\lambda \nu_{\tilde{A}}^L(x) + (1-\lambda)\, \nu_{\tilde{B}}^L(x), \lambda \nu_{\tilde{A}}^U(x) + (1-\lambda)\, \nu_{\tilde{B}}^U(x) \right] \end{aligned}$$

Corollary 5.3. *Assume the interval-valued picture fuzzy sets $\tilde{A}$, $\tilde{B}$, and $\tilde{D}$ on the reference set X and the real number $\lambda \in [0, 1]$, then,*

- *If $\lambda = 1$, then $\tilde{C}_\lambda(\tilde{A}, \tilde{B}) = \tilde{A}$, and if $\lambda = 0$, then $\tilde{C}_\lambda(\tilde{A}, \tilde{B}) = \tilde{B}$.*
- *If $\tilde{A} \subseteq \tilde{B}$, then $\tilde{A} \leq \tilde{C}_\lambda(\tilde{A}, \tilde{B}) \leq \tilde{B}$ (for any $\lambda \in [0, 1]$).*
- If $\tilde{B} \subseteq \tilde{A}$ *and $\lambda_1 \geq \lambda_2$, then $\tilde{C}_{\lambda_2}(\tilde{A}, \tilde{B}) \subseteq \tilde{C}_{\lambda_1}(\tilde{A}, \tilde{B})$.*
- $\tilde{C}_\lambda(\tilde{A} \cap \tilde{B}, \tilde{D}) = \tilde{C}_\lambda(\tilde{A}, \tilde{D}) \cap \tilde{C}_\lambda(\tilde{B}, \tilde{D})$.
- $\tilde{C}_\lambda(\tilde{A} \cup \tilde{B}, \tilde{D}) = \tilde{C}_\lambda(\tilde{A}, \tilde{D}) \cup \tilde{C}_\lambda(\tilde{B}, \tilde{D})$.

Definition 5.35 For the interval-valued picture fuzzy number $\tilde{A} = \left\{ \left[\mu_{\tilde{A}}^L, \mu_{\tilde{A}}^U \right], \left[\eta_{\tilde{A}}^L, \eta_{\tilde{A}}^U \right], \left[\nu_{\tilde{A}}^L, \nu_{\tilde{A}}^U \right] \right\}$, the below score function ($S(\tilde{A})$) and accuracy functions ($A(\tilde{A})$ and $G(\tilde{A})$) are defined.

$$S(\tilde{A}) = \frac{\mu_{\tilde{A}}^L + \mu_{\tilde{A}}^U - \eta_{\tilde{A}}^L - \eta_{\tilde{A}}^U - \nu_{\tilde{A}}^L - \nu_{\tilde{A}}^U}{2} \tag{5.28}$$

$$A(\tilde{A}) = \frac{\mu_{\tilde{A}}^L + \mu_{\tilde{A}}^U + \eta_{\tilde{A}}^L + \eta_{\tilde{A}}^U + \nu_{\tilde{A}}^L + \nu_{\tilde{A}}^U}{2} \tag{5.29}$$

$$G(\tilde{A}) = \mu_{\tilde{A}}^U + \eta_{\tilde{A}}^U + \nu_{\tilde{A}}^U + \mu_{\tilde{A}}^L + \eta_{\tilde{A}}^L + \nu_{\tilde{A}}^L \tag{5.30}$$

Now, for two interval-valued picture fuzzy numbers $\tilde{A}$ and $\tilde{B}$ the below rankings are obtained.

- If $S(\tilde{A}) > S(\tilde{B})$, then $\tilde{A} \succ \tilde{B}$,
- If $S(\tilde{A}) < S(\tilde{B})$, then $\tilde{A} \prec \tilde{B}$,
- If $S(\tilde{A}) = S(\tilde{B})$,
- If $A(\tilde{A}) > A(\tilde{B})$, then $\tilde{A} \succ \tilde{B}$,
- If $A(\tilde{A}) < A(\tilde{B})$, then $\tilde{A} \prec \tilde{B}$,
- If $A(\tilde{A}) = A(\tilde{B})$,
- If $G(\tilde{A}) > G(\tilde{B})$, then $\tilde{A} \succ \tilde{B}$,
- If $G(\tilde{A}) < G(\tilde{B})$, then $\tilde{A} \prec \tilde{B}$,
- If $G(\tilde{A}) = G(\tilde{B})$, then $\tilde{A} \approx \tilde{B}$.

5.5.1 Cubic Picture Fuzzy Sets

Definition 5.36 [2] Assume the reference set $X \neq \emptyset$. The set $\tilde{A} = \left\{ \left(x, \tilde{J}_{\tilde{A}}(x), \tilde{I}_{\tilde{A}}(x)\right) \middle| x \in X \right\}$ is called a cubic picture fuzzy set on X, where, $\tilde{J}_{\tilde{A}}(x) = \left\{ \left(x, \left[\mu^L_{\tilde{A}}(x), \mu^U_{\tilde{A}}(x)\right], \left[\eta^L_{\tilde{A}}(x), \eta^U_{\tilde{A}}(x)\right], \left[\nu^L_{\tilde{A}}(x), \nu^U_{\tilde{A}}(x)\right]\right) \middle| x \in X \right\}$ is an interval-valued picture fuzzy set on X and $\tilde{I}_{\tilde{A}}(x) = \left\{ \left(x, \mu_{\tilde{A}}(x), \eta_{\tilde{A}}(x), \nu_{\tilde{A}}(x)\right) \middle| x \in X \right\}$ is a picture fuzzy set on X. For more simplicity, a cubic picture fuzzy set is shown as $\tilde{A} = \langle \tilde{J}_{\tilde{A}}, \tilde{I}_{\tilde{A}} \rangle$.

Example 5.6 [2] Assume the reference set $X = \{x_1, x_2, x_3\}$. Now the below cubic picture fuzzy set is represented.

$$\tilde{A} = \left\{ \begin{array}{l|l} (x_1, [(0.3, 0.4), (0.0, 0.1), (0.4, 0.4)], (0.5, 0.1, 0.3)\,), & \\ (x_2, [(0.2, 0.3), (0.3, 0.4), (0.1, 0.2)], (0.4, 0.4, 0.1)\,), & x_1, x_2, x_3 \in X \\ (x_3, [(0.1, 0.3), (0.3, 0.3), (0.2, 0.3)], (0.2, 0.1, 0.6)\,) & \end{array} \right\}.$$

Definition 5.37 [2] Assume the reference set $X \neq \emptyset$. Furthermore, consider the cubic picture fuzzy sets $\tilde{A} = \langle \tilde{J}_{\tilde{A}}, \tilde{I}_{\tilde{A}} \rangle$ and $\tilde{B} = \langle \tilde{J}_{\tilde{B}}, \tilde{I}_{\tilde{B}} \rangle$, where,

$$\begin{aligned}
\tilde{J}_{\tilde{A}}(x) &= \left\{ \left(x, \left[\mu^L_{\tilde{A}}(x), \mu^U_{\tilde{A}}(x)\right], \left[\eta^L_{\tilde{A}}(x), \eta^U_{\tilde{A}}(x)\right], \left[\nu^L_{\tilde{A}}(x), \nu^U_{\tilde{A}}(x)\right]\right) \middle| x \in X \right\}, \\
\tilde{I}_{\tilde{A}}(x) &= \left\{ \left(x, \mu_{\tilde{A}}(x), \eta_{\tilde{A}}(x), \nu_{\tilde{A}}(x)\right) \middle| x \in X \right\}. \\
\tilde{J}_{\tilde{B}}(x) &= \left\{ \left(x, \left[\mu^L_{\tilde{B}}(x), \mu^U_{\tilde{B}}(x)\right], \left[\eta^L_{\tilde{B}}(x), \eta^U_{\tilde{B}}(x)\right], \left[\nu^L_{\tilde{B}}(x), \nu^U_{\tilde{B}}(x)\right]\right) \middle| x \in X \right\}, \\
\tilde{I}_{\tilde{B}}(x) &= \left\{ \left(x, \mu_{\tilde{B}}(x), \eta_{\tilde{B}}(x), \nu_{\tilde{B}}(x)\right) \middle| x \in X \right\}.
\end{aligned}$$

Then, the below logics are defined.

- $\tilde{A} \subseteq \tilde{B}$ if and only if for all $x \in X$, the conditions $\nu_{\tilde{A}}^{L}(x) \geq \nu_{\tilde{B}}^{L}(x)$, $\nu_{\tilde{A}}^{U}(x) \geq \nu_{\tilde{B}}^{U}(x)$, $\mu_{\tilde{A}}(x) \leq \mu_{\tilde{B}}(x)$, $\eta_{\tilde{A}}(x) \geq \eta_{\tilde{B}}(x)$, $\nu_{\tilde{A}}(x) \geq \nu_{\tilde{B}}(x)$, $\mu_{\tilde{A}}^{L}(x) \leq \mu_{\tilde{B}}^{L}(x)$, $\mu_{\tilde{A}}^{U}(x) \leq \mu_{\tilde{B}}^{U}(x)$, $\eta_{\tilde{A}}^{L}(x) \geq \eta_{\tilde{B}}^{L}(x)$, and $\eta_{\tilde{A}}^{U}(x) \geq \eta_{\tilde{B}}^{U}(x)$ are held.
- $\tilde{A} \cup \tilde{B} = \left\{ \left(x, \tilde{J}_{\tilde{A}\cup\tilde{B}}(x), \tilde{I}_{\tilde{A}\cup\tilde{B}}(x)\right) \middle| x \in X \right\}$, where,

$$\tilde{J}_{\tilde{A}\cup\tilde{B}}(x) = \left\{ \left(\begin{array}{l} x, \left[\max\left\{\mu_{\tilde{A}}^{L}(x), \mu_{\tilde{B}}^{L}(x)\right\}, \max\left\{\mu_{\tilde{A}}^{U}(x), \mu_{\tilde{B}}^{U}(x)\right\}\right], \\ \left[\min\left\{\eta_{\tilde{A}}^{L}(x), \eta_{\tilde{B}}^{L}(x)\right\}, \min\left\{\eta_{\tilde{A}}^{U}(x), \eta_{\tilde{B}}^{U}(x)\right\}\right], \\ \left[\min\left\{\nu_{\tilde{A}}^{L}(x), \nu_{\tilde{B}}^{L}(x)\right\}, \min\left\{\nu_{\tilde{A}}^{U}(x), \nu_{\tilde{B}}^{U}(x)\right\}\right] \end{array} \right) \middle| x \in X \right\},$$

and

$$\tilde{I}_{\tilde{A}\cup\tilde{B}}(x) = \left\{\left(x, \max\left\{\mu_{\tilde{A}}(x), \mu_{\tilde{B}}(x)\right\}, \min\left\{\eta_{\tilde{A}}(x), \eta_{\tilde{B}}(x)\right\}, \min\left\{\nu_{\tilde{A}}(x), \nu_{\tilde{B}}(x)\right\}\right)\right\},$$

- $\tilde{A} \cap \tilde{B} = \left\{ \left(x, \tilde{J}_{\tilde{A}\cap\tilde{B}}(x), \tilde{I}_{\tilde{A}\cap\tilde{B}}(x)\right) \middle| x \in X \right\}$, where,

$$\tilde{J}_{\tilde{A}\cap\tilde{B}}(x) = \left\{ \left(\begin{array}{l} x, \left[\min\left\{\mu_{\tilde{A}}^{L}(x), \mu_{\tilde{B}}^{L}(x)\right\}, \min\left\{\mu_{\tilde{A}}^{U}(x), \mu_{\tilde{B}}^{U}(x)\right\}\right], \\ \left[\min\left\{\eta_{\tilde{A}}^{L}(x), \eta_{\tilde{B}}^{L}(x)\right\}, \min\left\{\eta_{\tilde{A}}^{U}(x), \eta_{\tilde{B}}^{U}(x)\right\}\right], \\ \left[\max\left\{\nu_{\tilde{A}}^{L}(x), \nu_{\tilde{B}}^{L}(x)\right\}, \max\left\{\nu_{\tilde{A}}^{U}(x), \nu_{\tilde{B}}^{U}(x)\right\}\right] \end{array} \right) \middle| x \in X \right\},$$

and

$$\tilde{I}_{\tilde{A}\cap\tilde{B}}(x) = \left\{\left(x, \min\left\{\mu_{\tilde{A}}(x), \mu_{\tilde{B}}(x)\right\}, \min\left\{\eta_{\tilde{A}}(x), \eta_{\tilde{B}}(x)\right\}, \max\left\{\nu_{\tilde{A}}(x), \nu_{\tilde{B}}(x)\right\}\right)\right\},$$

- $\tilde{A}^{c} = \left\langle \tilde{J}_{\tilde{A}^{c}}, \tilde{I}_{\tilde{A}^{c}} \right\rangle$ where,

$$\tilde{J}_{\tilde{A}^{c}} = \left\{ \left(x, \left[\nu_{\tilde{A}}^{L}(x), \nu_{\tilde{A}}^{U}(x)\right], \left[\eta_{\tilde{A}}^{L}(x), \eta_{\tilde{A}}^{U}(x)\right], \left[\mu_{\tilde{A}}^{L}(x), \mu_{\tilde{A}}^{U}(x)\right]\right) \middle| x \in X \right\},$$

and

$$\tilde{I}_{\tilde{A}^{c}} = \left\{ \left(x, \nu_{\tilde{A}}(x), \eta_{\tilde{A}}(x), \mu_{\tilde{A}}(x)\right) \middle| x \in X \right\}.$$

Example 5.7 [2] Assume the set $X = \{x_1, x_2, x_3\}$ and the cubic picture fuzzy sets $\tilde{A} = \left\langle \tilde{J}_{\tilde{A}}, \tilde{I}_{\tilde{A}} \right\rangle$ and $\tilde{B} = \left\langle \tilde{J}_{\tilde{B}}, \tilde{I}_{\tilde{B}} \right\rangle$ defined by Table 5.1.

Now, the logics of Definition 5.37 for these two cubic picture fuzzy sets are obtained and reported by Tables 5.2, 5.3, and 5.4.

Definition 5.38 [2] Assume the reference set $X \neq \emptyset$. Furthermore, consider the cubic picture fuzzy sets $\tilde{A} = \left\langle \tilde{J}_{\tilde{A}}, \tilde{I}_{\tilde{A}} \right\rangle$ and $\tilde{B} = \left\langle \tilde{J}_{\tilde{B}}, \tilde{I}_{\tilde{B}} \right\rangle$, where,

$$\tilde{J}_{\tilde{A}}(x) = \left\{ \left(x, \left[M_{\tilde{A}}^{L}(x), M_{\tilde{A}}^{U}(x)\right], \left[L_{\tilde{A}}^{L}(x), L_{\tilde{A}}^{U}(x)\right], \left[N_{\tilde{A}}^{L}(x), N_{\tilde{A}}^{U}(x)\right]\right) \middle| x \in X \right\},$$

Table 5.1 Data of the cubic fuzzy sets of Example 5.7

X	$\tilde{J}_{\tilde{A}}$	$\tilde{I}_{\tilde{A}}$
x_1	([0.20, 0.30], [0.15, 0.26], [0.39, 0.41])	(0.51, 0.08, 0.40)
x_2	([0.14, 0.28], [0.29, 0.39], [0.18, 0.28])	(0.31, 0.43, 0.24)
x_3	([0.41, 0.43], [0.09, 0.18], [0.27, 0.33])	(0.26, 0.27, 0.28)
X	$\tilde{J}_{\tilde{B}}$	$\tilde{I}_{\tilde{B}}$
x_1	([0.20, 0.33], [0.05, 0.16], [0.39, 0.41])	(0.35, 0.19, 0.40)
x_2	([0.11, 0.29], [0.30, 0.40], [0.15, 0.25])	(0.32, 0.29, 0.24)
x_3	([0.33, 0.39], [0.15, 0.23], [0.27, 0.29])	(0.42, 0.27, 0.28)

Table 5.2 The values of $\tilde{A} \cup \tilde{B}$ for Example 5.7

X	$\tilde{J}_{\tilde{A}\cup\tilde{B}}$	$\tilde{I}_{\tilde{A}\cup\tilde{B}}$
x_1	([0.20, 0.33], [0.15, 0.26], [0.39, 0.41])	(0.51, 0.19, 0.40)
x_2	([0.14, 0.29], [0.30, 0.40], [0.18, 0.28])	(0.32, 0.43, 0.24)
x_3	([0.41, 0.43], [0.15, 0.23], [0.27, 0.33])	(0.42, 0.27, 0.28)

Table 5.3 The values of $\tilde{A} \cap \tilde{B}$ for Example 5.7

X	$\tilde{J}_{\tilde{A}\cap\tilde{B}}$	$\tilde{I}_{\tilde{A}\cap\tilde{B}}$
x_1	([0.20, 0.30], [0.05, 0.16], [0.39, 0.41])	(0.35, 0.08, 0.40)
x_2	([0.11, 0.28], [0.29, 0.39], [0.15, 0.25])	(0.13, 0.29, 0.24)
x_3	([0.33, 0.39], [0.09, 0.18], [0.27, 0.29])	(0.26, 0.27, 0.28)

Table 5.4 The values of $\tilde{A}^c$ for Example 5.7

X	$\tilde{J}_{\tilde{A}^c}$	$\tilde{I}_{\tilde{A}^c}$
x_1	([0.39, 0.41], [0.15, 0.26], [0.20, 0.30])	(0.40, 0.08, 0.51)
x_2	([0.18, 0.28], [0.29, 0.39], [0.14, 0.28])	(0.24, 0.43, 0.13)
x_3	([0.27, 0.33], [0.09, 0.18], [0.41, 0.43])	(0.28, 0.27, 0.26)

$$\tilde{I}_{\tilde{A}}(x) = \left\{ \left(x, \mu_{\tilde{A}}(x), \eta_{\tilde{A}}(x), \nu_{\tilde{A}}(x)\right) \middle| x \in X \right\}$$

$$\tilde{J}_{\tilde{B}}(x) = \left\{ \left(x, \left[M^L_{\tilde{B}}(x), M^U_{\tilde{B}}(x)\right], \left[L^L_{\tilde{B}}(x), L^U_{\tilde{B}}(x)\right], \left[N^L_{\tilde{B}}(x), N^U_{\tilde{B}}(x)\right]\right) \middle| x \in X \right\},$$

$$\tilde{I}_{\tilde{B}}(x) = \left\{ \left(x, \mu_{\tilde{B}}(x), \eta_{\tilde{B}}(x), \nu_{\tilde{B}}(x)\right) \middle| x \in X \right\}$$

Then, the below arithmetic operations are defined.

- $\tilde{A} \oplus \tilde{B} = \left\langle \tilde{J}_{\tilde{A}} \oplus \tilde{J}_{\tilde{B}}, \tilde{I}_{\tilde{A}} \oplus \tilde{I}_{\tilde{B}} \right\rangle$

  where

$$\begin{aligned}\tilde{J}_{\tilde{A}} \oplus \tilde{J}_{\tilde{B}} = \big(x|\big[&\mu_{\tilde{A}}^L(x) + \mu_{\tilde{B}}^L(x) - \mu_{\tilde{A}}^L(x)\,\mu_{\tilde{B}}^L(x),\\ &\mu_{\tilde{A}}^U(x) + \mu_{\tilde{B}}^U(x) - \mu_{\tilde{A}}^U(x)\,\mu_{\tilde{B}}^U(x)\big],\\ &\big[\eta_{\tilde{A}}^L(x)\ \eta_{\tilde{B}}^L(x), \eta_{\tilde{A}}^U(x)\ \eta_{\tilde{B}}^U(x)\big],\\ &\big[\nu_{\tilde{A}}^L(x)\ \nu_{\tilde{B}}^L(x), \nu_{\tilde{A}}^U(x)\ \nu_{\tilde{B}}^U(x)\big]\big)\end{aligned}$$

and

$$\begin{aligned}\tilde{I}_{\tilde{A}} \oplus \tilde{I}_{\tilde{B}} = \big\{x|\big(&\mu_{\tilde{A}}(x) + \mu_{\tilde{B}}(x)\\ &-\mu_{\tilde{A}}(x)\,\mu_{\tilde{B}}(x), \eta_{\tilde{A}}(x)\,\eta_{\tilde{B}}(x), \nu_{\tilde{A}}(x)\,\nu_{\tilde{B}}(x)\big)\big\}\end{aligned}$$

- $\tilde{A} \otimes \tilde{B} = \langle \tilde{J}_{\tilde{A}} \otimes \tilde{J}_{\tilde{B}}, \tilde{I}_{\tilde{A}} \otimes \tilde{I}_{\tilde{B}} \rangle$

  where

$$\begin{aligned}\tilde{J}_{\tilde{A}} \otimes \tilde{J}_{\tilde{B}} = \big(x|&\big[\mu_{\tilde{A}}^L(x)\,\mu_{\tilde{B}}^L(x), \mu_{\tilde{A}}^U(x)\,\mu_{\tilde{B}}^U(x)\big], \big[\eta_{\tilde{A}}^L(x)\,\eta_{\tilde{B}}^L(x), \eta_{\tilde{A}}^U(x)\,\eta_{\tilde{B}}^U(x)\big],\\ &\big[\nu_{\tilde{A}}^L(x) + \nu_{\tilde{B}}^L(x) - \nu_{\tilde{A}}^L(x)\,\nu_{\tilde{B}}^L(x), \nu_{\tilde{A}}^U(x) + \nu_{\tilde{B}}^U(x) - \nu_{\tilde{A}}^U(x)\,\nu_{\tilde{B}}^U(x)\big]\big)\end{aligned}$$

and

$$\tilde{I}_{\tilde{A}} \otimes \tilde{I}_{\tilde{B}} = \big\{x|\big(\mu_{\tilde{A}}(x)\mu_{\tilde{B}}(x), \eta_{\tilde{A}}(x)\eta_{\tilde{B}}(x), \nu_{\tilde{A}}(x) + \nu_{\tilde{B}}(x) - \nu_{\tilde{A}}(x)\nu_{\tilde{B}}(x)\big)\big\}$$

- $k\,\tilde{A} = \langle k\,\tilde{J}_{\tilde{A}}, k\,\tilde{I}_{\tilde{B}} \rangle,\ k \geq 0,$

  where

$$\begin{aligned}k\,\tilde{J}_{\tilde{A}} = \Big\{x|&\Big[1 - \big(1 - \mu_{\tilde{A}}^L(x)\big)^k, 1 - \big(1 - \mu_{\tilde{A}}^U(x)\big)^k\Big],\\ &\Big[\big(\eta_{\tilde{A}}^L(x)\big)^k, \big(\eta_{\tilde{A}}^U(x)\big)^k\Big], \Big[\big(\nu_{\tilde{A}}^L(x)\big)^k, \big(\nu_{\tilde{A}}^L(x)\big)^k\Big]\Big\}\end{aligned}$$

and

$$k\,\tilde{I}_{\tilde{B}} = \Big\{x|1 - \big(1 - \mu_{\tilde{B}}(x)\big)^k, k\,\eta_{\tilde{B}}(x), k\,\nu_{\tilde{B}}(x)\Big\}$$

- $\tilde{A}^k = \Big\langle \tilde{J}_{\tilde{A}}^k, \tilde{I}_{\tilde{A}}^k \Big\rangle,\ k \geq 0,$

where

$$\begin{aligned}\tilde{J}_{\tilde{A}}^k = \Big\{x|&\Big[\big(\mu_{\tilde{A}}^L(x)\big)^k, \big(\mu_{\tilde{A}}^L(x)\big)^k\Big], \Big[\big(\eta_{\tilde{A}}^L(x)\big)^k, \big(\eta_{\tilde{A}}^U(x)\big)^k\Big],\\ &\Big[1 - \big(1 - \nu_{\tilde{A}}^L(x)\big)^k, 1 - \big(1 - \nu_{\tilde{A}}^L(x)\big)^k\Big]\Big\}\end{aligned}$$

and

$$\tilde{I}_{\tilde{A}}^{k} = \left\{ x | \left(\mu_{\tilde{B}}(x)\right)^{k}, \left(\eta_{\tilde{B}}(x)\right)^{k}, 1 - \left(1 - \nu_{\tilde{B}}(x)\right)^{k} \right\}$$

Definition 5.39 [2] for the cubic picture fuzzy number $\tilde{A} = \langle \tilde{J}_{\tilde{A}}, \tilde{I}_{\tilde{A}} \rangle$ the below score and accuracy functions are defined.

$$S(\tilde{A}) = \frac{\mu_{\tilde{A}}^{L} + \mu_{\tilde{A}}^{U} - \eta_{\tilde{A}}^{L} - \eta_{\tilde{A}}^{U} - \nu_{\tilde{A}}^{L} - \nu_{\tilde{A}}^{U} + \mu_{\tilde{A}} - \eta_{\tilde{A}} - \nu_{\tilde{A}} + 3}{6}, \tag{5.31}$$

$$A(\tilde{A}) = \frac{\mu_{\tilde{A}}^{L} + \mu_{\tilde{A}}^{U} - \eta_{\tilde{A}}^{L} - \eta_{\tilde{A}}^{U} + \mu_{\tilde{A}} - \nu_{\tilde{A}}}{6}. \tag{5.32}$$

For two cubic picture fuzzy numbers $\tilde{A} = \langle \tilde{J}_{\tilde{A}}, \tilde{I}_{\tilde{A}} \rangle$ and $\tilde{B} = \langle \tilde{J}_{\tilde{B}}, \tilde{I}_{\tilde{B}} \rangle$, the below rankings are obtained.

- If $S(\tilde{A}) > S(\tilde{B}) \Rightarrow \tilde{A} \succ \tilde{B}$,
- If $S(\tilde{A}) < S(\tilde{B}) \Rightarrow \tilde{A} \prec \tilde{B}$,
- If $S(\tilde{A}) = S(\tilde{B})$,

$$\begin{aligned} A(\tilde{A}) > A(\tilde{B}) &\Rightarrow \tilde{A} \succ \tilde{B}, \\ A(\tilde{A}) < A(\tilde{B}) &\Rightarrow \tilde{A} \prec \tilde{B}, \\ A(\tilde{A}) > A(\tilde{B}) &\Rightarrow \tilde{A} \approx \tilde{B}. \end{aligned}$$

Definition 5.40 [2] The cubic picture fuzzy weighted geometric aggregator (CPFWGA) for the cubic picture fuzzy numbers $\tilde{A}_i = \langle \tilde{J}_{\tilde{A}_i}, \tilde{I}_{\tilde{A}_i} \rangle$, $i = 1, 2, \ldots, n$, is defined as below.

$$\begin{aligned} CPFWGA(\tilde{A}_1, \ldots, \tilde{A}_n) &= \prod_{i=1}^{n} (\tilde{A}_i)^{w_i} \\ &= \left(\begin{array}{l} \left(\begin{array}{l} \left[\prod_{i=1}^{n} \left(\mu_{\tilde{A}_i}^{L}(x)\right)^{w_i}, \prod_{i=1}^{n} \left(\mu_{\tilde{A}_i}^{U}(x)\right)^{w_i} \right], \\ \left[\prod_{i=1}^{n} \left(\eta_{\tilde{A}_i}^{L}(x)\right)^{w_i}, \prod_{i=1}^{n} \left(\eta_{\tilde{A}_i}^{U}(x)\right)^{w_i} \right], \\ \left[1 - \prod_{i=1}^{n} \left(1 - \nu_{\tilde{A}_i}^{L}(x)\right)^{w_i}, 1 - \prod_{i=1}^{n} \left(1 - \nu_{\tilde{A}_i}^{U}(x)\right)^{w_i} \right], \end{array} \right), \\ \left(\prod_{i=1}^{n} \left(\mu_{\tilde{A}_i}\right)^{w_i}, \prod_{i=1}^{n} \left(\eta_{\tilde{A}_i}\right)^{w_i}, 1 - \prod_{i=1}^{n} \left(1 - \nu_{\tilde{A}_i}\right)^{w_i} \right) \end{array} \right) \end{aligned} \tag{5.33}$$

Here, $w_i \in [0, 1]$ is the weight of $\tilde{A}_i = \langle \tilde{J}_{\tilde{A}_i}, \tilde{I}_{\tilde{A}_i} \rangle$, and $\sum_{i=1}^{n} w_i = 1$.

5.6 Possibility Theory for Picture Fuzzy Environment

Possibility theory is an important method for dealing with an uncertain event, especially for fuzzy-type uncertainty [12]. According to the decision maker's opinion, some chance constraints are to be satisfied in some confidence levels. In the case of fuzzy-type uncertainty, this method has been defined for various types of fuzzy numbers. Therefore, it is a very useful method to tackle the fuzzy objective function and fuzzy constraints of an optimization problem [9]. In this method, three measures for a fuzzy event, such as necessity measure, credibility measure, and possibility measure, are defined. As a shortcoming of the literature, this method has not yet been used for picture fuzzy numbers. In this section, the classical possibility theory is extended to an uncertain environment with picture fuzzy numbers [11].

The rest of this section presents the measures of possibility theory for picture fuzzy numbers.

Definition 5.41 Considering $\tilde{A}$ and $\tilde{B}$ as a pair of triangular picture fuzzy numbers with memberships $\mu_{\tilde{A}}$, $\mu_{\tilde{B}}$, $\eta_{\tilde{A}}$, $\eta_{\tilde{B}}$, $\upsilon_{\tilde{A}}$, and $\nu_{\tilde{B}}$, the possibility measure of a fuzzy event for the positive membership, neutral membership, and negative membership functions (Pos_{μ}, Pos_{η}, and Pos_{υ}) and their necessity measures (Nec_{μ}, Nec_{η}, and Nec_{υ}) are defined by the following equations.

$$Pos_{\mu}(\tilde{A} * \tilde{B}) = \sup\{\min(\mu_{\tilde{A}}, \mu_{\tilde{B}}), x, y \in \mathbb{R}, x * y\} \tag{5.34}$$

$$Pos_{\eta}(\tilde{A} * \tilde{B}) = \sup\{\min(\eta_{\tilde{A}}, \eta_{\tilde{B}}), x, y \in \mathbb{R}, x * y\} \tag{5.35}$$

$$Pos_{\upsilon}(\tilde{A} * \tilde{B}) = \sup\{\min(\nu_{\tilde{A}}, \nu_{\tilde{B}}), x, y \in \mathbb{R}, x * y\} \tag{5.36}$$

$$Nec_{\mu}(\tilde{A} * \tilde{B}) = \inf\{\max(\mu_{\tilde{A}}, \mu_{\tilde{B}}), x, y \in \mathbb{R}, x * y\} \tag{5.37}$$

$$Nec_{\eta}(\tilde{A} * \tilde{B}) = \inf\{\max(\eta_{\tilde{A}}, \eta_{\tilde{B}}), x, y \in \mathbb{R}, x * y\} \tag{5.38}$$

$$Nec_{\upsilon}(\tilde{A} * \tilde{B}) = \inf\{\max(\nu_{\tilde{A}}, \nu_{\tilde{B}}), x, y \in \mathbb{R}, x * y\} \tag{5.39}$$

In these equations, the notation * can be of the signs $\langle, \rangle, \leq, \geq, =$.

Definition 5.42 Considering $\tilde{A}$ as a triangular picture fuzzy number with the memberships $\mu_{\tilde{A}}$, $\eta_{\tilde{A}}$, and $\upsilon_{\tilde{A}}$, the credibility measure of its positive membership, neutral membership, and negative membership functions $(Cr_{\mu}, Cr_{\eta}, Cr_{\upsilon})$, are defined by the followings,

$$Cr_{\mu}\{\tilde{A}\} = \frac{Pos_{\mu}\{\tilde{A}\} + Nec_{\mu}\{\tilde{A}\}}{2}, \tag{5.40}$$

$$Cr_\eta\{\tilde{A}\} = \frac{Pos_\eta\{\tilde{A}\} + Nec_\eta\{\tilde{A}\}}{2}, \tag{5.41}$$

$$Cr_\upsilon\{\tilde{A}\} = \frac{Pos_\upsilon\{\tilde{A}\} + Nec_\upsilon\{\tilde{A}\}}{2}. \tag{5.42}$$

Definition 5.43 Considering the triangular picture fuzzy numbers $\tilde{A} = ((a_1, a_2, a_3, a_4; \alpha), (a_5, a_2, a_3, a_6; \beta), (a_7, a_2, a_3, a_8; \gamma))$, and $\tilde{B} = \big((b_1, b_2, b_3, b_4; \alpha'), (b_5, b_2, b_3, b_6; \beta'), (b_7, b_2, b_3, b_8; \gamma')\big)$, and Definition 5.41, the possibility measures of the event $\tilde{A} \le \tilde{B}$, for all of the membership functions are as below.

$$Pos_\mu\big(\tilde{A} \le \tilde{B}\big) = \begin{cases} t_1, & a_2 \le b_3, t_1 = \min\{\alpha, \alpha'\}, \\ \frac{(b_4-a_1)\alpha\alpha'}{(b_4-b_3)\alpha+(a_2-a_1)\alpha'}, & a_1 < b_4,\ a_2 > b_3, \\ 0, & b_4 \le a_1, \end{cases} \tag{5.43}$$

$$Pos_\eta\big(\tilde{A} \le \tilde{B}\big) = \begin{cases} t_2, & a_2 \le b_3, t_2 = \min\{\beta, \beta'\}, \\ \frac{(b_6-a_5)\beta\beta'}{(b_6-b_3)\beta+(a_2-a_4)\beta'}, & a_5 < b_6,\ a_2 > b_3, \\ 0, & b_6 \le a_5, \end{cases} \tag{5.44}$$

$$Pos_\upsilon\big(\tilde{A} \le \tilde{B}\big) = \begin{cases} 0, & a_2 \le b_3, \\ \frac{(a_2-b_3)\gamma\gamma'}{(b_8-b_3)\gamma+(a_2-a_7)\gamma'}, & a_2 > b_3,\ b_8 > a_7, \\ t_3, & b_8 \le a_7, t_3 = \max\big\{\gamma, \gamma'\big\}, \end{cases} \tag{5.45}$$

For the particular case of $\tilde{A} \le x$, according to Definition 5.38, the following possibility, necessity, and credibility measures are obtained.

$$Pos_\mu\big(\tilde{A} \le x\big) = \begin{cases} \alpha, & x \ge a_2, \\ \frac{\alpha(x-a_1)}{a_2-a_1}, & a_1 \le x \le a_2, \\ 0, & x \le a_1, \end{cases} \tag{5.46}$$

$$Pos_\eta\big(\tilde{A} \le x\big) = \begin{cases} \beta, & x \ge a_2, \\ \frac{\beta(x-a_5)}{a_2-a_5}, & a_5 \le x \le a_2, \\ 0, & x \le a_5, \end{cases} \tag{5.47}$$

$$Pos_\upsilon\big(\tilde{A} \le x\big) = \begin{cases} 0, & x \ge a_2, \\ \frac{\gamma(a_2-x)}{a_2-a_7}, & a_7 \le x \le a_2, \\ \gamma, & x \le a_7, \end{cases} \tag{5.48}$$

$$Nec_\mu\big(\tilde{A} \le x\big) = \begin{cases} 0, & x \le a_3, \\ \frac{\alpha(x-a_3)}{a_4-a_3}, & a_3 \le x \le a_4, \\ \alpha, & x \ge a_4, \end{cases} \tag{5.49}$$

$$Nec_\eta(\tilde{A} \le x) = \begin{cases} 0, & x \le a_3, \\ \frac{\beta(x-a_3)}{a_6-a_3}, & a_3 \le x \le a_6, \\ \beta, & x \ge a_6, \end{cases} \tag{5.50}$$

$$Nec_\upsilon(\tilde{A} \le x) = \begin{cases} \gamma, & x \le a_3, \\ \frac{\gamma(a_8-x)}{a_8-a_3}, & a_3 \le x \le a_8, \\ 0, & x \ge a_8. \end{cases} \tag{5.51}$$

Now from Definition 5.42 the credibility measure of $\tilde{A} \le x$, for positive membership, neutral membership, and negative membership functions are as follows:

$$Cr_\mu(\tilde{A} \le x) = \begin{cases} 0, & x \le a_1, \\ \frac{\alpha(x-a_1)}{2(a_2-a_1)}, & a_1 \le x \le a_2, \\ \frac{\alpha}{2}, & a_2 \le x \le a_3, \\ \frac{\alpha(x+a_4-2a_3)}{2(a_4-a_3)}, & a_3 \le x \le a_4, \\ \alpha, & x \ge a_4, \end{cases} \tag{5.52}$$

$$Cr_\eta(\tilde{A} \le x) = \begin{cases} 0, & x \le a_5, \\ \frac{\beta(x-a_5)}{2(a_2-a_5)}, & a_5 \le x \le a_2, \\ \frac{\beta}{2}, & a_2 \le x \le a_3, \\ \frac{\beta(x+a_6-2a_3)}{2(a_6-a_3)}, & a_3 \le x \le a_6, \\ \beta, & x \ge a_6, \end{cases} \tag{5.53}$$

$$Cr_\upsilon(\tilde{A} \le x) = \begin{cases} \gamma, & x \le a_7, \\ \frac{\gamma(2a_2-x-a_7)}{2(a_2-a_7)}, & a_7 \le x \le a_2, \\ \frac{\gamma}{2}, & a_2 \le x \le a_3, \\ \frac{\gamma(a_8-x)}{2(a_8-a_3)}, & a_3 \le x \le a_8, \\ 0, & x \ge a_8, \end{cases} \tag{5.54}$$

The Eqs. (5.52)–(5.54) are easily converted to the following equations, where $0 \le \lambda + \theta + \omega \le 1$.

$$Cr_\mu\{\tilde{A} \le x\} \ge \lambda \Leftrightarrow \begin{cases} \frac{\alpha(x-a_1)}{2(a_2-a_1)} \ge \lambda, & \lambda \in \left[0, \frac{\alpha}{2}\right], \\ \frac{\alpha(x+a_4-2a_3)}{2(a_4-a_3)} \ge \lambda, & \lambda \in \left(\frac{\alpha}{2}, \alpha\right], \end{cases} \tag{5.55}$$

$$Cr_\eta\{\tilde{A} \le x\} \ge \theta \Leftrightarrow \begin{cases} \frac{\beta(x-a_5)}{2(a_2-a_5)} \ge \theta & \theta \in \left[0, \frac{\beta}{2}\right] \\ \frac{\beta(x+a_6-2a_3)}{2(a_6-a_3)} \ge \theta & \theta \in \left(\frac{\beta}{2}, \beta\right] \end{cases} \tag{5.56}$$

$$Cr_\upsilon\{\tilde{A} \le x\} \le \omega \Leftrightarrow \begin{cases} \frac{\gamma(a_8-x)}{2(a_8-a_3)} \le \omega, & \omega \in \left[0, \frac{\gamma}{2}\right], \\ \frac{\gamma(2a_2-x-a_7)}{2(a_2-a_7)} \le \omega, & \omega \in \left(\frac{\gamma}{2}, \gamma\right]. \end{cases} \tag{5.57}$$

Considering the fact $0 \leq \lambda + \theta + \omega \leq 1$, the Eqs. (5.55)–(5.57) are converted to the below forms [11].

$$Cr_{\mu}\{\tilde{A} \leq x\} \geq \lambda \Leftrightarrow x \geq \left(1 - \frac{2\lambda}{\alpha}\right)a_1 + \frac{2\lambda}{\alpha}a_2, \quad \lambda \in \left[0, \frac{\alpha}{2}\right], \tag{5.58}$$

$$Cr_{\mu}\{\tilde{A} \leq x\} \geq \lambda \Leftrightarrow x \geq 2\left(1 - \frac{\lambda}{\alpha}\right)a_3 + \left(\frac{2\lambda}{\alpha} - 1\right)a_4, \quad \lambda \in \left(\frac{\alpha}{2}, \alpha\right], \tag{5.59}$$

$$Cr_{\eta}\{\tilde{A} \leq x\} \geq \theta \Leftrightarrow x \geq \frac{2\theta}{\beta}a_2 + \left(1 - \frac{2\theta}{\beta}\right)a_5, \quad \theta \in \left[0, \frac{\beta}{2}\right], \tag{5.60}$$

$$Cr_{\eta}\{\tilde{A} \leq x\} \geq \theta \Leftrightarrow x \geq 2\left(1 - \frac{\theta}{\beta}\right)a_3 + \left(\frac{2\theta}{\beta} - 1\right)a_6, \quad \theta \in \left(\frac{\beta}{2}, \beta\right], \tag{5.61}$$

$$Cr_{\upsilon}\{\tilde{A} \leq x\} \leq \omega \Leftrightarrow x \geq \frac{2\omega}{\gamma}a_3 + \left(1 - \frac{2\omega}{\gamma}\right)a_8, \quad \omega \in \left[0, \frac{\gamma}{2}\right]. \tag{5.62}$$

$$Cr_{\upsilon}\{\tilde{A} \leq x\} \leq \omega \Leftrightarrow x \geq \left(\frac{2\omega}{\gamma} - 1\right)a_7 + 2\left(1 - \frac{\omega}{\gamma}\right)a_2, \quad \omega \in \left(\frac{\gamma}{2}, \gamma\right]. \tag{5.63}$$

Applying the same procedure that resulted in the credibility measures (5.58)–(5.63), the credibility measures of $\tilde{A} \geq x$ for all of the membership functions are obtained as below,

$$Cr_{\mu}\{\tilde{A} \geq x\} \geq \lambda \Leftrightarrow x \leq \frac{2\lambda}{\alpha}a_3 + \left(1 - \frac{2\lambda}{\alpha}\right)a_4, \quad \lambda \in \left[0, \frac{\alpha}{2}\right], \tag{5.64}$$

$$Cr_{\mu}\{\tilde{A} \geq x\} \geq \lambda \Leftrightarrow x \leq \left(\frac{2\lambda}{\alpha} - 1\right)a_1 + 2\left(1 - \frac{\lambda}{\alpha}\right)a_2, \quad \lambda \in \left(\frac{\alpha}{2}, \alpha\right], \tag{5.65}$$

$$Cr_{\eta}\{\tilde{A} \geq x\} \geq \theta \Leftrightarrow x \leq \frac{2\theta}{\beta}a_3 + \left(1 - \frac{2\theta}{\beta}\right)a_6, \quad \theta \in \left[0, \frac{\beta}{2}\right], \tag{5.66}$$

$$Cr_{\eta}\{\tilde{A} \geq x\} \geq \theta \Leftrightarrow x \leq \left(\frac{2\theta}{\beta} - 1\right)a_5 + 2\left(1 - \frac{\theta}{\beta}\right)a_2, \quad \theta \in \left(\frac{\beta}{2}, \beta\right], \tag{5.67}$$

$$Cr_{\upsilon}\{\tilde{A} \geq x\} \leq \omega \Leftrightarrow x \leq \left(1 - \frac{2\omega}{\gamma}\right)a_7 + \frac{2\omega}{\gamma}a_2, \quad \omega \in \left[0, \frac{\gamma}{2}\right], \tag{5.68}$$

$$Cr_{\upsilon}\{\tilde{A} \geq x\} \leq \omega \Leftrightarrow x \leq 2\left(1 - \frac{\omega}{\gamma}\right)a_3 + \left(\frac{2\omega}{\gamma} - 1\right)a_8, \quad \omega \in \left(\frac{\gamma}{2}, \gamma\right]. \tag{5.69}$$

5.7 Application: BCC-IO Data Envelopment Analysis Model

Data envelopment analysis (DEA) is one of the non-parametric methods that is used to measure the performance efficiency of a decision-making unit (DMU) in existence of some input-based and output-based factors (criteria) [14]. In the existence of several DMUs in a system, a DEA model is solved for each DMU separately to decide whether the DMU is efficient or inefficient [13]. Although there are many DEA models in the literature, the BCC model of DEA is one of the most famous. This model in an input-oriented model (BCC-IO model) and is formulated as follows [3, 6],

$$\max \varphi_p = \sum_{r=1}^{s} y_{rp} U_r - u_0 \tag{5.70}$$

subject to

$$\sum_{i=1}^{m} x_{ip} V_i = 1, \tag{5.71}$$

$$\sum_{r=1}^{s} y_{rj} U_r - \sum_{i=1}^{m} x_{ij} V_i - u_0 \leq 0, \quad j = 1, \ldots, n, \tag{5.72}$$

$$U_r \geq 0, \quad r = 1, \ldots, s, \tag{5.73}$$

$$V_i \geq 0, \quad i = 1, \ldots, m. \tag{5.74}$$

In this model,

- n is the number of existing DMUs to be evaluated (parameter),
- m is the number of existing input criteria for evaluating the DMUs (parameter),
- s is the number of existing output criteria for evaluating the DMUs (parameter),
- j is the index for identifying the DMUs (index),
- p is the index of the DMU, which is under evaluation (index),
- i is the index used for the inputs (index),
- r is the index used for the outputs (index),
- x_{ij} is the performance value of DMU j in input i (parameter),
- y_{rj} is the performance value of DMU j in output r (parameter),
- V_i is the weight value for input i (variable),
- U_r is the weight value for output r (variable),
- u_0 is a free of sign variable (variable).

In the above-mentioned model, the objective function (5.70) maximizes the efficiency score of DMU p. The constraint (5.71) guarantees that the weighted sum of input values for the DMU is equal to 1. The constraints set (5.72) guarantees that the weighted sum of output values minus the weighted sum of input values and the free of sign variable, does not exceed zero. The non-negativity of the variable weight values are constrained by the constraints sets (5.73) and (5.74).

If the model (5.70)–(5.74) is solved for DMU_p, the optimal objective function value of φ_p^* is obtained, and the definition below is considered to decide its situation.

Definition 5.44 According to the BCC-IO model (5.70)–(5.74), DMU p is efficient if $\varphi_p^* = 1$, otherwise, it is inefficient.

In this section, the BCC-IO DEA model (5.70)–(5.74) is developed for a picture fuzzy environment, and a solution approach based on the possibility theory is proposed to solve the proposed fuzzy model.

The assumptions below have to be considered in advance to construct the formulation of the picture fuzzy BCC-IO model.

- The inputs and outputs of the classic BCC-IO model are considered to be of TrPFNs as below,

$$\tilde{x}_{ij} = \left(\left(x_{ij}^1, x_{ij}^2, x_{ij}^3, x_{ij}^4; \alpha_{ij}\right), \left(x_{ij}^5, x_{ij}^2, x_{ij}^3, x_{ij}^6; \beta_{ij}\right), \left(x_{ij}^7, x_{ij}^2, x_{ij}^3, x_{ij}^8; \gamma_{ij}\right)\right), \quad \begin{array}{l} i = 1, \ldots, m, \\ j = 1, \ldots, n, \end{array} \tag{5.75}$$

$$\tilde{y}_{rj} = \left(\left(y_{rj}^1, y_{rj}^2, y_{rj}^3, y_{rj}^4; \rho_{rj}\right), \left(y_{rj}^5, y_{rj}^2, y_{rj}^3, y_{rj}^6; \sigma_{rj}\right), \left(y_{rj}^7, y_{rj}^2, y_{rj}^3, y_{rj}^8; \varsigma_{rj}\right)\right), \quad \begin{array}{l} r = 1, \ldots, s, \\ j = 1, \ldots, n. \end{array} \tag{5.76}$$

- The constraint $\sum_{i=1}^{m} \tilde{x}_{ip} V_i = 1$, in the classical BCC-IO model is converted to the constraint $\sum_{i=1}^{m} \tilde{x}_{ip} V_i \leq 1$, in the fuzzy model [10].

Therefore, the below picture fuzzy BCC-IO model for measuring the efficiency of DMU_p is introduced here.

$$\max \varphi_p = \sum_{r=1}^{s} \tilde{y}_{rp} U_r - u_0 \tag{5.77}$$

subject to

$$\sum_{i=1}^{m} \tilde{x}_{ip} V_i \leq 1, \tag{5.78}$$

$$\sum_{r=1}^{s} \tilde{y}_{rj} U_r - \sum_{i=1}^{m} \tilde{x}_{ij} V_i - u_0 \leq 0, \quad j = 1, \ldots, n, \tag{5.79}$$

$$U_r \geq 0, \quad r = 1, \ldots, s, \tag{5.80}$$

$$V_i \geq 0, \quad i = 1, \ldots, m. \tag{5.81}$$

The picture fuzzy BCC-IO model presented by the formulation (5.77)–(5.81) is too complex to be solved directly as it has a fuzzy nature. To solve this problem, the model should first be converted to a crisp form. As mentioned earlier, the possibility theory is a well-known and effective method to tackle fuzzy optimization models, including DEA models. Among the measures used in the possibility theory, the credibility measure could be more logical and effective comparing to the others (the possibility and necessity measures). Therefore, this section applies a possibility theory approach with a credibility measure to solve the fuzzy BCC-IO model presented by the formulation (5.77)–(5.81). Notably, the credibility measure has not yet been applied to the picture fuzzy DEA models in the literature. Therefore, this study, applies the credibility measure to the fuzzy picture BCC-IO model.

First, the issues are considered in applying the possibility theory with credibility measures to the formulation (5.77)–(5.81).

- A continuous variable replaces the fuzzy objective function, say f_p, and the objective function is bounded by f_p as a constraint.
- The credibility measure of each fuzzy constraint is obtained by the relationships introduced by Definition 5.10 from Sect. 5.3.

According to these issues, the below credibility-based formulation is obtained for the picture fuzzy BCC-IO model (5.77)–(5.81).

$$\max f_p \tag{5.82}$$

subject to

$$Cr_\mu \left\{ \sum_{r=1}^{s} \tilde{y}_{rp} U_r - u_0 \geq f_p \right\} \geq \kappa_1, \tag{5.83}$$

$$Cr_\eta \left\{ \sum_{r=1}^{s} \tilde{y}_{rp} U_r - u_0 \geq f_p \right\} \geq \kappa_2, \tag{5.84}$$

$$Cr_\upsilon \left\{ \sum_{r=1}^{s} \tilde{y}_{rp} U_r - u_0 \geq f_p \right\} \leq \kappa_3, \tag{5.85}$$

$$Cr_\mu \left\{ \sum_{i=1}^{m} \tilde{x}_{ip} V_i \leq 1 \right\} \geq \pi_1, \tag{5.86}$$

$$Cr_\eta \left\{ \sum_{i=1}^{m} \tilde{x}_{ip} V_i \leq 1 \right\} \geq \pi_2, \tag{5.87}$$

$$Cr_{\upsilon}\left\{\sum_{i=1}^{m}\tilde{x}_{ip}V_i \leq 1\right\} \leq \pi_3, \tag{5.88}$$

$$Cr_{\mu}\left\{\sum_{r=1}^{s}\tilde{y}_{rj}U_r - \sum_{i=1}^{m}\tilde{x}_{ij}V_i - u_0 \leq 0\right\} \geq \lambda_j, \quad j = 1, \ldots, n, \tag{5.89}$$

$$Cr_{\eta}\left\{\sum_{r=1}^{s}\tilde{y}_{rj}U_r - \sum_{i=1}^{m}\tilde{x}_{ij}V_i - u_0 \leq 0\right\} \geq \theta_j, \quad j = 1, \ldots, n, \tag{5.90}$$

$$Cr_{\upsilon}\left\{\sum_{r=1}^{s}\tilde{y}_{rj}U_r - \sum_{i=1}^{m}\tilde{x}_{ij}V_i - u_0 \leq 0\right\} \leq \omega_j, \quad j = 1, \ldots, n, \tag{5.91}$$

$$U_r \geq 0, V_i \geq 0, \quad r = 1, \ldots, s, i = 1, \ldots, m. \tag{5.92}$$

In the above formulation, the parameters $\kappa_1, \kappa_2, \kappa_3, \pi_1, \pi_2, \pi_3, \lambda_j, \theta_j$, and ω_j are some confidence levels considered for the chance constraints of the model. These confidence levels should follow the below conditions.

- $0 \leq \kappa_1 + \kappa_2 + \kappa_3 \leq 1$,
- $0 \leq \pi_1 + \pi_2 + \pi_3 \leq 1$,
- $0 \leq \lambda_j + \theta_j + \omega_j \leq 1, \ (j = 1, 2, \ldots, n)$.

The formulation (5.82)–(5.92) is a complex formulation and cannot be solved by any optimization solver because of its chance constraints. Therefore, this model is converted to the below form,

$$\max f_p \tag{5.93}$$

subject to

$$\frac{2\kappa_1}{A_1}\sum_{r=1}^{s} y_{rp}^3 U_r + \left(1 - \frac{2\kappa_1}{\rho_{rp}}\right)\sum_{r=1}^{s} y_{rp}^4 U_r - u_0 \geq f_p, \quad \kappa_1 \in \left[0, \frac{A_1}{2}\right], \tag{5.94}$$

$$\left(\frac{2\kappa_1}{A_1} - 1\right)\sum_{r=1}^{s} y_{rp}^1 U_r + 2\left(1 - \frac{\kappa_1}{A_1}\right)\sum_{r=1}^{s} y_{rp}^2 U_r - u_0 \geq f_p, \quad \kappa_1 \in \left(\frac{A_1}{2}, A_1\right], \tag{5.95}$$

$$\frac{2\kappa_2}{A_2}\sum_{r=1}^{s} y_{rp}^3 U_r + \left(1 - \frac{2\kappa_2}{A_2}\right)\sum_{r=1}^{s} y_{rp}^6 U_r - u_0 \geq f_p, \quad \kappa_2 \in \left[0, \frac{A_2}{2}\right] \tag{5.96}$$

$$\left(\frac{2\kappa_2}{A_2} - 1\right)\sum_{r=1}^{s} y_{rp}^5 U_r + 2\left(1 - \frac{\kappa_2}{A_2}\right)\sum_{r=1}^{s} y_{rp}^2 U_r - u_0 \geq f_p, \quad \kappa_2 \in \left(\frac{A_2}{2}, A_2\right] \tag{5.97}$$

$$\left(1-\frac{2\kappa_3}{A_3}\right)\sum_{r=1}^{s} y_{rp}^{7}U_r+\frac{2\kappa_3}{A_3}\sum_{r=1}^{s} y_{rp}^{2}U_r-u_0\geq f_p,\quad \kappa_3\in\left[0,\frac{A_3}{2}\right] \tag{5.98}$$

$$2\left(1-\frac{\kappa_3}{A_3}\right)\sum_{r=1}^{s} y_{rp}^{3}U_r+\left(\frac{2\kappa_3}{A_3}-1\right)\sum_{r=1}^{s} y_{rp}^{8}U_r-u_0\geq f_p,\quad \kappa_3\in\left(\frac{A_3}{2},A_3\right] \tag{5.99}$$

$$\left(1-\frac{2\pi_1}{B_1}\right)\sum_{i=1}^{m} x_{ip}^{1}V_i+\frac{2\pi_1}{B_1}\sum_{i=1}^{m} x_{ip}^{2}V_i\leq 1,\quad \pi_1\in\left[0,\frac{B_1}{2}\right] \tag{5.100}$$

$$2\left(1-\frac{\pi_1}{B_1}\right)\sum_{i=1}^{m} x_{ip}^{3}V_i+\left(\frac{2\pi_1}{B_1}-1\right)\sum_{i=1}^{m} x_{ip}^{4}V_i\leq 1,\quad \pi_1\in\left(\frac{B_1}{2},B_1\right] \tag{5.101}$$

$$\frac{2\pi_2}{B_2}\sum_{i=1}^{m} x_{ip}^{2}V_i+\left(1-\frac{2\pi_2}{B_2}\right)\sum_{i=1}^{m} x_{ip}^{5}V_i\leq 1,\quad \pi_2\in\left[0,\frac{B_2}{2}\right] \tag{5.102}$$

$$2\left(1-\frac{\pi_2}{B_2}\right)\sum_{i=1}^{m} x_{ip}^{3}V_i+\left(\frac{2\pi_2}{B_2}-1\right)\sum_{i=1}^{m} x_{ip}^{6}V_i\leq 1,\quad \pi_2\in\left(\frac{B_2}{2},B_2\right] \tag{5.103}$$

$$\frac{2\pi_3}{B_3}\sum_{i=1}^{m} x_{ip}^{3}V_i+\left(1-\frac{2\pi_3}{B_3}\right)\sum_{i=1}^{m} x_{ip}^{8}V_i\leq 1,\quad \pi_3\in\left[0,\frac{B_3}{2}\right] \tag{5.104}$$

$$\left(\frac{2\pi_3}{B_3}-1\right)\sum_{i=1}^{m} x_{ip}^{7}V_i+2\left(1-\frac{\pi_3}{B_3}\right)\sum_{i=1}^{m} x_{ip}^{2}V_i\leq 1,\quad \pi_3\in\left(\frac{B_3}{2},B_3\right] \tag{5.105}$$

$$\begin{aligned}&\left(1-\frac{2\lambda_j}{C_{1j}}\right)\sum_{r=1}^{s} y_j^{1}U_r+\frac{2\lambda_j}{C_{1j}}\sum_{r=1}^{s} y_{rj}^{2}U_r && \lambda_j\in\left[0,\frac{C_{1j}}{2}\right],\\ &-\left(1-\frac{2\lambda_j}{C_{1j}}\right)\sum_{i=1}^{m} x_{ij}^{4}V_i-\frac{2\lambda_j}{C_{1j}}\sum_{i=1}^{m} x_{ij}^{3}V_i-u_0\leq 0, && j=1,2,\ldots,n^2\end{aligned} \tag{5.106}$$

$$\begin{aligned}&2\left(1-\frac{\lambda_j}{C_{1j}}\right)\sum_{r=1}^{s} y_{rj}^{3}U_r+\left(\frac{2\lambda_j}{C_{1j}}-1\right)\sum_{r=1}^{s} y_{rj}^{4}U_r && \lambda_j\in\left(\frac{C_{1j}}{2},C_{1j}\right],\\ &-2\left(1-\frac{\lambda_j}{C_{1j}}\right)\sum_{i=1}^{m} x_{ij}^{2}V_i-\left(\frac{2\lambda_j}{C_{1j}}-1\right)\sum_{i=1}^{m} x_{ij}^{1}V_i-u_0\leq 0, && j=1,2,\ldots,n\end{aligned} \tag{5.107}$$

$$\begin{aligned}&\left(1-\frac{2\theta_j}{C_{2j}}\right)\sum_{r=1}^{s} y_{rj}^{5}U_r+\frac{2\theta_j}{C_{2j}}\sum_{r=1}^{s} y_{rj}^{2}U_r && \theta_j\in\left[0,\frac{C_{2j}}{2}\right],\\ &-\left(1-\frac{2\theta_j}{C_{2j}}\right)\sum_{i=1}^{m} x_{ij}^{6}V_i-\frac{2\theta_j}{C_{2j}}\sum_{i=1}^{m} x_{ij}^{3}V_i-u_0\leq 0, && j=1,2,\ldots,n,\end{aligned} \tag{5.108}$$

$$\begin{array}{ll} 2\left(1-\frac{\theta_j}{C_{2j}}\right)\sum\limits_{r=1}^{s} y_{rj}^3 U_r+\left(\frac{2\theta_j}{C_{2j}}-1\right)\sum\limits_{r=1}^{s} y_{rj}^6 U_r & \theta_j\in\left(\frac{C_{2j}}{2},C_{2j}\right], \\ -2\left(1-\frac{\theta_j}{C_{2j}}\right)\sum\limits_{i=1}^{m} x_{ij}^2 V_i-\left(\frac{2\theta_j}{C_{2j}}-1\right)\sum\limits_{i=1}^{m} x_{ij}^5 V_i-u_0\le 0,\ j=1,2,\ldots,n, & \end{array} \tag{5.109}$$

$$\begin{array}{ll} \frac{2\omega_j}{C_{3j}}\sum\limits_{r=1}^{s} y_{rj}^3 U_r+\left(1-\frac{2\omega_j}{C_{3j}}\right)\sum\limits_{r=1}^{s} y_{rj}^8 U_r & \omega_j\in\left[0,\frac{C_{3j}}{2}\right], \\ -\frac{2\omega_j}{C_{3j}}\sum\limits_{i=1}^{m} x_{ij}^2 V_i-\left(1-\frac{2\omega_j}{C_{3j}}\right)\sum\limits_{i=1}^{m} x_{ij}^7 V_i-u_0\le 0,\ j=1,2,\ldots,n, & \end{array} \tag{5.110}$$

$$\begin{array}{ll} \left(\frac{2\omega_j}{C_{3j}}-1\right)\sum\limits_{r=1}^{s} y_{rj}^7 U_r+2\left(1-\frac{\omega_j}{C_{3j}}\right)\sum\limits_{r=1}^{s} y_{rj}^2 U_r & \omega_j\in\left(\frac{C_{3j}}{2},C_{3j}\right], \\ -\left(\frac{2\omega_j}{C_{3j}}-1\right)\sum\limits_{i=1}^{m} x_{ij}^8 V_i-2\left(1-\frac{\omega_j}{C_{3j}}\right)\sum\limits_{i=1}^{m} x_{ij}^3 V_i-u_0\le 0,\ j=1,2,\ldots,n, & \end{array} \tag{5.111}$$

$$U_r\ge 0,\ V_i\ge 0,\quad r=1,\ldots,s,\ i=1,\ldots,m, \tag{5.112}$$

where,

- $A_1=\sum\limits_{r=1}^{s}\rho_{rp}-\sum\limits_{r<t}\rho_{rp}\rho_{tp}+\sum\limits_{r<t<k}\rho_{rp}\rho_{tp}\rho_{kp}+\cdots+(-1)^{s-1}\prod\limits_{r=1}^{s}\rho_{rp}$
- $A_2=\prod\limits_{r=1}^{s}\sigma_{rp}$,
- $A_3=\prod\limits_{r=1}^{s}\varsigma_{rp}$,
- $B_1=\sum\limits_{i=1}^{m}\alpha_{ip}-\sum\limits_{i<g}\alpha_{ip}\alpha_{gp}+\sum\limits_{i<g<h}\alpha_{ip}\alpha_{gp}\alpha_{hp}+\cdots+(-1)^{m-1}\prod\limits_{i=1}^{m}\alpha_{ip}$,
- $B_2=\prod\limits_{i=1}^{m}\beta_{ip}$,
- $B_3=\prod\limits_{i=1}^{m}\gamma_{ip}$,
- $C_{1j}=\left(D_{1j}+D_{2j}-D_{1j}D_{2j}\right)$, for $j=1,2,\ldots,n$, where

$$D_{1j}=\sum_{r=1}^{s}\rho_{rj}-\sum_{r<t}\rho_{rj}\rho_{tj}+\sum_{r<t<k}\rho_{rj}\rho_{tj}\rho_{kj}+\cdots+(-1)^{s-1}\prod_{r=1}^{s}\rho_{rj},$$
$$D_{2j}=\sum_{i=1}^{m}\alpha_{ij}-\sum_{i<q}\alpha_{ij}\alpha_{qj}+\sum_{i<l<h}\alpha_{ij}\alpha_{lj}\alpha_{hj}+\cdots+(-1)^{m-1}\prod_{i=1}^{m}\alpha_{ij},$$

- $C_{2j=}\prod\limits_{r=1}^{s}\sigma_{rj}\prod\limits_{i=1}^{m}\beta_{ij}$, for $j=1,2,\ldots,n$,
- $C_{3j}=\prod\limits_{r=1}^{s}\varsigma_{rj}\prod\limits_{i=1}^{m}\gamma_{ij}$, for $j=1,2,\ldots,n$.

The formulation (5.93)–(5.112) is a general crisp form for the chance-constrained formulation (5.59)–(5.69). All possible ranges of confidence levels are included in the formulation. Here, the confidence levels $\kappa_1, \kappa_2, \kappa_3, \pi_1, \pi_2, \pi_3, \lambda_j, \theta_j$, and ω_j can take different values from 0 to 1, where the decision maker sets their values in the evaluation procedure. To evaluate the efficiency or inefficiency situation of a DMU (for example, DMU_p) using the crisp formulation (5.93)–(5.112), first, the decision maker sets the values of the confidence levels $\kappa_1, \kappa_2, \kappa_3, \pi_1, \pi_2, \pi_3, \lambda_j, \theta_j$, and ω_j according to his/her preferences. Then, some constraints of the formulation (5.93)–(5.112), which are extra according to the selected confidence level values, are removed and the net formulation is extracted. The extracted net formulation is solved to decide about the efficiency/inefficiency of DMU_p. According to the obtained results from the selected net formulation, the below definitions are proposed for DMU_p.

Definition 5.45 Given a set of confidence level values, DMU_p is called an efficient DMU (E) if its optimal f_p value be not less than 1. Otherwise, the DMU is inefficient (IE).

Definition 5.46 Considering all available sets of the confidence levels values, DMU_p is called a marginal-efficient DMU (E+) if its optimal f_p value for at least one set of the confidence levels values be not less than 1.

Definition 5.47 Considering all available sets of the confidence level values, DMU_p is called an ultra-efficient DMU (E++) if its optimal f_p value for all sets of the confidence levels values be not less than 1.

Definition 5.48 Considering all available sets of the confidence level values, DMU_p is and ultra-inefficient DMU (IE–) if it's optimal f_p value for all sets of the confidence levels values is less than 1.

References

1. Akram, M., Ullah, I., Allahviranloo, T.: A new method to solve linear programming problems in the environment of picture fuzzy sets. Iran. J. Fuzzy Syst. **19**(6), 29–49 (2022)
2. Ashraf, S., Abdullah, S., Muneeza: Some novel aggregation operators for cubic picture fuzzy information: application in multi-attribute decision support problem. Granul. Comput. **6**, 603–618 (2021)
3. Banker, R., Charnes, A., Cooper, W.: Some models for estimating technical and scale inefficiencies in data envelopment analysis. Manage. Sci. **30**, 1078–1092 (1984)
4. Cao, G.: A novel similarity measure between picture fuzzy sets based on transformation techniques and its applications in mobile multimedia healthcare. Multimed. Tools Appl. **83**, 58645–58688 (2024)
5. Cuong, B.C., Kreinovich, V.: Picture fuzzy sets. J. Comput. Sci. Cybern. **30**(4), 409–420 (2014)
6. Cuong, B.C.: Picture fuzzy sets—a new concept for computational intelligence problems. In: Proceedings of the 2013 Third World Congress on Information and Communication Technologies, Hanoi, Vietnam, pp. 1–6 (2013)

7. Dutta, P., Ganju, S.: Some aspects of picture fuzzy set. Trans. A. Razmadze Math. Inst. **172**(2), 164–175 (2018)
8. Ju, Y., Ju, D., Gonzalez, E.S., Giannakis, M., Wang, A.: Study of site selection of electric vehicle charging station based on extended GRP method on picture fuzzy sets. Comput. Ind. Eng. **135**, 1271–1285 (2019)
9. Mahmoodirad, A., Niroomand, S.: Uncertain location–allocation decisions for a bi-objective two-stage supply chain network design problem with environmental impacts. Expert. Syst. **37**(5), e12558 (2020)
10. Mahmoodirad, A., Pamucar, D., Niroomand, S.: A new intuitionistic fuzzy scheme of data envelopment analysis for evaluating rural comprehensive health service centers. Socioecon. Plann. Sci. **95**, 102004 (2024)
11. Mahmoodirad, A., Pamucar, D., Niroomand, S., Simic, V.: Data envelopment analysis based performance evaluation of hospitals—implementation of novel picture fuzzy BCC model. Expert Syst. Appl. **263**, 125775 (2025)
12. Mirzaei, N., Mahmoodirad, A., Niroomand, S.: An uncertain multi-objective assembly line balancing problem: a credibility-based fuzzy modeling approach. Int. J. Fuzzy Syst. **21**(8), 2392–2404 (2019)
13. Modhej, D., Sanei, M., Shoja, N., Hosseinzadehlotfi, F.: Integrating inverse data envelopment analysis and neural network to preserve relative efficiency values. J. Intell. Fuzzy Syst. **32**(6), 4047–4058 (2017)
14. Monzeli, A., Daneshian, B., Tohidi, G., Sanei, M., Razavian, S.: Efficiency study with undesirable inputs and outputs in DEA. J. Fuzzy Extens. Appl. **1**(1), 73–80 (2020)
15. Popa, L.: Centroid-induced ranking of triangular picture fuzzy numbers and applications in decision-making. Symmetry **16**, 1492 (2024)
16. Qiyas, M., Abdullah, S., Ashraf, S., Khan, S., Khan, A.: Triangular picture fuzzy linguistic induced ordered weighted aggregation operators and its application on decision making problems. Found. Comput. Math. **2**, 183–201 (2019)
17. Singh, P., Mishra, N.K.: Risk analysis of flood disaster based on similarity measures in picture fuzzy environment. Afr. Mat. **29**, 1019–1038 (2018)
18. Luo, M., Gao, J., Li, W.: Some new construction methods of similarity measure on picture fuzzy sets. Soft. Comput. **29**, 1273–1287 (2025)

Chapter 6
Neutrosophic Fuzzy Sets

6.1 Introduction

The neutrosophic fuzzy sets (NFS) for the first time introduced by [1, 2]. A neutrosophic fuzzy set in addition to the membership and non-membership degrees of an event, uses the neutral degree of the event for describing more uncertain and vague situations. In the case of neutrosophic fuzzy sets, for a set, truth-membership function, indeterminacy-membership function, and falsity-membership function are defined. It can be widely used for cases where the evaluations are determined by the experts. As an example, for a claim, an expert may believe that it can be "true", "no opinion", and "wrong". In such cases, the neutrosophic fuzzy sets can be applied to model the uncertain event.

Some concepts (taken from Smarandache [2] and applications of the neutrosophic fuzzy sets are reviewed in this chapter.

6.2 Neutrosophic Fuzzy Sets

In this section first, the non-standard analysis is explained, and then the main concepts of the neutrosophic fuzzy sets are presented.

A number such as a is a non-standard small infinity number if and only if for any positive discrete number such as n, the condition $|a| \leq \frac{1}{n}$ be true. The interval $\left]^{-}0, 1^{+}\right[$ is called the non-standard unit interval. Here, $1^{+} = 1 + \varepsilon$ is a set of finite non-standard number where 1 is its standard part and ε is its non-standard small infinity part. Also, $^{-}0 = 0 - \varepsilon$ is a set of finite non-standard number where 0 is its standard part and ε is its non-standard small infinity part. Therefore, the non-standard unit interval $\left]^{-}0, 1^{+}\right[$ includes 0, 1, the non-standard small infinity numbers less than 0, and the non-standard small infinity numbers greater than 1.

A. Mahmoodirad et al., *Recent Advances on Fuzzy Sets*, Studies in Fuzziness and Soft Computing 440, https://doi.org/10.1007/978-3-032-08776-8_6

Definition 6.1 Assume the reference set X. A neutrosophic fuzzy set such as $\tilde{N}$ on the reference set X is defined as $\tilde{N} = \{(x, T_{\tilde{N}}(x), I_{\tilde{N}}(x), F_{\tilde{N}}(x)), x \in X\}$. Here, $T_{\tilde{N}}(x) : X \to]^{-}0, 1^{+}[$ is the truth-membership function, $I_{\tilde{N}}(x) : X \to]^{-}0, 1^{+}[$ is the indeterminacy-membership function, and $T_{\tilde{N}}(x) : X \to]^{-}0, 1^{+}[$ is the falsity-membership function of $\tilde{N}$, where, $]^{-}0, 1^{+}[$ is a real non-standard interval. The truth-membership, indeterminacy-membership, and falsity-membership functions re independent, and $0^{-} \le sup\ T_{\tilde{N}}(x) + sup\ I_{\tilde{N}}(x) + sup\ F_{\tilde{N}}(x) \le 3^{+}$.

Definition 6.2 The complement of neutrosophic fuzzy set $\tilde{N}$ (on the reference set X) is shown by $\tilde{N}^{c}$ which is also a neutrosophic set on X with the following functions,

$$\begin{aligned} T_{\tilde{N}^c}(x) = \{1^{+}\} - T_{\tilde{N}}(x), I_{\tilde{N}^c}(x) = \{1^{+}\} - I_{\tilde{N}}(x), F_{\tilde{N}^c}(x) \\ = \{1^{+}\} - F_{\tilde{N}}(x) \end{aligned} \tag{6.1}$$

where, $\{1^{+}\} - S = \{x | x = 1^{+} - s, s \in S\}$.

A schematic comparison of the neutrosophic fuzzy sets with other fuzzy sets is represented by Fig. 6.1.

Example 6.1 Assume there are only two choices a and b, and 10 experts should vote for them. About the choice a, 5 experts vote "yes", 2 experts vote "no opinion", and 3 experts vote "no". About the choice b, 3 experts vote "yes", 5 experts vote "no opinion", and 2 experts vote "no". This information can be shown by the neutrosophic fuzzy set $\{(a, 0.5, 0.2, 0.3), (b, 03, 0.5, 0.2)\}$.

Example 6.2 Assume there is a claim such as "tomorrow is rainy" shown by p. An expert evaluates its truth, indeterminacy, and falsity degrees of this event

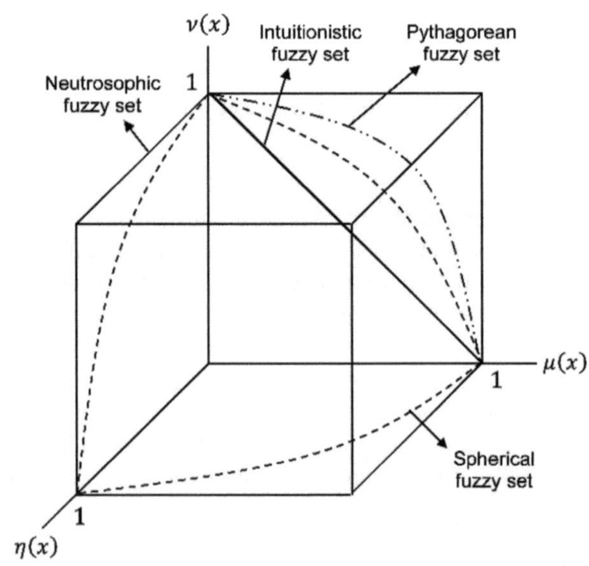

Fig. 6.1 Geometric representation of NFS and other fuzzy sets

as $[0, 0.4]$, $[0.1, 0.3]$, and $[0.5, 0.6]$ respectively. Another expert evaluates its truth, indeterminacy, and falsity degrees of this event as $[0.6, 0.8]$, $[0.2, 0.4]$, and $[0.3, 0.5]$ respectively. In this case, the neutrosophic fuzzy sets $\tilde{N}_1 = \{(p, [0, 0.4], [0.1, 0.3], [0.5, 0.6])\}$ and $\tilde{N}_2 = \{(p, [0.6, 0.8], [0.2, 0.4], [0.3, 0.5])\}$ can be defined. Even, these sets can be combined as $\tilde{N} = \{(p, [0, 0.4] \cup [0.6, 0.8], [0.1, 0.3] \cup [0.2, 0.4], [0.5, 0.6] \cup [0.3, 0.5])\}$.

Definition 6.3 Assume $\tilde{N}_1$ and $\tilde{N}_2$ as two neutrosophic fuzzy sets on the reference set X. The below issues are defined for these sets.

- $\tilde{N}_1$ is the subset of $\tilde{N}_2$, if and only if for any $x \in X$ there are,

$$\begin{aligned} inf\ T_{\tilde{N}_1}(x) \leq inf\ T_{\tilde{N}_2}(x), sup\ T_{\tilde{N}_1}(x) \leq sup\ T_{\tilde{N}_2}(x), \\ inf\ I_{\tilde{N}_1}(x) \geq inf\ I_{\tilde{N}_2}(x), sup\ I_{\tilde{N}_1}(x) \geq sup\ I_{\tilde{N}_2}(x), \\ inf\ F_{\tilde{N}_1}(x) \geq inf\ F_{\tilde{N}_2}(x), sup\ F_{\tilde{N}_1}(x) \geq sup\ F_{\tilde{N}_2}(x). \end{aligned} \tag{6.2}$$

- $\tilde{N}_1 \cup \tilde{N}_2$ is defined for any $x \in X$ as below,

$$\begin{aligned} T_{\tilde{N}_1 \cup \tilde{N}_2}(x) = T_{\tilde{N}_1}(x) + T_{\tilde{N}_2}(x) - T_{\tilde{N}_1}(x).T_{\tilde{N}_2}(x), \\ I_{\tilde{N}_1 \cup \tilde{N}_2}(x) = I_{\tilde{N}_1}(x) + I_{\tilde{N}_2}(x) - I_{\tilde{N}_1}(x).I_{\tilde{N}_2}(x), \\ F_{\tilde{N}_1 \cup \tilde{N}_2}(x) = F_{\tilde{N}_1}(x) + F_{\tilde{N}_2}(x) - F_{\tilde{N}_1}(x).F_{\tilde{N}_2}(x). \end{aligned} \tag{6.3}$$

- $\tilde{N}_1 \cap \tilde{N}_2$ is defined for any $x \in X$ as below,

$$\begin{aligned} T_{\tilde{N}_1 \cap \tilde{N}_2}(x) = T_{\tilde{N}_1}(x) T_{\tilde{N}_2}(x), \\ I_{\tilde{N}_1 \cap \tilde{N}_2}(x) = I_{\tilde{N}_1}(x) I_{\tilde{N}_2}(x), \\ F_{\tilde{N}_1 \cap \tilde{N}_2}(x) = F_{\tilde{N}_1}(x) F_{\tilde{N}_2}(x). \end{aligned} \tag{6.4}$$

- $\tilde{N}_1 \backslash \tilde{N}_2$ (difference of $\tilde{N}_1$ and $\tilde{N}_2$) is defined for any $x \in X$ as below,

$$\begin{aligned} T_{\tilde{N}_1 \backslash \tilde{N}_2}(x) = T_{\tilde{N}_1}(x) - T_{\tilde{N}_1}(x) T_{\tilde{N}_2}(x), \\ I_{\tilde{N}_1 \backslash \tilde{N}_2}(x) = I_{\tilde{N}_1}(x) - I_{\tilde{N}_1}(x) I_{\tilde{N}_2}(x), \\ F_{\tilde{N}_1 \backslash \tilde{N}_2}(x) = F_{\tilde{N}_1}(x) - F_{\tilde{N}_1}(x) F_{\tilde{N}_2}(x). \end{aligned} \tag{6.5}$$

In this definition, the operations are in standard or non-standard sets, meaning that,

$$\begin{aligned} S_1 + S_2 = \{s_1 + s_2 : s_1 \in S_1, s_2 \in S_2\} \\ S_1 - S_2 = \{s_1 - s_2 : s_1 \in S_1, s_2 \in S_2\} \\ S_1.S_2 = \{s_1.s_2 : s_1 \in S_1, s_2 \in S_2\} \end{aligned}$$

Example 6.3 Assume the reference set $X = \{a, b, c\}$ and the below neutrosophic fuzzy sets.

$$\tilde{N}_1 = \{(a, 0.4, 0.3, 0.5), (b, 0.5, 0.3, 0.1), (c, 0.2, 0.5, 0.9)\},$$
$$\tilde{N}_2 = \{(a, 0.3, 0.5, 0.5), (b, 0.1, 0.4, 0.6), (c, 0.5, 0.5, 0.5)\},$$
$$\tilde{N}_3 = \{(a, 0.4, 0.3, 0.2), (b, 0.6, 0.3, 0.1), (c, 0.4, 0.3, 0.8)\},$$

According to the above-mentioned definitions, the below results are obtained.

$$\tilde{N}_1^c = \{(a, 0.6, 0.7, 0.5), (b, 0.5, 0.7, 0.9), (c, 0.8, 0.5, 0.1)\},$$
$$\tilde{N}_1 \cap \tilde{N}_2 = \{(a, 0.12, 0.15, 0.25), (b, 0.05, 0.12, 0.06), (c, 0.1, 0.25, 0.45)\},$$
$$\tilde{N}_1 \cup \tilde{N}_2 = \{(a, 0.58, 0.65, 0.75), (b, 0.55, 0.58, 0.64), (c, 0.6, 0.75, 0.95)\},$$
$$\tilde{N}_1 \backslash \tilde{N}_2 = \{(a, 0.28, 0.15, 0.25), (b, 0.45, 0.18, 0.04), (c, 0.1, 0.25, 0.45)\},$$

As another result of this example, $\tilde{N}_1 \subseteq \tilde{N}_3$ and $\tilde{N}_1 \subsetneq \tilde{N}_2$.

Definition 6.4 [3] A single-valued neutrosophic fuzzy set (SVNFS) such as $\tilde{N}$ on the reference set X is a neutrosophic fuzzy set where $T_{\tilde{N}}(x), I_{\tilde{N}}(x), F_{\tilde{N}}(x) : X \to [0, 1]$ and $0 \le T_{\tilde{N}}(x) + I_{\tilde{N}}(x) + F_{\tilde{N}}(x) \le 3$. In other words, if $T_{\tilde{N}}(x)$, $I_{\tilde{N}}(x)$, and $F_{\tilde{N}}(x)$ be singletons in [0, 1], then, the set $\tilde{N} = \{(x, T_{\tilde{N}}(x), I_{\tilde{N}}(x), F_{\tilde{N}}(x)), x \in X\}$ is a SVNFS.

Definition 6.5 Assume the SVNFSs $\tilde{N}_1 = \{(x, T_{\tilde{N}_1}(x), I_{\tilde{N}_1}(x), F_{\tilde{N}_1}(x)), x \in X\}$ and $\tilde{N}_2 = \{(x, T_{\tilde{N}_2}(x), I_{\tilde{N}_2}(x), F_{\tilde{N}_2}(x)), x \in X\}$ on the reference set X. Then,

$\tilde{N}_1 \subseteq \tilde{N}_2$ if and only if $T_{\tilde{N}_1}(x) \le T_{\tilde{N}_2}(x)$, $I_{\tilde{N}_1}(x) \ge I_{\tilde{N}_2}(x)$, $F_{\tilde{N}_1}(x) \ge F_{\tilde{N}_2}(x)$, $\forall x \in X$, $\tilde{N}_1 = \tilde{N}_2$ if and only if $\tilde{N}_1 \subseteq \tilde{N}_2$, $\tilde{N}_2 \subseteq \tilde{N}_1$,

$$\tilde{N}_1^c = \{(x, F_{\tilde{N}_1}(x), 1 - I_{\tilde{N}_1}(x), T_{\tilde{N}_1}(x)), x \in X\}.$$

In the rest of this section for more simplicity, the set $\tilde{N} = \{(x, T_{\tilde{N}}(x), I_{\tilde{N}}(x), F_{\tilde{N}}(x)), x \in X\}$ is represented as $\tilde{N} = \{T_{\tilde{N}}(x), I_{\tilde{N}}(x), F_{\tilde{N}}(x)\}$.

Definition 6.6 [4] Assume the SVNFSs $\tilde{N}_1 = \{T_{\tilde{N}_1}(x), I_{\tilde{N}_1}(x), F_{\tilde{N}_1}(x)\}$ and $\tilde{N}_2 = \{T_{\tilde{N}_2}(x), I_{\tilde{N}_2}(x), F_{\tilde{N}_2}(x)\}$ on the reference set X. Then,

$$\tilde{N}_1 \cup \tilde{N}_2 = \{\max\{T_{\tilde{N}_1}(x), T_{\tilde{N}_2}(x)\}, \min\{I_{\tilde{N}_1}(x), I_{\tilde{N}_2}(x)\}, \min\{F_{\tilde{N}_1}(x), F_{\tilde{N}_2}(x)\}\},$$
$$\forall x \in X,$$
$$\tilde{N}_1 \cap \tilde{N}_2 = \{\min\{T_{\tilde{N}_1}(x), T_{\tilde{N}_2}(x)\}, \max\{I_{\tilde{N}_1}(x), I_{\tilde{N}_2}(x)\}, \max\{F_{\tilde{N}_1}(x), F_{\tilde{N}_2}(x)\}\},$$
$$\forall x \in X,$$
$$\tilde{N}_1 \oplus \tilde{N}_2 = \{T_{\tilde{N}_1}(x) + T_{\tilde{N}_2}(x) - T_{\tilde{N}_1}(x)T_{\tilde{N}_2}(x), I_{\tilde{N}_1}(x)I_{\tilde{N}_2}(x), F_{\tilde{N}_1}(x)F_{\tilde{N}_2}(x)\}, \forall x \in X,$$
$$\tilde{N}_1 \otimes \tilde{N}_2 = \left\{\begin{array}{l} T_{\tilde{N}_1}(x)T_{\tilde{N}_2}(x), I_{\tilde{N}_1}(x) + I_{\tilde{N}_2}(x) - I_{\tilde{N}_1}(x)I_{\tilde{N}_2}(x), \\ F_{\tilde{N}_1}(x) + F_{\tilde{N}_2}(x) - F_{\tilde{N}_1}(x)F_{\tilde{N}_2}(x) \end{array}\right\}, \forall x \in X,$$

$$\lambda\tilde{N}_1 = \left\{1 - \left(1 - T_{\tilde{N}_1}(x)\right)^{\lambda}, \left(I_{\tilde{N}_1}(x)\right)^{\lambda}, \left(F_{\tilde{N}_1}(x)\right)^{\lambda}\right\}, \lambda > 0,$$

$$\tilde{N}_1^{\lambda} = \left\{\left(T_{\tilde{N}_1}(x)\right)^{\lambda}, 1 - \left(1 - I_{\tilde{N}_1}(x)\right)^{\lambda}, 1 - \left(1 - F_{\tilde{N}_1}(x)\right)^{\lambda}\right\}, \lambda > 0.$$

Definition 6.7 [4] For the SVNFS $\tilde{N} = \{T_{\tilde{N}}(x), I_{\tilde{N}}(x), F_{\tilde{N}}(x)\}$, the below score and accuracy functions are defined respectively.

$$SV(\tilde{N}) = \frac{2 + T_{\tilde{N}} - I_{\tilde{N}} - F_{\tilde{N}}}{3}, SV(\tilde{N}) \in [0, 1], \tag{6.6}$$

$$AV(\tilde{N}) = T_{\tilde{N}} - F_{\tilde{N}}, AV(\tilde{N}) \in [-1, 1], \tag{6.7}$$

According to these functions, for the SVNFSs $\tilde{N}_1 = \{T_{\tilde{N}_1}(x), I_{\tilde{N}_1}(x), F_{\tilde{N}_1}(x)\}$ and $\tilde{N}_2 = \{T_{\tilde{N}_2}(x), I_{\tilde{N}_2}(x), F_{\tilde{N}_2}(x)\}$, the below ranking logics can be considered.

- $SV(\tilde{N}_1) < SV(\tilde{N}_2) \Rightarrow \tilde{N}_1 < \tilde{N}_2$,
- $SV(\tilde{N}_1) > SV(\tilde{N}_2) \Rightarrow \tilde{N}_1 > \tilde{N}_2$,
- $SV(\tilde{N}_1) = SV(\tilde{N}_2)$, then,
  - $AV(\tilde{N}_1) < AV(\tilde{N}_2) \Rightarrow \tilde{N}_1 < \tilde{N}_2$,
  - $AV(\tilde{N}_1) > AV(\tilde{N}_2) \Rightarrow \tilde{N}_1 > \tilde{N}_2$,
  - $AV(\tilde{N}_1) = AV(\tilde{N}_2) \Rightarrow \tilde{N}_1 = \tilde{N}_2$.

Remark 6.1 A neutrosophic fuzzy number (NFN) is obtained by considering a discrete neutrosophic fuzzy set in a continuous environment.

Definition 6.8 [4] Assume a set of single-valued neutrosophic fuzzy numbers shown as $\tilde{N}_j = \left(T_{\tilde{N}_j}, I_{\tilde{N}_j}, F_{\tilde{N}_j}\right), j = 1, \ldots, n$. Consider also the vector of importance weight values of these numbers as $W = (w_1, \ldots, w_n)^T$, where $w_j \in [0, 1]$ and $\sum_{j=1}^{n} w_j = 1$. Then the single-valued neutrosophic weighted arithmetic (SVNNWA) aggregator and the single-valued neutrosophic weighted geometric (SVNNWG) aggregator of these numbers are defined as below.

$$\begin{aligned} SVNNWA_w(\tilde{N}_1, \ldots, \tilde{N}_n) &= w_1\tilde{N}_1 + \ldots + w_n\tilde{N}_n \\ &= \left(1 - \prod_{j=1}^{n}\left(1 - T_{\tilde{N}_j}\right)^{w_j}, \prod_{j=1}^{n}\left(I_{\tilde{N}_j}\right)^{w_j}, \prod_{j=1}^{n}\left(F_{\tilde{N}_j}\right)^{w_j}\right), \end{aligned} \tag{6.8}$$

$$\begin{aligned} SVNNWG_w(\tilde{N}_1, \ldots, \tilde{N}_n) &= \prod_{j=1}^{n} \tilde{N}_j^{w_j} \\ &= \left(\prod_{j=1}^{n}\left(T_{\tilde{N}_j}\right)^{w_j}, 1 - \prod_{j=1}^{n}\left(1 - I_{\tilde{N}_j}\right)^{w_j}, \right. \end{aligned}$$

$$1-\prod_{j=1}^{n}\left(1-F_{\tilde{N}_j}\right)^{w_j}\Bigg). \tag{6.9}$$

6.2.1 Ranking of Neutrosophic Fuzzy Numbers

Definition 6.9 [4] Assume the neutrosophic fuzzy number $\tilde{N} = \left(T_{\tilde{N}}, I_{\tilde{N}}, F_{\tilde{N}}\right)$. For this number the below score and accuracy functions are defined respectively.

$$S\left(\tilde{N}\right) = T_{\tilde{N}} - I_{\tilde{N}} - F_{\tilde{N}}, S\left(\tilde{N}\right) \in [-2, 1], \tag{6.10}$$

$$A\left(\tilde{N}\right) = T_{\tilde{N}} + I_{\tilde{N}} + F_{\tilde{N}}, A\left(\tilde{N}\right) \in [0, 3], \tag{6.11}$$

In order to rank the neutrosophic fuzzy numbers $\tilde{N}_1$ and $\tilde{N}_2$, by applying the above functions the below logics are defined.

- $S\left(\tilde{N}_1\right) < S\left(\tilde{N}_2\right) \Rightarrow \tilde{N}_1 \prec \tilde{N}_2$,
- $S\left(\tilde{N}_1\right) > S\left(\tilde{N}_2\right) \Rightarrow \tilde{N}_1 \succ \tilde{N}_2$,
- $S\left(\tilde{N}_1\right) = S\left(\tilde{N}_2\right)$, then,
  - $A\left(\tilde{N}_1\right) < A\left(\tilde{N}_2\right) \Rightarrow \tilde{N}_1 \prec \tilde{N}_2$,
  - $A\left(\tilde{N}_1\right) > A\left(\tilde{N}_2\right) \Rightarrow \tilde{N}_1 \succ \tilde{N}_2$,
  - $A\left(\tilde{N}_1\right) = A\left(\tilde{N}_2\right) \Rightarrow \tilde{N}_1 \approx \tilde{N}_2$.

6.2.2 Distance of Neutrosophic Fuzzy Sets

Definition 6.10 [5] Consider two sets of single-valued neutrosophic fuzzy sets such as $\tilde{N}_1 = \left\{T_{\tilde{N}_1}(x_i), I_{\tilde{N}_1}(x_i), F_{\tilde{N}_1}(x_i)\right\}, i = 1, \ldots, n$ and $\tilde{N}_2 = \left\{T_{\tilde{N}_2}(x_i), I_{\tilde{N}_2}(x_i), F_{\tilde{N}_2}(x_i)\right\}, i = 1, \ldots, n$. Then the Hamming distance, Euclidean distance, and normalized Euclidean distance of these sets are defined as below respectively.

$$d_H\left(\tilde{N}_1, \tilde{N}_2\right) = \frac{\left|T_{\tilde{N}_1}(x_i) - T_{\tilde{N}_2}(x_i)\right| + \left|I_{\tilde{N}_1}(x_i) - I_{\tilde{N}_2}(x_i)\right| + \left|F_{\tilde{N}_1}(x_i) - F_{\tilde{N}_2}(x_i)\right|}{3}, \tag{6.12}$$

$$d_E\left(\tilde{N}_1, \tilde{N}_2\right) = \sqrt{\sum\nolimits_{i=1}^{n}\left(\begin{array}{c}\left(T_{\tilde{N}_1}(x_i) - T_{\tilde{N}_2}(x_i)\right)^2 + \left(I_{\tilde{N}_1}(x_i) - I_{\tilde{N}_2}(x_i)\right)^2 \\ + \left(F_{\tilde{N}_1}(x_i) - F_{\tilde{N}_2}(x_i)\right)^2\end{array}\right)}, \tag{6.13}$$

$$d_{NE}\left(\tilde{N}_1, \tilde{N}_2\right) = \sqrt{\frac{1}{3n}\sum_{i=1}^{n}\left(\begin{array}{r}\left(T_{\tilde{N}_1}(x_i) - T_{\tilde{N}_2}(x_i)\right)^2 + \left(I_{\tilde{N}_1}(x_i) - I_{\tilde{N}_2}(x_i)\right)^2 \\ + \left(F_{\tilde{N}_1}(x_i) - F_{\tilde{N}_2}(x_i)\right)^2\end{array}\right)} \tag{6.14}$$

6.3 Interval Valued Neutrosophic Fuzzy Sets

Definition 6.11 [4] An interval-valued neutrosophic fuzzy set (IVNS) such as $\tilde{N}^{IVNS}$ on the reference set X has the functions $T_{\tilde{N}}(x) = \left[T^L_{\tilde{N}}(x), T^U_{\tilde{N}}(x)\right] \subseteq [0, 1]$, $I_{\tilde{N}}(x) = \left[I^L_{\tilde{N}}(x), I^U_{\tilde{N}}(x)\right] \subseteq [0, 1]$, and $F_{\tilde{N}}(x) = \left[F^L_{\tilde{N}}(x), F^U_{\tilde{N}}(x)\right] \subseteq [0, 1]$, where $0 \leq T^U_{\tilde{N}}(x) + I^U_{\tilde{N}}(x) + F^U_{\tilde{N}}(x) \leq 3$. This IVNS is shown as below.

$$\begin{aligned}\tilde{N}^{IVNS} &= \left\{\left(x, T_{\tilde{N}}(x), I_{\tilde{N}}(x), F_{\tilde{N}}(x)\right), x \in X\right\} \\ &= \left\{\left(x, \left[T^L_{\tilde{N}}(x), T^U_{\tilde{N}}(x)\right], \left[I^L_{\tilde{N}}(x), I^U_{\tilde{N}}(x)\right], \left[F^L_{\tilde{N}}(x), F^U_{\tilde{N}}(x)\right]\right), x \in X\right\}\end{aligned} \tag{6.15}$$

Therefore, for $x \in X$, the interval-valued neutrosophic fuzzy number (IVNN) of $\tilde{N}^{IVNN} = \left\{\left[T^L_{\tilde{N}}(x), T^U_{\tilde{N}}(x)\right], \left[I^L_{\tilde{N}}(x), I^U_{\tilde{N}}(x)\right], \left[F^L_{\tilde{N}}(x), F^U_{\tilde{N}}(x)\right]\right\}$ is defined.

Example 6.4 Assume the reference set $X = \{x_1, x_2\}$, where x_1 is quality and x_2 is price that can be used to evaluate a product. Then, the characteristics of a certain product in terms of quality and price based on the opinions of an expert can be shown as an interval-valued neutrosophic fuzzy set as below.

$$\begin{aligned}\tilde{N}^{IVNS} = \{&(x_1, [0.1, 1.0], [0.1, 0.4], [0.4, 0.7]), \\ &(x_2, [0.6, 0.9], [0.8, 1.0], [0.4, 0.6])\}\end{aligned}$$

Definition 6.12 [6] Assume the IVNS $\tilde{N} = \left\{\left(x, \left[T^L_{\tilde{N}}(x), T^U_{\tilde{N}}(x)\right], \left[I^L_{\tilde{N}}(x), I^U_{\tilde{N}}(x)\right], \left[F^L_{\tilde{N}}(x), F^U_{\tilde{N}}(x)\right]\right), x \in X\right\}$. Then for any $x \in X$,

- $\tilde{\emptyset} = \{(x, [0, 0], [1, 1], [1, 1]), x \in X\}$,
- The set $\tilde{E} = \{(x, [1, 1], [0, 0], [0, 0]), x \in X\}$ is called reference set.
- The complement of $\tilde{N}$ is obtained as below.

$$\begin{aligned}\tilde{N}^c = \Big\{&x, \left[F^L_{\tilde{N}}(x), F^U_{\tilde{N}}(x)\right], \left[1 - I^U_{\tilde{N}}(x), 1 - I^L_{\tilde{N}}(x)\right], \\ &\left(\left[T^L_{\tilde{N}}(x), T^U_{\tilde{N}}(x)\right]\right), x \in X\Big\},\end{aligned} \tag{6.16}$$

Definition 6.13 [6] Consider the IVNSs $\tilde{A} = \left\{\left(x, \left[T^L_{\tilde{A}}(x), T^U_{\tilde{A}}(x)\right], \left[I^L_{\tilde{A}}(x), I^U_{\tilde{A}}(x)\right], \left[F^L_{\tilde{A}}(x), F^U_{\tilde{A}}(x)\right]\right), x \in X\right\}$ and $\tilde{B} = \left\{\left(x, \left[T^L_{\tilde{B}}(x), T^U_{\tilde{B}}(x)\right], \left[I^L_{\tilde{B}}(x), I^U_{\tilde{B}}(x)\right], \left[F^L_{\tilde{B}}(x), F^U_{\tilde{B}}(x)\right]\right), x \in X\right\}$. Then, the below logics and operations are defined.

$\tilde{A} \subseteq \tilde{B}$ if and only if $T_{\tilde{A}}^{L}(x) \leq T_{\tilde{B}}^{L}(x), T_{\tilde{A}}^{U}(x) \leq T_{\tilde{B}}^{U}(x), I_{\tilde{A}}^{L}(x) \geq I_{\tilde{B}}^{L}(x), I_{\tilde{A}}^{U}(x) \geq I_{\tilde{B}}^{U}(x)$,

$$F_{\tilde{A}}^{L}(x) \geq F_{\tilde{B}}^{L}(x), F_{\tilde{A}}^{U}(x) \geq F_{\tilde{B}}^{U}(x), \forall x \in X,$$

$$\tilde{A} \cap \tilde{B} = \{(x, [\min\{T_{\tilde{A}}^{L}(x), T_{\tilde{B}}^{L}(x)\}, \min\{T_{\tilde{A}}^{U}(x), T_{\tilde{B}}^{U}(x)\}],$$

$$[\max\{I_{\tilde{A}}^{L}(x), I_{\tilde{B}}^{L}(x)\}, \max\{I_{\tilde{A}}^{U}(x), I_{\tilde{B}}^{U}(x)\}]$$

$$[\max\{F_{\tilde{A}}^{L}(x), F_{\tilde{B}}^{L}(x)\}, \max\{F_{\tilde{A}}^{U}(x), F_{\tilde{B}}^{U}(x)\}]), x \in X\}$$

$$\tilde{A} \cup \tilde{B} = \{(x, [\max\{T_{\tilde{A}}^{L}(x), T_{\tilde{B}}^{L}(x)\}, \max\{T_{\tilde{A}}^{U}(x), T_{\tilde{B}}^{U}(x)\}],$$

$$[\min\{I_{\tilde{A}}^{L}(x), I_{\tilde{B}}^{L}(x)\}, \min\{I_{\tilde{A}}^{U}(x), I_{\tilde{B}}^{U}(x)\}]$$

$$[\min\{F_{\tilde{A}}^{L}(x), F_{\tilde{B}}^{L}(x)\}, \min\{F_{\tilde{A}}^{U}(x), F_{\tilde{B}}^{U}(x)\}]), x \in X\}$$

$$\tilde{A} \oplus \tilde{B} = \{(x, [T_{\tilde{A}}^{L}(x) + T_{\tilde{B}}^{L}(x) - T_{\tilde{A}}^{L}(x)T_{\tilde{B}}^{L}(x), T_{\tilde{A}}^{U}(x) + T_{\tilde{B}}^{U}(x) - T_{\tilde{A}}^{U}(x)T_{\tilde{B}}^{U}(x)],$$

$$[I_{\tilde{A}}^{L}(x)I_{\tilde{B}}^{L}(x), I_{\tilde{A}}^{U}(x)I_{\tilde{B}}^{U}(x)], [F_{\tilde{A}}^{L}(x)F_{\tilde{B}}^{L}(x), F_{\tilde{A}}^{U}(x)F_{\tilde{B}}^{U}(x)]), x \in X\}$$

$$\begin{aligned}\tilde{A} \otimes \tilde{B} = \{(x, &[T_{\tilde{A}}^{L}(x)T_{\tilde{B}}^{L}(x), T_{\tilde{A}}^{U}(x)T_{\tilde{B}}^{U}(x)], [I_{\tilde{A}}^{L}(x) + I_{\tilde{B}}^{L}(x) - I_{\tilde{A}}^{L}(x)I_{\tilde{B}}^{L}(x), \\ &I_{\tilde{A}}^{U}(x) + I_{\tilde{B}}^{U}(x) - I_{\tilde{A}}^{U}(x)I_{\tilde{B}}^{U}(x)], [F_{\tilde{A}}^{L}(x) + F_{\tilde{B}}^{L}(x) - F_{\tilde{A}}^{L}(x)F_{\tilde{B}}^{L}(x), \\ &F_{\tilde{A}}^{U}(x) + F_{\tilde{B}}^{U}(x) - F_{\tilde{A}}^{U}(x)F_{\tilde{B}}^{U}(x)]), x \in X\},\end{aligned}$$

For $k > 0$:

$$\begin{aligned}k\tilde{A} = \Big\{\Big(x, &\Big[1 - (1 - T_{\tilde{A}}^{L}(x))^{k}, 1 - (1 - T_{\tilde{A}}^{U}(x))^{k}\Big], \\ &\Big(\Big[(I_{\tilde{A}}^{L}(x))^{k}, (I_{\tilde{A}}^{U}(x))^{k}\Big], \Big[(F_{\tilde{A}}^{L}(x))^{k}, (F_{\tilde{A}}^{U}(x))^{k}\Big]\Big), x \in X\Big\},\end{aligned}$$

Definition 6.14 Assume the IVNN $\tilde{N} = \{[T_{\tilde{N}}^{L}, T_{\tilde{N}}^{U}], [I_{\tilde{N}}^{L}, I_{\tilde{N}}^{U}], [F_{\tilde{N}}^{L}, F_{\tilde{N}}^{U}]\}$. Some available score functions of the literature are mentioned below.

- Score function proposed by [7]:

$$S(\tilde{N}) = \frac{1}{8}(4 + T_{\tilde{N}}^{L} + T_{\tilde{N}}^{U} - F_{\tilde{N}}^{L} - F_{\tilde{N}}^{U} - 2I_{\tilde{N}}^{L} - 2I_{\tilde{N}}^{U}(4 - T_{\tilde{N}}^{L} - T_{\tilde{N}}^{U} - F_{\tilde{N}}^{L} - F_{\tilde{N}}^{U})), \tag{6.17}$$

- Score function proposed by [8]:

$$S(\tilde{N}) = \frac{1}{4}\left(2 + T_{\tilde{N}}^{L} + T_{\tilde{N}}^{U} - 2I_{\tilde{N}}^{L} - 2I_{\tilde{N}}^{U} - F_{\tilde{N}}^{L} - F_{\tilde{N}}^{U}\right), \tag{6.18}$$

- Score function proposed by [9]:

$$S(\tilde{N}) = \left(\frac{2}{3} + \frac{T_{\tilde{N}}^{L} + T_{\tilde{N}}^{U}}{6} - \frac{I_{\tilde{N}}^{L} + I_{\tilde{N}}^{U}}{6} - \frac{F_{\tilde{N}}^{L} + F_{\tilde{N}}^{U}}{6}\right), \tag{6.19}$$

- Score function proposed by [10]:

$$S(\tilde{N}) = \left(2 + \frac{T_{\tilde{N}}^{L} + T_{\tilde{N}}^{U}}{2} - \frac{I_{\tilde{N}}^{L} + I_{\tilde{N}}^{U}}{2} - \frac{F_{\tilde{N}}^{L} + F_{\tilde{N}}^{U}}{2}\right), \tag{6.20}$$

- Score function proposed by [11]:

$$S(\tilde{N}) = \left(\frac{T_{\tilde{N}}^{L} + T_{\tilde{N}}^{U}}{2} + \left(1 - \frac{I_{\tilde{N}}^{L} + I_{\tilde{N}}^{U}}{2}\right) I_{\tilde{N}}^{U}(x) - \left(\frac{F_{\tilde{N}}^{L} + F_{\tilde{N}}^{U}}{2}\right)\left(1 - F_{\tilde{N}}^{U}\right)\right), \tag{6.21}$$

Based on the above-mentioned score functions, the IVNNs $\tilde{N}_1$ and $\tilde{N}_2$ are ranked according to the below logics.

- $S(\tilde{N}_1) < S(\tilde{N}_2) \Rightarrow \tilde{N}_1 \prec \tilde{N}_2$,
- $S(\tilde{N}_1) > S(\tilde{N}_2) \Rightarrow \tilde{N}_1 \succ \tilde{N}_2$,

6.4 Trapezoidal Neutrosophic Fuzzy Numbers

Assume the trapezoidal neutrosophic fuzzy number $\tilde{N} = \left(w_{\tilde{N}}, u_{\tilde{N}}, v_{\tilde{N}}\right)$ on the set of real numbers. This number has the following functions.

$$T_{\tilde{N}}(x) = \begin{cases} T_{\tilde{N}}^{L}(x), a_1 \le x \le a_2, \\ 1, a_2 \le x \le a_3, \\ T_{\tilde{N}}^{U}(x), a_3 \le x \le a_4, \\ 0, \textit{otherwise} \end{cases} \tag{6.22}$$

$$I_{\tilde{N}}(x) = \begin{cases} I^L_{\tilde{N}}(x), b_1 \le x \le b_2, \\ 0, b_2 \le x \le b_3, \\ I^U_{\tilde{N}}(x), b_3 \le x \le b_4, \\ 1, otherwise \end{cases} \tag{6.23}$$

$$F_{\tilde{N}}(x) = \begin{cases} F^L_{\tilde{N}}(x), c_1 \le x \le c_2, \\ 0, c_2 \le x \le c_3, \\ F^U_{\tilde{N}}(x), c_3 \le x \le c_4, \\ 1, otherwise \end{cases} \tag{6.24}$$

In these functions, $0 \le sup\ T_{\tilde{N}}(x) + sup\ I_{\tilde{N}}(x) + sup\ F_{\tilde{N}}(x) \le 3$, $a_1 \le a_2 \le a_3 \le a_4$, $b_1 \le b_2 \le b_3 \le b_4$, and $c_1 \le c_2 \le c_3 \le c_4$ (for any $x \in X$). On the other hand, $T^L_{\tilde{N}}(x), I^L_{\tilde{N}}(x), F^L_{\tilde{N}}(x) \in [0, 1]$ are continuous monotonic increasing functions, and $T^U_{\tilde{N}}(x), I^U_{\tilde{N}}(x), F^U_{\tilde{N}}(x) \in [0, 1]$ are continuous monotonic decreasing functions.

Assume $w_{\tilde{N}}$, $u_{\tilde{N}}$, and $v_{\tilde{N}}$ from the interval [0, 1]. Further assume that $a, b, c, d \in \mathbb{R}$ and $a \le b \le c \le d$. An interval-valued trapezoidal neutrosophic fuzzy number (SVTrNN) is defined as a typical neutrosophic fuzzy set of the set of real numbers and shown as $\tilde{N} = \{(a, b, c, d); w_{\tilde{N}}, u_{\tilde{N}}, v_{\tilde{N}}\}$. This SVTrNN has the functions $T_{\tilde{N}} : \mathbb{R} \to [0, w_{\tilde{N}}]$, $I_{\tilde{N}} : \mathbb{R} \to [u_{\tilde{N}}, 1]$, and $F_{\tilde{N}} : \mathbb{R} \to [v_{\tilde{N}}, 1]$ detailed as below.

$$T_{\tilde{N}}(x) = \begin{cases} \frac{x-a}{b-a} w_{\tilde{N}}, a \le x \le b, \\ w_{\tilde{N}}, b \le x \le c, \\ \frac{d-x}{d-c} w_{\tilde{N}}, c \le x \le d, \\ 0, otherwise \end{cases} \tag{6.25}$$

$$I_{\tilde{N}}(x) = \begin{cases} \frac{(b-x)+(x-a)u_{\tilde{N}}}{b-a}, a \le x \le b, \\ u_{\tilde{N}}, b \le x \le c, \\ \frac{(x-c)+(d-x)u_{\tilde{N}}}{d-c}, c \le x \le d, \\ 1, otherwise \end{cases} \tag{6.26}$$

$$F_{\tilde{N}}(x) = \begin{cases} \frac{(b-x)+(x-a)v_{\tilde{N}}}{b-a}, a \le x \le b, \\ v_{\tilde{N}}, b \le x \le c, \\ \frac{(x-c)+(d-x)v_{\tilde{N}}}{d-c}, c \le x \le d, \\ 1, otherwise \end{cases} \tag{6.27}$$

Here, $w_{\tilde{N}}$ is maximum of the truth-membership function, $u_{\tilde{N}}$ is minimum of the indeterminacy-membership function, and $v_{\tilde{N}}$ is minimum of the falsity-membership function. This number is shown in Fig. 6.2.

Definition 6.15 [12] Assume the SVTrNNs $\tilde{N}_1 = \{(a_1, b_1, c_1, d_1); w_{\tilde{N}_1}, u_{\tilde{N}_1}, v_{\tilde{N}_1}\}$ and $\tilde{N}_2 = \{(a_2, b_2, c_2, d_2); w_{\tilde{N}_2}, u_{\tilde{N}_2}, v_{\tilde{N}_2}\}$, and the real number k. Then,

$$\tilde{N}_1 \oplus \tilde{N}_2 = (a_1 + a_2, b_1 + b_2, c_1 + c_2, d_1 + d_2;$$

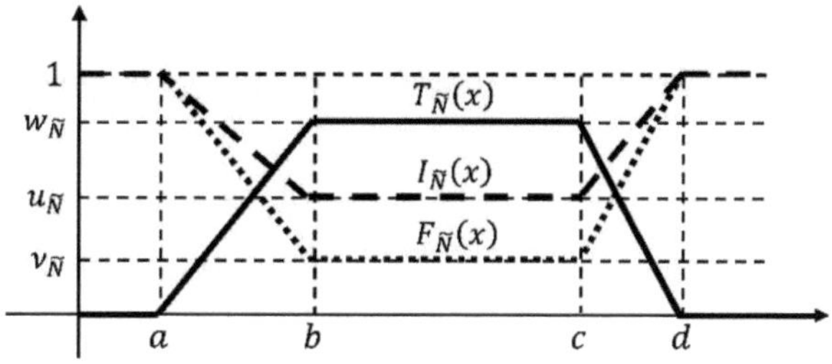

Fig. 6.2 Schematic representation of a SVTrNN

$$\min\{w_{\tilde{N}_1}, w_{\tilde{N}_2}\}, \max\{u_{\tilde{N}_1}, u_{\tilde{N}_2}\}, \max\{v_{\tilde{N}_1}, v_{\tilde{N}_2}\}), \tag{6.28}$$

$$\tilde{N}_1 \ominus \tilde{N}_2 = (a_1 - d_2, b_1 - c_2, c_1 - b_2, d_1 - a_2; \\ \min\{w_{\tilde{N}_1}, w_{\tilde{N}_2}\}, \max\{u_{\tilde{N}_1}, u_{\tilde{N}_2}\}, \max\{v_{\tilde{N}_1}, v_{\tilde{N}_2}\}), \tag{6.29}$$

$$\tilde{N}_1 \otimes \tilde{N}_2 = \begin{cases} \begin{pmatrix} a_1 a_2, b_1 b_2, c_1 c_2, d_1 d_2; \\ \min\{w_{\tilde{N}_1}, w_{\tilde{N}_2}\}, \max\{u_{\tilde{N}_1}, u_{\tilde{N}_2}\}, \max\{v_{\tilde{N}_1}, v_{\tilde{N}_2}\} \end{pmatrix}, a_1, a_2 > 0, \\ \begin{pmatrix} a_1 d_2, b_1 c_2, c_1 b_2, d_1 a_2; \\ \min\{w_{\tilde{N}_1}, w_{\tilde{N}_2}\}, \max\{u_{\tilde{N}_1}, u_{\tilde{N}_2}\}, \max\{v_{\tilde{N}_1}, v_{\tilde{N}_2}\} \end{pmatrix}, a_1 > 0, d_2 < 0, \\ \begin{pmatrix} d_1 d_2, c_1 c_2, b_1 b_2, a_1 a_2; \\ \min\{w_{\tilde{N}_1}, w_{\tilde{N}_2}\}, \max\{u_{\tilde{N}_1}, u_{\tilde{N}_2}\}, \max\{v_{\tilde{N}_1}, v_{\tilde{N}_2}\} \end{pmatrix}, a_1, a_2 < 0, \end{cases} \tag{6.30}$$

$$\frac{\tilde{N}_1}{\tilde{N}_2} = \left(\frac{a_1}{d_2}, \frac{b_1}{c_2}, \frac{c_1}{b_2}, \frac{d_1}{a_2}; \min\{w_{\tilde{N}_1}, w_{\tilde{N}_2}\}, \\ \max\{u_{\tilde{N}_1}, u_{\tilde{N}_2}\}, \max\{v_{\tilde{N}_1}, v_{\tilde{N}_2}\}\right), a_1, a_2 > 0, \tag{6.31}$$

$$k\tilde{N}_1 = \begin{cases} \{(ka_1, kb_1, kc_1, kd_1); w_{\tilde{N}_1}, u_{\tilde{N}_1}, v_{\tilde{N}_1}\}, k > 0, \\ \{(kd_1, kc_1, kb_1, ka_1); w_{\tilde{N}_1}, u_{\tilde{N}_1}, v_{\tilde{N}_1}\}, k < 0, \end{cases} \tag{6.32}$$

$$\tilde{N}_1^{-1} = \left\{\left(\frac{1}{d_1}, \frac{1}{c_1}, \frac{1}{b_1}, \frac{1}{a_1}\right); w_{\tilde{N}_1}, u_{\tilde{N}_1}, v_{\tilde{N}_1}\right\} \tag{6.33}$$

Remark 6.2 According to Definition 6.15, the below remarks are mentioned.

- If $w_{\tilde{N}_1} = 1$ and $u_{\tilde{N}_1} = v_{\tilde{N}_1} = 0$, then, we obtain a single-valued trapezoidal neutrosophic fuzzy number.
- If $b = c$, then, we obtain a single-valued triangular neutrosophic fuzzy number (SVTNN).

Example 6.5 For the SVTNNs $\tilde{N}_1 = \{(1, 3, 4); 0.3, 0.4, 0.6\}$ and $\tilde{N}_2 = \{(2, 6, 7); 0.5, 0.7, 0.9\}$ the below calculations are done.

$$\tilde{N}_1 \oplus \tilde{N}_2 = (3, 9, 11; 0.3, 0.7, 0.9),$$

$$\begin{aligned}\tilde{N}_1 \ominus \tilde{N}_2 &= (-6, -3, 2; 0.3, 0.7, 0.9),\\ \tilde{N}_1 \otimes \tilde{N}_2 &= (2, 18, 28; 0.3, 0.7, 0.9)\\ \frac{\tilde{N}_1}{\tilde{N}_2} &= \left(\frac{1}{7}, \frac{3}{6}, 2; 0.3, 0.7, 0.9\right),\\ 3\tilde{N}_1 &= (3, 9, 12; 0.3, 0.7, 0.9),\\ \tilde{N}_1^{-1} &= \left\{\left(\frac{1}{4}, \frac{1}{3}, 1\right); 0.3, 0.4, 0.6\right\}.\end{aligned}$$

Definition 6.16 [13] For the SVTrNN $\tilde{N} = \{(a, b, c, d); w_{\tilde{N}}, u_{\tilde{N}}, v_{\tilde{N}}\}$, the (α, β, γ)-cuts below are defined.

$$\tilde{N}_\alpha = \{x | T_{\tilde{N}}(x) \geq \alpha\} = \left[a + \frac{\alpha}{w_{\tilde{N}}}(b - a), d + \frac{\alpha}{w_{\tilde{N}}}(c - d)\right], 0 \leq \alpha \leq w_{\tilde{N}}, \tag{6.34}$$

$$\tilde{N}_\beta = \{x | I_{\tilde{N}}(x) \leq \beta\} = \left[a + \frac{1-\beta}{1-u_{\tilde{N}}}(b - a), d + \frac{1-\beta}{1-u_{\tilde{N}}}(c - d)\right], u_{\tilde{N}} \leq \beta \leq 1, \tag{6.35}$$

$$\tilde{N}_\gamma = \{x | F_{\tilde{N}}(x) \leq \gamma\} = \left[a + \frac{1-\gamma}{1-v_{\tilde{N}}}(b - a), d + \frac{1-\gamma}{1-v_{\tilde{N}}}(c - d)\right], v_{\tilde{N}} \leq \gamma \leq 1, \tag{6.36}$$

$$\begin{aligned}\tilde{N}_{\alpha,\beta,\gamma} &= \{x | T_{\tilde{N}}(x) \geq \alpha, I_{\tilde{N}}(x) \leq \beta, F_{\tilde{N}}(x) \leq \gamma\}\\ &= \left[a + \frac{\alpha}{w_{\tilde{N}}}(b - a), d + \frac{\alpha}{w_{\tilde{N}}}(c - d)\right]\\ &\cap \left[a + \frac{1-\beta}{1-u_{\tilde{N}}}(b - a), d + \frac{1-\beta}{1-u_{\tilde{N}}}(c - d)\right]\\ &\cap \left[a + \frac{1-\gamma}{1-v_{\tilde{N}}}(b - a), d + \frac{1-\gamma}{1-v_{\tilde{N}}}(c - d)\right],\end{aligned} \tag{6.37}$$

Here, $0 \leq \alpha \leq w_{\tilde{N}}$, $u_{\tilde{N}} \leq \beta \leq 1$, $v_{\tilde{N}} \leq \gamma \leq 1$, and $0 \leq \alpha + \beta + \gamma \leq 2 + w_{\tilde{N}}$.

Definition 6.17 [13] For the SVTrNN $\tilde{N} = \{(a, b, c, d); w_{\tilde{N}}, u_{\tilde{N}}, v_{\tilde{N}}\}$, the $(\alpha, \beta', \gamma')$-cuts below are defined.

$$\tilde{N}_{\beta'} = \{x | I_{\tilde{N}}(x) \geq \beta'\} = \left[a + \frac{\beta'}{1-u_{\tilde{N}}}(b - a), d + \frac{\beta'}{1-u_{\tilde{N}}}(c - d)\right], \quad 0 \leq \beta' \leq 1 - u_{\tilde{N}}, \tag{6.38}$$

$$\tilde{N}_{\gamma'} = \{x | F_{\tilde{N}}(x) \geq \gamma'\} = \left[a + \frac{\gamma'}{1-v_{\tilde{N}}}(b - a), d + \frac{\gamma'}{1-v_{\tilde{N}}}(c - d)\right],$$

$$0 \le \gamma' \le 1 - v_{\tilde{N}}, \tag{6.39}$$

$$\begin{aligned}\tilde{N}_{\alpha,\beta',\gamma'} &= \left\{x | T_{\tilde{N}}(x) \ge \alpha, I_{\tilde{N}}(x) \ge \beta', F_{\tilde{N}}(x) \ge \gamma'\right\} \\ &= \left[a + \frac{\alpha}{w_{\tilde{N}}}(b-a), d + \frac{\alpha}{w_{\tilde{N}}}(c-d)\right] \\ &\cap \left[a + \frac{\beta'}{1-u_{\tilde{N}}}(b-a), d + \frac{\beta'}{1-u_{\tilde{N}}}(c-d)\right] \\ &\cap \left[a + \frac{\gamma'}{1-v_{\tilde{N}}}(b-a), d + \frac{\gamma'}{1-v_{\tilde{N}}}(c-d)\right],\end{aligned} \tag{6.40}$$

Here, $0 \le \alpha \le w_{\tilde{N}}$, $0 \le \beta' \le 1 - u_{\tilde{N}}$, $0 \le \gamma' \le 1 - v_{\tilde{N}}$, and $0 \le \alpha + \beta' + \gamma' \le 2 + w_{\tilde{N}} - u_{\tilde{N}} - v_{\tilde{N}}$.

6.4.1 Ranking of Trapezoidal Neutrosophic Fuzzy Numbers

There are several score, accuracy, and ranking functions in the literature for trapezoidal neutrosophic fuzzy numbers (TrNFNs). In this sub-section, some of them are reviewed.

Definition 6.18 [14] Assume the SVTrNN $\tilde{N} = \left\{(a, b, c, d); w_{\tilde{N}}, u_{\tilde{N}}, v_{\tilde{N}}\right\}$. The below score and accuracy functions are defined for this number respectively.

$$S(\tilde{N}) = \frac{1}{12}(a + b + c + d)(2 + w_{\tilde{N}} - u_{\tilde{N}} - v_{\tilde{N}}), \tag{6.41}$$

$$A(\tilde{N}) = \frac{1}{12}(a + b + c + d)(2 + w_{\tilde{N}} - u_{\tilde{N}} + v_{\tilde{N}}), \tag{6.42}$$

Definition 6.19 [15] Assume the SVTrNN $\tilde{N} = \left\{(a, b, c, d); w_{\tilde{N}}, u_{\tilde{N}}, v_{\tilde{N}}\right\}$. The below ranking function is defined for this number.

$$\Re_{UU}(\tilde{N}) = \frac{(a + b + c + d)(2 + w_{\tilde{N}} - u_{\tilde{N}} - v_{\tilde{N}})}{12}, \tag{6.43}$$

Definition 6.20 [16] Assume the SVTrNN $\tilde{N} = \left\{(a, b, c, d); w_{\tilde{N}}, u_{\tilde{N}}, v_{\tilde{N}}\right\}$. The below ranking function is defined for this number.

$$\Re_K(\tilde{N}) = \frac{(c + d - a - d)(w_{\tilde{N}} - u_{\tilde{N}} - v_{\tilde{N}})}{2}, \tag{6.44}$$

Definition 6.21 Assume the SVTrNN $\tilde{N} = \left\{(a, b, c, d); w_{\tilde{N}}, u_{\tilde{N}}, v_{\tilde{N}}\right\}$. The below ranking functions are defined for this number for maximization and minimization problems respectively.

$$\Re(\tilde{N})\frac{|(3d-a+2b+2c)(w_{\tilde{N}}-u_{\tilde{N}}-\nu_{\tilde{N}})|}{4}_{\max} \tag{6.45}$$

$$\Re(\tilde{N})\frac{|(3b+3c-a-d)(w_{\tilde{N}}-u_{\tilde{N}}-\nu_{\tilde{N}})|}{4}_{\min} \tag{6.46}$$

Definition 6.22 [13] Assume the SVTrNN $\tilde{N} = \{(a,b,c,d); w_{\tilde{N}}, u_{\tilde{N}}, \nu_{\tilde{N}}\}$. The below ranking function in confidence levels not less than α, β', γ' is defined for this number.

$$\begin{aligned}\Re_{\alpha,\beta',\gamma'}(\tilde{N}) &= (a+d)(w_{\tilde{N}}-\alpha)+(b+c-a-d)\left(\frac{w_{\tilde{N}}^2-\alpha^2}{2w_{\tilde{N}}}\right)\\&+(a+d)(1-u_{\tilde{N}}-\beta')+(b+c-a-d)\left(\frac{(1-u_{\tilde{N}})^2-\beta'^2}{2(1-u_{\tilde{N}})}\right)\\&+(a+d)(1-\nu_{\tilde{N}}-\gamma')+(b+c-a-d)\left(\frac{(1-\nu_{\tilde{N}})^2-\gamma'^2}{2(1-\nu_{\tilde{N}})}\right)\end{aligned} \tag{6.47}$$

Theorem 6.1 [13] *Assume real number k, the SVTrNNs* $\tilde{N} = \{(a,b,c,d); w_{\tilde{N}}, u_{\tilde{N}}, \nu_{\tilde{N}}\}$, $\tilde{N}_1 = \{(a_1,b_1,c_1,d_1); w_{\tilde{N}_1}, u_{\tilde{N}_1}, \nu_{\tilde{N}_1}\}$, *and* $\tilde{N}_2 = \{(-d_2,-c_2,-b_2,-a_2); w_{\tilde{N}_2}, u_{\tilde{N}_2}, \nu_{\tilde{N}_2}\}$, *where,* $w_{\tilde{N}_1} = w_{\tilde{N}_2}$, $u_{\tilde{N}_1} = u_{\tilde{N}_2}$, *and* $\nu_{\tilde{N}_1} = \nu_{\tilde{N}_2}$. *Then, according to the ranking function* (6.47),

- $\Re_{\alpha,\beta',\gamma'}(\tilde{N}_1) = -\Re_{\alpha,\beta',\gamma'}(\tilde{N}_2)$
- $\Re_{\alpha,\beta',\gamma'}(\tilde{N}_1 \oplus \tilde{N}_2) = \Re_{\alpha,\beta',\gamma'}(\tilde{N}_1) + \Re_{\alpha,\beta',\gamma'}(\tilde{N}_2)$
- $\Re_{\alpha,\beta',\gamma'}(k\tilde{N}) = k\Re_{\alpha,\beta',\gamma'}(\tilde{N})$

Definition 6.23 [13] Assume the SVTrNNs $\tilde{N}_1 = \{(a_1,b_1,c_1,d_1); w_{\tilde{N}_1}, u_{\tilde{N}_1}, \nu_{\tilde{N}_1}\}$ and $\tilde{N}_2 = \{(a_2,b_2,c_2,d_2); w_{\tilde{N}_2}, u_{\tilde{N}_2}, \nu_{\tilde{N}_2}\}$. Then, for decision level greater than α ($0 \le \alpha < \min\{w_{\tilde{N}_1}, w_{\tilde{N}_2}\}$), and aspiration levels greater than β' and β' for $u_{\tilde{N}_2}$ and $\nu_{\tilde{N}_2}$ respectively which means that $0 \le \beta' < \min\{1-u_{\tilde{N}_1}, 1-u_{\tilde{N}_2}\}$ and $0 \le \gamma' < \min\{1-\nu_{\tilde{N}_1}, 1-\nu_{\tilde{N}_2}\}$, there is,

$$\tilde{N}_1 \prec_{\alpha,\beta',\gamma'} \tilde{N}_2 \Leftrightarrow \Re_{\alpha,\beta',\gamma'}(\tilde{N}_1) < \Re_{\alpha,\beta',\gamma'}(\tilde{N}_2), \tag{6.48}$$

$$\tilde{N}_1 \succ_{\alpha,\beta',\gamma'} \tilde{N}_2 \Leftrightarrow \Re_{\alpha,\beta',\gamma'}(\tilde{N}_1) > \Re_{\alpha,\beta',\gamma'}(\tilde{N}_2), \tag{6.49}$$

$$\tilde{N}_1 \sim_{\alpha,\beta',\gamma'} \tilde{N}_2 \Leftrightarrow \Re_{\alpha,\beta',\gamma'}(\tilde{N}_1) = \Re_{\alpha,\beta',\gamma'}(\tilde{N}_2). \tag{6.50}$$

Definition 6.24 [13] Assume the SVTrNNs $\tilde{N}_1 = \{(a_1, b_1, c_1, d_1); w_{\tilde{N}_1}, u_{\tilde{N}_1}, v_{\tilde{N}_1}\}$ and $\tilde{N}_2 = \{(a_2, b_2, c_2, d_2); w_{\tilde{N}_2}, u_{\tilde{N}_2}, v_{\tilde{N}_2}\}$. Then, considering $0 \le \alpha < \min\{w_{\tilde{N}_1}, w_{\tilde{N}_2}\}$, $0 \le \beta' < \min\{1 - u_{\tilde{N}_1}, 1 - u_{\tilde{N}_2}\}$, and $0 \le \gamma' < \min\{1 - v_{\tilde{N}_1}, 1 - v_{\tilde{N}_2}\}$, there is,

$$\tilde{N}_1 \prec_{\alpha,\beta',\gamma'} \tilde{N}_2 \Leftrightarrow \forall\beta', \gamma : \Re_{\alpha,\beta',\gamma'}(\tilde{N}_1) < \Re_{\alpha,\beta',\gamma'}(\tilde{N}_2), \tag{6.51}$$

$$\tilde{N}_1 \succ_{\alpha,\beta',\gamma'} \tilde{N}_2 \Leftrightarrow \forall\beta', \gamma : \Re_{\alpha,\beta',\gamma'}(\tilde{N}_1) > \Re_{\alpha,\beta',\gamma'}(\tilde{N}_2), \tag{6.52}$$

$$\tilde{N}_1 \sim_{\alpha,\beta',\gamma'} \tilde{N}_2 \Leftrightarrow \forall\beta', \gamma : \Re_{\alpha,\beta',\gamma'}(\tilde{N}_1) = \Re_{\alpha,\beta',\gamma'}(\tilde{N}_2). \tag{6.53}$$

Example 6.6 [13] Assume the SVTrNNs $\tilde{N}_1 = \{(1, 3, 4, 5); 0.8, 0.5, 0.4\}$ and $\tilde{N}_2 = \{(1, 2, 5, 6); 0.8, 0.5, 0.4\}$. These numbers are ranked for different values of α, β', and γ', in Tables 6.1, 6.2, 6.3 and 6.4.

Table 6.1 Ranking results of Example 6.6 for $\alpha = 0, 0.1, 0.2, 0.3, 0.4, 0.5$, $\beta' = 0$, and $\gamma' = 0$

SVTrNNs	$\alpha = 0$, $\beta' =$ $\gamma' = 0$	$\alpha = 0.1$, $\beta' =$ $\gamma' = 0$	$\alpha = 0.2$, $\beta' =$ $\gamma' = 0$	$\alpha = 0.3$, $\beta' =$ $\gamma' = 0$	$\alpha = 0.4$, $\beta' =$ $\gamma' = 0$	$\alpha = 0.5$, $\beta' =$ $\gamma' = 0$	Ranking results
$\tilde{N}_1$	12.35	11.74	11.12	10.49	9.85	9.19	$\tilde{N}_2 \succ \tilde{N}_1$
$\tilde{N}_2$	13.3	12.6	11.9	11.2	10.5	9.8	

Table 6.2 Ranking results of Example 6.6 for $\alpha = 0$, $\beta' = 0.1, 0.2, 0.3, 0.4, 0.5$, and $\gamma' = 0$

SVTrNNs	$\alpha = 0, \gamma' = 0 \beta' = 0.1$	$\alpha = 0, \gamma' = 0 \beta' = 0.2$	$\alpha = 0, \gamma' = 0 \beta' = 0.3$	$\alpha = 0, \gamma' = 0 \beta' = 0.4$	$\alpha = 0, \gamma' = 0 \beta' = 0.5$	Ranking results
$\tilde{N}_1$	11.74	11.11	10.46	9.79	9.1	$\tilde{N}_2 \succ \tilde{N}_1$
$\tilde{N}_2$	12.6	11.9	11.2	10.5	9.8	

Table 6.3 Ranking results of Example 6.6 for $\alpha = 0$, $\beta' = 0$, and $\gamma' = 0.1, 0.2, 0.3, 0.4, 0.5$

SVTNNs	$\alpha = 0, \beta' = 0 \gamma' = 0.1$	$\alpha = 0, \beta' = 0 \gamma' = 0.2$	$\alpha = 0, \beta' = 0 \gamma' = 0.3$	$\alpha = 0, \beta' = 0 \gamma' = 0.4$	$\alpha = 0, \beta' = 0 \gamma' = 0.5$	Ranking results
$\tilde{N}_1$	11.74	11.11	10.47	9.81	9.14	$\tilde{N}_2 \succ \tilde{N}_1$
$\tilde{N}_2$	12.6	11.9	11.2	10.5	9.8	

Table 6.4 Ranking results of Example 6.6 for $\alpha = \beta' = \gamma' = 0.1, 0.2, 0.3, 0.4, 0.5$

SVTNNs	$\alpha = \beta' = \gamma' = 0.1$	$\alpha = \beta' = \gamma' = 0.2$	$\alpha = \beta' = \gamma' = 0.3$	$\alpha = \beta' = \gamma' = 0.4$	$\alpha = \beta' = \gamma' = 0.5$	Ranking results
$\tilde{N}_1$	10.52	8.65	6.72	4.75	2.73	$\tilde{N}_2 \succ \tilde{N}_1$
$\tilde{N}_2$	11.2	9.1	7	4.9	2.8	

6.4.2 *Interval-Valued Trapezoidal Neutrosophic Numbers*

Definition 6.25 [17] Assume $\tilde{N}$ as a trapezoidal neutrosophic fuzzy number with the below truth-membership, indeterminacy-membership, and falsity-membership functions respectively.

$$T_{\tilde{N}}(x) = \begin{cases} \frac{x-a}{b-a} w_{\tilde{N}}, a \leq x \leq b, \\ w_{\tilde{N}}, b \leq x \leq c, \\ \frac{d-x}{d-c} w_{\tilde{N}}, c \leq x \leq d, \\ 0, O.W. \end{cases} \tag{6.54}$$

$$I_{\tilde{N}}(x) = \begin{cases} \frac{(b-x)+(x-a)u_{\tilde{N}}}{b-a}, a \leq x \leq b, \\ u_{\tilde{N}}, b \leq x \leq c, \\ \frac{(x-c)+(d-x)u_{\tilde{N}}}{d-c}, c \leq x \leq d, \\ 1, O.W. \end{cases} \tag{6.55}$$

$$F_{\tilde{N}}(x) = \begin{cases} \frac{(b-x)+(x-a)v_{\tilde{N}}}{b-a}, a \leq x \leq b, \\ v_{\tilde{N}}, b \leq x \leq c, \\ \frac{(x-c)+(d-x)v_{\tilde{N}}}{d-c}, c \leq x \leq d, \\ 1, O.W. \end{cases} \tag{6.56}$$

If here, $w_{\tilde{N}} = \left[w_{\tilde{N}}^L, w_{\tilde{N}}^U\right] \subseteq [0, 1]$, $u_{\tilde{N}} = \left[u_{\tilde{N}}^L, u_{\tilde{N}}^U\right] \subseteq [0, 1]$, and $v_{\tilde{N}} = \left[v_{\tilde{N}}^L, v_{\tilde{N}}^U\right] \subseteq [0, 1]$, then, $\tilde{N}$ is an interval-valued trapezoidal neutrosophic fuzzy number (IVTrNN) shown as $\tilde{N}^{IV} = \left\{(a, b, c, d); \left[w_{\tilde{N}}^L, w_{\tilde{N}}^U\right], \left[u_{\tilde{N}}^L, u_{\tilde{N}}^U\right], \left[v_{\tilde{N}}^L, v_{\tilde{N}}^U\right]\right\}$. If $a > 0$, then $\tilde{N}^{IV}$ is positive. If $b = c$, then $\tilde{N}^{IV}$ is an interval-valued triangular neutrosophic fuzzy number.

Another representation of an IVTrNN is $\tilde{N}^{IV} = \left\{\left((a, b, c, d); w_{\tilde{N}^{IV}}^{IV}\right), \left((e, f, g, h); u_{\tilde{N}^{IV}}^{IV}\right), \left((m, n, o, p); v_{\tilde{N}^{IV}}^{IV}\right)\right\}$, where, $w_{\tilde{N}^{IV}}^{IV} = \left[w_{\tilde{N}}^L, w_{\tilde{N}}^U\right] \subseteq [0, 1]$, $u_{\tilde{N}^{IV}}^{IV} = \left[u_{\tilde{N}}^L, u_{\tilde{N}}^U\right] \subseteq [0, 1]$, and $v_{\tilde{N}^{IV}}^{IV} = \left[v_{\tilde{N}}^L, v_{\tilde{N}}^U\right] \subseteq [0, 1]$.

6.4.3 Operational Laws on IVTrNNs

Assume the IVTrNN $\tilde{N}_1^{IV} = \left\{\left((a_1, b_1, c_1, d_1); w^{IV}_{\tilde{N}_1^{IV}}\right), \left((e_1, f_1, g_1, h_1); u^{IV}_{\tilde{N}_1^{IV}}\right), \left((m_1, n_1, o_1, p_1); v^{IV}_{\tilde{N}_1^{IV}}\right)\right\}$ and $\tilde{N}_2^{IV} = \left\{\left((a_2, b_2, c_2, d_2); w^{IV}_{\tilde{N}_2^{IV}}\right), \left((e_2, f_2, g_2, h_2); u^{IV}_{\tilde{N}_2^{IV}}\right), \left((m_2, n_2, o_2, p_2); v^{IV}_{\tilde{N}_2^{IV}}\right)\right\}$, where, $w^{IV}_{\tilde{N}_1^{IV}} = \left[w^{L}_{\tilde{N}_1^{IV}}, w^{U}_{\tilde{N}_1^{IV}}\right] \subseteq [0, 1]$, $u^{IV}_{\tilde{N}_1^{IV}} = \left[u^{L}_{\tilde{N}_1^{IV}}, u^{U}_{\tilde{N}_1^{IV}}\right] \subseteq [0, 1]$, $v^{IV}_{\tilde{N}_1^{IV}} = \left[v^{L}_{\tilde{N}_1^{IV}}, v^{U}_{\tilde{N}_1^{IV}}\right] \subseteq [0, 1]$, $w^{IV}_{\tilde{N}_2^{IV}} = \left[w^{L}_{\tilde{N}_2^{IV}}, w^{U}_{\tilde{N}_2^{IV}}\right] \subseteq [0, 1]$, $u^{IV}_{\tilde{N}_2^{IV}} = \left[u^{L}_{\tilde{N}_2^{IV}}, u^{U}_{\tilde{N}_2^{IV}}\right] \subseteq [0, 1]$, and $v^{IV}_{\tilde{N}_2^{IV}} = \left[v^{L}_{\tilde{N}_2^{IV}}, v^{U}_{\tilde{N}_2^{IV}}\right] \subseteq [0, 1]$. Then,

$$\tilde{N}_1^{IV} \oplus \tilde{N}_2^{IV} = \left\{\begin{array}{l} \left((a_1 + a_2, b_1 + b_2, c_1 + c_2, d_1 + d_2); \min\left\{w^{V}_{\tilde{N}_1^{IV}}, w^{IV}_{\tilde{N}_2^{IV}}\right\}\right), \\ \left((e_1 + e_2, f_1 + f_2, g_1 + g_2, h_1 + h_2); \max\left\{u^{IV}_{\tilde{N}_1^{IV}}, u^{IV}_{\tilde{N}_2^{IV}}\right\}\right), \\ \left((m_1 + m_2, n_1 + n_2, o_1 + o_2, p_1 + p_2); \max\left\{v^{IV}_{\tilde{N}_1^{IV}}, v^{IV}_{\tilde{N}_2^{IV}}\right\}\right) \end{array}\right\},$$

$$-\tilde{N}_1^{IV} = \left\{\left((-d_1, -c_1, -b_1, -a_1); w^{IV}_{\tilde{N}_1^{IV}}\right), \left((-h_1, -g_1, -f_1, -e_1); u^{IV}_{\tilde{N}_1^{IV}}\right), \left((-p_1, -o_1, -n_1, -m_1); v^{IV}_{\tilde{N}_1^{IV}}\right)\right\},$$

$$\tilde{N}_1^{IV} \ominus \tilde{N}_2^{IV} = \left\{\begin{array}{l} \left((a_1 - d_2, b_1 - c_2, c_1 - b_2, d_1 - a_2); \min\left\{w^{IV}_{\tilde{N}_1^{IV}}, w^{IV}_{\tilde{N}_2^{IV}}\right\}\right), \\ \left((e_1 - h_2, f_1 - g_2, g_1 - f_2, h_1 - e_2); \max\left\{u^{IV}_{\tilde{N}_1^{IV}}, u^{IV}_{\tilde{N}_2^{IV}}\right\}\right), \\ \left((m_1 - p_2, n_1 - o_2, o_1 - n_2, p_1 - m_2); \max\left\{v^{IV}_{\tilde{N}_1^{IV}}, v^{IV}_{\tilde{N}_2^{IV}}\right\}\right) \end{array}\right\},$$

$$k\tilde{N}_1^{IV} = \left\{\left((ka_1, kb_1, kc_1, kd_1); w^{IV}_{\tilde{N}_1^{IV}}\right), \left((ke_1, kf_1, kg_1, kh_1); u^{IV}_{\tilde{N}_1^{IV}}\right), \left((km_1, kn_1, ko_1, kp_1); v^{IV}_{\tilde{N}_1^{IV}}\right)\right\}, k > 0,$$

$$k\tilde{N}_1^{IV} = \left\{\left((kd_1, kc_1, kb_1, ka_1); w^{IV}_{\tilde{N}_1^{IV}}\right), \left((kh_1, kg_1, kf_1, ek_1); u^{IV}_{\tilde{N}_1^{IV}}\right), \left((kp_1, ko_1, kn_1, km_1); v^{IV}_{\tilde{N}_1^{IV}}\right)\right\}, k < 0,$$

$$\tilde{N}_1^{IV} \otimes \tilde{N}_2^{IV} = \left\{\begin{array}{l} \left((a_1a_2, b_1b_2, c_1c_2, d_1d_2); \min\left\{w^{IV}_{\tilde{N}_1^{IV}}, w^{IV}_{\tilde{N}_2^{IV}}\right\}\right), \\ \left((e_1e_2, f_1f_2, g_1g_2, h_1h_2); \max\left\{u^{IV}_{\tilde{N}_1^{IV}}, u^{IV}_{\tilde{N}_2^{IV}}\right\}\right), \\ \left((m_1m_2, n_1n_2, o_1o_2, p_1p_2); \max\left\{v^{IV}_{\tilde{N}_1^{IV}}, v^{IV}_{\tilde{N}_2^{IV}}\right\}\right) \end{array}\right\},$$
$$a_1 > 0, a_2 > 0, e_1 > 0, e_2 > 0, m_1 > 0, m_2 > 0,$$

$$\frac{\tilde{N}_1^{IV}}{\tilde{N}_2^{IV}} = \left\{ \begin{array}{l} \left(\left(\frac{a_1}{d_2}, \frac{b_1}{c_2}, \frac{c_1}{b_2} \frac{d_1}{a_2} \right); \min\left\{ w_{\tilde{N}_1^{IV}}^{IV}, w_{\tilde{N}_2^{IV}}^{IV} \right\} \right), \\ \left(\left(\frac{e_1}{h_2}, \frac{f_1}{g_2}, \frac{g_1}{f_2} \frac{h_1}{e_2} \right); \max\left\{ u_{\tilde{N}_1^{IV}}^{IV}, u_{\tilde{N}_2^{IV}}^{IV} \right\} \right), \\ \left(\left(\frac{m_1}{p_2}, \frac{n_1}{o_2}, \frac{o_1}{n_2} \frac{p_1}{m_2} \right); \max\left\{ v_{\tilde{N}_1^{IV}}^{IV}, v_{\tilde{N}_2^{IV}}^{IV} \right\} \right) \end{array} \right\},$$

$$a_1 > 0, a_2 > 0, e_1 > 0, e_2 > 0, m_1 > 0, m_2 > 0,$$

where $\min\left\{ w_{\tilde{N}_1^{IV}}^{IV}, w_{\tilde{N}_2^{IV}}^{IV} \right\} = \left[\min\left\{ w_{\tilde{N}_1^{IV}}^{L}, w_{\tilde{N}_2^{IV}}^{L} \right\}, \min\left\{ w_{\tilde{N}_1^{IV}}^{U}, w_{\tilde{N}_2^{IV}}^{U} \right\} \right]$, $\max\left\{ u_{\tilde{N}_1^{IV}}^{IV}, u_{\tilde{N}_2^{IV}}^{IV} \right\} = \left[\max\left\{ u_{\tilde{N}_1^{IV}}^{L}, u_{\tilde{N}_2^{IV}}^{L} \right\}, \max\left\{ u_{\tilde{N}_1^{IV}}^{U}, u_{\tilde{N}_2^{IV}}^{U} \right\} \right]$, $\max\left\{ v_{\tilde{N}_1^{IV}}^{IV}, v_{\tilde{N}_2^{IV}}^{IV} \right\} = \left[\max\left\{ v_{\tilde{N}_1^{IV}}^{L}, v_{\tilde{N}_2^{IV}}^{L} \right\}, \max\left\{ v_{\tilde{N}_1^{IV}}^{U}, v_{\tilde{N}_2^{IV}}^{U} \right\} \right]$.

Sahin [18] used the concept of score function to compare IVTrNNs. This means that an IVTrNN with higher score is greater. Assume the IVTrNN $\tilde{N}^{IV} = \left\{ \left((a, b, c, d); w_{\tilde{N}^{IV}}^{IV} \right), \left((e, f, g, h); u_{\tilde{N}^{IV}}^{IV} \right), \left((m, n, o, p); v_{\tilde{N}^{IV}}^{IV} \right) \right\}$. Then the score function $S\left(\tilde{N}^{IV}\right)$ and accuracy function $A\left(\tilde{N}^{IV}\right)$ are defined as below [19].

$$\begin{aligned} S\left(\tilde{N}^{IV}\right) = & \frac{1}{12}(8 + (a + b + c + d) - (e + f + g + h) - (m + n + o + p)) \\ & \times \left(2 + w_{\tilde{N}}^{L} + w_{\tilde{N}}^{U} - u_{\tilde{N}}^{L} - u_{\tilde{N}}^{U} - v_{\tilde{N}}^{L} - v_{\tilde{N}}^{U}\right), \end{aligned} \tag{6.57}$$

$$\begin{aligned} A\left(\tilde{N}^{IV}\right) = & \frac{1}{4}(a + b + c + d - (m + n + o + p)) \\ & \times \left(2 + w_{\tilde{N}}^{L} + w_{\tilde{N}}^{U} - u_{\tilde{N}}^{L} - u_{\tilde{N}}^{U} - v_{\tilde{N}}^{L} - v_{\tilde{N}}^{U}\right). \end{aligned} \tag{6.58}$$

Now, the IVTrNNs $\tilde{N}_1^{IV}$ and $\tilde{N}_2^{IV}$ are ranked by the below logics.

- $S\left(\tilde{N}_1^{IV}\right) < S\left(\tilde{N}_2^{IV}\right) \Rightarrow \tilde{N}_1^{IV} \prec \tilde{N}_2^{IV}$,
- $S\left(\tilde{N}_1^{IV}\right) > S\left(\tilde{N}_2^{IV}\right) \Rightarrow \tilde{N}_1^{IV} \succ \tilde{N}_2^{IV}$,
- $S\left(\tilde{N}_1^{IV}\right) = S\left(\tilde{N}_2^{IV}\right)$, then,
  - $A\left(\tilde{N}_1^{IV}\right) < A\left(\tilde{N}_2^{IV}\right) \Rightarrow \tilde{N}_1^{IV} \prec \tilde{N}_2^{IV}$
  - $A\left(\tilde{N}_1^{IV}\right) > A\left(\tilde{N}_2^{IV}\right) \Rightarrow \tilde{N}_1^{IV} \succ \tilde{N}_2^{IV}$
  - $A\left(\tilde{N}_1^{IV}\right) = A\left(\tilde{N}_2^{IV}\right) \Rightarrow \tilde{N}_1^{IV} \approx \tilde{N}_2^{IV}$

Example 6.7 Assume the IVTrNN $\tilde{N}_1^{IV} = \{((7, 11, 16, 21); [0.6, 0.8]), ((6, 10, 15, 20); [0.3, 0.4]), ((5, 9, 14, 19); [0.4, 0.6])\}$ and $\tilde{N}_2^{IV} = \{((6, 11, 13, 20); [0.7, 0.8]), ((5, 10, 12, 18); [0.4, 0.5]), ((3, 8, 11, 16); [0.5, 0.6])\}$. According to the above score and accuracy functions the values of $S\left(\tilde{N}_1^{IV}\right) = -4.95$, $S\left(\tilde{N}_2^{IV}\right) = -9.35$, $A\left(\tilde{N}_1^{IV}\right) = 5.1$, and $A\left(\tilde{N}_2^{IV}\right) = 4.87$ are obtained. As $S\left(\tilde{N}_1^{IV}\right) = -4.95 > S\left(\tilde{N}_2^{IV}\right) = -9.35$,, then $\tilde{N}_1^{IV} \succ \tilde{N}_2^{IV}$.

6.5 Type-2 Neutrosophic Fuzzy Numbers

Definition 6.26 [20] Considering the reference set X, a type-2 neutrosophic (T2NN) is defined as $\tilde{N} = \{x, (\tilde{T}_{\tilde{N}}(x), \tilde{I}_{\tilde{N}}(x), \tilde{F}_{\tilde{N}}(x)), x \in X\}$ where $\tilde{T}_{\tilde{N}} : X \to [0, 1]$ is its truth-membership function, $\tilde{I}_{\tilde{N}} : X \to [0, 1]$ is its indeterminacy-membership function, and $\tilde{F}_{\tilde{N}} : X \to [0, 1]$ is its falsity-membership function. For any $x \in X$, the condition $0 \le (\tilde{T}_{\tilde{N}}(x))^3 + (\tilde{I}_{\tilde{N}}(x))^3 + (\tilde{F}_{\tilde{N}}(x))^3 \le 3$ should be held.

For more simplicity, a T2NN is shown as $\tilde{N} = \{(T_{T_{\tilde{N}}}, T_{I_{\tilde{N}}}, T_{F_{\tilde{N}}}), (I_{T_{\tilde{N}}}, I_{I_{\tilde{N}}}, I_{F_{\tilde{N}}}), (F_{T_{\tilde{N}}}, F_{I_{\tilde{N}}}, F_{F_{\tilde{N}}})\}$.

Example 6.8 The number $\tilde{N} = \{(0.90, 0.85, 0.95), (0.10, 0.15, 0.10), (0.05, 0.05, 0.10)\}$ is a T2NN because the condition $(0.9)^3 + (0.85)^3 + (0.95)^3 + (0.1)^3 + (0.15)^3 + (0.1)^3 + (0.05)^3 + (0.05)^3 + (0.1)^3 = 2.21$ is held.

Definition 6.27 [20] Consider the T2NNs $\tilde{N}_1 = \left\{\left(T_{T_{\tilde{N}_1}}, T_{I_{\tilde{N}_1}}, T_{F_{\tilde{N}_1}}\right), \left(I_{T_{\tilde{N}_1}}, I_{I_{\tilde{N}_1}}, I_{F_{\tilde{N}_1}}\right), \left(F_{T_{\tilde{N}_1}}, F_{I_{\tilde{N}_1}}, F_{F_{\tilde{N}_1}}\right)\right\}$ and $\tilde{N}_2 = \left\{\left(T_{T_{\tilde{N}_2}}, T_{I_{\tilde{N}_2}}, T_{F_{\tilde{N}_2}}\right), \left(I_{T_{\tilde{N}_2}}, I_{I_{\tilde{N}_2}}, I_{F_{\tilde{N}_2}}\right), \left(F_{T_{\tilde{N}_2}}, F_{I_{\tilde{N}_2}}, F_{F_{\tilde{N}_2}}\right)\right\}$. Some arithmetic operations are defined below.

$$\begin{aligned}
\tilde{N}_1 \oplus \tilde{N}_2 = \Big\{ & \Big(\Big(T_{T_{\tilde{N}_1}} + T_{T_{\tilde{N}_2}} - T_{T_{\tilde{N}_1}} T_{T_{\tilde{N}_2}}\Big), \Big(T_{I_{\tilde{N}_1}} + T_{I_{\tilde{N}_2}} - T_{I_{\tilde{N}_1}} T_{I_{\tilde{N}_2}}\Big), \\
& \Big(T_{F_{\tilde{N}_1}} + T_{F_{\tilde{N}_2}} - T_{F_{\tilde{N}_1}} T_{F_{\tilde{N}_2}}\Big)\Big), \Big(\Big(I_{T_{\tilde{N}_1}} I_{T_{\tilde{N}_2}}\Big), \Big(I_{I_{\tilde{N}_1}} I_{I_{\tilde{N}_2}}\Big), \Big(I_{F_{\tilde{N}_1}} I_{F_{\tilde{N}_2}}\Big)\Big), \\
& \Big(\Big(F_{T_{\tilde{N}_1}} F_{T_{\tilde{N}_2}}\Big), \Big(F_{I_{\tilde{N}_1}} F_{I_{\tilde{N}_2}}\Big), \Big(F_{F_{\tilde{N}_1}} F_{F_{\tilde{N}_2}}\Big)\Big)\Big\}
\end{aligned} \tag{6.59}$$

$$\begin{aligned}
\tilde{N}_1 \otimes \tilde{N}_2 = \Big\{ & \Big(\Big(T_{T_{\tilde{N}_1}} T_{T_{\tilde{N}_2}}\Big), \Big(T_{I_{\tilde{N}_1}} T_{I_{\tilde{N}_2}}\Big), \Big(T_{F_{\tilde{N}_1}} T_{F_{\tilde{N}_2}}\Big)\Big), \\
& \Big(\Big(T_{T_{\tilde{N}_1}} + T_{T_{\tilde{N}_2}} - T_{T_{\tilde{N}_1}} T_{T_{\tilde{N}_2}}\Big), \Big(I_{I_{\tilde{N}_1}} + I_{I_{\tilde{N}_2}} - I_{I_{\tilde{N}_1}} I_{I_{\tilde{N}_2}}\Big), \\
& \Big(F_{F_{\tilde{N}_1}} + F_{F_{\tilde{N}_2}} - F_{F_{\tilde{N}_1}} F_{F_{\tilde{N}_2}}\Big)\Big), \Big(\Big(F_{T_{\tilde{N}_1}} + F_{T_{\tilde{N}_2}} - F_{T_{\tilde{N}_1}} F_{T_{\tilde{N}_2}}\Big), \\
& \Big(F_{I_{\tilde{N}_1}} + F_{I_{\tilde{N}_2}} - F_{I_{\tilde{N}_1}} F_{I_{\tilde{N}_2}}\Big), \Big(F_{F_{\tilde{N}_1}} + F_{F_{\tilde{N}_2}} - F_{F_{\tilde{N}_1}} F_{F_{\tilde{N}_2}}\Big)\Big)\Big\},
\end{aligned} \tag{6.60}$$

$$\begin{aligned}
k\tilde{N}_1 = \Big\{ & \Big(1 - \Big(1 - T_{T_{\tilde{N}_1}}\Big)^k, 1 - \Big(1 - T_{I_{\tilde{N}_1}}\Big)^k, 1 - \Big(1 - T_{F_{\tilde{N}_1}}\Big)^k\Big), \\
& \Big(I^k_{T_{\tilde{N}_1}}, I^k_{I_{\tilde{N}_1}}, I^k_{F_{\tilde{N}_1}}\Big), \Big(F^k_{T_{\tilde{N}_1}}, F^k_{I_{\tilde{N}_1}}, F^k_{F_{\tilde{N}_1}}\Big)\Big\},
\end{aligned} \tag{6.61}$$

$$\tilde{N}_1^k = \left\{\begin{array}{l}
\Big(T^k_{T_{\tilde{N}_1}}, T^k_{I_{\tilde{N}_1}}, T^k_{F_{\tilde{N}_1}}\Big), \Big(1 - \Big(1 - I_{T_{\tilde{N}_1}}\Big)^k, 1 - \Big(1 - I_{I_{\tilde{N}_1}}\Big)^k, 1 - \Big(1 - I_{F_{\tilde{N}_1}}\Big)^k\Big), \\
\Big(1 - \Big(1 - F_{T_{\tilde{N}_1}}\Big)^k, 1 - \Big(1 - F_{I_{\tilde{N}_1}}\Big)^k, 1 - \Big(1 - F_{F_{\tilde{N}_1}}\Big)^k\Big)
\end{array}\right\},$$
$$k > 0. \tag{6.62}$$

Definition 6.28 [20] Consider the T2NNs $\tilde{N}_1, ..., \tilde{N}_k$ and their associated weight values shown by the vector $W = (w_1, ..., w_n)^T$ where $w_i \in [0, 1]$ and $\sum_{i=1}^{n} w_i = 1$. The type-2 neutrosophic fuzzy number arithmetic mean of these T2NNs are obtained by the below aggregator.

$$
\begin{aligned}
& T2NNWA_w(\tilde{N}_1, ..., \tilde{N}_k) \\
& = w_1\tilde{N}_1 + ... + w_k\tilde{N}_k \\
& = \left\{ \begin{array}{l} \left(\left(1 - \prod_{j=1}^{k} \left(1 - T_{T_{\tilde{N}_j}}\right)^{w_j}, 1 - \prod_{j=1}^{k} \left(1 - T_{I_{\tilde{N}_j}}\right)^{w_j}, 1 - \prod_{j=1}^{k} \left(1 - T_{F_{\tilde{N}_j}}\right)^{w_j} \right) \right), \\ \left(\prod_{j=1}^{k} I_{T_{\tilde{N}_j}}^{w_j}, \prod_{j=1}^{k} I_{I_{\tilde{N}_j}}^{w_j}, \prod_{j=1}^{k} I_{F_{\tilde{N}_j}}^{w_j} \right), \left(\prod_{j=1}^{k} F_{T_{\tilde{N}_j}}^{w_j}, \prod_{j=1}^{k} F_{I_{\tilde{N}_j}}^{w_j}, \prod_{j=1}^{k} F_{F_{\tilde{N}_j}}^{w_j} \right) \end{array} \right\}
\end{aligned}
\tag{6.63}
$$

Definition 6.29 [20] For the T2NN $\tilde{N} = \left\{\left(T_{T_{\tilde{N}}}, T_{I_{\tilde{N}}}, T_{F_{\tilde{N}}}\right), \left(I_{T_{\tilde{N}}}, I_{I_{\tilde{N}}}, I_{F_{\tilde{N}}}\right), \left(F_{T_{\tilde{N}}}, F_{I_{\tilde{N}}}, F_{F_{\tilde{N}}}\right)\right\}$, the below score and accuracy functions are proposed respectively.

$$
S(\tilde{N}) = \frac{1}{12}\left(8 + T_{T_{\tilde{N}}} + 2T_{I_{\tilde{N}}} + T_{F_{\tilde{N}}} - I_{T_{\tilde{N}}} - 2I_{I_{\tilde{N}}} - I_{F_{\tilde{N}}} - F_{T_{\tilde{N}}} - 2F_{I_{\tilde{N}}} - F_{F_{\tilde{N}}}\right),
\tag{6.64}
$$

$$
A(\tilde{N}) = \frac{1}{4}\left(T_{T_{\tilde{N}}} + 2T_{I_{\tilde{N}}} + T_{F_{\tilde{N}}} - F_{T_{\tilde{N}}} - 2F_{I_{\tilde{N}}} - F_{F_{\tilde{N}}}\right).
\tag{6.65}
$$

Now, the T2NNs $\tilde{N}_1$ and $\tilde{N}_2$ are ranked by the below logics.

- $S(\tilde{N}_1) > S(\tilde{N}_2), \Leftrightarrow \tilde{N}_1 \succ \tilde{N}_2$,
- $S(\tilde{N}_1) < S(\tilde{N}_2), \Leftrightarrow \tilde{N}_1 \prec \tilde{N}_2$,
- $S(\tilde{N}_1) = S(\tilde{N}_2)$, then,
  - $A(\tilde{N}_1) > A(\tilde{N}_2), \Leftrightarrow \tilde{N}_1 \succ \tilde{N}_2$,
  - $A(\tilde{N}_1) < A(\tilde{N}_2), \Leftrightarrow \tilde{N}_1 \prec \tilde{N}_2$,
  - $A(\tilde{N}_1) = A(\tilde{N}_2), \Leftrightarrow \tilde{N}_1 \approx \tilde{N}_2$.

Definition 6.30 [21] For the triangular T2NN $\tilde{N} = \left\{\left(\tilde{N}_1, \tilde{N}_2, \tilde{N}_3\right); \left(T_{\tilde{N}}, I_{\tilde{N}}, F_{\tilde{N}}\right)\right\} = \left\{ \begin{array}{l} \left\langle (a_1, b_1, c_1); w_{\tilde{N}_1}, u_{\tilde{N}_1}, v_{\tilde{N}_1} \right\rangle, \left\langle (a_2, b_2, c_2); w_{\tilde{N}_2}, u_{\tilde{N}_2}, v_{\tilde{N}_2} \right\rangle, \\ \left\langle (a_3, b_3, c_3); w_{\tilde{N}_3}, u_{\tilde{N}_3}, v_{\tilde{N}_3} \right\rangle; \left(w_{\tilde{N}}, u_{\tilde{N}}, v_{\tilde{N}}\right) \end{array} \right\}$, considering the values $w_{\tilde{N}} = \min\left\{w_{\tilde{N}_1}, w_{\tilde{N}_2}, w_{\tilde{N}_3}\right\}, u_{\tilde{N}} = \max\left\{u_{\tilde{N}_1}, u_{\tilde{N}_2}, u_{\tilde{N}_3}\right\}$, and $v_{\tilde{N}} = \max\left\{v_{\tilde{N}_1}, v_{\tilde{N}_2}, v_{\tilde{N}_3}\right\}$, the below ranking function is proposed.

$$
\begin{aligned}
\Re(\tilde{N}) = & \left(\frac{w_{\tilde{N}} + u_{\tilde{N}} + v_{\tilde{N}}}{3}\right)\left(\frac{1}{3}\right) \\
& \left(\frac{w_{\tilde{N}_1} + w_{\tilde{N}_2} + w_{\tilde{N}_3}}{3} + \frac{u_{\tilde{N}_1} + u_{\tilde{N}_2} + u_{\tilde{N}_3}}{3} + \frac{v_{\tilde{N}_1} + v_{\tilde{N}_2} + v_{\tilde{N}_3}}{3}\right),
\end{aligned}
\tag{6.66}
$$

Table 6.5 The logics proposed by [22] to convert linguistic evaluations to T2NN

Linguistic variable	Equivalent T2NN
Very low	$\{(0.20, 0.20, 0.10), (0.65, 0.80, 0.85), (0.45, 0.80, 0.70)\}$
Low	$\{(0.35, 0.35, 0.10), (0.50, 0.75, 0.80), (0.50, 0.75, 0.65)\}$
Medium low	$\{(0.40, 0.30, 0.35), (0.50, 0.45, 0.60), (0.45, 0.40, 0.60)\}$
Medium	$\{(0.50, 0.45, 0.50), (0.40, 0.35, 0.50), (0.35, 0.30, 0.45)\}$
Medium high	$\{(0.60, 0.45, 0.50), (0.20, 0.15, 0.25), (0.10, 0.25, 0.15)\}$
High	$\{(0.70, 0.75, 0.80), (0.15, 0.20, 0.25), (0.10, 0.15, 0.20)\}$
Very high	$\{(0.95, 0.90, 0.95), (0.10, 0.10, 0.05), (0.05, 0.05, 0.05)\}$

In addition, T2NNs can be used to convert linguistic evaluations to a numerical form. This issue is very useful in optimization problems. For this aim, the conversion logics of Table 6.5 have been proposed in the literature.

6.6 Possibility Theory of Single-Valued Neutrosophic Fuzzy Numbers

This section presents the measures of possibility theory for single-valued neutrosophic fuzzy numbers (SVNNs).

Definition 6.31 Consider $\tilde{A} = \left(T_{\tilde{A}}, I_{\tilde{A}}, F_{\tilde{A}}\right)$ and $\tilde{B} = \left(T_{\tilde{B}}, I_{\tilde{B}}, F_{\tilde{B}}\right)$ as a pair of SVNNs with truth-membership functions $T_{\tilde{A}}(x)$ and $T_{\tilde{B}}(x)$, indeterminacy-membership functions $I_{\tilde{A}}(x)$ and $I_{\tilde{B}}(x)$, and falsity-membership functions $F_{\tilde{A}}(x)$ and $F_{\tilde{B}}(x)$, respectively. Then, the possibility measure of a fuzzy event for the truth-membership, indeterminacy-membership, and falsity-membership functions (Pos_T, Pos_I, Pos_F) and their necessity measures (Nec_T, Nec_I, Nec_F) are defined by the following equations.

$$
\begin{aligned}
Pos_T\left(\tilde{A} * \tilde{B}\right) &= sup\left\{\min\left(T_{\tilde{A}}(x), T_{\tilde{B}}(y)\right), x, y \in \mathbb{R}, x * y\right\},\\
Pos_I\left(\tilde{A} * \tilde{B}\right) &= sup\left\{\min\left(I_{\tilde{A}}(x), I_{\tilde{B}}(y)\right), x, y \in \mathbb{R}, x * y\right\},\\
Pos_F\left(\tilde{A} * \tilde{B}\right) &= sup\left\{\min\left(F_{\tilde{A}}(x), F_{\tilde{B}}(y)\right), x, y \in \mathbb{R}, x * y\right\},\\
Nec_T\left(\tilde{A} * \tilde{B}\right) &= inf\left\{\max\left(T_{\tilde{A}}(x), T_{\tilde{B}}(y)\right), x, y \in \mathbb{R}, x * y\right\},\\
Nec_I\left(\tilde{A} * \tilde{B}\right) &= inf\left\{\max\left(I_{\tilde{A}}(x), I_{\tilde{B}}(y)\right), x, y \in \mathbb{R}, x * y\right\},\\
Nec_F\left(\tilde{A} * \tilde{B}\right) &= inf\left\{\max\left(F_{\tilde{A}}(x), F_{\tilde{B}}(y)\right), x, y \in \mathbb{R}, x * y\right\},
\end{aligned}
$$

In these equations, the notation * can be of the signs <, >, ≤, ≥, =.

The dual relationship of possibility and necessity shows that,

$$
Nec_T\left(\tilde{A} * \tilde{B}\right) = 1 - Pos_T\left(\overline{\tilde{A} * \tilde{B}}\right),
$$

$$Nec_I(\tilde{A} * \tilde{B}) = 1 - Pos_I\left(\overline{\tilde{A} * \tilde{B}}\right),$$
$$Nec_F(\tilde{A} * \tilde{B}) = 1 - Pos_F\left(\overline{\tilde{A} * \tilde{B}}\right),$$

where $\overline{\tilde{A} * \tilde{B}}$ represents complement of the event $\tilde{A} * \tilde{B}$.

Definition 6.32 Considering $\tilde{A}$ as a SVNN with the truth-membership function $T_{\tilde{A}}(x)$, indeterminacy-membership function $I_{\tilde{A}}(x)$, and falsity-membership function $F_{\tilde{A}}(x)$, the *Me* measure of its truth-membership functions, indeterminacy-membership function, and falsity-membership function are represented as (Me_T, Me_I, Me_F) and defined by the followings,

$$Me_T(\tilde{A}) = \lambda Pos_T(\tilde{A}) + (1 - \lambda) Nec_T(\tilde{A}),$$
$$Me_I(\tilde{A}) = \lambda Pos_I(\tilde{A}) + (1 - \lambda) Nec_I(\tilde{A}),$$
$$Me_F(\tilde{A}) = \lambda Pos_F(\tilde{A}) + (1 - \lambda) Nec_F(\tilde{A}),$$

where $\lambda \in [0, 1]$ is the optimistic-pessimistic parameter to determine the combined attitude of a decision maker.

If $\lambda = 0$, then $Me_T(\tilde{A}) = Nec_T(\tilde{A}), Me_I(\tilde{A}) = Nec_I(\tilde{A}), Me_F(\tilde{A}) = Nec_F(\tilde{A})$,
If $\lambda = 0.5$, then $Me_T(\tilde{A}) = Cr_T(\tilde{A}), Me_I(\tilde{A}) = Cr_I(\tilde{A}), Me_F(\tilde{A}) = Cr_F(\tilde{A})$,
If $\lambda = 1$, then $Me_T(\tilde{A}) = Pos_T(\tilde{A}), Me_I(\tilde{A}) = Pos_I(\tilde{A}), Me_F(\tilde{A}) = Pos_F(\tilde{A})$,
Here, the sign *Cr* is credibility measure.

Definition 6.33 Let $\tilde{A} = \{(a_1, a_2, a_3, a_4); w_{\tilde{A}}, u_{\tilde{A}}, v_{\tilde{A}}\}$ and $\tilde{B} = \{(b_1, b_2, b_3, b_4); w_{\tilde{B}}, u_{\tilde{B}}, v_{\tilde{B}}\}$ be two SVNNs. From Definition 6.31, the possibility measures of the event $\tilde{A} \leq \tilde{B}$ for all of the membership functions are as below.

$$Pos_T(\tilde{A} \leq \tilde{B}) = \begin{cases} \min\{w_{\tilde{A}}, w_{\tilde{B}}\}, & a_2 \leq b_3, \\ \frac{(b_4 - a_4)\min(w_A, w_B)}{(b_4 - b_3)w_A + (a_2 - a_1)w_{\tilde{B}}}, & a_1 < b_4, a_2 > b_3 \\ 0, & b_4 \leq a_1 \end{cases} \tag{6.67}$$

$$Pos_I(\tilde{A} \leq \tilde{B}) = \begin{cases} \max\{u_{\tilde{A}}, u_{\tilde{B}}\}, & a_2 \leq b_3, \\ \frac{(b_4 u_B - b_3)(u_A - 1) - (a_2 - a_1 u_A)(1 - u_B)}{(b_4 - b_3)(u_A - 1) - (a_2 - a_1)(1 - u_{\tilde{B}})}, & a_2 < b_3, a_1 > b_4 \\ 1, & b_4 \leq a_1 \end{cases} \tag{6.68}$$

$$Pos_F(\tilde{A} \leq \tilde{B}) = \begin{cases} \max\{v_{\tilde{A}}, v_{\tilde{B}}\}, & a_2 \leq b_3, \\ \frac{(b_4 v_B - b_3)(v_A - 1) - (a_2 - a_1 v_A)(1 - v_B)}{(b_4 - b_3)(v_A - 1) - (a_2 - a_1)(1 - v_{\tilde{B}})}, & a_2 < b_3, a_1 > b_4 \\ 1, & b_4 \leq a_1 \end{cases} \tag{6.69}$$

For the particular case of $\tilde{A} \leq x$, according to Definition 6.32, the following possibility, measures are obtained.

$$Pos_T\left(\tilde{A} \leq x\right) = \begin{cases} w_{\tilde{A}} & a_2 \leq x \\ \frac{w_{\tilde{A}}(x-a_1)}{a_2-a_1} & a_1 \leq x \leq a_2 \\ 0 & x \leq a_1 \end{cases} \tag{6.70}$$

$$Pos_I\left(\tilde{A} \leq x\right) = \begin{cases} u_{\tilde{A}} & x \geq a_2 \\ \frac{(a_2-x)+u_{\tilde{A}}(x-a_1)}{a_2-a_1} & a_1 \leq x \leq a_2 \\ 0 & x \leq a_1 \end{cases} \tag{6.71}$$

$$Pos_F\left(\tilde{A} \leq x\right) = \begin{cases} v_{\tilde{A}} & x \geq a_2 \\ \frac{(a_2-x)+v_{\tilde{A}}(x-a_1)}{a_2-a_1} & a_1 \leq x \leq a_2 \\ 0 & x \leq a_1 \end{cases} \tag{6.72}$$

Now, from Definition 6.31, possibility measure of $\tilde{A} \geq \tilde{B}$ for truth-membership functions, indeterminacy-membership functions and falsity-membership functions, are follows.

$$Pos_T(\tilde{A} \geq \tilde{B}) = \begin{cases} \min\left\{w_{\tilde{A}}, w_{\tilde{B}}\right\}, & a_3 \geq b_2, \\ \frac{(a_4-b_1)\min(w_{\mathcal{A}}, w_B)}{(b_2-b_1)w_{\mathcal{A}}+(a_4-a_3)w_{\tilde{B}}}, & a_3 < b_2, a_4 > b_1 \\ 0, & a_4 \leq b_1 \end{cases} \tag{6.73}$$

$$Pos_I(\tilde{A} \geq \tilde{B}) = \begin{cases} \max\left\{u_{\tilde{A}}, u_{\tilde{B}}\right\}, & a_3 \geq b_2, \\ \frac{(b_2-b_1u_B)(1-u_A)-(a_4u_A-a_3)(u_B-1)}{(b_2-b_1)(1-u_A)+(a_4-a_3)\left(u_{\tilde{B}}-1\right)}, & a_3 < b_2, a_4 > b_1 \\ 0, & a_4 \leq b_1 \end{cases} \tag{6.74}$$

$$Pos_F(\tilde{A} \geq \tilde{B}) = \begin{cases} \max\left\{v_{\tilde{A}}, v_{\tilde{B}}\right\}, & a_3 \geq b_2, \\ \frac{(b_2-b_1v_B)(1-v_A)-(a_4v_A-a_3)(v_B-1)}{(b_2-b_1)(1-v_A)+(a_4-a_B)(v_B-1)}, & a_3 < b_2, a_4 > b_1 \\ 0, & a_4 \leq b_1 \end{cases} \tag{6.75}$$

For the particular case of $\tilde{A} \geq x$, according to Definition 6.32, the following possibility, measures are obtained.

$$Pos_T\left(\tilde{A} \geq x\right) = \begin{cases} w_{\tilde{A}}, & x \leq a_3, \\ \frac{a_4-x}{a_4-a_3}w_{\tilde{A}}, & a_3 \leq x \leq a_4 \\ 0, & x \geq a_4 \end{cases} \tag{6.76}$$

$$Pos_I\left(\tilde{A} \geq x\right) = \begin{cases} u_{\tilde{A}}, & x \leq a_3, \\ \frac{(x-a_3)+(a_4-x)u_{\tilde{A}}}{a_4-a_3}, & a_3 \leq x \leq a_4 \\ 1, & x \geq a_4 \end{cases} \tag{6.77}$$

$$Pos_F\left(\tilde{A} \geq x\right) = \begin{cases} v_{\tilde{A}}, & x \leq a_3, \\ \frac{(x-a_3)+(a_4-x)v_{\tilde{A}}}{a_4-a_3}, & a_3 \leq x \leq a_4 \\ 1, & x \geq a_4 \end{cases} \tag{6.78}$$

Lemma 6.1 *Assume the SVNN* $\tilde{A} = \{(a_1, a_2, a_3, a_4); w_{\tilde{A}}, u_{\tilde{A}}, v_{\tilde{A}}\}$. *Then,*

$$Pos_T(\tilde{A} \le x) \ge \xi_1 \Leftrightarrow \frac{w_{\tilde{A}}(x - a_1)}{a_2 - a_1} \ge \xi_1 \Leftrightarrow x \ge a_1 + \frac{(a_2 - a_1)\xi_1}{w_{\tilde{A}}}, \tag{6.79}$$

$$\begin{aligned} Pos_I(\tilde{A} \le x) \le \xi_2 \Leftrightarrow \frac{(a_2 - x) + u_{\tilde{A}}(x - a_1)}{a_2 - a_1} &\le \xi_2 \Leftrightarrow x \\ \ge \frac{(\xi_2 - u_{\tilde{A}})a_1 + (1 - \xi_2)a_2}{1 - u_{\tilde{A}}}&, \end{aligned} \tag{6.80}$$

$$\begin{aligned} Pos_F(\tilde{A} \le x) \le \xi_3 \Leftrightarrow \frac{(a_2 - x) + v_{\tilde{A}}(x - a_1)}{a_2 - a_1} &\le \xi_3 \Leftrightarrow x \\ \ge \frac{(\xi_3 - v_{\tilde{A}})a_1 + (1 - \xi_3)a_2}{1 - v_{\tilde{A}}}&. \end{aligned} \tag{6.81}$$

Lemma 6.2 *Assume the SVNN* $\tilde{A} = \{(a_1, a_2, a_3, a_4); w_{\tilde{A}}, u_{\tilde{A}}, v_{\tilde{A}}\}$. *Then,*

$$Pos_T(\tilde{A} \ge x) \ge \varsigma_1 \Leftrightarrow \frac{a_4 - x}{a_4 - a_3} w_{\tilde{A}} \ge \varsigma_1 \Leftrightarrow x \le a_4 - \frac{(a_4 - a_3)\xi_1}{w_{\tilde{A}}}, \tag{6.82}$$

$$\begin{aligned} Pos_I(\tilde{A} \ge x) \le \varsigma_2 \Leftrightarrow \frac{(x - a_3) + (a_4 - x)u_{\tilde{A}}}{a_4 - a_3} &\le \varsigma_2 \Leftrightarrow x \\ \le \frac{(1 - \xi_2)a_3 + (\varsigma_2 - u_{\tilde{A}})a_4}{1 - u_{\tilde{A}}}&, \end{aligned} \tag{6.83}$$

$$\begin{aligned} Pos_F(\tilde{A} \ge x) \le \varsigma_3 \Leftrightarrow \frac{(x - a_3) + (a_4 - x)v_{\tilde{A}}}{a_4 - a_3} &\le \varsigma_3 \Leftrightarrow x \\ \le \frac{(1 - \varsigma_3)a_3 + (\varsigma_3 - v_{\tilde{A}})a_4}{1 - v_{\tilde{A}}}&. \end{aligned} \tag{6.84}$$

6.6.1 Possibility Expected Value of Single-Valued Trapezoidal Neutrosophic Fuzzy Numbers

Definition 6.34 [23] Assume $\tilde{A}$ as a SVNN. The possibility expected value of this number for its truth-membership, indeterminacy-membership, and falsity-membership functions are defined as below.

$$E_T(\tilde{A}) = \frac{1}{2}\left(\int_0^{w_{\tilde{A}}} \left(Pos_T(\tilde{A} \le x) \ge \alpha\right) d\alpha + \int_0^{w_{\tilde{A}}} \left(Pos_T(\tilde{A} \ge x) \ge \alpha\right) d\alpha\right),$$

$$E_I(\tilde{A}) = \frac{1}{2}\left(\int_{u_{\tilde{A}}}^{1} (Pos_I(\tilde{A} \leq x) \leq \beta)d\beta + \int_{u_{\tilde{A}}}^{1} (Pos_I(\tilde{A} \geq x) \leq \beta)d\beta\right),$$

$$E_F(\tilde{A}) = \frac{1}{2}\left(\int_{\nu_{\tilde{A}}}^{1} (Pos_F(\tilde{A} \leq x) \leq \gamma)d\gamma + \int_{u_{\tilde{A}}}^{1} (Pos_I(\tilde{A} \geq x) \leq \gamma)d\gamma\right),$$

For the particular case of SVTrNN such as $\tilde{A} = \{(a, b, c, d); w_{\tilde{A}}, u_{\tilde{A}}, \nu_{\tilde{A}}\}$, the possibility expected value of this number for its truth-membership, indeterminacy-membership, and falsity-membership functions are defined as below.

$$E_T(\tilde{A}) = \frac{w_{\tilde{A}}}{4}(a + b + c + d),$$
$$E_I(\tilde{A}) = \frac{1 - u_{\tilde{A}}}{4}(a + b + c + d),$$
$$E_F(\tilde{A}) = \frac{1 - \nu_{\tilde{A}}}{4}(a + b + c + d),$$

Therefore, the possibility expected value of a SVTrNN is obtained as below.

$$E(\tilde{A}) = \frac{a + b + c + d}{4}\left(\eta w_{\tilde{A}} + (1 - \eta)(1 - u_{\tilde{A}}) + (1 - \eta)(1 - \nu_{\tilde{A}})\right),$$
$$\forall \eta \in [0, 1]. \tag{6.85}$$

Theorem 6.2 [23] *Assume the SVTrNN* $\tilde{A}_i = \{(a_i, b_i, c_i, d_i); w_{\tilde{A}_i}, u_{\tilde{A}_i}, \nu_{\tilde{A}_i}\}$ *where* $i = 1, \ldots, k$, *and the set of real numbers* $\theta_1, \ldots, \theta_k$. *Then, for any* $\eta \in [0, 1]$,

$$E\left(\sum_{i=1}^{k} \theta_i \tilde{A}_i\right) = \frac{\sum_{i=1}^{k} \theta_i(a_i + b_i + c_i + d_i)}{4}\left(\begin{array}{l}\eta \min\limits_{1 \leq i \leq k} w_{\tilde{A}_i} + (1 - \eta)\left(1 - \max\limits_{1 \leq i \leq k} u_{\tilde{A}_i}\right) \\ +(1 - \eta)\left(1 - \max\limits_{1 \leq i \leq k} \nu_{\tilde{A}_i}\right)\end{array}\right),$$

6.7 Application: Neutrosophic Compromise Optimization Approach

In this sub-section the fuzzy programming approach of Bellman and Zadeh [24] is extended to a neutrosophic environment.

According to Bellman and Zadeh [24], the concept of fuzzy decision (D) includes fuzzy goal (G) and fuzzy constraints (C). Therefore, it is defined as $D = G \cap C$.

Now, neutrosophic fuzzy decision set (D_N) includes neutrosophic fuzzy goal and neutrosophic fuzzy constraints and shown as below.

$$D_N = \left(\bigcap_{k=1}^{K} G_k\right)\left(\bigcap_{i=1}^{M} C_i\right) = (x, T_D(x), I_D(x), F_D(x)) \tag{6.86}$$

Here, for any x there are,

$$T_D(x) = \min\{T_{G_1}(x), \ldots, T_{G_K}(x); T_{C_1}(x), \ldots, T_{C_m}(x)\}, \tag{6.87}$$

$$I_D(x) = \min\{I_{G_1}(x), \ldots, I_{G_K}(x); I_{C_1}(x), \ldots, I_{C_m}(x)\}, \tag{6.88}$$

$$F_D(x) = \max\{F_{G_1}(x), \ldots, F_{G_K}(x); F_{C_1}(x), \ldots, F_{C_m}(x)\}. \tag{6.89}$$

where, $T_D(x)$, $I_D(x)$, and $F_D(x)$ are truth-membership, indeterminacy-membership, and falsity-membership functions under the neutrosophic fuzzy decision set.

In order to explain the proposed neutrosophic fuzzy decision set, first the below multi-objective formulation is considered and steps of the proposed approach are described.

$$\begin{aligned} \text{Max } Z_p(x) &= (z_1, \ldots, z_t), p = 1, \ldots, p_1, \\ \text{Min } Z_p(x) &= (z_1, \ldots, z_t), p = p_2, \ldots, P, \end{aligned} \tag{6.90}$$

Subject to

$$\begin{aligned} g(x) &\geq b_i, i = 1, \ldots, m, \\ x &= (x_1, \ldots, x_n) \geq 0. \end{aligned}$$

In this formulation, z_k shows the k -th objective function, $g(x) \geq b_i$ is the i -th constraint, and $x = (x_1, \ldots, x_n)$ is the set of variables.

In order to obtain a Pareto-optimal solution for the above multi-objective problem a neutrosophic fuzzy programming approach is proposed here. The steps of this approach are represented below.

Step 1. The possible maximum and minimum of each objective function are obtained as below.

$$L_p = \min\{Z_p(X_p)\}, \quad U_p = \max\{Z_p(X_p)\}, \quad p = 1, \ldots, P, \tag{6.91}$$

Step 2. The following lower and upper bounds are obtained,

$$\begin{aligned} &L_p^T = L_p, U_p^T = U_p, \text{ for truth membership}, \\ &L_p^I = L_p^T, U_p^I = L_p^T + q_p\left(U_p^T - L_p^T\right), \text{ for indeterminacy membership}, \\ &L_p^F = L_p^T + r_p\left(U_p^T - L_p^T\right), U_p^F = U_p^T, \text{ for falsity membership}, \end{aligned} \tag{6.92}$$

where, $q_p, r_p \in (0, 1)$ are determined by decision makers in advance.

Step 3. The following linear membership functions in a neutrosophic fuzzy environment for minimization type objective functions are constructed.

$$T_p(Z_p(x)) = \begin{cases} 1, & Z_p(x) \le L_p^T, \\ 1 - \frac{Z_p(x) - L_p^T}{U_p^T - L_p^T}, & L_p^T \le Z_p(x) \le U_p^T, \\ 0, & Z_p(x) \ge U_p^T, \end{cases}$$

$$I_p(Z_p(x)) = \begin{cases} 1, & Z_p(x) \le L_p^I, \\ 1 - \frac{Z_p(x) - L_p^I}{U_p^I - L_p^I}, & L_p^I \le Z_p(x) \le U_p^I, \\ 0, & Z_p(x) \ge U_p^I, \end{cases}$$

$$F_p(Z_p(x)) = \begin{cases} 1, & Z_p(x) \ge U_p^F, \\ 1 - \frac{U_p^F - Z_p(x)}{U_p^F - L_p^F}, & L_p^F \le Z_p(x) \le U_p^F, \\ 0, & Z_p(x) \le L_p^F, \end{cases} \tag{6.93}$$

The following linear membership functions in a neutrosophic fuzzy environment for maximization type objective functions are constructed.

$$T_p(Z_p(x)) = \begin{cases} 0, & Z_p(x) \le L_p^T, \\ 1 - \frac{U_p^T - Z_p(x)}{U_p^T - L_p^T}, & L_p^T \le Z_p(x) \le U_p^T, \\ 1, & Z_p(x) \ge U_p^T, \end{cases}$$

$$I_p(Z_p(x)) = \begin{cases} 0, & Z_p(x) \le L_p^I, \\ 1 - \frac{U_p^I - Z_p(x)}{U_p^I - L_p^I}, & L_p^I \le Z_p(x) \le U_p^I, \\ 1, & Z_p(x) \ge U_p^I, \end{cases}$$

$$F_p(Z_p(x)) = \begin{cases} 0, & Z_p(x) \ge U_p^F, \\ 1 - \frac{Z_p(x) - L_p^F}{U_p^F - L_p^F}, & L_p^F \le Z_p(x) \le U_p^F, \\ 1, & Z_p(x) \le L_p^F, \end{cases} \tag{6.94}$$

For all objective functions the relation $\boldsymbol{U_p^{(.)}} \ne \boldsymbol{L_p^{(.)}}$ should be true. If for an objective function such as p, there is $\boldsymbol{U_p^{(.)}} = \boldsymbol{L_p^{(.)}}$, then, this membership value is considered as 1.

Step 4. The below decision-making model in neutrosophic fuzzy environment is proposed.

$$\max \min_{p=1,\dots,P} T_p(Z_p(x))$$

$$\max \min_{p=1,\dots,P} I_p(Z_p(x))$$

$$
\begin{aligned}
&\min \max_{p=1,\ldots,P} F_p\big(Z_p(x)\big)\\
&\text{Subject to}\\
&g(x) \geq b_i, i = 1, \ldots, m,\\
&x = (x_1, \ldots, x_n) \geq 0.
\end{aligned} \tag{6.95}
$$

The model (6.95) is converted to the below model using some new variables.

$$
\begin{aligned}
&\max\ \alpha\\
&\max\ \beta\\
&\min\ \gamma
\end{aligned}
$$

Subject to

$$
\begin{aligned}
&T_p\big(Z_p(x)\big) \geq \alpha,\\
&I_p\big(Z_p(x)\big) \geq \beta,\\
&F_p\big(Z_p(x)\big) \leq \gamma,
\end{aligned} \tag{6.96}
$$

$$
\begin{aligned}
&\alpha \geq \beta,\\
&\alpha \geq \gamma,
\end{aligned}
$$

$$
\begin{aligned}
&\alpha + \beta + \gamma \leq 3,\\
&\alpha, \beta, \gamma \in (0, 1),\\
&g(x) \geq b_i, i = 1, \ldots, m,\\
&x = (x_1, \ldots, x_n) \geq 0.
\end{aligned}
$$

For more simplicity, the model (6.96) is converted to the below model.

$$
\max\ \alpha + \beta - \gamma
$$

Subject to

$$
\begin{aligned}
&T_p\big(Z_p(x)\big) \geq \alpha,\\
&I_p\big(Z_p(x)\big) \geq \beta\\
&F_p\big(Z_p(x)\big) \leq \gamma,\\
&\alpha \geq \beta,\\
&\alpha \geq \gamma,\\
&\alpha + \beta + \gamma \leq 3,\\
&\alpha, \beta, \gamma \in (0, 1),
\end{aligned} \tag{6.97}
$$

$$
g(x) \geq b_i, i = 1, \ldots, m,
$$

$$x = (x_1, \ldots, x_n) \geq 0.$$

Step 5. The model (6.97) can be easily solved by any optimization solver. The obtained solution is a Pareto-optimal solution of the model (6.90).

Example 6.9 [25] The below data is for a multi-objective transportation problem with neutrosophic fuzzy data where its data have been defuzzified by one of the proposed methods of this chapter.

	$p=1$				$p=2$				
	1	2	3	4	1	2	3	4	Supplies
1	1	2	7	7	4	4	3	4	a_1
2	1	9	3	4	5	8	9	10	a_2
3	8	9	4	6	6	2	5	1	a_3
Demand	11	3	14	16	11	3	14	16	

By solving the problem for each objective function separately, the solutions $X_1 = (5, 3, 0, 0, 6, 0, 0, 13, 0, 0, 14, 3)$ and $X_2 = (0, 0, 8, 0, 11, 2, 0, 6, 0, 0, 1, 16)$, and the objective function values $Z_1(X_1) = 143$, $Z_1(X_2) = 208$, $Z_2(X_1) = 167$, and $Z_2(X_2) = 265$ are obtained. Therefore, the below calculations are done for both objective functions.

$$\begin{aligned}
&U_{z_1}^T = 208, L_{z_1}^T = 143, U_{z_1}^I = 143 + 65q_1, L_{z_1}^I = 143,\\
&U_{z_1}^F = U_{z_1}^T = 208, L_{z_1}^F = 143 + 65r_1,\\
&U_{z_2}^T = 265, L_{z_2}^T = 167, U_{z_2}^I = 167 + 98q_2, L_{z_2}^I = 167,\\
&U_{z_2}^F = U_{z_2}^T = 265, L_{z_2}^F = 167 + 98r_2,
\end{aligned}$$

Then, the below membership functions are obtained for the objective functions.

$$\begin{aligned}
T_1(Z_1(x)) &= \begin{cases} 1, & Z_1(x) \leq 143, \\ 1 - \frac{Z_1(x)-143}{208-143}, & 143 \leq Z_1(x) \leq 208, \\ 0, & Z_1(x) \geq 208, \end{cases}\\
I_1(Z_1(x)) &= \begin{cases} 1, & Z_1(x) \leq 143, \\ 1 - \frac{Z_1(x)-143}{65q_1}, & 143 \leq Z_1(x) \leq 143 + 65q_1, \\ 0, & Z_1(x) \geq 143 + 65q_1, \end{cases}\\
F_1(Z_1(x)) &= \begin{cases} 1, & Z_1(x) \geq 208, \\ 1 - \frac{208-Z_1(x)}{208-143-65r_1}, & 143 + 65r_1 \leq Z_1(x) \leq 208, \\ 0, & Z_1(x) \leq 143 + 65r_1, \end{cases}
\end{aligned} \tag{6.98}$$

$$T_2(Z_2(x)) = \begin{cases} 1, & Z_1(x) \le 167, \\ 1 - \frac{Z_2(x)-167}{265-167}, & 167 \le Z_1(x) \le 265, \\ 0, & Z_1(x) \ge 265, \end{cases}$$

$$I_2(Z_2(x)) = \begin{cases} 1, & Z_2(x) \le 167, \\ 1 - \frac{Z_1(x)-167}{98q_2}, & 167 \le Z_2(x) \le 167 + 98q_2, \\ 0, & Z_2(x) \ge 167 + 98q_1, \end{cases}$$

$$F_2(Z_2(x)) = \begin{cases} 1, & Z_2(x) \ge 265, \\ 1 - \frac{265-Z_2(x)}{265-167-98r_2}, & 167 + 98r_2 \le Z_2(x) \le 265, \\ 0, & Z_2(x) \le 267 + 98r_2, \end{cases} \quad (6.99)$$

In continue, the following decision-making model in neutrosophic fuzzy environment is constructed.

$$\max \ \alpha + \beta - \gamma$$

Subject to

$$\begin{aligned} & T_1(Z_1(x)) \ge \alpha, \\ & T_2(Z_2(x)) \ge \alpha, \\ & I_1(Z_1(x)) \ge \beta, \\ & I_2(Z_2(x)) \ge \beta, \\ & F_1(Z_1(x)) \le \gamma, \\ & F_2(Z_2(x)) \le \gamma, \\ & \alpha \ge \beta, \\ & \alpha \ge \gamma, \\ & \alpha + \beta + \gamma \le 3, \\ & \alpha, \beta, \gamma \in (0, 1), \end{aligned} \quad (6.100)$$

$$\sum_{j=1}^{4} x_{ij} = s_i, i = 1, 2, 3,$$

$$\sum_{i=1}^{3} x_{ij} = d_j, j = 1, 2, 3, 4,$$

$$x_{ij} \ge 0, i = 1, 2, 3, j = 1, 2, 3, 4.$$

The above model is solved and the objective function values of $Z_1 = 160$ and $Z_2 = 195$ are obtained.

References

1. Smarandache, F.: Neutrosophy Neutrosophic Probability: Set, and Logic, pp. 12–20. American Research Press, Rehoboth (1998)
2. Smarandache, F.: A unifying field in logics: neutrosophic logic. Neutrosophy, neutrosophic set, neutrosophic probability. Infinite Study (2003)
3. Wang, H., Smarandache, F., Zhang, Y., Sunderraman, R.: Single valued neutrosophic sets. Multispace Multistructure **4**, 410–413 (2010)
4. Peng, J.J., Wang, J.Q., Wang, J., Zhang, H.Y., Chen, X.H.: Simplified neutrosophic sets and their applications in multi-criteria group decision-making problems. Int. J. Syst. Sci. **47**(10), 2342–2358 (2016)
5. Majumdar, P., Samanta, S.K.: On similarity and entropy of neutrosophic sets. J. Intell. Fuzzy Syst. **26**(3), 1245–1252 (2014)
6. Wang, H., Smarandache, F., Zhang, Y.Q., Sunderraman, R.: Interval neutrosophic sets and logic: theory and applications in computing. In: Neutrosophic Book Series, vol. 5. Hexis, Arizona (2005)
7. Nancy, Garg, H.: An improved score function for ranking neutrosophic sets and its application to decision making process. Int. J. Uncertainty Quantification **6**(5), 377–385 (2016)
8. Şahin, R., Liu, P.: Maximizing deviation method for neutrosophic multiple attribute decision making with incomplete weight information. Neural Comput. Appl. **27**(7), 2017–2029 (2016)
9. Peng, X., Dai, J.: Algorithms for interval neutrosophic multiple attribute decision-making based on MABAC, similarity measure, and EDAS. Int. J. Uncertainty Quantification **7**(5), 395–421 (2017)
10. Liu, P., You, X.: Interval neutrosophic muirhead mean operators and their applications in multiple-attribute group decision making. Int. J. Uncertainty Quantification **7**, 303–334 (2017)
11. Bolturk, E., Kahraman, C.: A novel interval-valued neutrosophic AHP with cosine similarity measure. Soft. Comput. **22**(15), 4941–4958 (2018)
12. Giri, B.K., Roy, S.K.: Neutrosophic multi-objective green four-dimensional fixed-charge transportation problem. Int. J. Mach. Learn. Cybern. **13**(10), 3089–3112 (2022)
13. Farnam, M., Darehmiraki, M., Behdani, Z.: Neutrosophic data envelopment analysis based on parametric ranking method. Appl. Soft Comput. **153**, 111297 (2024)
14. Broumi, S., Bakali, A., Talea, M., Smarandache, F., Vladareanu, L.: Computation of shortest path problem in a network with svtrapezoidal neutrosophic numbers. In: International Conference on Advanced Mechatronic Systems, pp. 417–422 (2016)
15. Umamageswari, R., Uthra, G.: Generalized single valued neutrosophic trapezoidal numbers and their application to solve transportation problem. Stud. Rosenthaliana **11**(1), 164–170 (2020)
16. Kalaivani, K.: A neutrosophic approach to the transportation problem using single-valued trapezoidal neutrosophic numbers. Proyecciones J. Math. **42**(2), 533–547 (2023)
17. Biswas, P., Pramanik, S., Giri, B.C.: Distance measure based MADM strategy with interval trapezoidal neutrosophic numbers. Neutrosophic Sets Syst. **19**, 40–46 (2018)
18. Şahin, R.: Fuzzy multicriteria decision making method based on the improved accuracy function for interval-valued intuitionistic fuzzy sets. Soft. Comput. **20**(7), 2557–2563 (2016)
19. Ye, J.: Multiple-attribute group decision-making method under a neutrosophic number environment. J. Intell. Fuzzy Syst. **25**(3), 377–386 (2016)
20. Abdel-Basset, M., Saleh, M., Gamal, A., Smarandache, F.: An approach of TOPSIS technique for developing supplier selection with group decision making under type-2 neutrosophic number. Appl. Soft Comput. **77**, 438–452 (2019)
21. Das, S.K., Yu, V.F., Roy, S.K., Weber, G.W.: Loceation-allocation problem for green efficient two-stage vehicle-based logistics system: a type-2 neutrosophic multi-objective modeling approach. Expert Syst. Appl. **238**, 122174 (2024)
22. Simic, V., Gokasar, I., Deveci, M., Karakurt, A.: An integrated CRITIC and MABAC based type-2 neutrosophic model for public transportation pricing system selection. Socio-Econ. Plan. Sci. **80**, 101157 (2022)

23. Samanta, S., Chakraborty, D., Jana, D. K.: Neutrosophic multi-period two stage four-dimensional transportation problem for breakable items. Expert Syst. Appl. **246**, 123266 (2024)
24. Bellman, R. E., Zadeh, L. A.: Decision-making in a fuzzy environment. Manag. Sci. **17**(4), B-141 (1970)
25. Rizk-Allah, R.M., Hassanien, A.E., Elhoseny, M.: A multi-objective transportation model under neutrosophic environment. Comput. Electr. Eng. **69**, 705–719 (2018)

Chapter 7
Spherical Fuzzy Sets

7.1 Introduction

As an extension of the classic fuzzy sets, Gündoğdu and Kahraman [1] introduced the concepts of spherical fuzzy sets (SFSs). The SFSs are a direct extension of picture fuzzy sets where the sum of squares of membership, non-membership, and neutral degrees must not exceed 1. As an example, the membership degree of 0.7, neutral degree of 0.5, and non-membership degree of 0.5, cannot be presented as a picture fuzzy set as $0.7 + 0.5 + 0.5 \geq 1$. But this event can be presented as a spherical fuzzy set as $(0.7)^2 + (0.5)^2 + (0.5)^2 = 0.99 \leq 1$. In fact, spherical fuzzy sets contain most of the advantages of other extensions of the classical fuzzy sets [2].

7.2 Spherical Fuzzy Set

Definition 7.1 [1] The spherical fuzzy set $\tilde{A}_S$ on the reference set X is defined as $\tilde{A}_S = \left\{ \left(x, \mu_{\tilde{A}_S}(x), \nu_{\tilde{A}_S}(x), \pi_{\tilde{A}_S}(x)\right) \middle| x \in X \right\}$, where $\mu_{\tilde{A}_S}(x) \in [0, 1]$ is the membership degree, $\pi_{\tilde{A}_S}(x) \in [0, 1]$ is the hesitancy degree, and $\nu_{\tilde{A}_S}(x) \in [0, 1]$ is the non-membership degree of x in $\tilde{A}_S$. The condition $0 \leq \mu^2_{\tilde{A}_S}(x) + \nu^2_{\tilde{A}_S}(x) + \pi^2_{\tilde{A}_S}(x) \leq 1$ should be held for any $x \in X$. Also, the degree of refusal-membership is defined as $\rho_{\tilde{A}_S} = \sqrt{1 - \left(\mu_{\tilde{A}_S}(x)\right)^2 - \left(\nu_{\tilde{A}_S}(x)\right)^2 - \left(\pi_{\tilde{A}_S}(x)\right)^2}$.

Definition 7.2 [2] consider the SFSs $\tilde{A}_S = \left\{ \left(x, \mu_{\tilde{A}_S}(x), \nu_{\tilde{A}_S}(x), \pi_{\tilde{A}_S}(x)\right) \middle| x \in X \right\}$ and $\tilde{B}_S = \left\{ \left(x, \mu_{\tilde{B}_S}(x), \nu_{\tilde{B}_S}(x), \pi_{\tilde{B}_S}(x)\right) \middle| x \in X \right\}$. Then,

$$\tilde{A}_S \cup \tilde{B}_S = \left\{ \left(\begin{array}{l} x, \max\left\{\mu_{\tilde{A}_S}, \mu_{\tilde{B}_S}\right\}, \min\left\{\nu_{\tilde{A}_S}, \nu_{\tilde{B}_S}\right\}, \\ \min\left\{1 - \left(\left(\max\left\{\mu_{\tilde{A}_S}, \mu_{\tilde{B}_S}\right\}\right)^2 + \left(\min\left\{\nu_{\tilde{A}_S}, \nu_{\tilde{B}_S}\right\}\right)^2\right), \max\left\{\pi_{\tilde{A}_S}, \pi_{\tilde{B}_S}\right\}\right\} \end{array} \right), x \in X \right\}, \tag{7.1}$$

A. Mahmoodirad et al., *Recent Advances on Fuzzy Sets*, Studies in Fuzziness and Soft Computing 440, https://doi.org/10.1007/978-3-032-08776-8_7

$$\tilde{A}_S \cap \tilde{B}_S = \left\{ \left(\begin{array}{l} x, \min\left\{\mu_{\tilde{A}_S}, \mu_{\tilde{B}_S}\right\}, \max\left\{\nu_{\tilde{A}_S}, \nu_{\tilde{B}_S}\right\}, \\ \min\left\{1 - \left(\left(\min\left\{\mu_{\tilde{A}_S}, \mu_{\tilde{B}_S}\right\}\right)^2 + \left(\max\left\{\nu_{\tilde{A}_S}, \nu_{\tilde{B}_S}\right\}\right)^2\right), \min\left\{\pi_{\tilde{A}_S}, \pi_{\tilde{B}_S}\right\}\right\} \end{array} \right), x \in X \right\}, \tag{7.2}$$

For the SFS $\tilde{A}_S, \tilde{A}_S = \left\{\left(\mu_{\tilde{A}_S}(x), \nu_{\tilde{A}_S}(x), \pi_{\tilde{A}_S}(x)\right)\right\}$ is a spherical fuzzy number (SFN) and is shown as $\tilde{A}_S = \left(\mu_{\tilde{A}_S}, \nu_{\tilde{A}_S}, \pi_{\tilde{A}_S}\right)$ for more simplicity.

Definition 7.3 [2] Assume the SFNs $\tilde{A}_S = \left(\mu_{\tilde{A}_S}, \nu_{\tilde{A}_S}, \pi_{\tilde{A}_S}\right)$ and $\tilde{B}_S = \left(\mu_{\tilde{B}_S}, \nu_{\tilde{B}_S}, \pi_{\tilde{B}_S}\right)$. Then,

$$\begin{aligned} \tilde{A}_S \oplus \tilde{B}_S = \Bigg(& \sqrt{\mu^2_{\tilde{A}_S} + \mu^2_{\tilde{B}_S} - \mu^2_{\tilde{A}_S}\mu^2_{\tilde{B}_S}}, \nu_{\tilde{A}_S}\nu_{\tilde{B}_S}, \\ & \sqrt{\left(1 - \mu^2_{\tilde{B}_S}\right)\pi^2_{\tilde{A}_S} + \left(1 - \mu^2_{\tilde{A}_S}\right)\pi^2_{\tilde{B}_S} - \pi^2_{\tilde{A}_S}\pi^2_{\tilde{B}_S}} \Bigg), \end{aligned} \tag{7.3}$$

$$k\tilde{A}_S = \left(\sqrt{1 - \left(1 - \mu^2_{\tilde{A}_S}\right)^k}, \nu^k_{\tilde{A}_S}, \sqrt{\left(1 - \mu^2_{\tilde{A}_S}\right)^k - \left(1 - \mu^2_{\tilde{A}_S} - \pi^2_{\tilde{A}_S}\right)^k} \right), \quad k > 0, \tag{7.4}$$

$$k\tilde{A}_S = \left(\sqrt{1 - \left(1 - \mu^2_{\tilde{A}_S}\right)^{-k}}, \nu^{-k}_{\tilde{A}_S}, \sqrt{\left(1 - \mu^2_{\tilde{A}_S}\right)^{-k} - \left(1 - \mu^2_{\tilde{A}_S} - \pi^2_{\tilde{A}_S}\right)^{-k}} \right), \quad k < 0, \tag{7.5}$$

$$\begin{aligned} \tilde{A}_S \otimes \tilde{B}_S = \Bigg(& \mu_{\tilde{A}_S}\mu_{\tilde{B}_S}, \sqrt{\nu^2_{\tilde{A}_S} + \nu^2_{\tilde{B}_S} - \nu^2_{\tilde{A}_S}\nu^2_{\tilde{B}_S}}, \\ & \sqrt{\left(1 - \nu^2_{\tilde{B}_S}\right)\pi^2_{\tilde{A}_S} + \left(1 - \nu^2_{\tilde{A}_S}\right)\pi^2_{\tilde{B}_S} - \pi^2_{\tilde{A}_S}\pi^2_{\tilde{B}_S}} \Bigg), \end{aligned} \tag{7.6}$$

$$\tilde{A}^k_S = \left(\mu^k_{\tilde{A}_S}, \sqrt{1 - \left(1 - \nu^2_{\tilde{A}_S}\right)^k}, \sqrt{\left(1 - \nu^2_{\tilde{A}_S}\right)^k - \left(1 - \nu^2_{\tilde{A}_S} - \pi^2_{\tilde{A}_S}\right)^k} \right), \quad k > 0, \tag{7.7}$$

$$\tilde{A}^k_S = \left(\mu^{-k}_{\tilde{A}_S}, \sqrt{1 - \left(1 - \nu^2_{\tilde{A}_S}\right)^{-k}}, \sqrt{\left(1 - \nu^2_{\tilde{A}_S}\right)^{-k} - \left(1 - \nu^2_{\tilde{A}_S} - \pi^2_{\tilde{A}_S}\right)^{-k}} \right), \quad k < 0, \tag{7.8}$$

Definition 7.4 [3] The conjugate of the SFN $\tilde{A}_S = \left(\mu_{\tilde{A}_S}, \nu_{\tilde{A}_S}, \pi_{\tilde{A}_S}\right)$ is defined as $\tilde{A}_S = \left(\nu_{\tilde{A}_S}, \mu_{\tilde{A}_S}, \pi_{\tilde{A}_S}\right)$.

Definition 7.5 [1] Assume the SFNs $\tilde{A}_S = \left(\mu_{\tilde{A}_S}, \nu_{\tilde{A}_S}, \pi_{\tilde{A}_S}\right)$ and $\tilde{B}_S = \left(\mu_{\tilde{B}_S}, \nu_{\tilde{B}_S}, \pi_{\tilde{B}_S}\right)$, and $k, k_1, k_2 > 0$. Then,

$$\tilde{A}_S \oplus \tilde{B}_S = \tilde{B}_S \oplus \tilde{A}_S, \tag{7.9}$$

$$\tilde{A}_S \otimes \tilde{B}_S = \tilde{B}_S \otimes \tilde{A}_S, \tag{7.10}$$

$$k\left(\tilde{A}_S \oplus \tilde{B}_S\right) = k\tilde{A}_S \oplus k\tilde{B}_S, \tag{7.11}$$

$$k_1\tilde{A}_S \oplus k_2\tilde{A}_S = (k_1 + k_2)\tilde{A}_S, \tag{7.12}$$

$$\left(\tilde{A}_S \otimes \tilde{B}_S\right)^k = \tilde{A}_S^k \otimes \tilde{B}_S^k, \tag{7.13}$$

$$\tilde{A}_S^{k_1} \otimes \tilde{A}_S^{k_2} = \tilde{A}_S^{k_1+k_2}, \tag{7.14}$$

7.3 Comparison Between Spherical Fuzzy Numbers and Other Types of Fuzzy Numbers

As the condition for a SFN such as $\tilde{A}_S = \left(\mu_{\tilde{A}_S}, \nu_{\tilde{A}_S}, \pi_{\tilde{A}_S}\right)$, is $0 \leq \mu_{\tilde{A}_S}^2 + \nu_{\tilde{A}_S}^2 + \pi_{\tilde{A}_S}^2 \leq 1$, therefore a spherical space is obtained as a feasible space of this condition. A schematic representation of this feasible space compared to the similar space obtained by intuitionistic, picture, and neutrosophic fuzzy numbers is given by Fig. 7.1.

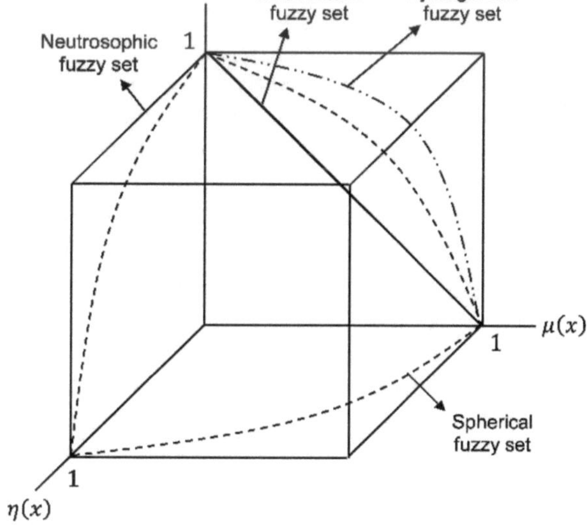

Fig. 7.1 Comparison of spherical fuzzy numbers with other types of fuzzy numbers

7.4 Ranking of Spherical Fuzzy Numbers

Definition 7.6 [1] The score function and accuracy function of the SFN $\tilde{A}_S = \left(\mu_{\tilde{A}_S}, \nu_{\tilde{A}_S}, \pi_{\tilde{A}_S}\right)$ is defined as below.

$$S\left(\tilde{A}_S\right) = \left(\mu_{\tilde{A}_S} - \pi_{\tilde{A}_S}\right)^2 - \left(\nu_{\tilde{A}_S} - \pi_{\tilde{A}_S}\right)^2, \quad (7.15)$$

$$A\left(\tilde{A}_S\right) = \mu_{\tilde{A}_S}^2 + \nu_{\tilde{A}_S}^2 + \pi_{\tilde{A}_S}^2, \quad (7.16)$$

Therefore, the SFNs $\tilde{A}_S = \left(\mu_{\tilde{A}_S}, \nu_{\tilde{A}_S}, \pi_{\tilde{A}_S}\right)$ and $\tilde{B}_S = \left(\mu_{\tilde{B}_S}, \nu_{\tilde{B}_S}, \pi_{\tilde{B}_S}\right)$ can be ranked based on the rules given below.

If $S\left(\tilde{A}_S\right) < S\left(\tilde{B}_S\right)$, then $\tilde{A}_S \prec \tilde{B}_S$,

If $S\left(\tilde{A}_S\right) > S\left(\tilde{B}_S\right)$, then $\tilde{A}_S \succ \tilde{B}_S$,

If $S\left(\tilde{A}_S\right) = S\left(\tilde{B}_S\right)$ and $A\left(\tilde{A}_S\right) < A\left(\tilde{B}_S\right)$, then $\tilde{A}_S \prec \tilde{B}_S$.

In addition, the below score functions are defined by Gündoğdu and Kahraman [4, 5].

$$S\left(\tilde{A}_S\right) = \left(2\mu_{\tilde{A}_S} - \pi_{\tilde{A}_S}\right)^2 - \left(\nu_{\tilde{A}_S} - \pi_{\tilde{A}_S}\right)^2, \quad (7.17)$$

$$S\left(\tilde{A}_S\right) = 4\mu_{\tilde{A}_S}^2 - \nu_{\tilde{A}_S}^2 - \frac{\pi_{\tilde{A}_S}^2}{2}, \quad (7.18)$$

$$S\left(\tilde{A}_S\right) = \sqrt{\left|100\left(3\mu_{\tilde{A}_S} - \frac{\pi_{\tilde{A}_S}}{2}\right)^2 - \left(\frac{\nu_{\tilde{A}_S}}{2} - \pi_{\tilde{A}_S}\right)^2\right|}. \quad (7.19)$$

Example 7.1 Consider the SFSs $\tilde{A}_{S_1} = (0.73, 0.3, 0.31)$, $\tilde{A}_{S_2} = (0.71, 0.54, 0.45)$, and $\tilde{A}_{S_3} = (0.81, 0.46, 0.26)$. Based on the score values calculated from the score functions (7.15), (7.17), (7.18), and (7.19), the common ranking of $\tilde{A}_{S_3} \succ \tilde{A}_{S_1} \succ \tilde{A}_{S_2}$ is obtained.

The above-mentioned functions may not be quietly appropriate for ranking spherical fuzzy numbers as they may result in zero or negative score values. For example, for the SFSs $\tilde{A}_S = (0.5, 0.5, 0.5)$ and $\tilde{B}_S = (0.4, 0.6, 0.4)$, the score values of $S\left(\tilde{A}_S\right) = \left(\mu_{\tilde{A}_S} - \pi_{\tilde{A}_S}\right)^2 - \left(\nu_{\tilde{A}_S} - \pi_{\tilde{A}_S}\right)^2 = 0$ and $S\left(\tilde{B}_S\right) = \left(\mu_{\tilde{B}_S} - \pi_{\tilde{B}_S}\right)^2 - \left(\nu_{\tilde{B}_S} - \pi_{\tilde{B}_S}\right)^2 = -0.16$ are obtained respectively. Therefore, the below ranking function is introduced to reduce the problems of the above-mentioned score functions.

$$PF\left(\tilde{A}_S\right) = \mu_{\tilde{A}_S}\left(1 - \nu_{\tilde{A}_S}\right)\left(1 - \pi_{\tilde{A}_S}\right), \quad (7.20)$$

7.5 Distance of Spherical Fuzzy Sets

Definition 7.7 [1] The below distances are defined for the SFSs $\tilde{A}_S = \left\{ \left(x, \mu_{\tilde{A}_S}(x), \nu_{\tilde{A}_S}(x), \pi_{\tilde{A}_S}(x)\right) \middle| x \in X \right\}$ and $\tilde{B}_S = \left\{ \left(x, \mu_{\tilde{B}_S}(x), \nu_{\tilde{B}_S}(x), \pi_{\tilde{B}_S}(x)\right) \middle| x \in X \right\}$.

- Euclidean distance:

$$dis_E\left(\tilde{A}_S, \tilde{B}_S\right) = \sqrt{\left(\mu^2_{\tilde{A}_S} - \mu^2_{\tilde{B}_S}\right)^2 + \left(\nu^2_{\tilde{A}_S} - \nu^2_{\tilde{B}_S}\right)^2 + \left(\pi^2_{\tilde{A}_S} - \pi^2_{\tilde{B}_S}\right)^2}, \quad (7.21)$$

- Xu and Zhang's distance:

$$dis_{XZ}\left(\tilde{A}_S, \tilde{B}_S\right) = \frac{1}{2}\left(\left(\mu^2_{\tilde{A}_S} - \mu^2_{\tilde{B}_S}\right)^2 + \left(\nu^2_{\tilde{A}_S} - \nu^2_{\tilde{B}_S}\right)^2 + \left(\pi^2_{\tilde{A}_S} - \pi^2_{\tilde{B}_S}\right)^2\right), \quad (7.22)$$

- Spherical distance:

$$dis_{SP}\left(\tilde{A}_S, \tilde{B}_S\right) = \frac{2}{\pi} \arccos\left(\mu_{\tilde{A}_S}\mu_{\tilde{B}_S} + \nu_{\tilde{A}_S}\nu_{\tilde{B}_S} + \pi_{\tilde{A}_S}\pi_{\tilde{B}_S}\right), \quad (7.23)$$

$$dis_{SP}\left(\tilde{A}_S, \tilde{B}_S\right) = \frac{2}{\pi}\sum_{i=1}^{n} \arccos\left(1 - \frac{1}{2}\left(\left(\mu_{\tilde{A}_S} - \mu_{\tilde{B}_S}\right)^2 + \left(\nu_{\tilde{A}_S} - \nu_{\tilde{B}_S}\right)^2 + \left(\pi_{\tilde{A}_S} - \pi_{\tilde{B}_S}\right)^2\right)\right), \quad (7.24)$$

- Normalized spherical distance

$$dis_{SP}\left(\tilde{A}_S, \tilde{B}_S\right) = \frac{2}{n\pi}\sum_{i=1}^{n} \arccos\left(\mu_{\tilde{A}_S}\mu_{\tilde{B}_S} + \nu_{\tilde{A}_S}\nu_{\tilde{B}_S} + \pi_{\tilde{A}_S}\pi_{\tilde{B}_S}\right), \quad (7.25)$$

$$dis_{SP}\left(\tilde{A}_S, \tilde{B}_S\right) = \frac{2}{n\pi}\sum_{i=1}^{n} \arccos\left(1 - \frac{1}{2}\left(\left(\mu_{\tilde{A}_S} - \mu_{\tilde{B}_S}\right)^2 + \left(\nu_{\tilde{A}_S} - \nu_{\tilde{B}_S}\right)^2 + \left(\pi_{\tilde{A}_S} - \pi_{\tilde{B}_S}\right)^2\right)\right), \quad (7.26)$$

Example 7.2 For the SFNs $\tilde{A}_{S_1} = (0.73, 0.3, 0.31)$, $\tilde{A}_{S_2} = (0.71, 0.54, 0.45)$, and $\tilde{A}_{S_3} = (0.81, 0.46, 0.26)$, the Euclidean distances of $dis_E\left(\tilde{A}_{S_1}, \tilde{A}_{S_2}\right) = 0.2786$, $dis_E\left(\tilde{A}_{S_1}, \tilde{A}_{S_3}\right) = 0.1857$, and $dis_E\left(\tilde{A}_{S_2}, \tilde{A}_{S_3}\right) = 0.2291$, the Xu and Zhang's distance of $dis_{XZ}\left(\tilde{A}_{S_1}, \tilde{A}_{S_2}\right) = 0.1687$, $dis_{XZ}\left(\tilde{A}_{S_1}, \tilde{A}_{S_3}\right) = 0.1367$, and $dis_{XZ}\left(\tilde{A}_{S_2}, \tilde{A}_{S_3}\right) = 0.1835$, and the spherical distance of $dis_{SP}\left(\tilde{A}_{S_1}, \tilde{A}_{S_2}\right) = 0.3882$, $dis_{SP}\left(\tilde{A}_{S_1}, \tilde{A}_{S_3}\right) = 0.3990$, and $dis_{SP}\left(\tilde{A}_{S_2}, \tilde{A}_{S_3}\right) = 0.2207$ are calculated.

Definition 7.8 [6] The below similarity measures are defined for the SFSs $\tilde{A}_S = \{(x, \mu_{\tilde{A}_S}(x), \nu_{\tilde{A}_S}(x), \pi_{\tilde{A}_S}(x)) | x \in X\}$ and $\tilde{B}_S = \{(x, \mu_{\tilde{B}_S}(x), \nu_{\tilde{B}_S}(x), \pi_{\tilde{B}_S}(x)) | x \in X\}$.

$$S_1(\tilde{A}_S, \tilde{B}_S) = \frac{1}{m}\sum_{i=1}^{m} \frac{\mu_{\tilde{A}_S}^2(x_i)\mu_{\tilde{B}_S}^2(x_i) + \nu_{\tilde{A}_S}^2(x_i)\nu_{\tilde{B}_S}^2(x_i) + \pi_{\tilde{A}_S}^2(x_i)\pi_{\tilde{B}_S}^2(x_i)}{\sqrt{\mu_{\tilde{A}_S}^4(x_i) + \nu_{\tilde{A}_S}^4(x_i) + \pi_{\tilde{A}_S}^4(x_i)}\sqrt{\mu_{\tilde{B}_S}^4(x_i) + \nu_{\tilde{B}_S}^4(x_i) + \pi_{\tilde{B}_S}^4(x_i)}}, \tag{7.27}$$

$$S_2(\tilde{A}_S, \tilde{B}_S) = \frac{1}{m}\sum_{i=1}^{m} \cos\Big(\frac{\pi}{2}\max\Big\{\Big|\mu_{\tilde{A}_S}^2(x_i) - \mu_{\tilde{B}_S}^2(x_i)\Big|, \Big|\nu_{\tilde{A}_S}^2(x_i) - \nu_{\tilde{B}_S}^2(x_i)\Big|, \Big|\pi_{\tilde{A}_S}^2(x_i) - \pi_{\tilde{B}_S}^2(x_i)\Big|\Big\}\Big), \tag{7.28}$$

$$S_3(\tilde{A}_S, \tilde{B}_S) = \frac{1}{m}\sum_{i=1}^{m} \cos\Big(\frac{\pi}{4}\Big(\Big|\mu_{\tilde{A}_S}^2(x_i) - \mu_{\tilde{B}_S}^2(x_i)\Big| + \Big|\nu_{\tilde{A}_S}^2(x_i) - \nu_{\tilde{B}_S}^2(x_i)\Big| + \Big|\pi_{\tilde{A}_S}^2(x_i) - \pi_{\tilde{B}_S}^2(x_i)\Big|\Big)\Big), \tag{7.29}$$

$$S_4(\tilde{A}_S, \tilde{B}_S) = \frac{1}{m}\sum_{i=1}^{m} \cot\Big(\frac{\pi}{4} + \frac{\pi}{4}\max\Big\{\Big|\mu_{\tilde{A}_S}^2(x_i) - \mu_{\tilde{B}_S}^2(x_i)\Big|, \Big|\nu_{\tilde{A}_S}^2(x_i) - \nu_{\tilde{B}_S}^2(x_i)\Big|, \Big|\pi_{\tilde{A}_S}^2(x_i) - \pi_{\tilde{B}_S}^2(x_i)\Big|\Big\}\Big) \tag{7.30}$$

$$S_5(\tilde{A}_S, \tilde{B}_S) = \frac{1}{m}\sum_{i=1}^{m} \cot\Big(\frac{\pi}{4} + \frac{\pi}{8}\Big(\Big|\mu_{\tilde{A}_S}^2(x_i) - \mu_{\tilde{B}_S}^2(x_i)\Big| + \Big|\nu_{\tilde{A}_S}^2(x_i) - \nu_{\tilde{B}_S}^2(x_i)\Big| + \Big|\pi_{\tilde{A}_S}^2(x_i) - \pi_{\tilde{B}_S}^2(x_i)\Big|\Big)\Big), \tag{7.31}$$

$$S_6(\tilde{A}_S, \tilde{B}_S) = \frac{1}{m}\sum_{i=1}^{m} \cos\Big(\frac{\pi}{2}\max\Big\{\Big|\mu_{\tilde{A}_S}^2(x_i) - \mu_{\tilde{B}_S}^2(x_i)\Big|, \Big|\nu_{\tilde{A}_S}^2(x_i) - \nu_{\tilde{B}_S}^2(x_i)\Big|, \Big|\pi_{\tilde{A}_S}^2(x_i) - \pi_{\tilde{B}_S}^2(x_i)\Big|, \Big|\rho_{\tilde{A}_S}^2(x_i) - \rho_{\tilde{B}_S}^2(x_i)\Big|\Big\}\Big), \tag{7.32}$$

$$S_7(\tilde{A}_S, \tilde{B}_S) = \frac{1}{m}\sum_{i=1}^{m} \cos\Big(\frac{\pi}{4}\Big(\Big|\mu_{\tilde{A}_S}^2(x_i) - \mu_{\tilde{B}_S}^2(x_i)\Big| + \Big|\nu_{\tilde{A}_S}^2(x_i) - \nu_{\tilde{B}_S}^2(x_i)\Big| + \Big|\pi_{\tilde{A}_S}^2(x_i) - \pi_{\tilde{B}_S}^2(x_i)\Big| + \Big|\rho_{\tilde{A}_S}^2(x_i) - \rho_{\tilde{B}_S}^2(x_i)\Big|\Big)\Big), \tag{7.33}$$

$$S_8(\tilde{A}_S, \tilde{B}_S) = \frac{1}{m}\sum_{i=1}^{m} \cot\Big(\frac{\pi}{4} + \frac{\pi}{4}\max\Big\{\Big|\mu_{\tilde{A}_S}^2(x_i) - \mu_{\tilde{B}_S}^2(x_i)\Big|,$$

$$\left|\nu^2_{\tilde{A}_S}(x_i) - \nu^2_{\tilde{B}_S}(x_i)\right|, \left|\pi^2_{\tilde{A}_S}(x_i) - \pi^2_{\tilde{B}_S}(x_i)\right|, \left|\rho^2_{\tilde{A}_S}(x_i) - \rho^2_{\tilde{B}_S}(x_i)\right|\Big\}\Big), \tag{7.34}$$

$$S_9(\tilde{A}_S, \tilde{B}_S) = \frac{1}{m}\sum_{i=1}^{m} \cot\left(\frac{\pi}{4} + \frac{\pi}{8}\left(\begin{array}{c}\left|\mu^2_{\tilde{A}_S}(x_i) - \mu^2_{\tilde{B}_S}(x_i)\right| + \left|\nu^2_{\tilde{A}_S}(x_i) - \nu^2_{\tilde{B}_S}(x_i)\right| + \\ \left|\pi^2_{\tilde{A}_S}(x_i) - \pi^2_{\tilde{B}_S}(x_i)\right| + \left|\rho^2_{\tilde{A}_S}(x_i) - \rho^2_{\tilde{B}_S}(x_i)\right|\end{array}\right)\right), \tag{7.35}$$

$$S_{10}(\tilde{A}_S, \tilde{B}_S) = \frac{\sum_{i=1}^{m}\left(\mu^2_{\tilde{A}_S}(x_i)\mu^2_{\tilde{B}_S}(x_i) + \nu^2_{\tilde{A}_S}(x_i)\nu^2_{\tilde{B}_S}(x_i) + \pi^2_{\tilde{A}_S}(x_i)\pi^2_{\tilde{B}_S}(x_i)\right)}{\sum_{i=1}^{m}\left(\max\left\{\mu^4_{\tilde{A}_S}(x_i), \mu^4_{\tilde{B}_S}(x_i)\right\} + \max\left\{\nu^4_{\tilde{A}_S}(x_i), \nu^4_{\tilde{B}_S}(x_i)\right\} + \max\left\{\pi^4_{\tilde{A}_S}(x_i), \pi^4_{\tilde{B}_S}(x_i)\right\}\right)}, \tag{7.36}$$

Definition 7.9 [7] The below similarity measure and distance are defined for the SFSs $\tilde{A}_S = \left\{\left(x, \mu_{\tilde{A}_S}(x), \nu_{\tilde{A}_S}(x), \pi_{\tilde{A}_S}(x)\right) \middle| x \in X\right\}$ and $\tilde{B}_S = \left\{\left(x, \mu_{\tilde{B}_S}(x), \nu_{\tilde{B}_S}(x), \pi_{\tilde{B}_S}(x)\right) \middle| x \in X\right\}$.

$$d(\tilde{A}_S, \tilde{B}_S) = 1 - \frac{\sum_{i=1}^{m}\left(\mu^2_{\tilde{A}_S}(x_i)\mu^2_{\tilde{B}_S}(x_i) + \nu^2_{\tilde{A}_S}(x_i)\nu^2_{\tilde{B}_S}(x_i) + \pi^2_{\tilde{A}_S}(x_i)\pi^2_{\tilde{B}_S}(x_i)\right)}{\sum_{i=1}^{m}\left(\max\left\{\mu^4_{\tilde{A}_S}(x_i), \mu^4_{\tilde{B}_S}(x_i)\right\} + \max\left\{\nu^4_{\tilde{A}_S}(x_i), \nu^4_{\tilde{B}_S}(x_i)\right\} + \max\left\{\pi^4_{\tilde{A}_S}(x_i), \pi^4_{\tilde{B}_S}(x_i)\right\}\right)}, \tag{7.37}$$

Example 7.3 [7] Consider the reference set $X = \{x_1, \ldots, x_5\}$ and the below SFSs.

$$\tilde{A}_S = \left\{\frac{(0.7, 0.2, 0.1)}{x_1, \frac{(0.7,0.2,0.2)}{x_2,} \frac{(0.2,0.7,0.1)}{x_3,} \frac{(0.9,0.2,0.1)}{x_4,} \frac{(0.2,0.6,0.1)}{x_5}}\right\}$$

$$\tilde{B}_S = \left\{\frac{(0.3, 0.4, 0.2)}{x_1, \frac{(0.5,0.1,0.2)}{x_2,} \frac{(0.1,0.7,0.1)}{x_3,} \frac{(0.4,0.3,0.1)}{x_4,} \frac{(0.1,0.7,0.1)}{x_5}}\right\}$$

Then, based on the previous definition, the below similarity measure and distance values are calculated for these sets.

$$S_{10}(\tilde{A}_S, \tilde{B}_S) = \frac{0.0509 + 0.1245 + 0.2406 + 0.1333 + 0.1769}{0.2673 + 0.2433 + 0.2418 + 0.6643 + 0.2418} = \frac{0.61}{0.91} = 0.6696$$

$$d(\tilde{A}_S, \tilde{B}_S) = 1 - \frac{0.0509 + 0.1245 + 0.2406 + 0.1333 + 0.1769}{0.2673 + 0.2433 + 0.2418 + 0.6643 + 0.2418} = 1 - \frac{0.61}{0.91} = 1 - 0.6696 = 0.3304.$$

Definition 7.10 [1] Assume the set of SFNs $\tilde{A}_{S_i} = \left(\mu_{\tilde{A}_{S_i}}, \nu_{\tilde{A}_{S_i}}, \eta_{\tilde{A}_{S_i}}\right), i = 1, 2, \ldots, n$ and their associated weight values $W = (w_1, \ldots, w_n)^T$, where, $w_i \in [0, 1]$ and $\sum_{i=1}^{n} w_i = 1$. The spherical weighted arithmetic mean (SWAM) and the spherical weighted geometric mean (SWGM) aggregators of these numbers are defined as below.

$$SWAM_w(\tilde{A}_{S_1}, \ldots, \tilde{A}_{S_n}) = w_1\tilde{A}_{S_1} + \cdots + w_n\tilde{A}_{S_n} = \tag{7.38}$$

$$\left(\sqrt{1-\prod_{i=1}^{n}\left(1-\mu_{\tilde{A}_{S_i}}^2\right)^{w_i}}, \prod_{i=1}^{n}\nu_{\tilde{A}_{S_i}}^{w_i}, \sqrt{\prod_{i=1}^{n}\left(1-\mu_{\tilde{A}_{S_i}}^2\right)^{w_i} - \prod_{i=1}^{n}\left(1-\mu_{\tilde{A}_{S_i}}^2 - \pi_{\tilde{A}_{S_i}}^2\right)^{w_i}}\right), \tag{7.39}$$

$$SWGM_w(\tilde{A}_{S_1}, \ldots, \tilde{A}_{S_n}) = \tilde{A}_{S_1}^{w_1} \otimes \cdots \otimes \tilde{A}_{S_n}^{w_n} = \tag{7.40}$$

$$\left(\prod_{i=1}^{n}\mu_{\tilde{A}_{S_i}}^{w_i}, \sqrt{1-\prod_{i=1}^{n}\left(1-\nu_{\tilde{A}_{S_i}}^2\right)^{w_i}}, \sqrt{\prod_{i=1}^{n}\left(1-\nu_{\tilde{A}_{S_i}}^2\right)^{w_i} - \prod_{i=1}^{n}\left(1-\nu_{\tilde{A}_{S_i}}^2 - \pi_{\tilde{A}_{S_i}}^2\right)^{w_i}}\right). \tag{7.41}$$

7.6 Trapezoidal Spherical Fuzzy Numbers

In this section the concepts of trapezoidal spherical fuzzy numbers (TrSFNs) are represented. All of the concepts of this section are taken from the study of Dhanalakshmi [8].

Definition 7.11 [8] A trapezoidal spherical fuzzy number is shown as $\tilde{A}_S = (a_1, a_2, a_3, a_4; \alpha_{\tilde{A}_S}, \beta_{\tilde{A}_S}, \gamma_{\tilde{A}_S})$, where, $a_1 \le a_2 \le a_3 \le a_4$, $\mu_{\tilde{A}_S}, \nu_{\tilde{A}_S}, \pi_{\tilde{A}_S} \in [0, 1]$, and $0 \le \alpha_{\tilde{A}_S}^2 + \beta_{\tilde{A}_S}^2 + \gamma_{\tilde{A}_S}^2 \le 1$. The below functions are defined for this number.

$$\mu_{\tilde{A}_S}(x) = \begin{cases} \frac{x-a_1}{a_2-a_1}\alpha_{\tilde{A}_S}, & a_1 \le x \le a_2, \\ \alpha_{\tilde{A}_S}, & a_2 \le x \le a_3, \\ \frac{a_4-x}{a_4-a_3}\alpha_{\tilde{A}_S}, & a_3 \le x \le a_4, \\ 0, & otherwise. \end{cases} \tag{7.42}$$

$$\nu_{\tilde{A}_S}(x) = \begin{cases} \frac{(a_2-x)+(x-a_1)}{a_2-a_1}\beta_{\tilde{A}_S}, & a_1 \le x \le a_2, \\ \beta_{\tilde{A}_S}, & a_2 \le x \le a_3, \\ \frac{(x-a_3)+(a_4-x)}{a_4-a_3}\beta_{\tilde{A}_S}, & a_3 \le x \le a_4, \\ 1, & otherwise. \end{cases} \tag{7.43}$$

$$\pi_{\tilde{A}_S}(x) = \begin{cases} \frac{x-a_1}{a_2-a_1}\gamma_{\tilde{A}_S}, & a_1 \le x \le a_2, \\ \gamma_{\tilde{A}_S}, & a_2 \le x \le a_3, \\ \frac{a_4-x}{a_4-a_3}\gamma_{\tilde{A}_S}, & a_3 \le x \le a_4, \\ 0, & otherwise. \end{cases} \tag{7.44}$$

Definition 7.12 [8] Assume the TrSFNs $\tilde{A}_S = (a_1, a_2, a_3, a_4; \alpha_{\tilde{A}_S}, \beta_{\tilde{A}_S}, \gamma_{\tilde{A}_S})$ and $\tilde{B}_S = (b_1, b_2, b_3, b_4; \alpha_{\tilde{B}_S}, \beta_{\tilde{B}_S}, \gamma_{\tilde{B}_S})$, and $k \in \mathbb{R}$. The below arithmetic operations are defined for these numbers where the output of each operation is a SFN.

$$\tilde{A}_S \oplus \tilde{B}_S = \begin{pmatrix} a_1 + b_1, a_2 + b_2, a_3 + b_3; \\ \sqrt{\alpha^2_{\tilde{A}_S} + \alpha^2_{\tilde{B}_S} - \alpha^2_{\tilde{A}_S}\alpha^2_{\tilde{B}_S}}, \beta_{\tilde{A}_S}\beta_{\tilde{B}_S}, \sqrt{\left(1 - \alpha^2_{\tilde{B}_S}\right)\alpha^2_{\tilde{A}_S} + \left(1 - \alpha^2_{\tilde{A}_S}\right)\gamma^2_{\tilde{B}_S} - \gamma^2_{\tilde{A}_S}\gamma^2_{\tilde{B}_S}} \end{pmatrix}, \tag{7.45}$$

$$k\tilde{A}_S = \begin{pmatrix} ka_1, ka_2, ka_3, ka_4; \sqrt{1 - \left(1 - \alpha^2_{\tilde{A}_S}\right)^k}, \beta^k_{\tilde{A}_S}, \\ \sqrt{\left(1 - \alpha^2_{\tilde{A}_S}\right)^k - \left(1 - \alpha^2_{\tilde{A}_S} - \gamma^2_{\tilde{A}_S}\right)^k} \end{pmatrix}, k > 0, \tag{7.46}$$

$$k\tilde{A}_S = \begin{pmatrix} ka_4, ka_3, ka_2, ka_1; \sqrt{1 - \left(1 - \alpha^2_{\tilde{A}_S}\right)^{-k}}, \beta^{-k}_{\tilde{A}_S}, \\ \sqrt{\left(1 - \alpha^2_{\tilde{A}_S}\right)^{-k} - \left(1 - \alpha^2_{\tilde{A}_S} - \gamma^2_{\tilde{A}_S}\right)^{-k}} \end{pmatrix}, k < 0, \tag{7.47}$$

$$\tilde{A}_S \otimes \tilde{B}_S = \begin{pmatrix} \min\{c_{1,4}\}, \min\{c_{2,3}\}, \max\{c_{2,3}\}, \max\{c_{1,4}\}; \\ \alpha_{\tilde{A}_S}\alpha_{\tilde{B}_S}, \sqrt{\beta^2_{\tilde{A}_S} + \beta^2_{\tilde{B}_S} - \beta^2_{\tilde{A}_S}\beta^2_{\tilde{B}_S}}, \sqrt{\left(1 - \beta^2_{\tilde{B}_S}\right)\gamma^2_{\tilde{A}_S} + \left(1 - \beta^2_{\tilde{A}_S}\right)\gamma^2_{\tilde{B}_S} - \gamma^2_{\tilde{A}_S}\gamma^2_{\tilde{B}_S}} \end{pmatrix}, \tag{7.48}$$

where $c_{i,j} = \{a_i b_i, a_i b_j, a_j b_i, a_j b_j\}$,

$$\tilde{A}^k_S = \left(a^k_1, a^k_2, a^k_3, a^k_4; \alpha^k_{\tilde{A}_S}, \sqrt{1 - \left(1 - \beta^2_{\tilde{A}_S}\right)^k}, \sqrt{\left(1 - \beta^2_{\tilde{A}_S}\right)^k - \left(1 - \beta^2_{\tilde{A}_S} - \gamma^2_{\tilde{A}_S}\right)^k}\right), k > 0, \tag{7.49}$$

$$\tilde{A}^k_S = \left(a^k_4, a^k_3, a^k_2, a^k_1; \alpha^{-k}_{\tilde{A}_S}, \sqrt{1 - \left(1 - \beta^2_{\tilde{A}_S}\right)^{-k}}, \sqrt{\left(1 - \beta^2_{\tilde{A}_S}\right)^{-k} - \left(1 - \beta^2_{\tilde{A}_S} - \gamma^2_{\tilde{A}_S}\right)^{-k}}\right), k < 0, \tag{7.50}$$

Theorem 7.1 [8] *Assume the TrSFNs* $\tilde{A}_{S_1}, \ldots, \tilde{A}_{S_n}$ *where* $\tilde{A}_{S_i} = \left(a^i_1, a^i_2, a^i_3, a^i_4; \alpha^i_{\tilde{A}_S}, \beta^i_{\tilde{A}_S}, \gamma^i_{\tilde{A}_S}\right)$. *The sum of these numbers is obtained as below.*

$$\sum_{i=1}^{n} \tilde{A}_{S_i} = \begin{pmatrix} \sum_{i=1}^{n} a^i_1, \sum_{i=1}^{n} a^i_2, \sum_{i=1}^{n} a^i_3, \sum_{i=1}^{n} a^i_4; \\ \sqrt{1 - \prod_{i=1}^{n}\left(1 - \alpha^2_{\tilde{A}_{S_i}}\right)}, \prod_{i=1}^{n} \beta_{\tilde{A}_{S_i}}, \sqrt{\prod_{i=1}^{n}\left(1 - \alpha^2_{\tilde{A}_{S_i}}\right) - \prod_{i=1}^{n}\left(1 - \alpha^2_{\tilde{A}_{S_i}} - \gamma^2_{\tilde{A}_{S_i}}\right)} \end{pmatrix}, \tag{7.51}$$

Theorem 7.2 [8] *Assume the TrSFNs* $\tilde{A}_{S_1}, \ldots, \tilde{A}_{S_n}$ *where* $\tilde{A}_{S_i} = \left(a^i_1, a^i_2, a^i_3, a^i_4; \alpha^i_{\tilde{A}_S}, \beta^i_{\tilde{A}_S}, \gamma^i_{\tilde{A}_S}\right)$. *The product of these numbers is obtained below where* $\boldsymbol{a}^i_1 \geq 0$.

$$\prod_{i=1}^{n}\tilde{A}_{S_i} = \left(\frac{\prod_{i=1}^{n} a_1^i, \prod_{i=1}^{n} a_2^i, \prod_{i=1}^{n} a_3^i, \prod_{i=1}^{n} a_4^i;}{\prod_{i=1}^{n}\alpha_{\tilde{A}_{S_i}}, \sqrt{1-\prod_{i=1}^{n}\left(1-\beta^2_{\tilde{A}_{S_i}}\right)}, , \sqrt{\prod_{i=1}^{n}\left(1-\beta^2_{\tilde{A}_{S_i}}\right) - \prod_{i=1}^{n}\left(1-\beta^2_{\tilde{A}_{S_i}}-\gamma^2_{\tilde{A}_{S_i}}\right)}} \right), \tag{7.52}$$

Definition 7.13 [9] Consider the TrSFN $\tilde{A}_S = \left(a_1, a_2, a_3, a_4; \alpha_{\tilde{A}_S}, \beta_{\tilde{A}_S}, \gamma_{\tilde{A}_S}\right)$. Then, the λ, δ, θ -cut of this number is defined as $A_S^{\lambda,\delta,\theta} = \left\{x|\alpha_{\tilde{A}_S} \geq \lambda, \beta_{\tilde{A}_S} \leq \delta, \gamma_{\tilde{A}_S} \leq \theta\right\}$ (where $0 \leq \lambda \leq \alpha_{\tilde{A}_S}, \beta_{\tilde{A}_S} \leq \delta \leq 1$, and $\gamma_{\tilde{A}_S} \leq \theta \leq 1$) and shown as $A_S^{\lambda,\delta,\theta} = \left(A_S^{\lambda}, A_S^{\delta}, A_S^{\theta}\right)$. The values of this cut is obtained as below.

$$A_S^{\lambda} = \left[L_{\tilde{A}_S}(\lambda), U_{\tilde{A}_S}(\lambda)\right] = \left[a_1 + \frac{\lambda}{\alpha_{\tilde{A}_S}}(a_2 - a_1), a_4 - \frac{\lambda}{\alpha_{\tilde{A}_S}}(a_4 - a_3)\right], \tag{7.53}$$

$$A_S^{\delta} = \left[L_{\tilde{A}_S}(\delta), U_{\tilde{A}_S}(\delta)\right] = \left[\frac{\left(\delta - \nu_{\tilde{A}_S}\right)a_1 + (1-\delta)a_2}{1-\nu_{\tilde{A}_S}}, \frac{\left(\delta - \nu_{\tilde{A}_S}\right)a_4 + (1-\delta)a_3}{1-\nu_{\tilde{A}_S}}\right], \tag{7.54}$$

$$A_S^{\gamma} = \left[L_{\tilde{A}_S}(\theta), U_{\tilde{A}_S}(\theta)\right] = \left[\frac{\left(\theta - \pi_{\tilde{A}_S}\right)a_1 + (1-\theta)a_2}{1-\pi_{\tilde{A}_S}}, \frac{\left(\theta - \pi_{\tilde{A}_S}\right)a_4 + (1-\theta)a_3}{1-\pi_{\tilde{A}_S}}\right], \tag{7.55}$$

Remark 7.1 Any real number such as a, can be represented as a TrSFN such as $a = (a, a, a, a; 1, 0, 0)$.

7.6.1 Ranking of Trapezoidal Spherical Fuzzy Numbers

Definition 7.14 For the $\tilde{A}_S = \left(a_1, a_2, a_3, a_4; \alpha_{\tilde{A}_S}, \beta_{\tilde{A}_S}, \gamma_{\tilde{A}_S}\right)$, the below ranking function is defined where $0 \leq \mathbf{w} \leq 1$ is a risk preference value.

$$\Re_w\left(\tilde{A}_S\right) = \frac{1}{2}\left[\int_0^{\alpha_{\tilde{A}_S}} \left((1-w)L_{\tilde{A}_S}(x) + wR_{\tilde{A}_S}(x)\right)dx + \int_0^{\sqrt{1-\beta_{\tilde{A}_S}-\gamma_{\tilde{A}_S}}} \left((1-w)L_{\tilde{A}_S}(x) + wR_{\tilde{A}_S}(x)\right)dx\right. \tag{7.56}$$

For the case of $w = 0.5$, the below ranking function is obtained for $\tilde{A}_S = \left(a_1, a_2, a_3, a_4; \alpha_{\tilde{A}_S}, \beta_{\tilde{A}_S}, \gamma_{\tilde{A}_S}\right)$.

$$\Re_w(\tilde{A}_S) = \frac{1}{4\alpha_{\tilde{A}_S}}\Big[\alpha_{\tilde{A}_S}(a_1 + a_4)\Big(\alpha_{\tilde{A}_S} + \sqrt{1 - \beta_{\tilde{A}_S} - \gamma_{\tilde{A}_S}}\Big) - \frac{1}{2}(a_1 - a_2 - a_3 + a_4)\big(1 + \alpha_{\tilde{A}_S} - \beta_{\tilde{A}_S} - \gamma_{\tilde{A}_S}\big)\Big] \quad (7.57)$$

Here, for the TrSFNs $\tilde{A}_S$ and $\tilde{B}_S$, the order of $\tilde{A}_S < \tilde{B}_S$ is obtained if and only if,

- $\Re(\tilde{A}_S) < \Re(\tilde{B}_S)$, or
- $\Re(\tilde{A}_S) = \Re(\tilde{B}_S)$ and $\delta_{\tilde{A}_S} > \delta_{\tilde{B}_S}$ where $\delta_{\tilde{A}_S} = \sqrt{1 - \alpha^2_{\tilde{A}_S} - \beta^2_{\tilde{A}_S} - \gamma^2_{\tilde{A}_S}}$ is the refusal degree of the TrSFN $\tilde{A}_S$.

7.6.2 Interval-Valued Spherical Fuzzy Sets

In this section, spherical fuzzy sets are extended to interval-valued (IVSFS) form and some basic concepts, arithmetic operations, and defuzzification methods of the interval-valued spherical fuzzy sets (IVSFSs) are explained.

Definition 7.15 [2] The interval-valued spherical fuzzy set (IVSFS) $\tilde{A}_S$ on the reference set X, is defined as below.

$$\tilde{A}_S = \Big\{\Big(x, \big[\mu^L_{\tilde{A}_S}(x), \mu^U_{\tilde{A}_S}(x)\big], \big[\nu^L_{\tilde{A}_S}(x), \nu^U_{\tilde{A}_S}(x)\big], \big[\pi^L_{\tilde{A}_S}(x), \pi^U_{\tilde{A}_S}(x)\big]\Big)\Big| x \in X\Big\} \quad (7.58)$$

For any $x \in X$, the values $\mu^L_{\tilde{A}_S}(x)$, $\nu^L_{\tilde{A}_S}(x)$, and $\pi^L_{\tilde{A}_S}(x)$ are upper values for the membership, non-membership, and resistance degrees respectively. The values $\mu^U_{\tilde{A}_S}(x)$, $\nu^U_{\tilde{A}_S}(x)$, and $\pi^U_{\tilde{A}_S}(x)$ are lower values for the membership, non-membership, and resistance degrees respectively. The conditions $0 \leq \mu^L_{\tilde{A}_S}(x) \leq \mu^U_{\tilde{A}_S}(x) \leq 1$, $0 \leq \nu^L_{\tilde{A}_S}(x) \leq \nu^U_{\tilde{A}_S}(x) \leq 1$, $0 \leq \pi^L_{\tilde{A}_S}(x) \leq \pi^U_{\tilde{A}_S}(x) \leq 1$, and $0 \leq \big(\mu^U_{\tilde{A}_S}(x)\big)^2 + \big(\nu^U_{\tilde{A}_S}(x)\big)^2 + \big(\pi^U_{\tilde{A}_S}(x)\big)^2 \leq 1$ are held. In addition, for the case of $\mu^L_{\tilde{A}_S}(x) = \mu^U_{\tilde{A}_S}(x)$, $\nu^L_{\tilde{A}_S}(x) = \nu^U_{\tilde{A}_S}(x)$, and $\pi^L_{\tilde{A}_S}(x) = \pi^U_{\tilde{A}_S}(x)$ for any $x \in X$, the IVSFS is converted to a SFS.

For the IVSFS $\tilde{A}_S$, the set $\Big\{\big[\mu^L_{\tilde{A}_S}(x), \mu^U_{\tilde{A}_S}(x)\big], \big[\nu^L_{\tilde{A}_S}(x), \nu^U_{\tilde{A}_S}(x)\big], \big[\pi^L_{\tilde{A}_S}(x), \pi^U_{\tilde{A}_S}(x)\big]\Big\}$ is an interval-valued spherical fuzzy number (IVSFN) and for more simplicity is denoted by $\{[a, b], [c, d], [e, f]\}$, where $[a, b] \subset [0, 1]$, $[c, d] \subset [0, 1]$, $[e, f] \subset [0, 1]$, and $0 \leq b^2 + d^2 + f^2 \leq 1$.

Clearly, $\{[1, 1], [0, 0], [0, 0]\}$ is the largest IVSFN, $\{[0, 0], [1, 1], [0, 0]\}$ is the smallest IVSFN, and $\{[0, 0], [0, 0], [1, 1]\}$ is between the largest and smallest IVSFNs.

Definition 7.16 [2] Consider the IVSFNs $\tilde{A}_S = \{[a, b], [c, d], [e, f]\}$, $\tilde{A}_{S_1} = \{[a_1, b_1], [c_1, d_1], [e_1, f_1]\}$, and $\tilde{A}_{S_2} = \{[a_2, b_2], [c_2, d_2], [e_2, f_2]\}$. Then the below operations are defined.

$$\tilde{A}_{S_1} \cup \tilde{A}_{S_2} = \left\{ \begin{array}{l} [\max\{a_1, a_2\}, \max\{b_1, b_2\}], [\min\{c_1, c_2\}, \min\{d_1, d_2\}], \\ [\min\{e_1, e_2\}, \min\{f_1, f_2\}] \end{array} \right\}, \quad (7.59)$$

$$\tilde{A}_{S_1} \cap \tilde{A}_{S_2} = \left\{ \begin{array}{l} [\min\{a_1, a_2\}, \min\{b_1, b_2\}], [\max\{c_1, c_2\}, \max\{d_1, d_2\}], \\ [\max\{e_1, e_2\}, \max\{f_1, f_2\}] \end{array} \right\}, \quad (7.60)$$

$$\tilde{A}_{S_1} \oplus \tilde{A}_{S_2} = \left(\begin{array}{l} \left[\sqrt{a_1^2 + a_2^2 - a_1^2 a_2^2}, \sqrt{b_1^2 + b_2^2 - b_1^2 b_2^2}\right], [c_1 c_2, d_1 d_2], \\ \left[\sqrt{\left(1 - a_2^2\right)e_1^2 + \left(1 - a_1^2\right)e_2^2 - e_1^2 e_2^2}, \sqrt{\left(1 - b_2^2\right)f_1^2 + \left(1 - b_1^2\right)f_2^2 - f_1^2 f_2^2}\right] \end{array} \right), \quad (7.61)$$

$$\tilde{A}_{S_1} \otimes \tilde{A}_{S_2} = \left(\begin{array}{l} [a_1 a_2, b_1 b_2], \left[\sqrt{c_1^2 + c_2^2 - c_1^2 c_2^2}, \sqrt{d_1^2 + d_2^2 - d_1^2 d_2^2}\right], \\ \left[\sqrt{\left(1 - c_2^2\right)e_1^2 + \left(1 - c_1^2\right)e_2^2 - e_1^2 e_2^2}, \sqrt{\left(1 - d_2^2\right)f_1^2 + \left(1 - d_1^2\right)f_2^2 - f_1^2 f_2^2}\right] \end{array} \right), \quad (7.62)$$

$$k\tilde{A}_S = \left(\begin{array}{l} \left[\sqrt{1 - \left(1 - a^2\right)^k}, \sqrt{1 - \left(1 - b^2\right)^k}\right], \left[c^k, d^k\right], \\ \left[\sqrt{\left(1 - a^2\right)^k - \left(1 - a^2 - e^2\right)^k}, \sqrt{\left(1 - b^2\right)^k - \left(1 - b^2 - f^2\right)^k}\right] \end{array} \right), \quad (7.63)$$

$$\tilde{A}_S^k = \left(\begin{array}{l} \left[a^k, b^k\right], \left[\sqrt{1 - \left(1 - c^2\right)^k}, \sqrt{1 - \left(1 - d^2\right)^k}\right], \\ \left[\sqrt{\left(1 - c^2\right)^k - \left(1 - c^2 - d^2\right)^k}, \sqrt{\left(1 - d^2\right)^k - \left(1 - d^2 - f^2\right)^k}\right] \end{array} \right). \quad (7.64)$$

***Remark** 7.2* Considering the real value k, the below special cases can be mentioned.

- If $\tilde{A}_S = \{[1, 1], [0, 0], [0, 0]\}$, then $k\tilde{A}_S = \tilde{A}_S$ and $\tilde{A}_S^k = \tilde{A}_S$.
- If $\tilde{A}_S = \{[0, 0], [1, 1], [0, 0]\}$, then $k\tilde{A}_S = \tilde{A}_S$ and $\tilde{A}_S^k = \tilde{A}_S$.
- If $\tilde{A}_S = \{[0, 0], [0, 0], [1, 1]\}$, then $k\tilde{A}_S = \tilde{A}_S$ and $\tilde{A}_S^k = \tilde{A}_S$.

Theorem 7.3 [2] *Consider the IVSFNs* $\tilde{A}_S = \{[a, b], [c, d], [e, f]\}$, $\tilde{A}_{S_1} = \{[a_1, b_1], [c_1, d_1], [e_1, f_1]\}$, *and* $\tilde{A}_{S_2} = \{[a_2, b_2], [c_2, d_2], [e_2, f_2]\}$. *Then, for the real values* k, k_1, *and* k_2, *the below equalities are held.*

$$\tilde{A}_S \oplus \tilde{B}_S = \tilde{B}_S \oplus \tilde{A}_S, \quad (7.65)$$

$$\tilde{A}_S \otimes \tilde{B}_S = \tilde{B}_S \otimes \tilde{A}_S, \quad (7.66)$$

$$k\left(\tilde{A}_S \oplus \tilde{B}_S\right) = k\tilde{A}_S \oplus k\tilde{B}_S, \quad (7.67)$$

$$k_1\tilde{A}_S \oplus k_2\tilde{A}_S = (k_1 + k_2)\tilde{A}_S, \quad (7.68)$$

$$\left(\tilde{A}_S \otimes \tilde{B}_S\right)^k = \tilde{A}_S^k \otimes \tilde{B}_S^k, \tag{7.69}$$

$$\tilde{A}_S^{k_1} \otimes \tilde{A}_S^{k_2} = \tilde{A}_S^{k_1+k_2}, \tag{7.70}$$

Definition 7.17 [2] Assume the set of IVSFNs $\tilde{A}_{S_i} = \left([a_i, b_i], [c_i, d_i], \left[e_i, f_i\right]\right)$, $i = 1, 2, \ldots, n$ and their associated weight values $W = (w_1, \ldots, w_n)^T$, where, $w_i \in [0, 1]$ and $\sum_{i=1}^{n} w_i = 1$. The interval-valued spherical weighted arithmetic mean (IVSWAM) and the interval-valued spherical weighted geometric mean (IVSWGM) aggregators of these numbers are defined as below.

$$IVSWAM_w\left(\tilde{A}_1, \ldots, \tilde{A}_n\right) = w_1\tilde{A}_1 + \cdots + w_n\tilde{A}_n =$$
$$\left(\begin{array}{c} \left[\sqrt{1-\prod_{i=1}^{n}\left(1-a_i^2\right)^{w_i}}, \sqrt{1-\prod_{i=1}^{n}\left(1-b_i^2\right)^{w_i}}\right], \left[\prod_{i=1}^{n} c_i^{w_i}, \prod_{i=1}^{n} d_i^{w_i}\right], \\ \left[\sqrt{\prod_{i=1}^{n}\left(1-a_i^2\right)^{w_i} - \prod_{i=1}^{n}\left(1-a_i^2-e_i^2\right)^{w_i}}, \sqrt{\prod_{i=1}^{n}\left(1-b_i^2\right)^{w_i} - \prod_{i=1}^{n}\left(1-b_i^2-f_i^2\right)^{w_i}}\right] \end{array}\right), \tag{7.71}$$

$$IVSWGM_w\left(\tilde{A}_1, \ldots, \tilde{A}_n\right) = \tilde{A}_1^{w_1} \otimes \cdots \otimes \tilde{A}_n^{w_n} =$$
$$\left(\begin{array}{c} \left[\prod_{i=1}^{n} a_i^{w_i}, \prod_{i=1}^{n} b_i^{w_i}\right], \left[\sqrt{1-\prod_{i=1}^{n}\left(1-c_i^2\right)^{w_i}}, \sqrt{1-\prod_{i=1}^{n}\left(1-d_i^2\right)^{w_i}}\right], \\ \left[\sqrt{\prod_{i=1}^{n}\left(1-c_i^2\right)^{w_i} - \prod_{i=1}^{n}\left(1-c_i^2-e_i^2\right)^{w_i}}, \sqrt{\prod_{i=1}^{n}\left(1-d_i^2\right)^{w_i} - \prod_{i=1}^{n}\left(1-d_i^2-f_i^2\right)^{w_i}}\right] \end{array}\right), \tag{7.72}$$

Definition 7.17 [2] For the IVSFN $\tilde{A}_S = \left\{[a, b], [c, d], \left[e, f\right]\right\}$, the below score and accuracy functions are defined.

$$S\left(\tilde{A}_S\right) = \frac{a^2 + b^2 - c^2 - d^2 - \left(\frac{e}{2}\right)^2 - \left(\frac{f}{2}\right)^2}{2}, \quad S\left(\tilde{A}_S\right) \in [-1, 1], \tag{7.73}$$

$$A\left(\tilde{A}_S\right) = \frac{a^2 + b^2 + c^2 + d^2 + e^2 + f^2}{2}, \quad A\left(\tilde{A}_S\right) \in [0, 1]. \tag{7.74}$$

The above-mentioned functions are used to rank the IVSFNs $\tilde{A}_{S_1}$ and $\tilde{A}_{S_2}$ as below.
If $S\left(\tilde{A}_{S_1}\right) < S\left(\tilde{A}_{S_2}\right)$ then $\tilde{A}_{S_1} \prec \tilde{A}_{S_2}$,
If $S\left(\tilde{A}_{S_1}\right) > S\left(\tilde{A}_{S_2}\right)$ then $\tilde{A}_{S_1} \succ \tilde{A}_{S_2}$,
If $S\left(\tilde{A}_{S_1}\right) = S\left(\tilde{A}_{S_2}\right)$ then $\tilde{A}_{S_1} \approx \tilde{A}_{S_2}$,
$A\left(\tilde{A}_{S_1}\right) < A\left(\tilde{A}_{S_2}\right)$ then $\tilde{A}_{S_1} \prec \tilde{A}_{S_2}$,
$A\left(\tilde{A}_{S_1}\right) > A\left(\tilde{A}_{S_2}\right)$ then $\tilde{A}_{S_1} \succ \tilde{A}_{S_2}$,

$A(\tilde{A}_{S_1}) = A(\tilde{A}_{S_2})$ then $\tilde{A}_{S_1} \approx \tilde{A}_{S_2}$.

***Example** 7.4* For the IVSFNs $\tilde{A}_{S_1} = \{[0.85, 0.95], [0.1, 0.15], [0.05, 0.15]\}$, $\tilde{A}_{S_2} = \{[0.2, 0.25], [0.65, 0.75], [0.2, 0.25]\}$, and $\tilde{A}_{S_3} = \{[0.55, 0.65], [0.25, 0.3], [0.25, 0.3]\}$, the score values $S(\tilde{A}_{S_1}) = 0.79$, $S(\tilde{A}_{S_2}) = -0.46$, and $S(\tilde{A}_{S_3}) = 0.25$, and therefore, the ranking of $\tilde{A}_{S_1} \succ \tilde{A}_{S_3} \succ \tilde{A}_{S_2}$ are obtained.

Definition 7.18 [2] For the IVSFNs $\tilde{A}_{S_1} = \{[a_1, b_1], [c_1, d_1], [e_1, f_1]\}$ and $\tilde{A}_{S_2} = \{[a_2, b_2], [c_2, d_2], [e_2, f_2]\}$, the below distance is obtained.

$$d(\tilde{A}_{S_1}, \tilde{A}_{S_2}) = \frac{1}{4}\big(|a_1^2 - a_2^2| + |b_1^2 - b_2^2| + |c_1^2 - c_2^2| + |d_1^2 - d_2^2| + |e_1^2 - e_2^2| + |f_1^2 - f_2^2|\big)$$

7.7 Application: Spherical Fuzzy Multi-objective Linear Programming Problem

In this section, the classical fuzzy programming approach is extended to a spherical fuzzy programming approach and is used to solve a general multi-objective linear programming approach.

Consider the below deterministic multi-objective linear programming model.

$\max Z^p(x) = (z_1, \ldots, z_t)p = 1, \ldots, p_1,$

$$\min Z^p(x) = (z_1, \ldots, z_t)p = p_2, \ldots, P, \tag{7.75}$$

subject to

$$g_i(x) \geq b_i, i = 1, \ldots, m,$$

$$x = (x_1, \ldots, x_n) \geq 0.$$

Here, z_j is j -th objective function, $g_i(x) \geq b_i$ is i -th constraint, and $x = (x_1, \ldots, x_n) \geq 0$ is the vector of decision variables.

According to Bellman and Zadeh [10], fuzzy decision (D) consists of fuzzy goals (G) and fuzzy constraints (C). Therefore, $D = G \cap C$. In the spherical fuzzy environment, in existence of spherical fuzzy goals (G_S) and spherical fuzzy constraints (C_S), the fuzzy decision of $D_s = G_s \cap C_s = (x, T_D(x), F_D(x), I_D(x))$ is obtained. Here, for any x, the membership ($T_D(x)$), non-membership ($F_D(x)$), and hesitancy ($I_D(x)$) functions are of a spherical fuzzy decision set, and defined as below respectively.

$$T_D(x) = \{T_{D_1}(x), \ldots, T_{D_P}(x), T_{C_1}(x), \ldots, T_{C_m}(x)\}, \tag{7.76}$$

$$F_D(x) = \{F_{D_1}(x), \ldots, F_{D_P}(x), F_{C_1}(x), \ldots, F_{C_m}(x)\}, \tag{7.77}$$

$$I_D(x) = \{I_{D_1}(x), \ldots, I_{D_P}(x), I_{C_1}(x), \ldots, I_{C_m}(x)\}, \tag{7.78}$$

An algorithm to find a Pareto-optimal solution for the problem (7.75) in a spherical fuzzy environment is described by the below steps.

Step 1. Solve each of the objective functions 1 to P separately and obtain the solutions X_1 to X_P respectively. For each solution calculate the value of all objective functions and construct the pay-off matrix shown by Table 7.1.

Step 2. Determine the lower bound (L_p) and upper bound (U_p) of each objective function as below.

$$L_p = \min\{Z^p(X_p)\},\ U_p = \max\{Z^p(X_p)\},\quad p = 1, \ldots, P, \tag{7.79}$$

Step 3. Determine the lower and upper bounds for the membership, non-membership, and hesitancy functions as below.

$$\begin{aligned} &L_p^T = L_p, U_p^T = U_p, \\ &L_p^F = L_p^T + r_p\left(U_p^T - L_p^T\right), U_p^F = U_p^T, \\ &L_p^I = L_p^T, U_p^I = L_p^T + q_p\left(U_p^T - L_p^T\right), \end{aligned} \tag{7.80}$$

Here, the values $q_p, r_p \in (0, 1)$ are determined by the decision maker in advance.

Step 4. For the minimization type objective functions the below functions are defined.

$$T_D(x) = \begin{cases} 1; & Z_p(x) \le L_p^T, \\ 1 - \frac{Z_p(x) - L_p^T}{U_p^T - L_p^T}, & L_p^T \le Z_p(x) \le U_p^T, \\ 0; & Z_p(x) \ge U_p^T, \end{cases}$$

Table 7.1 The pay-off table

	Z^1	Z^2	…	Z^P
X_1	$Z^1(X_1)$	$Z^2(X_1)$	…	$Z^P(X_1)$
X_2	$Z^1(X_2)$	$Z^2(X_2)$	…	$Z^P(X_2)$
…	…	…	…	…
X_P	$Z^1(X_P)$	$Z^2(X_P)$	…	$Z^P(X_P)$

$$F_D(x) = \begin{cases} 1; & Z_p(x) \geq U_p^F, \\ 1 - \frac{U_p^F - Z_p(x)}{U_p^F - L_p^F}, & L_p^F \leq Z_p(x) \leq U_p^F, \\ 0; & Z_p(x) \leq L_p^F, \end{cases}$$

$$I_D(x) = \begin{cases} 1; & Z_p(x) \leq L_p^I, \\ 1 - \frac{Z_p(x) - L_p^I}{U_p^I - L_p^I}, & L_p^I \leq Z_p(x) \leq U_p^I, \\ 0; & Z_p(x) \geq U_p^I, \end{cases} \tag{7.81}$$

For the maximization type objective functions the below functions are defined.

$$T_D(x) = \begin{cases} 0; & Z_p(x) \leq L_p^T, \\ 1 - \frac{U_p^T - Z_p(x)}{U_p^T - L_p^T}, & L_p^T \leq Z_p(x) \leq U_p^T, \\ 1; & Z_p(x) \geq U_p^T, \end{cases}$$

$$F_D(x) = \begin{cases} 0 & Z_p(x) \geq U_p^F \\ 1 - \frac{Z_p(x) - L_p^F}{U_p^F - L_p^F}, & L_p^F \leq Z_p(x) \leq U_p^F \\ 1 & Z_p(x) \leq L_p^F \end{cases}$$

$$I_D(x) = \begin{cases} 0 & Z_p(x) \leq L_p^I \\ 1 - \frac{U_p^I - Z_p(x)}{U_p^I - L_p^I}, & L_p^I \leq Z_p(x) \leq U_p^I \\ 1 & Z_p(x) \geq U_p^I \end{cases} \tag{7.82}$$

Step 5. The below multi-objective spherical fuzzy model which is an extension of the model of Bellman and Zadeh [10] is proposed as below.

$$\begin{aligned} &\max \min_{p=1,\ldots p} T_p(Z_p(x))^2 \\ &\min \max_{p=1} F_p(Z_p(x))^2 \\ &\min \max_{p=1,\ldots p} I_p(Z_p(x))^2 \end{aligned} \tag{7.83}$$

subject to

$$T_p(Z_p(x))^2 \geq F_p(Z_p(x))^2,$$

$$T_p(Z_p(x))^2 \geq I_p(Z_p(x))^2,$$

$$0 \leq T_p(Z_p(x))^2 + F_p(Z_p(x))^2 + I_p(Z_p(x))^2 \leq 1,$$

$$g(x) \geq b_i, i = 1, \ldots, m,$$

$$x = (x_1, \ldots, x_n) \geq 0$$

The model is converted to the below model by changing some variables.

$$\begin{aligned} &\max \alpha^2 \\ &\min \beta^2 \\ &\min \gamma^2 \end{aligned} \tag{7.84}$$

subject to

$$T_p\big(Z_p(x)\big)^2 \geq \alpha^2,$$

$$F_p\big(Z_p(x)\big)^2 \leq \beta^2,$$

$$I_p\big(Z_p(x)\big)^2 \leq \gamma^2$$

$$\alpha^2 \geq \beta^2,$$

$$\alpha^2 \geq \gamma^2,$$

$$0 \leq \alpha^2 + \beta^2 + \gamma^2 \leq 1$$

$$g(x) \geq b_i, i = 1, \ldots, m,$$

$$x = (x_1, \ldots, x_n) \geq 0$$

Finally, the below single-objective model is obtained.

$$\max\big(\alpha^2 - \beta^2 - \gamma^2\big) \tag{7.85}$$

subject to

$$T_p\big(Z_p(x)\big)^2 \geq \alpha^2$$

$$F_p\big(Z_p(x)\big)^2 \leq \beta^2,$$

$$I_p\big(Z_p(x)\big)^2 \leq \gamma^2,$$

$$\alpha^2 \geq \beta^2,$$

$$\alpha^2 \geq \gamma^2,$$

$$0 \leq \alpha^2 + \beta^2 + \gamma^2 \leq 1,$$

$$g(x) \geq b_i, i = 1, \ldots, m,$$

$$x = (x_1, \ldots, x_n) \geq 0.$$

Step 6. The single-objective model (7.83) is solved optimally, and the obtained solution is a Pareto-optimal solution for the multi-objective model (7.75).

Example 7.5 [11] Consider the multi-objective linear programming model below.

$$\min z_1 = 19x_1 + 33x_2 + 58x_3$$

$$\min z_2 = 25x_1 + 68x_2 + 72x_3$$

$$\max z_3 = 17x_1 + 35x_2 + 63x_3$$

Subject to

$$7x_1 + 2x_2 + 5x_3 \geq 125$$

$$4x_1 + 5x_2 + 3x_3 \geq 116$$

$$6x_1 + 9x_2 + 5x_3 \geq 214$$

$$X_1, X_2, X_3 \geq 0$$

By implementing the above-mentioned steps, the below bounds are obtained for the first objective function.

$$L_1^T = 657.68,\ U_1^T = 1436.87,$$

$$L_1^F = 657.68 + r_1(1425.87 - 657.68),\ U_1^F = 1425.87,$$

$$L_1^I = 657.68,\ U_1^I = 657.68 + q_1(1425.87 - 657.68),$$

In addition, the membership, non-membership, and hesitancy functions for the first objective function are obtained as below.

$$T_D(z_1(x)) = \begin{cases} 1 & Z_1(x) \leq L_1^T \\ 1 - \frac{Z_1(x) - L_1^T}{U_1^T - L_1^T}, & L_1^T \leq Z_1(x) \leq U_1^T \\ 0 & Z_1(x) \geq U_1^T \end{cases}$$

$$F_D(z_1(x)) = \begin{cases} 1 & Z_1(x) \geq U_1^F \\ 1 - \frac{U_1^F - Z_1(x)}{U_1^F - L_1^F}, & L_1^F \leq Z_1(x) \leq U_1^F \\ 0 & Z_1(x) \leq L_1^F \end{cases}$$

$$I_D(z_1(x)) = \begin{cases} 1 & Z_1(x) \leq L_1^I \\ 1 - \frac{Z_1(x) - L_1^I}{U_1^I - L_1^I}, & L_1^I \leq Z_1(x) \leq U_1^I \\ 0 & Z_1(x) \geq U_1^I \end{cases}$$

By implementing the above-mentioned steps, the below bounds are obtained for the second objective function.

$$L_2^T = 1754.36,\ U_2^T = 2254.56,$$

$$L_2^F = 1754.36 + r_2(2254.56 - 1754.36),\ U_2^F = 2254.56,$$

$$L_2^I = 1754.36,\ U_2^I = 1754.36 + q_2(2254.56 - 1754.36),$$

In addition, the membership, non-membership, and hesitancy functions for the second objective function are obtained as below.

$$T_D(z_2(x)) = \begin{cases} 1 & Z_2(x) \leq L_2^T, \\ 1 - \frac{Z_2(x) - L_2^T}{U_2^T - L_2^T}, & L_2^T \leq Z_2(x) \leq U_2^T, \\ 0 & Z_2(x) \geq U_2^T, \end{cases}$$

$$F_D(z_2(x)) = \begin{cases} 1 & Z_2(x) \geq U_2^F, \\ 1 - \frac{U_2^F - Z_2(x)}{U_2^F - L_2^F}, & L_2^F \leq Z_2(x) \leq U_2^F, \\ 0 & Z_2(x) \leq L_2^F, \end{cases}$$

$$I_D(z_2(x)) = \begin{cases} 1 & Z_2(x) \leq L_2^I, \\ 1 - \frac{Z_2(x) - L_2^I}{U_2^I - L_2^I}, & L_2^I \leq Z_2(x) \leq U_2^I, \\ 0 & Z_2(x) \geq U_2^I, \end{cases}$$

By implementing the above-mentioned steps, the below bounds are obtained for the third objective function.

$$L_3^T = 1125.77, U_3^T = 1927.85,$$

$$L_3^F = 1125.77 + r_3(1927.85 - 1125.77), U_3^F = 1927.85,$$

$$L_3^I = 1125.77, U_3^I = 1125.77 + q_3(1927.85 - 1125.77),$$

In addition, the membership, non-membership, and hesitancy functions for the third objective function are obtained as below.

$$T_D(Z_3(x)) = \begin{cases} 0 & Z_3(x) \le L_3^T, \\ 1 - \frac{U_3^T - Z_3(x)}{U_3^T - L_3^T}, & L_3^T \le Z_3(x) \le U_3^T, \\ 1 & Z_3(x) \ge U_3^T, \end{cases}$$

$$F_D(z_3(x)) = \begin{cases} 0 & Z_3(x) \ge U_3^F, \\ 1 - \frac{Z_{3,}(x) - L_3^F}{U_3^F - L_3^F}, & L_3^F \le Z_3(x) \le U_3^F, \\ 1 & Z_3(x) \le L_3^F, \end{cases}$$

$$I_D(Z_3(x)) = \begin{cases} 0 & Z_3(x) \le L_3^I, \\ 1 - \frac{U_3^I - Z_3(x)}{U_3^I - L_3^I}, & L_3^I \le Z_3(x) \le U_3^I, \\ 1 & Z_3(x) \ge U_3^I, \end{cases}$$

Then, the below spherical fuzzy optimization model is constructed.

$$\max\left(\alpha^2 - \beta^2 - \gamma^2\right)$$

Subject to

$$T_1(Z_1(x))^2 \ge \alpha^2, T_2(Z_2(x))^2 \ge \alpha^2, T_3(Z_3(x))^2 \ge \alpha^2,$$

$$F_1(Z_1(x))^2 \le \beta^2, F_2(Z_2(x))^2 \le \beta^2, F_3(Z_3(x))^2 \le \beta^2,$$

$$I_1(Z_1(x))^2 \le \gamma^2, I_2(Z_2(x))^2 \le \gamma^2, I_3(Z_3(x))^2 \le \gamma^2$$

$$\alpha^2 \ge \beta^2,$$

$$\alpha^2 \ge \gamma^2,$$

$$0 \leq \alpha^2 + \beta^2 + \gamma^2 \leq 1,$$

$$7x_1 + 2x_2 + 5x_3 \geq 125$$

$$4x_1 + 5x_2 + 3x_3 \geq 116$$

$$6x_1 + 9x_2 + 5x_3 \geq 214$$

$$x_1, x_2, x_3 \geq 0.$$

By solving the above model, the optimal solution of $x_1 = 12.83$, $x_2 = 7.38$, and $x_3 = 15.29$ is obtained which is a Pareto-optimal solution of the multi-objective model of the example with $z_1 = 1374.13$, $z_2 = 1923.47$, and $z_3 = 1439.68$ objective function values.

References

1. Gündoğdu, F.K., Kahraman, C.: Spherical fuzzy sets and spherical fuzzy TOPSIS method. J. Intell. Fuzzy Syst. **36**(1), 337–352 (2019)
2. Gündoğdu, F.K., Kahraman, C.: A novel fuzzy TOPSIS method using emerging interval-valued spherical fuzzy sets. Eng. Appl. Artif. Intell. **85**, 307–323 (2019)
3. Ashraf, S., Abdullah, S., Mahmood, T., Ghani, F., Mahmood, T.: Spherical fuzzy sets and their applications in multi-attribute decision making problems. J. Intell. Fuzzy Syst. **36**, 2829–2844 (2019)
4. Gündoğdu, F.K., Kahraman, C.: A novel spherical fuzzy QFD method and its application to the linear delta robot technology development. Eng. Appl. Artif. Intell. **87**, 103348 (2020)
5. Gündoğdu, F.K., Kahraman, C.: A novel spherical fuzzy analytic hierarchy process and its renewable energy application. Soft. Comput. **24**, 4607–4621 (2020)
6. Rafiq, M., Ashraf, S., Abdullah, S., Mahmood, T., Muhammad, S.: The cosine similarity measures of spherical fuzzy sets and their applications in decision making. J. Intell. Fuzzy Syst. 6059–6073 (2019)
7. Jabir Khan, M., Kumam, P., Deebani, W., Kumam, W., Shah, Z.: Distance and similarity measures for spherical fuzzy sets and their applications in selecting mega projects. Math. **8**(4), 519 (2020)
8. Dhanalakshmi, V.: Trapezoidal spherical fuzzy numbers and its application to fuzzy risk analysis. Mapana-J. Sci. **22**(2), 109–125 (2023)
9. Mohanta, K.K., Sharanappa, D.S., Dabke, D., Mishra, L.N., Mishra, V.N.: Data envelopment analysis in the context of spherical fuzzy inputs and outputs. Eur. J. Pure Appl. Math. **15**(3), 1158–1179 (2022)
10. Bellman, R.E., Zadeh, L.A.: Decision-making in a fuzzy environment. Manage. Sci. **17**(4), 140–164 (1970)
11. Ahmad, F., Adhami, A.Y.: Spherical fuzzy multiobjective linear programming problem. In: Kahraman, C., Kutlu Gündoğdu, F. (eds) Decision Making with Spherical Fuzzy Sets. Studies in Fuzziness and Soft Computing, vol. 392. Springer, Cham (2021)

Chapter 8
Z-Numbers

8.1 Introduction

According to Zadeh [11], a Z-number consists of two parts such as constraint and reliability of the constraint. Such a number simultaneously covers uncertainty and its probability. Therefore, using Z-numbers, the uncertainty of an event can be closer to real-world situations compared to the classical fuzzy numbers. On the other hand, Z-numbers can be a very good choice in order to reflect the uncertain nature of real-world decision-making problems [6].

8.2 Z-Numbers

Definition 8.1 [11] Assume $\tilde{A}$ as a fuzzy set in the space of U, and V as a random variable. The probability measure of event $\tilde{A}$ is obtained as below.

$$Prob(\tilde{A}) = \int_U p(u)\mu_{\tilde{A}}(u)du, \tag{8.1}$$

Here, $\mu_{\tilde{A}}(u) : U \to [0, 1]$ is the membership function of $\tilde{A}$ and $p(u)$ is the probability density function of V on U.

Definition 8.2 [11] A Z-number is shown as $Z = (\tilde{A}, \tilde{B})$, and is an ordered pair of fuzzy numbers. Its first element $\tilde{A}$ has the role to apply the limitations of the uncertain variable, where its second element $\tilde{B}$ reflect the reliability of the first element.

In general, the values of $\tilde{A}$ and $\tilde{B}$ are determined exactly by their membership functions such as $\mu_{\tilde{A}}(u) : U \to [0, 1]$ and $\mu_{\tilde{B}}(t) : [0, 1] \to [0, 1]$. As V is a random variable, the probability of occurring $(V, \tilde{A})$ is considered as the reliability value $\tilde{B}$.

A. Mahmoodirad et al., *Recent Advances on Fuzzy Sets*, Studies in Fuzziness and Soft Computing 440, https://doi.org/10.1007/978-3-032-08776-8_8

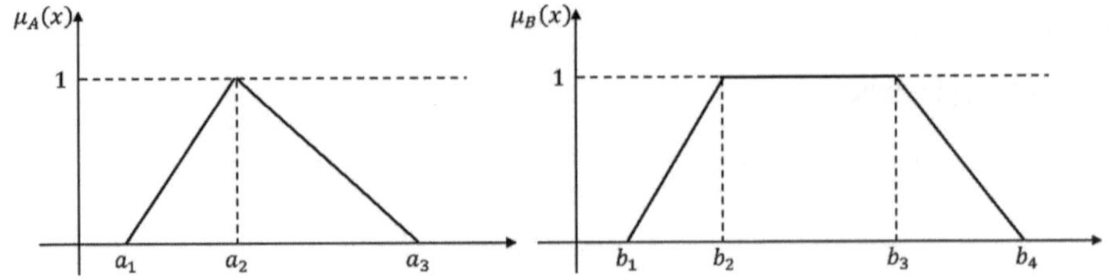

Fig. 8.1 A schematic representation for the Z-number $Z = (\tilde{A}, \tilde{B})$

Therefore, $\tilde{B}$ is an answer to the question "How much is the probability of V is equal to $\tilde{A}$?". The elements $\tilde{A}$ and $\tilde{B}$ can have any membership functions such as triangular, trapezoidal, etc.

Example 8.1 Assume the variable x as "attendance of people in an election". The Z-number for this variable can be shown as $Z(50000000, \textit{High probability})$. This means that, $\tilde{A}$ is approximately 50,000,000 people and $\tilde{B}$ means that with high probability it happens.

Example 8.2 In order to predict the future economic situation, the experts usually do not report a certain value. Instead, they use some linguistic terms to predict such situation. As a term, the term (Strong growth, Probable), can describe such a situation. This can be considered as a Z-number such as $Z = (\tilde{A}, \tilde{B})$ where $\tilde{A}$ represents the term "Strong growth" as a fuzzy set, and $\tilde{B}$ shows its probability degree.

Figure 8.1 represents a Z-number with a triangular membership function which plays the role to apply the limitations of the uncertain variable, and a trapezoidal membership function which reflects the reliability of the membership function.

Z-numbers are very useful to describe uncertain variables in real-world situations. But, as a shortcoming, operations of these numbers are difficult, and several steps should be followed for this aim.

Consider the Z-numbers $\tilde{Z}_1 = (\tilde{A}_1, \tilde{B}_1)$ and $\tilde{Z}_2 = (\tilde{A}_2, \tilde{B}_2)$, which present some incomplete information about the variables X_1 and X_2 respectively. The addition operator of these numbers is done in the below steps [7].

Step 1. The first elements of these numbers are summed up as $\tilde{A}_1 \oplus \tilde{A}_2 = \tilde{A}_{12}$.

Step 2. Apply the below conditions in order to find the probability density functions p_1 and p_2 for the variables X_1 and X_2 respectively. These conditions are defined based on the membership functions $\tilde{B}_1$ and $\tilde{B}_2$.

$$\mu_{\tilde{B}_k}\left(\int\limits_U p_k(u)\mu_{\tilde{A}_k}(u)du\right),$$

$$\int\limits_U p_k(u)du = 1,$$

$$\int\limits_U up_k(u)du = \frac{\int_U u\mu_{\tilde{A}_k}(u)du}{\int_U \mu_{\tilde{A}_k}(u)du}, \quad k = 1, 2,$$

Step 3. Calculate $p_1 + p_2$ according to the operational laws of random variables.

Step 4. The second elements of the given Z-numbers are summed up as below.

$$\begin{aligned}\tilde{B}_{12} &= \mu_{\tilde{B}_{12}}\left(\int\limits_U p_{12}(u)\mu_{\tilde{A}_{12}}(u)du\right)\\ &= \sup\left(\min_{k=1,2}\left(\mu_{\tilde{B}_k}\left(\int\limits_U p_k(u)\mu_{\tilde{A}_k}(u)du\right)\right)\right).\end{aligned}$$

As a result, the value of $\tilde{Z}_1 \oplus \tilde{Z}_2$ is obtained as $(\tilde{A}_{12}, \tilde{B}_{12})$. Other arithmetic operations can be done in a similar way.

As another method, Kang et al. [5] have proposed to reduce a Z-number to a fuzzy number and then apply the arithmetic operations of the fuzzy numbers on the reduced Z-number. In this method a fuzzy expected value is assigned to the second element of any Z-number, and the Z-number is converted to a classical fuzzy number. This can be done for a Z-number such as $\tilde{Z} = (\tilde{A}, \tilde{B})$ by the following steps.

Step 1. The second element $\tilde{B}$ is converted to a crisp form by the below formula (Fig. 8.2).

$$\alpha = \frac{\int x\mu_{\tilde{B}}(x)dx}{\int \mu_{\tilde{B}}(x)dx}, \tag{8.2}$$

Step 2. The weighted form of the first element of the Z-number $\tilde{Z} = (\tilde{A}, \tilde{B})$ is obtained as $Z^\alpha = \{(x, \mu_{\tilde{A}^\alpha}(x)) | \mu_{\tilde{A}^\alpha}(x) = \alpha\mu_{\tilde{A}}(x)\}$ and shown by Fig. 8.3.

Step 3. As the obtained Z^α is an irregular classic fuzzy number, it is converted to a regular classic fuzzy number a $Z' = \{(x, \mu_{\tilde{A}'}(x)) | \mu_{\tilde{A}'}(x) = \mu_{\tilde{A}}(\frac{x}{\alpha})\}$ which is a classical form of the Z-number $\tilde{Z} = (\tilde{A}, \tilde{B})$. It is shown in Fig. 8.4.

Theorem 8.1 [5] *Assume $\tilde{A}$ as a fuzzy number of the reference set X, and its fuzzy expected value $E_{\tilde{A}}(x)$. Then,*

- $E_{\tilde{A}^\alpha}(x) = \alpha E_{\tilde{A}}(x), \quad x \in X,$

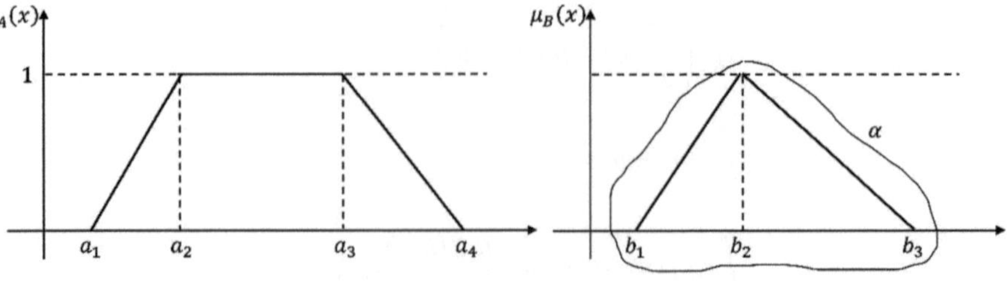

Fig. 8.2 Conversion of the second element of the Z-number $\tilde{Z} = (\tilde{A}, \tilde{B})$

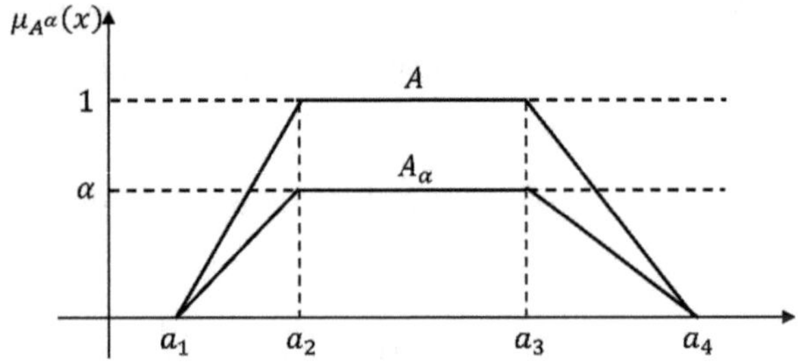

Fig. 8.3 The weighted classical form of the Z-number $\tilde{Z} = (\tilde{A}, \tilde{B})$

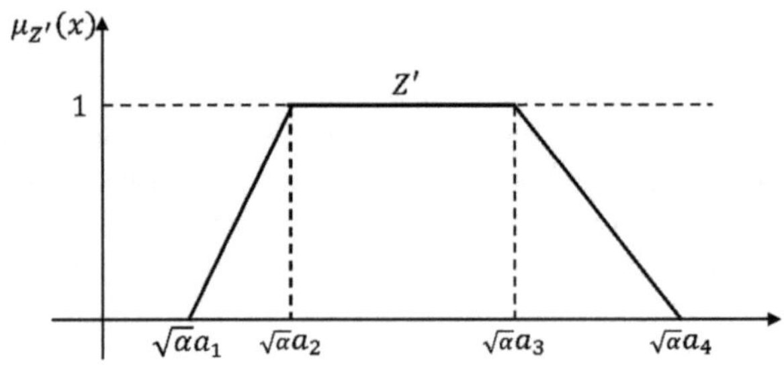

Fig. 8.4 The classical fuzzy form of the Z-number $\tilde{Z} = (\tilde{A}, \tilde{B})$

- $\mu_{\tilde{A}^\alpha}(x) = \alpha\mu_{\tilde{A}}(x), \quad x \in X,$
- $E_{\tilde{Z}'}(x) = \alpha E_{\tilde{A}}(x), \quad x \in \sqrt{\alpha}X,$
- $\mu_{\tilde{Z}'}(x) = \mu_{\tilde{A}}\left(\frac{x}{\alpha}\right), \quad x \in \sqrt{\alpha}X,$
- $E_{\tilde{Z}'}(x) = E_{\tilde{A}^\alpha}(x), \quad x \in X,$

Example 8.3 [5] The opinion and its reliability by an expert is determined as $\tilde{A} = (0.25, 0.35, 0.45, 1)$ and $\tilde{B} = (0.8, 0.9, 1)$ which result in the Z-number $\tilde{Z} = (\tilde{A}, \tilde{B}) = (0.25, 0.35, 0.45, 1; 0.8, 0.9, 1)$. In order to obtain the classical form of this Z-number, first the below certain value for the second element is obtained.

$$\alpha = \frac{\int x\mu_{\tilde{B}}(x)dx}{\int \mu_{\tilde{B}}(x)dx} = \frac{\int_{0.8}^{0.9} x\left(\frac{x-0.8}{0.9-0.8}\right)dx + \int_{0.9}^{1} x\left(\frac{1-x}{1-0.9}\right)dx}{\int_{0.8}^{0.9}\left(\frac{x-0.8}{0.9-0.8}\right)dx + \int_{0.9}^{1}\left(\frac{1-x}{1-0.9}\right)dx} = 0.9, \quad (8.3)$$

Then, the weighted Z-number of $\tilde{Z}^{\alpha} = (0.25, 0.35, 0.45, 1; 0.9)$ and after that the classical fuzzy value as below is obtained.

$$\tilde{Z}\prime = \left(\sqrt{0.9} \times 0.25, \sqrt{0.9} \times 0.35, \sqrt{0.9} \times 0.45, \sqrt{0.9} \times 1\right)$$
$$= (0.2372, 0.3320, 0.4270, 0.9487).$$

Here, two methods were proposed in order to perform the operations of Z-numbers. The first method has some computational difficulties and increases the entropy. On the other hand, the second methos [5] although is simple to perform, during its procedure some information may be lost as the second element is converted to a certain value [1].

8.3 Distance Between Z-Numbers

In order to measure the distance between two Z-numbers, first the distance of their reliability functions are measured using the concept of α-cut, and then the distance of the Z-numbers is measured. The rest of this sub-section describes these concepts.

8.3.1 *Distance of Reliability Functions*

Definition 8.2 [10] Assume the continuous Z-numbers $\tilde{Z}_i = \left(\tilde{A}_i, \tilde{B}_i\right)$ $(i = 1, 2)$ with the reliability membership function of $\mu_{\tilde{B}_i}$. Also assume the α-cut of $[B_{iL}(\alpha), B_{iR}(\alpha)]$ for $\tilde{B}_i$ where $0 \le \alpha \le 1$. Then, the center-to-center distance of $\tilde{B}_1$ and $\tilde{B}_2$ is calculated as below.

$$D_1^{\alpha}\left(\tilde{B}_1, \tilde{B}_2\right) = \left|\frac{B_{1L}(\alpha) + B_{1R}(\alpha)}{2} - \frac{B_{2L}(\alpha) + B_{2R}(\alpha)}{2}\right|, \tag{8.4}$$

Furthermore, the deviation of $[B_{1L}(\alpha), B_{1R}(\alpha)]$ and $[B_{2L}(\alpha), B_{2R}(\alpha)]$ in any cut-set level is calculated as below.

$$ZRMD\left(\tilde{B}_1, \tilde{B}_2\right) = \frac{1}{2}\left(\int_0^1 \left(D_1^{\alpha}\left(\tilde{B}_1, \tilde{B}_2\right) + D_2^{\alpha}\left(\tilde{B}_1, \tilde{B}_2\right)\right)d\alpha\right), \tag{8.5}$$

Here, $ZRMD\left(\tilde{B}_1, \tilde{B}_2\right)$ is a distance.

Now, consider the discrete Z-numbers $\tilde{\mathbf{Z}}_1$ and $\tilde{\mathbf{Z}}_2$ with the membership functions $\mu_{\tilde{B}_i} : \left\{x_{i1}, \ldots, x_{in_i}\right\} \to [0, 1]$, $\left\{x_{i1}, \ldots, x_{in_i}\right\} \subset [0, 1]$, $i = 1, 2$. The α-cut of $\tilde{B}_i^{\alpha} = \left\{x_{ij} \middle| \mu_{\tilde{B}_i}\left(x_{ij}\right) \ge \alpha, j = 1, 2, \ldots, n_i\right\}$ for $\tilde{B}_i$ is defined, and $(B_{iL}(\alpha), B_{iR}(\alpha))$ is obtained as below.

$$B_{iL}(\alpha) = \begin{cases} \min\{x_{ij} | x_{ij} \in B_i^{\alpha}\}, & \alpha > 0, \\ \min\{x_{ij} | x_{ij} \in \sup p(\tilde{B})\}, & \alpha = 0, \end{cases}$$

$$B_{iR}(\alpha) = \begin{cases} \max\{x_{ij} | x_{ij} \in B_i^{\alpha}\}, & \alpha > 0, \\ \max\{x_{ij} | x_{ij} \in \sup p(\tilde{B})\}, & \alpha = 0, \end{cases}$$

Furthermore, the deviation of $[B_{1L}(\alpha), B_{1R}(\alpha)]$ and $[B_{2L}(\alpha), B_{2R}(\alpha)]$ in any cut-set level is calculated as below.

$$ZRMD(\tilde{B}_1, \tilde{B}_2) = \frac{1}{2}\left(\int_0^1 (D_1^{\alpha}(\tilde{B}_1, \tilde{B}_2) + D_2^{\alpha}(\tilde{B}_1, \tilde{B}_2))d\alpha\right), \tag{8.6}$$

Example 8.4 [10] Assume the Z-numbers $\tilde{Z}_1 = (\tilde{A}_1, \tilde{B}_1)$ and $\tilde{Z}_2 = (\tilde{A}_2, \tilde{B}_2)$ with the reliability elements of $\tilde{B}_1 = \left\{\frac{0}{0.4}, \frac{0.5}{0.5}, \frac{1}{0.6}, \frac{0.6}{0.7}, \frac{0}{0.8}\right\}$ and $\tilde{B}_2 = \left\{\frac{0}{0.5}, \frac{0.6}{0.6}, \frac{1}{0.7}, \frac{0.4}{0.8}, \frac{0}{0.9}\right\}$ respectively. First the values of $B_{iL}(\alpha)$ and $B_{iR}(\alpha)$ for $i = 1, 2$ are calculated as below.

$$B_{1L}(\alpha) = \begin{cases} 0.5, \ 0 \le \alpha \le 0.5, \\ 0.6, \ 0.5 < \alpha \le 1, \end{cases}, \quad B_{1R}(\alpha) = \begin{cases} 0.7, \ 0 \le \alpha \le 0.6, \\ 0.6, \ 0.6 < \alpha \le 1, \end{cases}$$

$$B_{2L}(\alpha) = \begin{cases} 0.6, \ 0 \le \alpha \le 0.6, \\ 0.7, \ 0.6 < \alpha \le 1, \end{cases}, \quad B_{2R}(\alpha) = \begin{cases} 0.8, \ 0 \le \alpha \le 0.4, \\ 0.7, \ 0.4 < \alpha \le 1, \end{cases}$$

Then, the below results are obtained.

$$[B_{1L}(\alpha), B_{1U}(\alpha)] = \begin{cases} [0.5, 0.7], \ 0 \le \alpha \le 0.5, \\ [0.6, 0.7], \ 0.5 < \alpha \le 0.6, \\ [0.6, 0.6], \ 0.6 < \alpha \le 1, \end{cases}$$

$$[B_{2L}(\alpha), B_{2U}(\alpha)] = \begin{cases} [0.6, 0.8], \ 0 \le \alpha \le 0.4, \\ [0.6, 0.7], \ 0.4 < \alpha \le 0.6, \\ [0.7, 0.7], \ 0.6 < \alpha \le 1, \end{cases}$$

The center-to-center distance of $\tilde{B}_1$ and $\tilde{B}_2$ for any α-cut is obtained by the below formula.

$$D_1^{\alpha}(\tilde{B}_1, \tilde{B}_2) = \begin{cases} 0.1, & 0 \le \alpha \le 0.4, \\ 0.05, & 0.4 < \alpha \le 0.5, \\ 0, & 0.5 < \alpha \le 0.6, \\ 0.1, & 0.6 < \alpha \le 1, \end{cases}$$

Therefore,

$$ZRMD(\tilde{B}_1, \tilde{B}_2) = \frac{1}{2}\left(\int_0^1 (D_1^\alpha(\tilde{B}_1, \tilde{B}_2) + D_2^\alpha(\tilde{B}_1, \tilde{B}_2))d\alpha\right)$$
$$= \frac{1}{2}(0.085 + 0.01) = 0.0475.$$

8.3.2 Distance Measure of Underlying Probability Distributions

Definition 8.3 [9] The Hellinger distance of the discrete probability distributions $P = (p_1, \ldots, p_k)$ and $Q = (q_1, \ldots, q_k)$ is defined as below.

$$\boldsymbol{H}(\boldsymbol{P}, \boldsymbol{Q}) = \frac{1}{\sqrt{2}}\sqrt{\sum_{i=1}^{k}\left(\sqrt{\boldsymbol{p}_i} - \sqrt{\boldsymbol{q}_i}\right)^2}. \tag{8.7}$$

Assume $P_{\tilde{Z}_i}$ as the underlying probability distribution of the Z-number $\tilde{Z}_i = (\tilde{A}_i, \tilde{B}_i)$ where $i = 1, 2$, which satisfies the below conditions.

$$\int_{\mathbb{R}} P_{\tilde{Z}_i}(x)dx = 1, \tag{8.8}$$

$$\int_{\mathbb{R}} xP_{\tilde{Z}_i}(x)dx = \frac{\int_{\mathbb{R}} x\mu_{\tilde{A}_i}(x)dx}{\int_{\mathbb{R}} \mu_{\tilde{A}_i}(x)dx}, \tag{8.9}$$

$$\int_{\mathbb{R}} P_{\tilde{Z}_i}(x)\mu_{\tilde{A}_i}(x)dxis\tilde{B}_i. \tag{8.10}$$

Assume G_i as the set of probability distributions satisfying the above-mentioned conditions (8.8)–(8.10). Then, the Hellinger Z-number distance (ZHD) for the underlying probability distributions is as below.

$$ZHD(P_{\tilde{Z}_1}, P_{\tilde{Z}_2}) = \inf_{P_{\tilde{Z}_i}\in G_i}\left\{\sqrt{1 - \int_{\mathbb{R}}\sqrt{P_{\tilde{Z}_1}(x)P_{\tilde{Z}_2}(x)}dx}\right\}, \tag{8.11}$$

Notably, $ZHD(P_{\tilde{Z}_1}, P_{\tilde{Z}_2})$ satisfied the conditions of distance function [10].

Assuming the discrete Z-numbers $\tilde{Z}_i = (\tilde{A}_i, \tilde{B}_i), i = 1, 2$, and their discrete probability distributions $P_{\tilde{Z}_i}$, $i = 1, 2$, probability distribution of the set G_i is obtained as below.

$$G_i = \left\{ P_{\tilde{Z}_i}(x) \middle| \sum_{j=1}^{n} P_{\tilde{Z}_i}(x_j) = 1, \sum_{j=1}^{n} P_{\tilde{Z}_i}(x_j)\mu_{\tilde{A}_i}(x_j) \ is \ \tilde{B}_i, \right. \tag{8.12}$$

$$\left. \sum_{j=1}^{n} x_j P_{\tilde{Z}_i}(x_j) = \frac{\sum_{j=1}^{n} x_j \mu_{\tilde{A}_i}(x_j)}{\sum_{j=1}^{n} \mu_{\tilde{A}_i}(x_j)} \right\}$$

Then, the below Hellinger distance is obtained.

$$ZHD(P_{\tilde{Z}_1}, P_{\tilde{Z}_2}) = \inf_{P_{\tilde{Z}_i} \in G_i} \left\{ \frac{1}{\sqrt{2}} \sqrt{\sqrt{P_{\tilde{Z}_{1j}}} - \sqrt{P_{\tilde{Z}_{2j}}}} \right\}. \tag{8.13}$$

Example 8.5 [10] Consider the below probability distributions for G_1 and G_2.

$$G_1 = \left\{ P_{1j} \middle| \sum_{i=1}^{5} P_{1j}(x_i) = 1, \sum_{j=1}^{n} x_i P_{1j}(x_i) = 2.9, \right.$$

$$\left. \sum_{i=1}^{5} P_{1j}(x_i)\mu_{\tilde{A}_1}(x_i) is \{0.2, 0.4, 0.5, 0.6, 0.7\} \right\}$$

$$G_2 = \left\{ P_{2j} \middle| \sum_{i=1}^{5} P_{2j}(x_i) = 1, \sum_{j=1}^{n} x_i P_{2j}(x_i) = 3.6, \right.$$

$$\left. \sum_{i=1}^{5} P_{2j}(x_i)\mu_{\tilde{A}_2}(x_i) is \{0.4, 0.5, 0.6, 0.7, 0.8\} \right\}$$

Then,

$$P_{11} = \frac{0.351}{1} + \frac{0.145}{2} + \frac{0.044}{3} + \frac{0.173}{4} + \frac{0.287}{5}$$

$$P_{12} = \frac{0.222}{1} + \frac{0.208}{2} + \frac{0.199}{3} + \frac{0.191}{4} + \frac{0.18}{5}$$

$$P_{13} = \frac{0.157}{1} + \frac{0.239}{2} + \frac{0.276}{3} + \frac{0.201}{4} + \frac{0.127}{5}$$

$$P_{14} = \frac{0.093}{1} + \frac{0.271}{2} + \frac{0.354}{3} + \frac{0.209}{4} + \frac{0.073}{5}$$

$$P_{15} = \frac{0.028}{1} + \frac{0.302}{2} + \frac{0.432}{3} + \frac{0.219}{4} + \frac{0.019}{5}$$

Table 8.1 The Hellinger distances obtained for Example 8.5

	P_{21}	P_{22}	P_{23}	P_{24}	P_{25}
P_{11}	0.1878	0.2377	0.300	0.3654	0.4017
P_{12}	0.1268	0.1476	0.1968	0.2606	0.3042
P_{13}	0.1457	0.1404	0.1688	0.2243	0.2707
P_{14}	0.1925	0.1663	0.1652	0.2012	0.2477
P_{15}	0.2738	0.2346	0.2049	0.2070	0.2449

$$P_{21} = \frac{0.089}{1} + \frac{0.148}{2} + \frac{0.182}{3} + \frac{0.236}{4} + \frac{0.345}{5}$$

$$P_{22} = \frac{0.054}{1} + \frac{0.136}{2} + \frac{0.249}{3} + \frac{0.279}{4} + \frac{0.282}{5}$$

$$P_{23} = \frac{0.019}{1} + \frac{0.123}{2} + \frac{0.316}{3} + \frac{0.324}{4} + \frac{0.218}{5}$$

$$P_{24} = \frac{0.002}{1} + \frac{0.084}{2} + \frac{0.379}{3} + \frac{0.381}{4} + \frac{0.154}{5}$$

$$P_{25} = \frac{0.006}{1} + \frac{0.021}{2} + \frac{0.428}{3} + \frac{0.459}{4} + \frac{0.086}{5}$$

According to the relation (8.7), the Hellinger distance of $H(P_{11}, P_{12}) = 0.1878$ is calculated and all Hellinger distances are calculated and reported by Table 8.1.

Therefore, using the relation (8.12) the value of ZHD for G_1 and G_2 is obtained as $ZHD(P_{\tilde{Z}_1}, P_{\tilde{Z}_2}) = 0.1286$.

8.3.3 Comprehensive Weighted Distance Measure of Z-Numbers

In order to obtain distance of two Z-numbers, the above-defined distances ZRMD and ZHD should be considered together. Shen et al. [10] proposed the below distance for the Z-numbers $\tilde{Z}_1 = (\tilde{A}_1, \tilde{B}_1)$ and $\tilde{Z}_2 = (\tilde{A}_2, \tilde{B}_2)$,

$$D_w(\tilde{Z}_1, \tilde{Z}_2) = wZRMD(\tilde{B}_1, \tilde{B}_2) + (1 - w)ZHD(P_{\tilde{Z}_1}, P_{\tilde{Z}_2}) \tag{8.14}$$

where $0 < w < 1$. Therefore,

- if $w = 0$, then $D_w(\tilde{Z}_1, \tilde{Z}_2) = ZHD(P_{\tilde{Z}_1}, P_{\tilde{Z}_2})$,
- if $w = 0.5$, then $D_w(\tilde{Z}_1, \tilde{Z}_2) = \frac{1}{2}(ZRMD(\tilde{B}_1, \tilde{B}_2) + ZHD(P_{\tilde{Z}_1}, P_{\tilde{Z}_2}))$,
- if $w = 1$, then $D_w(\tilde{Z}_1, \tilde{Z}_2) = ZRMD(\tilde{B}_1, \tilde{B}_2)$.

In the case of $w = 0.5$, the obtained distance is shown as $GD(\tilde{Z}_1, \tilde{Z}_2)$ which is a general distance and respect to the conditions of distance function.

Example 8.6 [10] Consider the continuous Z-numbers $\tilde{Z}_1 = (\tilde{A}_1, \tilde{B}_1)$ and $\tilde{Z}_2 = (\tilde{A}_2, \tilde{B}_2)$ where $\tilde{A}_1 = (5, 6, 7)$, $\tilde{B}_1 = (0.7, 0.8, 0.9)$, $\tilde{A}_2 = (7, 8, 9)$, and $\tilde{B}_2 = (0.5, 0.6, 0.7)$. First, the below sets probability distributions are obtained for these Z-numbers respectively.

$$G_1 = \left\{ P_{\tilde{Z}_1}(x) \middle| \int_{\mathbb{R}} xP_{\tilde{Z}_1}(x)dx = 6, \int_{\mathbb{R}} P_{\tilde{Z}_1}(x)dx = 1, \int_{\mathbb{R}} P_{\tilde{Z}_1}(x)\mu_{\tilde{A}_1}(x)dxis[0.7, 0.9] \right\},$$

$$G_2 = \left\{ P_{\tilde{Z}_2}(x) \middle| \int_{\mathbb{R}} xP_{\tilde{Z}_2}(x)dx = 8, \int_{\mathbb{R}} P_{\tilde{Z}_2}(x)dx = 1, \int_{\mathbb{R}} P_{\tilde{Z}_2}(x)\mu_{\tilde{A}_2}(x)dxis[0.5, 0.7] \right\},$$

Then, according to the relation (8.13) $ZHD(P_{\tilde{Z}_1}, P_{\tilde{Z}_2}) = 0.5455$, and according to the relation (8.6) $ZRMD(\tilde{B}_1, \tilde{B}_2) = 0.1$. Therefore, $GD(\tilde{Z}_1, \tilde{Z}_2) = 0.323$.

8.4 Ranking of Z-Numbers

In this sub-section the Sigmoid function is used to rank a set of Z-numbers [4].

Definition 8.4 [4] The Sigmoid function is defined as below.

$$S(n) = \frac{1}{1 + e^{-n}}, S(n) \in [0, 1], \tag{8.15}$$

Now, in order to rank the set of Z-numbers $\tilde{Z}_i = (\tilde{A}_i, \tilde{B}_i)$, $i = 1, \ldots, n$, the bellow steps are applied.

Step 1. Calculate α for the set of Z-numbers $\tilde{Z}_i = (\tilde{A}_i, \tilde{B}_i)$, $i = 1, \ldots, n$ using the relation (8.2).

Step 2. For $\tilde{Z}_i = (\tilde{A}_i, \tilde{B}_i)$, $i = 1, \ldots, n$, convert $\tilde{A}_i = (a_{i1}, a_{i2}, a_{i3}, a_{i4}; w_{\tilde{A}_i})$ to a standard fuzzy number such as $\tilde{A}_i^* = (a_{i1}^*, a_{i2}^*, a_{i3}^*, a_{i4}^*; w_{\tilde{A}_i})$ where $a_{ij}^* = \frac{a_{ij}}{\gamma}, j = 1, \ldots, 4$ and $\gamma = \max\{a_{ij}, 1\}$.

Step 3. For any i, calculate $\eta_{\tilde{A}_i^*} = \frac{a_{i1}^* + a_{i2}^* + a_{i3}^* + a_{i4}^*}{4}$ where $\eta_{\tilde{A}_i^*} \in [-1, 1]$.

Step 4. Calculate the spread rank of $\tilde{Z}_i$ using the Sigmoid function as below.

$$\Re(\tilde{Z}_i) = \begin{cases} \frac{1}{1+e^{-\alpha|\eta_{\tilde{A}_i^*}|}}, & \eta_{\tilde{A}_i^*} > 0, \\ 0, & \eta_{\tilde{A}_i^*} = 0, \\ \frac{-1}{1+e^{-\alpha|\eta_{\tilde{A}_i^*}|}}, & \eta_{\tilde{A}_i^*} < 0, \end{cases} \tag{8.16}$$

Here, $\Re(\tilde{Z}_i) \in (-1, 1)$ and the Z-number with greater $\Re(\tilde{Z}_i)$ has the better ranking.

Example 8.7 Consider $\tilde{Z} = (\tilde{A}, \tilde{B})$. For the case of $\tilde{A} = \tilde{B} = (0, 0, 0, 0; 0)$ the value of $\Re(\tilde{Z}) = 0$, and for the case of $\tilde{A} = \tilde{B} = (1, 1, 1, 1; 1)$ the value of $\Re(\tilde{Z}) = 0.731$ are obtained.

As another method, in order to rank the set of Z-numbers $\tilde{Z}_i = (\tilde{A}_i, \tilde{B}_i), i = 1, \ldots, n$, the bellow steps are applied. This method is based on the concept of convex combination.

Step 1. Calculate α for the set of Z-numbers $\tilde{Z}_i = (\tilde{A}_i, \tilde{B}_i), i = 1, \ldots, n$ using the relation (8.2).

Step 2. For $\tilde{Z}_i = (\tilde{A}_i, \tilde{B}_i), i = 1, \ldots, n$, convert $\tilde{A}_i = (a_{i1}, a_{i2}, a_{i3}, a_{i4}; w_{\tilde{A}_i})$ to a standard fuzzy number such as $\tilde{A}_i^* = (a_{i1}^*, a_{i2}^*, a_{i3}^*, a_{i4}^*; w_{\tilde{A}_i})$ where $a_{ij}^* = \frac{a_{ij}}{\gamma}, j = 1, \ldots, 4$ and $\gamma = \max\{a_{ij}, 1\}$.

Step 3. For any i, calculate $\xi_{\tilde{A}_i^*} = \frac{\lambda(a_{i2}^*+a_{i3}^*)+(1-\lambda)(a_{i1}^*+a_{i4}^*)}{2}$ where $\xi_{\tilde{A}_i^*} \in [-1, 1]$. Here, the concept of convex combination is used.

Step 4. Calculate the following spread value.

$$STD_{\tilde{A}_i^*} = \sqrt{\lambda\left(\left(a_{i2}^* - \xi_{\tilde{A}_i^*}\right)^2 + \left(a_{i3}^* - \xi_{\tilde{A}_i^*}\right)^2\right) + (1-\lambda)\left(\left(a_{i1}^* - \xi_{\tilde{A}_i^*}\right)^2 + \left(a_{i4}^* - \xi_{\tilde{A}_i^*}\right)^2\right)} \tag{8.17}$$

Step 5. Calculate the rank of $\tilde{Z}_i$ using the Sigmoid function as below.

$$\Re(\tilde{Z}_i) = \begin{cases} \frac{1}{1+e^{-K}}, & \xi_{\tilde{A}_i^*} > 0, \\ 0, & \xi_{\tilde{A}_i^*} = 0, \\ \frac{-1}{1+e^{-\alpha K}}, & \xi_{\tilde{A}_i^*} < 0, \end{cases} \tag{8.18}$$

Here, $K = \left(\xi_{\tilde{A}_i^*} \times |\alpha|\right) + STD_{\tilde{A}_i^*}$ and $\Re(\tilde{Z}_i) \in (-1, 1)$. The Z-number with greater $\Re(\tilde{Z}_i)$, has the better ranking.

Example 8.8 [4] Assume the Z-numbers $\tilde{Z}_1 = (\tilde{A}, \tilde{B}_1)$ and $\tilde{Z}_2 = (\tilde{A}, \tilde{B}_2)$ where $\tilde{A} = (0.1, 0.3, 0.3, 0.5; 1)$, $\tilde{B}_1 = (0.1, 0.2, 0.4, 0.5; 1)$, and $\tilde{B}_2 = (0.3, 0.3, 0.3, 0.3; 1)$. Applying the method based on the Sigmoid function, the values of $\alpha_1 = \alpha_2 = 0.3$, $\eta_{\tilde{A}^*} = 0.6$, and $\Re(\tilde{Z}_1) = \Re(\tilde{Z}_2) = 0.522$ are obtained, which mean $\tilde{Z}_1 \approx \tilde{Z}_2$. On the

Table 8.2 The results of Example 8.8 using the convex combination based method

λ	Convex combination	$\xi_{\tilde{A}_i^*}$	$STD_{\tilde{A}^*}$	$\Re(\tilde{Z}_1)$	$\Re(\tilde{Z}_2)$
0	$\lambda(a_1^* + a_2^*) + (1 - \lambda)(a_3^* + a_4^*)$	0.4	0.14	0.5564	0.55649
0.5		0.3	0.2	0.5501	0.5501
1		0.2	0.14	0.5719	0.5719
0	$\lambda(a_1^* + a_3^*) + (1 - \lambda)(a_2^* + a_4^*)$	0.4	0.14	0.5649	0.5649
0.5		0.3	0.2	0.5501	0.5501
1		0.2	0.14	0.5719	0.5719
0	$\lambda(a_1^* + a_4^*) + (1 - \lambda)(a_2^* + a_3^*)$	0.3	0	0.5224	0.5224
0.5		0.3	0.2	0.5921	0.5921
1		0.3	0.28	0.5719	0.5719
0	$\lambda(a_{i2}^* + a_{i3}^*) + (1 - \lambda)(a_{i1}^* + a_{i4}^*)$	0.3	0.28	0.5921	0.5921
0.5		0.3	0.2	0.5224	0.5224
1		0.3	0	0.5719	0.5719
0	$\lambda(a_{i2}^* + a_{i4}^*) + (1 - \lambda)(a_{i1}^* + a_{i3}^*)$	0.2	0.31	0.5501	0.5501
0.5		0.3	0.2	0.5649	0.5649
1		0.4	0.31	0.5719	0.5719
0	$\lambda(a_{i3}^* + a_{i4}^*) + (1 - \lambda)(a_{i1}^* + a_{i2}^*)$	0.4	0.14	0.5649	0.5649
0.5		0.3	0.2	0.5501	0.5501
1		0.2	0.14	0.5719	0.5719

other hand, the method based on the convex combination also is used here, and the results of Table 8.2 is obtained.

According to the results of Table 8.2 the final ranking is $\tilde{Z}_1 \approx \tilde{Z}_2$.

8.5 Application: Z-Number Data Envelopment Analysis

The following CCR model [3] in a Z-number environment is considered [2].

$$\begin{aligned}
&\text{Max} \sum_{r=1}^{s} u_r \tilde{Y}_{ro} \\
&\text{subject to} \\
&\sum_{i=1}^{m} v_i \tilde{X}_{io} = 1, \\
&\sum_{r=1}^{s} u_r \tilde{Y}_{rj} - \sum_{i=1}^{m} v_i \tilde{X}_{ij} \le 0,\ j = 1, \ldots, n, \\
&u_r, v_j \ge 0,\ i = 1, \ldots, m, j = 1, \ldots, n.
\end{aligned} \tag{8.19}$$

Here, $j = 1, \ldots, n$ show the DMUs, $i = 1, \ldots, m$ show the inputs, and $r = 1, \ldots, s$ show the outputs. The i -th input and r -th output vector of j -th DMU are shown by $\tilde{X}_{ij} = \left(X\tilde{A}_{ij}, X\tilde{B}_{ij}\right)$ and $\tilde{Y}_{rj} = \left(Y\tilde{A}_{rj}, Y\tilde{B}_{rj}\right)$ respectively, where $X\tilde{A}_{ij} = \left(xa_{ij}^1, xa_{ij}^2, xa_{ij}^3, xa_{ij}^4\right)$, $X\tilde{B}_{ij} = \left(xb_{ij}^1, xb_{ij}^2, xb_{ij}^3, xb_{ij}^4\right)$, $Y\tilde{A}_{rj} = \left(ya_{rj}^1, ya_{rj}^2, ya_{rj}^3, ya_{rj}^4\right)$, and $Y\tilde{B}_{rj} = \left(yb_{rj}^1, yb_{rj}^2, yb_{rj}^3, yb_{rj}^4\right)$.

Now, first the below reductions are done.

$$\alpha_{io} = \frac{xb_{io}^1 + xb_{io}^2 + xb_{io}^3 + xb_{io}^4}{4}, \tag{8.20}$$

$$\alpha_{ij} = \frac{xb_{ij}^1 + xb_{ij}^2 + xb_{ij}^3 + xb_{ij}^4}{4}, \tag{8.21}$$

$$\beta_{ro} = \frac{yb_{ro}^1 + yb_{ro}^2 + yb_{ro}^3 + yb_{ro}^4}{4}, \tag{8.22}$$

$$\beta_{rj} = \frac{yb_{rj}^1 + yb_{rj}^2 + yb_{rj}^3 + yb_{rj}^4}{4}, \tag{8.23}$$

Therefore, the below model is obtained.

$$\begin{aligned}
&\text{Max} \sum_{r=1}^{s} u_r \sqrt{\beta_{ro}} Y\tilde{A}_{ro} \\
&\text{Subject to} \\
&\sum_{i=1}^{m} v_i \sqrt{\alpha_{io}} X\tilde{A}_{io} = 1, \\
&\sum_{r=1}^{s} u_r \sqrt{\beta_{rj}} Y\tilde{A}_{rj} - \sum_{i=1}^{m} v_i \sqrt{\alpha_{ij}} X\tilde{A}_{ij} \le 0, \quad j = 1, \ldots, n, \\
&u_r, v_j \ge 0, \quad i = 1, \ldots, m, j = 1, \ldots, n.
\end{aligned} \tag{8.24}$$

Then, considering the conversions $\sqrt{\beta_{ro}}Y\tilde{A}_{ro} = \tilde{y}_{ro}$, $\sqrt{\alpha_{io}}X\tilde{A}_{io} = \tilde{x}_{io}$, $\sqrt{\alpha_{ij}}X\tilde{A}_{ij} = \tilde{x}_{ij}$, and $\sqrt{\beta_{rj}}Y\tilde{A}_{rj} = \tilde{y}_{rj}$, the below fuzzy CCR model with trapezoidal parameters is obtained.

$$\begin{aligned}
&\text{Max} \sum_{r=1}^{s} u_r \tilde{y}_{ro} \\
&\text{Subject to} \\
&\sum_{i=1}^{m} v_i \tilde{x}_{io} = 1, \\
&\sum_{r=1}^{s} u_r \tilde{y}_{rj} - \sum_{i=1}^{m} v_i \tilde{x}_{ij} \leq 0, \quad j = 1, \ldots, n, \\
&u_r, v_j \geq 0, u_r, v_j \geq 0, quadi = 1, \ldots, m, j = 1, \ldots, n.
\end{aligned} \tag{8.25}$$

The model 98.250 can be used by and available methods of the literature (see [8]).

References

1. Ahmadi, F.R., Rafiei, H., Akbarzadeh-Totonchi, M.R.: An overview of Z-numbers and its applications. Fuzzy Syst. Appl. **6**(2), 1–50 (2024)
2. Azadeh, A., Kokabi, R.: Z-number DEA: a new possibilistic DEA in the context of Z-numbers. Adv. Eng. Inform. **30**(3), 604–617 (2016)
3. Charnes, A., Cooper, W.W., Rhodes, E.: Measuring the efficiency of decision making units. Eur. J. Oper. Res. **2**(6), 429–444 (1978)
4. Ezadi, S., Allahviranloo, T., Mohammadi, S.: Two new methods for ranking of Z-numbers based on sigmoid function and sign method. Int. J. Intell. Syst. **33**(7), 1476–1487 (2018)
5. Kang, B., Wei, D., Li, Y., Deng, Y.: A method of converting z-number to classical fuzzy number. J. Inf. Comput. Sci. **9**(3), 703–709 (2012)
6. Liao, H., Liu, F., Xiao, Y., Wu, Z., Zavadskas, E.K.: A survey on Z-number-based decision analysis methods and applications: What's going on and how to go further? Inf. Sci. **663**, 120234 (2024)
7. Li, Y., Kou, G., Peng, Y., Morente-Molinera, J.A.: Z-number linguistic term set for multi-criteria group decision-making and its application in predicting the acceptance of academic papers. Appl. Intell. **54**(21), 10962–10981 (2024)
8. Lotfi, F.H., Sanei, M., Hosseinzadeh, A.A., Niroomand, S., Mahmoodirad, A.: Uncertainty in data envelopment analysis: fuzzy and belief degree-based uncertainties. Elsevier (2023)
9. Nasir, E.A., Pan, R.: Simulation-based Bayesian optimal ALT designs for model discrimination. Reliab. Eng. Syst. Saf. **134**, 1–9 (2015)
10. Shen, K.W., Wang, J.Q.: Z-VIKOR method based on a new comprehensive weighted distance measure of Z-number and its application. IEEE Trans. Fuzzy Syst. **26**(6), 3232–3245 (2018)
11. Zadeh, L.A.: A note on z-numbers. Inf. Sci. **181**(14), 2923–2932 (2011)

Chapter 9
Hesitant Fuzzy Sets

9.1 Introduction

In classical fuzzy sets, for an element of a fuzzy set only one membership value is assigned which is between 0 and 1. Torra [1] introduced the hesitant fuzzy sets (HFSs) where more than one membership function value can be assigned to an element. This issue is very useful for decision makers in order to determine the membership values. HFSs can be used when the experts face hesitancy for making decisions. As an example assume that an expert considers the alternatives "average" and "good" for assessing the quality on an automobile where each of these alternatives have several membership function between 0.6 and 0.8.

9.2 Hesitant Fuzzy Set

Hesitant fuzzy sets (HFS) are used for the situations that an expert expresses his/her hesitancy degree by a finite set of values. Therefore, in a HFS instead of the interval [0, 1], a finite set of its elements is considered for degrees of hesitancy.

Definition 9.1 [1] Assume $\boldsymbol{X}$ as a reference set. A hesitant fuzzy set on $\boldsymbol{X}$, is a set of ordered pairs as below.

$$\tilde{H} = \left\{ \left(x, h_{\tilde{H}}(x)\right) : x \in X \right\} \tag{9.1}$$

Here, $h_{\tilde{H}} : X \to P([0, 1])$, and $P([0, 1])$ is the set of all sub-sets of [0, 1].

The variable $h_{\tilde{H}}(x)$ is shown as a hesitant fuzzy element (HFE). If a HFS be convex and defined on a limited and closed interval, then it is a hesitant fuzzy number (HFN).

Example 9.1 Assume the reference set $X = \{x_1, x_2, x_3, x_4\}$, and the HFEs $h_{\tilde{H}}(x_1) = \{0.2, 0.4\}$, $h_{\tilde{H}}(x_2) = \{0.1, 0.3, 0.4\}$, $h_{\tilde{H}}(x_3) = \{0.5\}$, and $h_{\tilde{H}}(x_4) = \{0.4, 0.6, 0.7\}$.

A. Mahmoodirad et al., *Recent Advances on Fuzzy Sets*, Studies in Fuzziness and Soft Computing 440, https://doi.org/10.1007/978-3-032-08776-8_9

The HFS $\tilde{H}$ can be defined as below.

$$\tilde{H} = \{(x_1, \{0.2, 0.4\}), (x_2, \{0.1, 0.3, 0.4\}), (x_3, \{0.5\}), (x_4, \{0.4, 0.6, 0.7\})\}.$$

Definition 9.2 [2] Assume the HFEs $h = \bigcup_{\gamma \in h} \{\gamma\}$, $h_1 = \bigcup_{\gamma_1 \in h_1} \{\gamma_1\}$, and $h_2 = \bigcup_{\gamma_2 \in h_2} \{\gamma_2\}$. Then, the following operations are defined.

1. **Complement**: $h^c = \bigcup_{\gamma \in h} \{1 - \gamma\}$,
2. **Union**: $h_1 \cup h_2 = \bigcup_{\gamma_1 \in h_1, \gamma_2 \in h_2} \max\{\gamma_1, \gamma_2\}$,
3. **Intersection**: $h_1 \cap h_2 = \bigcup_{\gamma_1 \in h_1, \gamma_2 \in h_2} \min\{\gamma_1, \gamma_2\}$.

Definition 9.3 [3] Assume the HFEs $h = \bigcup_{\gamma \in h} \{\gamma\}$, $h_1 = \bigcup_{\gamma_1 \in h_1} \{\gamma_1\}$, and $h_2 = \bigcup_{\gamma_2 \in h_2} \{\gamma_2\}$. Then, considering $\lambda \geq 0$, the below operations are defined.

1. $h^\lambda = \bigcup_{\gamma \in h} \{\gamma^\lambda\}$,
2. $\lambda h = \bigcup_{\gamma \in h} \{1 - (1 - \gamma)^\lambda\}$,
3. $h_1 \oplus h_2 = \bigcup_{\gamma_1 \in h_1, \gamma_2 \in h_2} \{\gamma_1 + \gamma_2 - \gamma_1 \gamma_2\}$,
4. $h_1 \ominus h_2 = \bigcup_{\substack{\gamma_1 \in h_1, \\ \gamma_2 \in h_2}} \{\gamma^*\}$, where $\gamma^* = \begin{cases} \frac{\gamma_1 - \gamma_2}{1 - \gamma_2}, \gamma_1 \geq \gamma_2, \gamma_2 \neq 0, \\ 0, O.W. \end{cases}$
5. $h_1 \otimes h_2 = \bigcup_{\gamma_1 \in h_1, \gamma_2 \in h_2} \{\gamma_1 \gamma_2\}$,
6. $h_1 \div h_2 = \bigcup_{\gamma_1 \in h_1, \gamma_2 \in h_2} \{\gamma^+\}$, where $\gamma^+ = \begin{cases} \frac{\gamma_1}{\gamma_2}, \gamma_1 \leq \gamma_2, \gamma_2 \neq 0, \\ 1, O.W. \end{cases}$

Example 9.2 Consider the HFEs $h_1 = \{0.1, 0.2\}$ and $h_2 = \{0.1, 0.3, 0.5\}$, and the value of $\lambda = 2$. Then,

1. $h_1^2 = \{0.01, 0.04\}$,
2. $2 \times h_1 = \{0.19, 0.36\}$,
3. $h_1 \oplus h_2 = \{0.19, 0.37, 0.55, 0.28, 0.44, 0.60\}$,
4. $h_1 \ominus h_2 = \{0.11\}$,
5. $h_1 \otimes h_2 = \{0.01, 0.03, 0.05, 0.02, 0.06, 0.10\}$,
6. $h_1 \div h_2 = \{0.2, 0.33, 0.4, 0.67, 1\}$.

Definition 9.4 [3] Assume the HFE $h = \bigcup_{\gamma \in h} \{\gamma\}$. Then, the score function of h is obtained as $S(h) = \frac{1}{l(h)} \sum_{\gamma \in h} \gamma$, where $l(h)$ is the number of elements of h.

Definition 9.5 [4] Standard deviation degree of the HFE h is shown by $D_s(h)$ and is obtained as below.

$$D_s(h) = \frac{1}{l(h)} \sqrt{\sum_{\gamma_i, \gamma_j \in h} (\gamma_i - \gamma_j)^2} \tag{9.2}$$

Definition 9.6 [5] Deviation degree of the HFE h is shown by $D(h)$ and is obtained as below.

$$D(h) = \sqrt{\frac{1}{l(h)} \sum_{\gamma \in h} (\gamma - S(h))^2} \tag{9.3}$$

Definition of $S(h)$ is stated in Definition 9.4.

Definition 9.7 Assume the HFEs $h_1 = \bigcup_{\gamma_1 \in h_1} \{\gamma_1\}$ and $h_2 = \bigcup_{\gamma_2 \in h_2} \{\gamma_2\}$ with the score functions $S(h_1)$ and $S(h_2)$, and deviation degrees $D(h_1)$ and $D(h_2)$ respectively. Then,

- If $S(h_1) < S(h_2) \Rightarrow h_1 \prec h_2$,
- If $S(h_1) > S(h_2) \Rightarrow h_1 \succ h_2$,
- If $S(h_1) = S(h_2)$, then,
  - If $D(h_1) < D(h_2) \Rightarrow h_1 \succ h_2$,
  - If $D(h_1) > D(h_2) \Rightarrow h_1 \prec h_2$,
  - If $D(h_1) = D(h_2) \Rightarrow h_1 \approx h_2$.

Also, Liao and Xu [4] applied the standard deviation degree in order to rank HFEs as below,

- If $D_s(h_1) < D_s(h_2) \Rightarrow h_1 \succ h_2$,
- If $D_s(h_1) > D_s(h_2) \Rightarrow h_1 \prec h_2$,
- If $D_s(h_1) = D_s(h_2) \Rightarrow h_1 \approx h_2$.

Definition 9.8 [6] Assume the HFN $h = \bigcup_{\gamma \in h} \{\gamma\} = \{\gamma_j | j = 1, 2, \ldots, l(h)\}$. Then its score function is defined as,

$$S(h) = \frac{\sum_{j=1}^{l(h)} \delta(j) \gamma_j}{\sum_{j=1}^{l(h)} \delta(j)}, \tag{9.4}$$

where, $\{\delta(\gamma_j) | j = 1, 2, \ldots, l(h)\}$ is a positive-valued monotonic increasing sequence of index j. The logic of $S(h_1) < S(h_2) \Rightarrow h_1 \prec h_2$ can be used to rank the HFNs h_1 and h_2.

The function (9.4) assumes that the values of the HFNs are sorted increasingly. If two HFNs are with different lengths, then the shorter one should be extended by repeating its last value (largest value) as long as the longer one in advance.

Example 9.3 [7] Assume the HFNs $h_1 = \{0.5\}$, $h_2 = \{0.2, 0.8\}$, and $h_3 = \{0.2, 0.5, 0.8\}$. Then, $S(h_1) = 0.50$, $S(h_2) = 0.70$, and $S(h_3) = 0.60$ (if h_2 is extended to $\{0.2, 0.8, 0.8\}$). Now, as $S(h_1) < S(h_3) < S(h_2)$, then, $h_1 < h_3 < h_2$.

Definition 9.9 [6] Assume the HFSs $\tilde{H}_1$ and $\tilde{H}_2$. Then,

$$\tilde{H}_1 \prec \tilde{H}_2 \Leftrightarrow S(\tilde{H}_1) < S(\tilde{H}_2), \tag{9.5}$$

where, $S(\tilde{H}_1) = \frac{1}{n}\sum_{i=1}^{n}\left(\frac{1}{l_{x_i}}\sum_{j=1}^{l_{x_i}}\gamma_{\tilde{H}_1}^{\sigma(j)}(x_i)\right)$, and $\{\delta(\gamma_j)|j = 1, 2, \ldots, l_{x_i}\}$ is a positive-valued monotonic increasing sequence of index j.

The function (9.5) assumes that the values of $\{h_i\}$ in the HFSs are sorted increasingly. If two $\{h_i\}$ are with different lengths, then the shorter one should be extended by repeating its last value (largest value) as long as the longer one in advance.

Example 9.4 [8] Assume the HFSs $\tilde{H}_1 = \{(x_1, \{0.4\}), (x_2, \{0.3, 0.4\}), (x_3, \{0.3, 0.2, 0.5, 0.6\})\}$ and $\tilde{H}_2 = \{(x_1, \{0.2, 0.5\}), (x_2, \{0.3\}), (x_3, \{0.3, 0.2\})\}$ on the reference set $X = \{x_1, x_2, x_3\}$. Then, $\tilde{H}_1$ and $\tilde{H}_2$ are rewritten as below.

$$\tilde{H}_1 = \{(x_1, \{0.4, 0.4\}), (x_2, \{0.3, 0.4\}), (x_3, \{0.2, 0.3, 0.5, 0.6\})\}$$

$$\tilde{H}_2 = \{(x_1, \{0.2, 0.5\}), (x_2, \{0.3, 0.3\}), (x_3, \{0.2, 0.3, 0.3, 0.3\})\}$$

Then, according to Definition 9.9, there is,

$$S(\tilde{H}_1) = \frac{1}{3}\left(\frac{0.4 + 0.4}{2} + \frac{0.3 + 0.4}{2} + \frac{0.2 + 0.3 + 0.5 + 0.6}{4}\right) = 0.3833,$$

$$S(\tilde{H}_2) = \frac{1}{3}\left(\frac{0.2 + 0.5}{2} + \frac{0.3 + 0.3}{2} + \frac{0.2 + 0.3 + 0.3 + 0.3}{4}\right) = 0.3083,$$

As $S(\tilde{H}_2) < S(\tilde{H}_1)$, then $\tilde{H}_2 \prec \tilde{H}_1$.

Definition 9.10 [9, 10] Assume the HFNs $\tilde{H}_1 = (a, h_{\tilde{H}_1})$ and $\tilde{H}_2 = (b, h_{\tilde{H}_2})$ where $a, b \in \mathbb{R}$, $h_{\tilde{H}_1}, h_{\tilde{H}_2} \in [0, 1]$, $|h_{\tilde{H}_1}| = k$, and $|h_{\tilde{H}_2}| = l$. The Euclidean distance (d_E) and Hamming distance (d_H) of these HFNs are defined as below.

$$d_E(\tilde{H}_1, \tilde{H}_2) = \sqrt{\frac{1}{1 + k \times l}\left(|a - b|^2 + \sum_{\substack{\gamma_i \in h_{\tilde{H}_1}, \\ \gamma_j \in h_{\tilde{H}_2}}} |\gamma_i - \gamma_j|^2\right)}, \quad (9.6)$$

$$d_H = \frac{1}{1 + k \times l}\left(|a - b| + \sum_{\substack{\gamma_i \in h_{\tilde{H}_1}, \\ \gamma_j \in h_{\tilde{H}_2}}} |\gamma_i - \gamma_j|\right). \quad (9.7)$$

Definition 9.11 Assume the HFEs $h_1, \ldots, h_n$, each of them with l elements. The vector $w = (w_1, \ldots, w_n)$ with the conditions $\sum_{j=1}^{n} w_j = 1$ and $w_j \in [0, 1]$ represents their associated weight values. Then, the adjusted hesitant fuzzy weighted average (AHFWA) and the adjusted hesitant fuzzy weighted geometric (AHFWG) operators are defined below.

$$AHFWA(h_1, \ldots, h_n) = \oplus_{j=1}^{n} w_j h_j = \left\{ 1 - \prod_{j=1}^{n} \left(1 - h_j^{\sigma(t)}\right)^{w_j} \middle| t = 1, \ldots, l \right\}, \quad (9.8)$$

$$AHFWG(h_1, \ldots, h_n) = \otimes_{j=1}^{n} \left(h_j\right)^{w_j} = \left\{ \prod_{j=1}^{n} \left(h_j^{\sigma(t)}\right)^{w_j} \middle| t = 1, \ldots, l \right\}. \quad (9.9)$$

9.3 Trapezoidal Hesitant Fuzzy Set

Definition 9.12 [11] The trapezoidal hesitant fuzzy set $\tilde{A}^H$ on the set X is shown as $\tilde{A}^H = \left\{ \left(x, h_{\tilde{A}^H}\right) \middle| x \in X \right\}$ where $h_{\tilde{A}^H}$ is known as a trapezoidal hesitant fuzzy element (HTrFE) and shows the possible membership degrees of $x \in X$ to $\tilde{A}^H$. It is shown as,

$$\tilde{A}^H = \left\{ \left(x, \bigcup_{(a,b,c,d) \in h_{\tilde{A}^H}} \{(a, b, c, d)\} \right) \middle| x \in X \right\} \quad (9.10)$$

where,

$$(a, b, c, d) = \begin{cases} \frac{t-a}{b-a}, a \le t \le b, \\ 1, b \le t \le c, \\ \frac{d-t}{d-c}, c \le t \le d. \end{cases} \quad (9.11)$$

Clearly, the condition $0 \le a, b, c, d \le 1$ should be met.

Definition 9.13 [11] Assume the reference set X, $\xi_i \in [0, 1]$ (where $i \in I = \{1, \ldots, n\} or \{1, \ldots, m\} or..$), $a_i, b_i, c_i, d_i \in \mathbb{R}$, and $a_i \le b_i \le c_i \le d_i$ (for $i \in I$). Then the generalized trapezoidal hesitant fuzzy set is defined as,

$$\tilde{A}^H = \{ (x, \{((a_i, b_i, c_i, d_i); \xi_i) : i \in I\}) | x \in X \} \quad (9.12)$$

where, $\{((a_i, b_i, c_i, d_i); \xi_i) : i \in I\}$ is a set of generalized trapezoidal fuzzy numbers in the set or real numbers where show the possible membership function values of $x \in X$.

If $a = a_i$, $b = b_i$, $c = c_i$, and $d = d_i$, then the generalized trapezoidal hesitant fuzzy number is converted to the single-valued generalized trapezoidal hesitant fuzzy number. Meaning that,

$$\xi_{GTHFN} = \{(a, b, c, d); \{\xi_i : \xi_i \in \xi(x)\}\},$$
$$\xi(x) \text{is a set of some values in } [0, 1], \tag{9.13}$$

is a particular hesitant fuzzy set on the set of real numbers with the below membership functions.

$$\mu_i(x) = \begin{cases} \frac{x-a}{b-a}\xi_i, a \le x \le b, \\ \xi_i, b \le x \le c, \\ \frac{d-x}{d-c}\xi_i, c \le x \le d. \end{cases}$$

If $a = a_i$, $b = b_i = c = c_i$, and $d = d_i$, then the generalized trapezoidal hesitant fuzzy number is converted to a generalized hesitant triangular fuzzy number.

Definition 9.14 [12] assume the generalized trapezoidal hesitant fuzzy numbers $\xi_{GTHFN} = \{(a, b, c, d); \{\xi(x)\}\}$, $\xi^1_{GTHFN} = \{(a_1, b_1, c_1, d_1); \{\xi_1 = \xi_1(x)\}\}$, and $\xi^2_{GTHFN} = \{(a_2, b_2, c_2, d_2); \{\xi_2 = \xi_2(x)\}\}$. For the real number $\gamma \neq 0$,

- $\xi^1_{GTHFN} \oplus \xi^2_{GTHFN} = \{(a_1 + a_2, b_1 + b_2, c_1 + c_2, d_1 + d_2); \{\xi_1 + \xi_2 - \xi_1\xi_2\}\}$,
- $\xi^1_{GTHFN} \ominus \xi^2_{GTHFN} = \left\{(a_1 - d_2, b_1 - c_2, c_1 - b_2, d_1 - a_2); \left\{\bigcup_{\xi_1 \in \xi^1, \xi_2 \in \xi^2} \{\xi^*\},\right\}\right\}$,
  where $\xi^* = \begin{cases} \frac{\xi_1 - \xi_2}{1 - \xi_2}, \xi_1 \ge \xi_2, \xi_2 \neq 0, \\ 0, O.W. \end{cases}$
- $\xi^1_{GTHFN} \div \xi^2_{GTHFN} = \left\{\left(\frac{a_1}{d_2}, \frac{b_1}{c_2}, \frac{c_1}{b_2}, \frac{d_1}{a_2}\right); \left\{\bigcup_{\xi_1 \in \xi^1, \xi_2 \in \xi^2} \{\xi_1\xi_2\},\right\}\right\}, a_1 > 0, a_2 > 0$,
- $\xi^1_{GTHFN} \ominus \xi^2_{GTHFN} = \left\{(a_1 - d_2, b_1 - c_2, c_1 - b_2, d_1 - a_2); \left\{\bigcup_{\xi_1 \in \xi^1, \xi_2 \in \xi^2} \{\xi^+\},\right\}\right\}$,
  where $\xi^+ = \begin{cases} \frac{\xi_1}{\xi_2}, \xi_1 \le \xi_2, \xi_2 \neq 0, \\ 1, O.W. \end{cases}$,
- $k\xi_{GTHFN} = \left\{(ka, kb, kc, kd); \left\{\bigcup_{\xi \in \xi(x)} \{1 - (1 - \xi)^k\},\right\}\right\}, k \ge 0$,
- $(\xi_{GTHFN})^k = \left\{\left(a^k, b^k, c^k, d^k\right); \left\{\bigcup_{\xi \in \xi(x)} \{\xi^k\}\right\}\right\}, k \ge 0$.

Definition 9.15 [12] Assume $\xi_{GTHFN} = \{(a, b, c, d); \{\xi(x)\}\}$ as a generalized trapezoidal hesitant fuzzy number where $l(h)$ is the number of elements of ξ_{GTHFN}. Then, the score function, standard deviation degree, and deviation degree of ξ_{GTHFN} are obtained respectively as below.

$$S(\xi_{GTHFN}) = \frac{c^2 + d^2 - a^2 - b^2}{2l(h)} \times \sum_{\xi \in \xi_{GTHFN}} \xi, \tag{9.14}$$

$$D_s(\xi_{GTHFN}) = \frac{c^2 + d^2 - a^2 - b^2}{2l(h)} \times \sqrt{\sum_{\xi_i, \xi_j \in \xi_{GTHFN}} \left(\xi_i - \xi_j\right)^2}, \tag{9.15}$$

$$D(\xi_{GTHFN}) = \frac{c^2+d^2-a^2-b^2}{2l(h)} \times \sqrt{\frac{1}{l(h)} \sum_{\xi \in \xi_{GTHFN}} (\xi - S(\xi_{GTHFN}))^2}, \quad (9.16)$$

Example 9.5 [12] Assume the generalized trapezoidal hesitant fuzzy number $\xi_{GTHFN} = \{(2,4,6,8); \{0.5, 0.7, 0.2, 0.2\}\}$. Then,

$$S(\xi_{GTHFN}) = \frac{6^2+8^2-2^2-4^2}{2\times 4} \times (0.5+0.7+0.2+0.2) = 16,$$

$$D_s(\xi_{GTHFN}) = \frac{6^2+8^2-2^2-4^2}{2\times 4} \times \sqrt{0.04+0.25+0.09+0.09+0.25+0.09+0.25+0.09+0.04+0.25} = 12,$$

$$D(\xi_{GTHFN}) = \frac{6^2+8^2-2^2-4^2}{2\times 4} \times \sqrt{240.25+234.09+249.64+249.64} = 624.05,$$

Definition 9.16 [12] Assume the generalized trapezoidal hesitant fuzzy numbers ξ^1_{GTHFN} and ξ^2_{GTHFN} with score functions $S(\xi^1_{GTHFN})$ and $S(\xi^2_{GTHFN})$, and deviation degrees $D(\xi^1_{GTHFN})$ and $D(\xi^2_{GTHFN})$ respectively. Then,

If $S(\xi^1_{GTHFN}) < S(\xi^2_{GTHFN}) \Rightarrow \xi^1_{GTHFN} \prec \xi^2_{GTHFN}$,
If $S(\xi^1_{GTHFN}) > S(\xi^2_{GTHFN}) \Rightarrow \xi^1_{GTHFN} \succ \xi^2_{GTHFN}$,
If $S(\xi^1_{GTHFN}) = S(\xi^2_{GTHFN})$, then
If $D(\xi^1_{GTHFN}) < D(\xi^2_{GTHFN}) \Rightarrow \xi^1_{GTHFN} \succ \xi^2_{GTHFN}$,
If $D(\xi^1_{GTHFN}) > D(\xi^2_{GTHFN}) \Rightarrow \xi^1_{GTHFN} \prec \xi^2_{GTHFN}$,
If $D(\xi^1_{GTHFN}) = D(\xi^2_{GTHFN}) \Rightarrow \xi^1_{GTHFN} \approx \xi^2_{GTHFN}$.

Example 9.6 [12] Assume the generalized trapezoidal hesitant fuzzy numbers $\xi^1_{GTHFN} = \{(1,2,3,4); \{0.9, 0.7\}\}$, $\xi^2_{GTHFN} = \{(2,4,6,8); \{0.1, 0.2, 0.2, 0.3\}\}$, and $\xi^3_{GTHFN} = \{(0,5,6,8); \{0.8, 0.1, 0.4, 1.0, 0, 9\}\}$. As $S(\xi^1_{GTHFN}) = 8$, $S(\xi^2_{GTHFN}) = 8$, and $S(\xi^3_{GTHFN}) = 24$, then $\xi^1_{GTHFN}, \xi^2_{GTHFN} < \xi^3_{GTHFN}$. On the other hand, as $D(\xi^1_{GTHFN}) = 50.91$ and $D(\xi^2_{GTHFN}) = 156$, then $\xi^2_{GTHFN} < \xi^1_{GTHFN} < \xi^3_{GTHFN}$.

Theorem 9.1 [12] *Consider the generalized trapezoidal hesitant fuzzy numbers* ξ^1_{GTHFN} *and* ξ^2_{GTHFN}, *and* $k > 0$. *Then,*

$$k\left(\xi^1_{GTHFN} \oplus \xi^2_{GTHFN}\right) = k\xi^1_{GTHFN} \oplus k\xi^2_{GTHFN}, \quad (9.17)$$

$$k\left(\xi^1_{GTHFN} \ominus \xi^2_{GTHFN}\right) = k\xi^1_{GTHFN} \ominus k\xi^2_{\mathrm{GTHFN}}, \text{ if } \xi^1_1 \in \xi^1_{\mathrm{GTHFN}} \geq \xi^2_1 \in \xi^2_{\mathrm{GTHFN}}, \xi^2_1 \neq 1, \quad (9.18)$$

$$\left(\xi^1_{GTHFN} \otimes \xi^2_{GTHFN}\right)^k = \left(\xi^1_{GTHFN}\right)^k \otimes \left(\xi^2_{GTHFN}\right)^k, \quad (9.19)$$

$$\left(\xi^1_{GTHFN} \div \xi^2_{GTHFN}\right)^k = \left(\xi^1_{GTHFN}\right)^k \div \left(\xi^2_{GTHFN}\right)^k,$$
$$\text{if } \xi^1_1 \in \xi^1_{GTHFN} \leq \xi^2_1 \in \xi^2_{GTHFN}, \xi^2_1 \neq 0, \tag{9.20}$$

Definition 9.17 [12] Assume Ω as the set of all generalized trapezoidal hesitant fuzzy numbers on the set of real numbers. Assume $\xi^j_{GTHFN} \in \Omega$ where $j \in I_n$ is a set of generalized trapezoidal hesitant fuzzy numbers. Then, $H^{\tilde{A}}_w : \Omega^n \to \Omega$, is a n dimensional generalized trapezoidal hesitant fuzzy weighted arithmetic operator defined as,

$$\begin{aligned} H^{\tilde{A}}_w\left(\xi^1_{GTHFN}, \xi^2_{GTHFN}, \ldots, \xi^n_{GTHFN}\right) &= \oplus^n_{j=1}\left(w_j \xi^j_{GTHFN}\right) \\ &= \left\{\left(\sum_{j=1}^n w_j a_j, \sum_{j=1}^n w_j b_j, \sum_{j=1}^n w_j c_j, \sum_{j=1}^n w_j d_j\right);\right. \\ &\left.\left\{\bigcup_{\xi^1_1 \in \xi^1, \xi^2_1 \in \xi^2, \ldots, \xi^n_1 \in \xi^n} \left\{1 - \prod_{j=1}^n \left(1 - \xi^j_1\right)^{w_j}\right\}\right\}\right\}, \end{aligned} \tag{9.21}$$

where, $w = (w_1, \ldots, w_n)^T$ is the weight vector for $\xi^j_{GTHFN}, j \in I_n$, $w_j \in [0, 1]$, and $\sum_{j=1}^n w_j = 1$.

Definition 9.18 [12] Assume Ω as the set of all generalized trapezoidal hesitant fuzzy numbers on the set of real numbers. Assume $\xi^j_{GTHFN} \in \Omega$ where $j \in I_n$ is a set of generalized trapezoidal hesitant fuzzy numbers. Then, $H^{\tilde{G}}_w : \Omega^n \to \Omega$ is a n dimensional generalized trapezoidal hesitant fuzzy weighted geometric operator defined as,

$$\begin{aligned} H^{\tilde{G}}_w\left(\xi^1_{GTHFN}, \xi^2_{GTHFN}, \ldots, \xi^n_{GTHFN}\right) &= \otimes^n_{j=1}\left(\xi^j_{GTHFN}\right)^{w_j} \\ &= \left\{\left(\prod_{j=1}^n a_j^{w_j}, \prod_{j=1}^n b_j^{w_j}, \prod_{j=1}^n c_j^{w_j}, \prod_{j=1}^n d_j^{w_j}\right);\right. \\ &\left.\left\{\bigcup_{\xi^1_1 \in \xi^1, \xi^2_1 \in \xi^2, \ldots, \xi^n_1 \in \xi^n} \left\{\left(\prod_{j=1}^n \xi^j_1\right)^{w_j}\right\}\right\}\right\}, \end{aligned} \tag{9.22}$$

where, $w = (w_1, \ldots, w_n)^T$ is the weight vector for $\xi^j_{GTHFN}, j \in I_n$, $w_j \in [0, 1]$, and $\sum_{j=1}^n w_j = 1$.

Example 9.7 [12] Assume the generalized trapezoidal hesitant fuzzy numbers $\xi^1_{GTHFN} = \{(2, 3, 4, 6); \{0.5, 0.6, 0.2\}\}$, $\xi^2_{GTHFN} = \{(1, 5, 6, 8); \{0.3, 0.4\}\}$, and $\xi^3_{GTHFN} = \{(5, 6, 8, 10); \{0.2, 0.6\}\}$, where the weight vector of $w = (0.25, 0.35, 0.40)^T$ is given. Then,

$$H_w^{\tilde{A}}\left(\xi_{GTHFN}^1, \xi_{GTHFN}^2, \xi_{GTHFN}^3\right) = \oplus_{j=1}^3\left(w_j\xi_{GTHFN}^j\right)$$
$$= \{(2.85, 4.9, 6.30, 8.30); \{0.71, 0.55, 0.68, 0.50, 0.78, 0.66, 0.76, 0.63, 0.77, 0.64, 0.75, 0.60\}\},$$

$$H_w^{\tilde{G}}\left(\xi_{GTHFN}^1, \xi_{GTHFN}^2, \xi_{GTHFN}^3\right) = \otimes_{j=1}^3\left(\xi_{GTHFN}^j\right)^{w_j}$$
$$= \{(2.26, 4.73, 6.08, 8.14); \{0.29, 0.45, 0.32, 0.50, 0.30, 0.47, 0.34, 0.52, 0.23, 0.36, 0.25, 0.40\}\},$$

Theorem 9.2 [12] *Assume Ω as the set of all generalized trapezoidal hesitant fuzzy numbers on the set of real numbers. Assume $\xi_{GTHFN}^j \in \Omega$ where $j \in I_n$ is a set of generalized trapezoidal hesitant fuzzy numbers. Then,*

$$H_w^{\tilde{G}}\left(\xi_{GTHFN}^1, \xi_{GTHFN}^2, \ldots, \xi_{GTHFN}^n\right) \le H_w^{\tilde{A}}\left(\xi_{GTHFN}^1, \xi_{GTHFN}^2, \ldots, \xi_{GTHFN}^n\right), \quad (9.23)$$

9.4 Interval-Valued Hesitant Fuzzy Sets

In this sub-section the hesitant fuzzy sets are extended to an interval form where the membership function degrees are of interval values.

Definition 9.19 [13] Assume the reference set X and $D[0, 1]$ as all the closed sub-sets of the interval $[0, 1]$. An interval-valued hesitant fuzzy set (IVHFS) on X, is defined as $\tilde{H} = \left\{\left(x_i, \tilde{h}_{\tilde{H}}(x_i)\right) : x_i \in X, i = 1, \ldots, n\right\}$ where $\tilde{h}_{\tilde{H}}(x_i) : X \to D[0, 1]$ represents all interval-valued membership degrees of $x_i \in X$ to the set $\tilde{H}$. For simplicity, $\tilde{h}_{\tilde{H}}(x_i)$ is called an interval-valued hesitant fuzzy element (IVHFE) where $\tilde{h}_{\tilde{H}}(x_i) = \left\{\left[\tilde{\gamma}^L, \tilde{\gamma}^U\right] | \tilde{\gamma}^L = \widetilde{\inf \gamma}, \tilde{\gamma}^U = \widetilde{\sup \gamma}, \tilde{\gamma} \in h_{\tilde{H}}(x_i)\right\}$. Here, $\tilde{\gamma}^L$ and $\tilde{\gamma}^U$ are lower bound and upper bound of $\tilde{\gamma}$ respectively.

Example 9.8 [13] Consider the reference set $X = \{x_1, x_2\}$ and the IVHFEs $\tilde{h}_{\tilde{H}}(x_1) = \{[0.1, 0.3], [0.4, 0.5]\}$ and $\tilde{h}_{\tilde{H}}(x_2) = \{[0.1, 0.2], [0.3, 0.5], [0.7, 0.9]\}$. Then, the below IVHFS is considered.

$$\tilde{H} = \{(x_1, \{[0.1, 0.3], [0.4, 0.5]\}), (x_2, \{[0.1, 0.2], [0.3, 0.5], [0.7, 0.9]\})\}$$

Definition 9.20 [14] Assume the intervals $A = \left[a^L, a^U\right]$ and $B = \left[b^L, b^U\right]$ where $l_A = a^U - a^L$ and $l_B = b^U - b^L$. The possibility degree of $A \ge B$ is defined as below.

$$P(A \ge B) = \max\left\{1 - \max\left\{\frac{b^U - a^L}{l_A + l_B}, 0\right\}, 0\right\}, \quad (9.24)$$

Here, $0 \le P(A \ge B) \le 1$ and $P(A \ge B) + P(B \ge A) = 1$.

Definition 9.21 [13] Assume the IVHFEs $\tilde{h}$, $\tilde{h}_1$, and $\tilde{h}_2$. Then,

- $\tilde{h}^c = \{[1-\gamma^U, 1-\gamma^L] | \gamma \in \tilde{h}\}$,
- $\tilde{h}_1 \cup \tilde{h}_2 = \{[\max\{\gamma_1^L, \gamma_2^L\}, \max\{\gamma_1^U, \gamma_2^U\}] | \gamma_1 \in \tilde{h}_1, \gamma_2 \in \tilde{h}_2\}$,
- $\tilde{h}_1 \cap \tilde{h}_2 = \{[\min\{\gamma_1^L, \gamma_2^L\}, \min\{\gamma_1^U, \gamma_2^U\}] | \gamma_1 \in \tilde{h}_1, \gamma_2 \in \tilde{h}_2\}$,
- $\tilde{h}^\lambda = \left\{\left[(\gamma^L)^\lambda, (\gamma^U)^\lambda\right] \middle| \gamma \in \tilde{h}\right\}, \lambda > 0$,
- $\lambda\tilde{h} = \left\{\left[1-(1-\gamma^L)^\lambda, 1-(1-\gamma^U)^\lambda\right] \middle| \gamma \in \tilde{h}\right\}, \lambda > 0$,
- $\tilde{h}_1 \oplus \tilde{h}_2 = \{[\gamma_1^L + \gamma_2^L - \gamma_1^L\gamma_2^L, \gamma_1^U + \gamma_2^U - \gamma_1^U\gamma_2^U] | \gamma_1 \in \tilde{h}_1, \gamma_2 \in \tilde{h}_2\}$,
- $\tilde{h}_1 \otimes \tilde{h}_2 = \{[\gamma_1^L\gamma_2^L, \gamma_1^U\gamma_2^U] | \gamma_1 \in \tilde{h}_1, \gamma_2 \in \tilde{h}_2\}$,
- $\tilde{h}_1 \ominus \tilde{\mathrm{h}}_2 = \left\{\left[\max\left\{0, \frac{\gamma_1^{\mathrm{L}} - \gamma_2^{\mathrm{L}}}{1-\gamma_2^L}\right\}, \max\left\{0, \frac{\gamma_1^{\mathrm{U}} - \gamma_2^{\mathrm{U}}}{1-\gamma_2^U}\right\}\right] \middle| \gamma_1 \in \tilde{\mathrm{h}}_1, \gamma_2 \in \tilde{\mathrm{h}}_2\right\}$,
- $\tilde{h}_1 \div \tilde{h}_2 = \left\{\left[\min\left\{1, \frac{\gamma_1^L}{\gamma_2^L}\right\}, \min\left\{1, \frac{\gamma_1^U}{\gamma_2^U}\right\}\right] \middle| \gamma_1 \in \tilde{h}_1, \gamma_2 \in \tilde{h}_2\right\}$.

Theorem 9.3 [13] *Assume the IVHFEs $\tilde{h}$, $\tilde{h}_1$, and $\tilde{h}_2$. Then,*

- $\tilde{h}_1 \oplus \tilde{h}_2 = \tilde{h}_2 \oplus \tilde{h}_1$,
- $\tilde{h}_1 \otimes \tilde{h}_2 = \tilde{h}_2 \otimes \tilde{h}_1$,
- $k(\tilde{h}_1 \oplus \tilde{h}_2) = k\tilde{h}_1 \oplus k\tilde{h}_2, k > 0$,
- $(\tilde{h}_1 \otimes \tilde{h}_2)^k = \tilde{h}_1^k \otimes \tilde{h}_2^k, k > 0$,
- $k_1\tilde{h} \oplus k_2\tilde{h} = (k_1 + k_2)\tilde{h}, k_1, k_2 > 0$,
- $\tilde{h}^{k_1} \otimes \tilde{h}^{k_2} = \tilde{h}^{(k_1+k_2)}, k_1, k_2 > 0$.

Theorem 9.4 [13] *Assume the IVHFEs $\tilde{h}$, $\tilde{h}_1$, and $\tilde{h}_2$. For $k > 0$,*

- $\tilde{h}_1^c \cup \tilde{h}_2^c = (\tilde{h}_1 \cap \tilde{h}_2)^c$,
- $\tilde{h}_1^c \cap \tilde{h}_2^c = (\tilde{h}_1 \cup \tilde{h}_2)^c$,
- $(\tilde{h}^c)^k = (k\tilde{h})^c$,
- $k(\tilde{h}^c) = (\tilde{h}^k)^c$,
- $\tilde{h}_1^c \oplus \tilde{h}_2^c = (\tilde{h}_1 \otimes \tilde{h}_2)^c$,
- $\tilde{h}_1^c \otimes \tilde{h}_2^c = (\tilde{h}_1 \oplus \tilde{h}_2)^c$.

Definition 9.22 [13] Assume the IVHFEs $\tilde{\alpha}$ and $\tilde{\beta}$. Consider $l = \max\{l_{\tilde{\alpha}}, l_{\tilde{\beta}}\}$ where $l_{\tilde{\alpha}}$ and $l_{\tilde{\beta}}$ are the length of $\tilde{\alpha}$ and $\tilde{\beta}$, respectively. The elements of $\tilde{\alpha}$ and $\tilde{\beta}$ are re-ordered in an increasing sequence. Also assume $\tilde{\alpha}_{\sigma(i)}, \tilde{\beta}_{\sigma(i)}, i = 1, \ldots, l$ as the i -th least value of $\tilde{\alpha}$ and $\tilde{\beta}$ respectively. Then, the below Hamming and Euclidean distances of these elements are obtained as below respectively.

$$d_H(\tilde{\alpha}, \tilde{\beta}) = \frac{1}{2l}\sum_{i=1}^{l}\left(\left|\tilde{\alpha}_{\sigma(i)}^L - \tilde{\beta}_{\sigma(i)}^L\right| + \left|\tilde{\alpha}_{\sigma(i)}^U - \tilde{\beta}_{\sigma(i)}^U\right|\right), \tag{9.25}$$

$$d_E(\tilde{\alpha}, \tilde{\beta}) = \sqrt{\frac{1}{2l}\sum_{i=1}^{l}\left(\left|\tilde{\alpha}_{\sigma(i)}^L - \tilde{\beta}_{\sigma(i)}^L\right|^2 + \left|\tilde{\alpha}_{\sigma(i)}^U - \tilde{\beta}_{\sigma(i)}^U\right|^2\right)}. \tag{9.26}$$

Definition 9.23 [13] For the IVFHEs $\tilde{h}_j, j = 1, \ldots, n$ and the weight vector of $w = (w_1, \ldots, w_n)$, the interval-valued hesitant fuzzy weighted arithmetic and geometric operators (which are functions from $\tilde{H}^n \to \tilde{H}$) are defined as below respectively.

$$IVHFWA\left(\tilde{h}_1, \ldots, \tilde{h}_n\right) = \oplus_{j=1}^{n}\left(w_j \tilde{h}_j\right) = \left\{\left[1 - \prod_{j=1}^{n}\left(1 - \gamma_j^L\right)^{w_j}, 1 - \prod_{j=1}^{n}\left(1 - \gamma_j^U\right)^{w_j}, \right] \middle| \gamma_1 \in \tilde{h}_1, \ldots, \gamma_n \in \tilde{h}_n\right\}, \tag{9.27}$$

$$IVHFWG\left(\tilde{h}_1, \ldots, \tilde{h}_n\right) = \otimes_{j=1}^{n}\left(\tilde{h}_j\right)^{w_j} = \left\{\left[\prod_{j=1}^{n}\left(\gamma_j^L\right)^{w_j}, \prod_{j=1}^{n}\left(\gamma_j^U\right)^{w_j}, \right] \middle| \gamma_1 \in \tilde{h}_1, \ldots, \gamma_n \in \tilde{h}_n\right\}, \tag{9.28}$$

Here, $w_j \in [0, 1]$, $\sum_{j=1}^{n} w_j = 1$, and $\lambda > 0$.

9.5 Intuitionistic Hesitant Fuzzy Sets

Definition 9.24 Assume the reference set X. An intuitionistic hesitant fuzzy set (IHFS) such as $\tilde{A}$ on X based on the membership and non-membership functions $\mu_{\tilde{A}}(x)$ and $\nu_{\tilde{A}}(x)$ where $\mu_{\tilde{A}}(x), \nu_{\tilde{A}}(x) \in [0, 1]$, is defined as below.

$$\tilde{A} = \left\{\left(x, \mu_{\tilde{A}}(x), \nu_{\tilde{A}}(x)\right) \middle| x \in X\right\} \tag{9.29}$$

Here, for $x \in X$ if we assume $\mu_{\tilde{A}}(x) = \bigcup_{\gamma_{\tilde{A}} \in \mu_{\tilde{A}}(x)} \left\{\gamma_{\tilde{A}}\right\}$, $\nu_{\tilde{A}}(x) = \bigcup_{\eta_{\tilde{A}} \in \nu_{\tilde{A}}(x)} \left\{\eta_{\tilde{A}}\right\}$, $\gamma_{\tilde{A}}^+ \in \mu_{\tilde{A}}^+ = \bigcup_{x \in X} \max_{\gamma_{\tilde{A}} \in \mu_{\tilde{A}}(x)} \left\{\gamma_{\tilde{A}}\right\}$, $\eta_{\tilde{A}}^+ \in \nu_{\tilde{A}}^+ = \bigcup_{x \in X} \max_{\eta_{\tilde{A}} \in \nu_{\tilde{A}}(x)} \left\{\eta_{\tilde{A}}\right\}$, then $0 \le \gamma_{\tilde{A}}, \eta_{\tilde{A}} \le 1$, $0 \le \gamma_{\tilde{A}}^+ + \eta_{\tilde{A}}^+ \le 1$. Therefore, an IHFS can be shown as below too.

$$\tilde{A} = \left\{\left(x, \bigcup_{\gamma_{\tilde{A}} \in \mu_{\tilde{A}}} \left\{\gamma_{\tilde{A}}\right\}, \bigcup_{\eta_{\tilde{A}} \in \nu_{\tilde{A}}} \left\{\eta_{\tilde{A}}\right\}\right) \middle| x \in X\right\}. \tag{9.30}$$

For more simplicity, in continue the term $h_{\tilde{A}} = \left(\mu_{\tilde{A}}(x), \nu_{\tilde{A}}(x)\right) = \left(\bigcup_{\gamma_{\tilde{A}} \in \mu_{\tilde{A}}} \left\{\gamma_{\tilde{A}}\right\}, \bigcup_{\eta_{\tilde{A}} \in \nu_{\tilde{A}}} \left\{\eta_{\tilde{A}}\right\}\right)$ represents an intuitionistic hesitant fuzzy element (IHFE).

Example 9.9 Assume the reference set $X = \{x_1, x_2\}$, and the IHFEs $h_{\tilde{A}}(x_1) = (\{0.2, 0.5\}, \{0.3\})$ and $h_{\tilde{A}}(x_2) = (\{0.3, 0.4\}, \{0.1, 0.6\})$. As,

$$\gamma^+_{\tilde{A}}(x_1) = 0.5, \eta^+_{\tilde{A}}(x_1) = 0.3, \gamma^+_{\tilde{A}}(x_2) = 0.4, \eta^+_{\tilde{A}}(x_2) = 0.6,$$

$$\begin{aligned}\gamma^+_{\tilde{A}}(x_1) + \eta^+_{\tilde{A}}(x_1) = 0.5 + 0.3 = 0.8 \leq 1, \gamma^+_{\tilde{A}}(x_2) + \eta^+_{\tilde{A}}(x_2)\\ = 0.4 + 0.6 = 1 \leq 1,\end{aligned}$$

then, $\tilde{A}$ is an IHFS shown as below.

$$\tilde{A} = \{(x_1, \{0.2, 0.5\}, \{0.3\}), (x_2, \{0.3, 0.4\}, \{0.1, 0.6\})\}.$$

Definition 9.25 Assume the IHFEs $h_{\tilde{A}} = \left(\mu_{\tilde{A}}(x), \nu_{\tilde{A}}(x)\right)$, $h_{\tilde{A}_1} = \left(\mu_{\tilde{A}_1}(x), \nu_{\tilde{A}_1}(x)\right)$, and $h_{\tilde{A}_2} = \left(\mu_{\tilde{A}_2}(x), \nu_{\tilde{A}_2}(x)\right)$. Then,

$$h^c_{\tilde{A}} = \left(\nu_{\tilde{A}}(x), \mu_{\tilde{A}}(x)\right), \tag{9.31}$$

$$h_{\tilde{A}_1} \cup h_{\tilde{A}_2} = \left(\mu_{\tilde{A}_1} \cup \mu_{\tilde{A}_2}, \nu_{\tilde{A}_1} \cap \nu_{\tilde{A}_2}\right), \tag{9.32}$$

$$h_{\tilde{A}_1} \cap h_{\tilde{A}_2} = \left(\mu_{\tilde{A}_1} \cap \mu_{\tilde{A}_2}, \nu_{\tilde{A}_1} \cup \nu_{\tilde{A}_2}\right), \tag{9.33}$$

$$h_{\tilde{A}_1} \oplus h_{\tilde{A}_2} = \left(\mu_{\tilde{A}_1} \oplus \mu_{\tilde{A}_2}, \nu_{\tilde{A}_1} \otimes \nu_{\tilde{A}_2}\right), \tag{9.34}$$

$$h_{\tilde{A}_1} \otimes h_{\tilde{A}_2} = \left(\mu_{\tilde{A}_1} \otimes \mu_{\tilde{A}_2}, \nu_{\tilde{A}_1} \oplus \nu_{\tilde{A}_2}\right), \tag{9.35}$$

where $0 \leq \gamma^+_{\tilde{A}} + \eta^+_{\tilde{A}} \leq 1, 0 \leq \gamma^+_{\tilde{A}_1} + \eta^+_{\tilde{A}_1} \leq 1, 0 \leq \gamma^+_{\tilde{A}_2} + \eta^+_{\tilde{A}_2} \leq 1$.

Definition 9.26 Assume the IHFSs $\tilde{A}_1$ and $\tilde{A}_2$. Then,

$$\begin{aligned}\tilde{A}_1 \cup \tilde{A}_2 &= \bigcup_{h_{\tilde{A}_1} \in \tilde{A}_1, h_{\tilde{A}_2} \in \tilde{A}_2} h_{\tilde{A}_1} \cup h_{\tilde{A}_2}\\ &= \left\{ \left(x, \mu_{\tilde{A}_1}(x) \cup \mu_{\tilde{A}_2}(x), \nu_{\tilde{A}_1}(x) \cap \nu_{\tilde{A}_2}(x)\right) \middle| x \in X \right\},\end{aligned} \tag{9.36}$$

$$\begin{aligned}\tilde{A}_1 \cap \tilde{A}_2 &= \bigcup_{h_{\tilde{A}_1} \in \tilde{A}_1, h_{\tilde{A}_2} \in \tilde{A}_2} h_{\tilde{A}_1} \cap h_{\tilde{A}_2}\\ &= \left\{ \left(x, \mu_{\tilde{A}_1}(x) \cap \mu_{\tilde{A}_2}(x), \nu_{\tilde{A}_1}(x) \cup \nu_{\tilde{A}_2}(x)\right) \middle| x \in X \right\},\end{aligned} \tag{9.37}$$

$$\tilde{A}_1 \oplus \tilde{A}_2 = \bigcup_{h_{\tilde{A}_1} \in \tilde{A}_1, h_{\tilde{A}_2} \in \tilde{A}_2} h_{\tilde{A}_1} \oplus h_{\tilde{A}_2}$$

$$= \left\{ \left(x, \mu_{\tilde{A}_1}(x) \oplus \mu_{\tilde{A}_2}(x), \nu_{\tilde{A}_1}(x) \otimes \nu_{\tilde{A}_2}(x) \right) \middle| x \in X \right\}, \tag{9.38}$$

$$\tilde{A}_1 \otimes \tilde{A}_2 = \bigcup_{h_{\tilde{A}_1} \in \tilde{A}_1, h_{\tilde{A}_2} \in \tilde{A}_2} h_{\tilde{A}_1} \otimes h_{\tilde{A}_2}$$
$$= \left\{ \left(x, \mu_{\tilde{A}_1}(x) \otimes \mu_{\tilde{A}_2}(x), \nu_{\tilde{A}_1}(x) \oplus \nu_{\tilde{A}_2}(x) \right) \middle| x \in X \right\}. \tag{9.39}$$

9.5.1 Interval-Valued Intuitionistic Hesitant Fuzzy Sets

Definition 9.27 Assume the reference set X. An interval-valued intuitionistic hesitant fuzzy set (IVIHFS) such as $\tilde{A}$ on X based on the membership and non-membership functions $\mu_{\tilde{A}}(x)$ and $\nu_{\tilde{A}}(x)$ is defined as below.

$$\tilde{A} = \left\{ \left(x, \mu_{\tilde{A}}(x), \nu_{\tilde{A}}(x) \right) \middle| x \in X \right\} \tag{9.40}$$

Here, $\mu_{\tilde{A}}(x)$ and $\nu_{\tilde{A}}(x)$ are interval values from [0, 1].

In this definition, if for any $x \in X$ is assumed that,

$$\mu_{\tilde{A}}(x) = \bigcup_{\left[\gamma_{\tilde{A}}^L, \gamma_{\tilde{A}}^U\right] \in \mu_{\tilde{A}}(x)} \left\{ \left[\gamma_{\tilde{A}}^L, \gamma_{\tilde{A}}^U \right] \right\}, \nu_{\tilde{A}}(x) = \bigcup_{\left[\eta_{\tilde{A}}^L, \eta_{\tilde{A}}^U\right] \in \nu_{\tilde{A}}(x)} \left\{ \left[\eta_{\tilde{A}}^L, \eta_{\tilde{A}}^U \right] \right\},$$

$$\gamma_{\tilde{A}}^{U+} \in \bigcup_{x \in X} \max_{\gamma_{\tilde{A}} \in \mu_{\tilde{A}}(x)} \left\{ \gamma_{\tilde{A}}^U(x) \right\}, \eta_{\tilde{A}}^{U+} \in \bigcup_{x \in X} \max_{\eta_{\tilde{A}} \in \nu_{\tilde{A}}(x)} \left\{ \eta_{\tilde{A}}^U(x) \right\},$$

then, $0 \le \gamma_{\tilde{A}}^L, \gamma_{\tilde{A}}^U, \eta_{\tilde{A}}^L, \eta_{\tilde{A}}^U \le 1, 0 \le \gamma_{\tilde{A}}^{U+} + \eta_{\tilde{A}}^{U+} \le 1$. Therefore, an IVIHFS is shown as,

$$\tilde{A} = \left\{ \left(x, \bigcup_{\left[\gamma_{\tilde{A}}^L, \gamma_{\tilde{A}}^U\right] \in \mu_{\tilde{A}}(x)} \left\{ \left[\gamma_{\tilde{A}}^L, \gamma_{\tilde{A}}^U \right] \right\}, \bigcup_{\left[\eta_{\tilde{A}}^L, \eta_{\tilde{A}}^U\right] \in \nu_{\tilde{A}}(x)} \left\{ \left[\eta_{\tilde{A}}^L, \eta_{\tilde{A}}^U \right] \right\} \right) \middle| x \in X \right\} \tag{9.41}$$

In addition, for more simplicity, an interval-valued intuitionistic fuzzy element (IVIHFE) is shown as below.

$$h_{\tilde{A}} = \left(\mu_{\tilde{A}}(x), \nu_{\tilde{A}}(x) \right) = \left(\bigcup_{\left[\gamma_{\tilde{A}}^L, \gamma_{\tilde{A}}^U\right] \in \mu_{\tilde{A}}(x)} \left\{ \left[\gamma_{\tilde{A}}^L, \gamma_{\tilde{A}}^U \right] \right\}, \bigcup_{\left[\eta_{\tilde{A}}^L, \eta_{\tilde{A}}^U\right] \in \nu_{\tilde{A}}(x)} \left\{ \left[\eta_{\tilde{A}}^L, \eta_{\tilde{A}}^U \right] \right\} \right)$$

Example 9.10 Assume the reference set $X = \{x_1, x_2\}$, and the IVIHFEs $h_{\tilde{A}}(x_1) = (\{[0.2, 0.4], [0.3, 0.3]\}, \{[0.2, 0.5]\})$ and $h_{\tilde{A}}(x_2) = (\{[0.3, 0.4], [0.1, 0.5]\}, \{[0.2, 0.4]\})$. As, $\gamma_{\tilde{A}}^{U+}(x_1) = 0.4, \eta_{\tilde{A}}^{U+}(x_1) = 0.5$, $\gamma_{\tilde{A}}^{U+}(x_2) = 0.5, \eta_{\tilde{A}}^{U+}(x_2) = 0.4$ then $\gamma_{\tilde{A}}^{U+}(x_1) = 0.4, \eta_{\tilde{A}}^{U+}(x_1) = 0.5$, $\gamma_{\tilde{A}}^{U+}(x_2) = 0.5, \eta_{\tilde{A}}^{U+}(x_2) = 0.4$. Therefore, $\tilde{A}$ is an IVIHFS shown as below.

$$\tilde{A} = \{(x_1, \{[0.2, 0.4], [0.3, 0.3]\}, \{[0.2, 0.5]\}), \\ (x_2, \{[0.3, 0.4], [0.1, 0.5]\}, \{[0.2, 0.4]\})\}.$$

Definition 9.28 Assume the IVIHFEs $h_{\tilde{A}} = \left(\bigcup_{[\gamma_{\tilde{A}}^L, \gamma_{\tilde{A}}^U] \in \mu_{\tilde{A}}(x)} \left\{\left[\gamma_{\tilde{A}}^L, \gamma_{\tilde{A}}^U\right]\right\}, \bigcup_{[\eta_{\tilde{A}}^L, \eta_{\tilde{A}}^U] \in \nu_{\tilde{A}}(x)} \left\{\left[\eta_{\tilde{A}}^L, \eta_{\tilde{A}}^U\right]\right\}\right)$, $h_{\tilde{A}_1} = \left(\bigcup_{[\gamma_{\tilde{A}_1}^L, \gamma_{\tilde{A}_1}^U] \in \mu_{\tilde{A}_1}(x)} \left\{\left[\gamma_{\tilde{A}_1}^L, \gamma_{\tilde{A}_1}^U\right]\right\}, \bigcup_{[\eta_{\tilde{A}_1}^L, \eta_{\tilde{A}_1}^U] \in \nu_{\tilde{A}_1}(x)} \left\{\left[\eta_{\tilde{A}_1}^L, \eta_{\tilde{A}_1}^U\right]\right\}\right)$, and $h_{\tilde{A}_2} = \left(\bigcup_{[\gamma_{\tilde{A}}^L, \gamma_{\tilde{A}_2}^U] \in \mu_{\tilde{A}_2}(x)} \left\{\left[\gamma_{\tilde{A}_2}^L, \gamma_{\tilde{A}_2}^U\right]\right\}, \bigcup_{[\eta_{\tilde{A}_2}^L, \eta_{\tilde{A}_2}^U] \in \nu_{\tilde{A}_2}(x)} \left\{\left[\eta_{\tilde{A}_2}^L, \eta_{\tilde{A}_2}^U\right]\right\}\right)$. Then, for $k > 0$,

$$h_{\tilde{A}_1} \cup h_{\tilde{A}_2} = \begin{pmatrix} \bigcup_{[\gamma_{\tilde{A}_1}^L, \gamma_{\tilde{A}_1}^U] \in \mu_{\tilde{A}_1}, [\gamma_{\tilde{A}_2}^L, \gamma_{\tilde{A}_2}^U] \in \mu_{\tilde{A}_2}} \left[\max\left\{\gamma_{\tilde{A}_1}^L, \gamma_{\tilde{A}_2}^L\right\}, \max\left\{\gamma_{\tilde{A}_1}^U, \gamma_{\tilde{A}_2}^U\right\}\right], \\ \bigcup_{[\eta_{\tilde{A}_1}^L, \eta_{\tilde{A}_1}^U] \in \nu_{\tilde{A}_1}, [\eta_{\tilde{A}_2}^L, \eta_{\tilde{A}_2}^U] \in \nu_{\tilde{A}_2}} \left[\min\left\{\eta_{\tilde{A}_1}^L, \eta_{\tilde{A}_2}^L\right\}, \min\left\{\eta_{\tilde{A}_1}^U, \eta_{\tilde{A}_2}^U\right\}\right] \end{pmatrix}, \tag{9.42}$$

$$h_{\tilde{A}_1} \cap h_{\tilde{A}_2} = \begin{pmatrix} \bigcup_{[\gamma_{\tilde{A}_1}^L, \gamma_{\tilde{A}_1}^U] \in \mu_{\tilde{A}_1}, [\gamma_{\tilde{A}_2}^L, \gamma_{\tilde{A}_2}^U] \in \mu_{\tilde{A}_2}} \left[\min\left\{\gamma_{\tilde{A}_1}^L, \gamma_{\tilde{A}_2}^L\right\}, \min\left\{\gamma_{\tilde{A}_1}^U, \gamma_{\tilde{A}_2}^U\right\}\right], \\ \bigcup_{[\eta_{\tilde{A}_1}^L, \eta_{\tilde{A}_1}^U] \in \nu_{\tilde{A}_1}, \{\eta_{\tilde{A}_2}^L, \eta_{\tilde{A}_2}] \in \nu_{\tilde{A}_2}} \left[\max\left\{\eta_{\tilde{A}_1}^L, \eta_{\tilde{A}_2}^L\right\}, \max\left\{\eta_{\tilde{A}_1}^U, \eta_{\tilde{A}_2}^U\right\}\right] \end{pmatrix}, \tag{9.43}$$

$$h_{\tilde{A}_1} \oplus h_{\tilde{A}_2} = \begin{pmatrix} \bigcup_{[r_{\tilde{A}_1}^L, r_{\tilde{A}_1}^U] \in \mu_{\tilde{A}_1} [r_{\tilde{A}_2}^L, \gamma_{\tilde{A}_2}^U] \in \mu_{\tilde{A}_2}} \left\{\left[\gamma_{\tilde{A}_1}^L + \gamma_{\tilde{A}_2}^L - \gamma_{\tilde{A}_1}^L \gamma_{\tilde{A}_2}^L, \gamma_{\tilde{A}_1}^U + \gamma_{\tilde{A}_2}^U - \gamma_{\tilde{A}_1}^U \gamma_{\tilde{A}_2}^U\right]\right\}, \\ \bigcup_{[\eta_{\tilde{A}_1}^L, \eta_{\tilde{A}_1}^U] \in \nu_{\tilde{A}_1} [\eta_{\tilde{A}_2}^L, \eta_{\tilde{A}_2}^U] \in \nu_{\tilde{A}_2}} \left\{\left[\eta_{\tilde{A}_1}^L \times \eta_{\tilde{A}_2}^L, \eta_{\tilde{A}_1}^U \times \eta_{\tilde{A}_2}^U\right]\right\} \end{pmatrix}, \tag{9.44}$$

$$h_{\tilde{A}_1} \otimes h_{\tilde{A}_2} = \begin{pmatrix} \bigcup_{[\gamma_{\tilde{A}_1}^L, \gamma_{\tilde{A}_1}^U] \in \mu_{\tilde{A}_1}, [\gamma_{\tilde{A}_2}^L, \gamma_{\tilde{A}_2}^U] \in \mu_{\tilde{A}_2}} \left\{\left[\gamma_{\tilde{A}_1}^L \gamma_{\tilde{A}_2}^L, \gamma_{\tilde{A}_1}^U \gamma_{\tilde{A}_2}^U\right]\right\}, \\ \bigcup_{[\eta_{\tilde{A}_1}^L, \eta_{\tilde{A}_1}^U] \in \nu_{\tilde{A}_1}^U, \eta_{\tilde{A}_2}^L, \eta_{\tilde{A}_2}^U] \in \nu_{\tilde{A}_2}} \left\{\left[\eta_{\tilde{A}_1}^L + \eta_{\tilde{A}_2}^L - \eta_{\tilde{A}_1}^L \eta_{\tilde{A}_2}^L, \eta_{\tilde{A}_1}^U + \eta_{\tilde{A}_2}^U - \eta_{\tilde{A}_1}^U \eta_{\tilde{A}_2}^U\right]\right\} \end{pmatrix}, \tag{9.45}$$

$$kh_{\tilde{A}} = \begin{pmatrix} \bigcup_{\left[\gamma_{\tilde{A}}^L, \gamma_{\tilde{A}}^U\right] \in \mu_{\tilde{A}}(x)} \left\{ \left[1 - \left(1 - \gamma_{\tilde{A}}^L\right)^k, 1 - \left(1 - \gamma_{\tilde{A}}^U\right)^k \right] \right\}, \\ \bigcup_{\left[\eta_{\tilde{A}}^L, \eta_{\tilde{A}}^U\right] \in \nu_{\tilde{A}}(x)} \left\{ \left[\left(\eta_{\tilde{A}}^L\right)^k, \left(\eta_{\tilde{A}}^U\right)^k \right] \right\} \end{pmatrix}, \tag{9.46}$$

$$\left(h_{\tilde{A}}\right)^k = \begin{pmatrix} \bigcup_{\left[\gamma_{\tilde{A}}^L, \gamma_{\tilde{A}}^U\right] \in \mu_{\tilde{A}}(x)} \left\{ \left[\left(\gamma_{\tilde{A}}^L\right)^k, \left(\gamma_{\tilde{A}}^U\right)^k \right] \right\}, \\ \bigcup_{\left[\eta_{\tilde{A}}^L, \eta_{\tilde{A}}^U\right] \in \nu_{\tilde{A}}(x)} \left\{ \left[1 - \left(1 - \eta_{\tilde{A}}^L\right)^k, 1 - \left(1 - \eta_{\tilde{A}}^U\right)^k \right] \right\} \end{pmatrix}. \tag{9.47}$$

9.6 Neutrosophic Hesitant Fuzzy Sets

In a neutrosophic hesitant fuzzy set, for any element the values of truth-membership, indeterminacy-membership, and falsity-membership functions are considered. Some concepts about this type of fuzzy numbers are explained by the rest of this sub-section.

Definition 9.29 [15] A single-valued neutrosophic hesitant fuzzy set (SVNHFS) on the reference set X is defined as below,

$$\tilde{A} = \left\{ \left(x, \mu_{\tilde{A}}(x), w_{\tilde{A}}(x), \nu_{\tilde{A}}(x)\right) \middle| x \in X \right\} \tag{9.48}$$

where, $\mu_{\tilde{A}}(x)$, $w_{\tilde{A}}(x)$, and $\nu_{\tilde{A}}(x)$ are respectively the truth-membership function, indeterminacy-membership function, and falsity-membership function of $x \in X$, and $\mu_{\tilde{A}}(x), w_{\tilde{A}}(x), \nu_{\tilde{A}}(x) \in [0, 1]$.

Therefore, a SVNHFS is shown as below too.

$$\tilde{A} = \left\{ \left(x, \bigcup_{\gamma_{\tilde{A}} \in \mu_{\tilde{A}}} \{\gamma_{\tilde{A}}\}, \bigcup_{\delta_{\tilde{A}} \in w_{\tilde{A}}} \{\delta_{\tilde{A}}\}, \bigcup_{\eta_{\tilde{A}} \in \nu_{\tilde{A}}} \{\eta_{\tilde{A}}\} \right) \middle| x \in X \right\}. \tag{9.49}$$

For more simplicity, a single-valued neutrosophic hesitant fuzzy element (SVNHFE) is shown as $h_{\tilde{A}} = \left(\mu_{\tilde{A}}(x), w_{\tilde{A}}(x), \nu_{\tilde{A}}(x)\right)$.

Example 9.11 Assume the reference set $X = \{x_1, x_2\}$, and the SVNHFEs $h_{\tilde{A}}(x_1) = (\{0.1, 0.5\}, \{0.2, 0.3\}, \{0.4, 0.6\})$ and $h_{\tilde{A}}(x_2) = (\{0.2\}, \{0.5\}, \{0.3, 0.5, 0.6\})$. Therefore, $\tilde{A}$ is an SVNHFS shown as below.

$$\begin{aligned} \tilde{A} = \{&(x_1, \{0.1, 0.5\}, \{0.2, 0.3\}, \{0.4, 0.6\}), \\ &(x_2, \{0.2\}, \{0.5\}, \{0.3, 0.5, 0.6\})\}. \end{aligned}$$

Definition 9.30 [8, 16] Assume the SVNHFEs $h_{\tilde{A}} = \left(\bigcup_{\gamma_{\tilde{A}} \in \mu_{\tilde{A}}} \{\gamma_{\tilde{A}}\}, \bigcup_{\delta_{\tilde{A}} \in w_{\tilde{A}}} \{\delta_{\tilde{A}}\}, \bigcup_{\eta_{\tilde{A}} \in v_{\tilde{A}}} \{\eta_{\tilde{A}}\}\right)$, $h_{\tilde{A}_1} = \left(\bigcup_{\gamma_{\tilde{A}_1} \in \mu_{\tilde{A}_1}} \{\gamma_{\tilde{A}_1}\}, \bigcup_{\delta_{\tilde{A}_1} \in w_{\tilde{A}_1}} \{\delta_{\tilde{A}_1}\}, \bigcup_{\eta_{\tilde{A}_1} \in v_{\tilde{A}_1}} \{\eta_{\tilde{A}_1}\}\right)$, and $h_{\tilde{A}_2} = \left(\bigcup_{\gamma_{\tilde{A}_2} \in \mu_{\tilde{A}_2}} \{\gamma_{\tilde{A}_2}\}, \bigcup_{\delta_{\tilde{A}_2} \in w_{\tilde{A}_2}} \{\delta_{\tilde{A}_2}\}, \bigcup_{\eta_{\tilde{A}_2} \in v_{\tilde{A}_2}} \{\eta_{\tilde{A}_2}\}\right)$. Then, for $k > 0$,

$$h_{\tilde{A}_1} \cup h_{\tilde{A}_2} = \left(\bigcup_{\substack{\gamma_{\tilde{A}_1} \in \mu_{\tilde{A}_1} \\ \gamma_{\tilde{A}_2} \in \mu_{\tilde{A}_2}}} \max\{\gamma_{\tilde{A}_1}, \gamma_{\tilde{A}_2}\}, \bigcup_{\substack{\delta_{\tilde{A}_1} \in w_{\tilde{A}_1} \\ \delta_{\tilde{A}_2} \in w_{\tilde{A}_2}}} \min\{\delta_{\tilde{A}_1}, \delta_{\tilde{A}_2}\}, \bigcup_{\substack{\eta_{\tilde{A}_1} \in v_{\tilde{A}_1} \\ \eta_{\tilde{A}_2} \in v_{\tilde{A}_2}}} \min\{\eta_{\tilde{A}_1}, \eta_{\tilde{A}_2}\} \right), \tag{9.50}$$

$$h_{\tilde{A}_1} \cap h_{\tilde{A}_2} = \left(\bigcup_{\substack{\gamma_{\tilde{A}_1} \in \mu_{\tilde{A}_1} \\ \gamma_{\tilde{A}_2} \in \mu_{\tilde{A}_2}}} \min\{\gamma_{\tilde{A}_1}, \gamma_{\tilde{A}_2}\}, \bigcup_{\substack{\delta_{\tilde{A}_1} \in w_{\tilde{A}_1} \\ \delta_{\tilde{A}_2} \in w_{\tilde{A}_2}}} \max\{\delta_{\tilde{A}_1}, \delta_{\tilde{A}_2}\}, \bigcup_{\substack{\eta_{\tilde{A}_1} \in v_{\tilde{A}_1} \\ \eta_{\tilde{A}_2} \in v_{\tilde{A}_2}}} \max\{\eta_{\tilde{A}_1}, \eta_{\tilde{A}_2}\} \right), \tag{9.51}$$

$$h_{\tilde{A}_1} \oplus h_{\tilde{A}_2} = \left(\bigcup_{\substack{\gamma_{\tilde{A}_1} \in \mu_{\tilde{A}_1} \\ \gamma_{\tilde{A}_2} \in \mu_{\tilde{A}_2}}} \{\gamma_{\tilde{A}_1} + \gamma_{\tilde{A}_2} - \gamma_{\tilde{A}_1} \times \gamma_{\tilde{A}_2}\}, \bigcup_{\substack{\delta_{\tilde{A}_1} \in w_{\tilde{A}_1} \\ \delta_{\tilde{A}_2} \in w_{\tilde{A}_2}}} \{\delta_{\tilde{A}_1} \times \delta_{\tilde{A}_2}\}, \bigcup_{\substack{\eta_{\tilde{A}_1} \in v_{\tilde{A}_1} \\ \eta_{\tilde{A}_2} \in v_{\tilde{A}_2}}} \{\eta_{\tilde{A}_1} \times \eta_{\tilde{A}_2}\} \right), \tag{9.52}$$

$$h_{\tilde{A}_1} \otimes h_{\tilde{A}_2} = \left(\bigcup_{\substack{\gamma_{\tilde{A}_1} \in \mu_{\tilde{A}_1} \\ \gamma_{\tilde{A}_2} \in \mu_{\tilde{A}_2}}} \{\gamma_{\tilde{A}_1} \times \gamma_{\tilde{A}_2}\}, \bigcup_{\substack{\delta_{\tilde{A}_1} \in w_{\tilde{A}_1} \\ \delta_{\tilde{A}_2} \in w_{\tilde{A}_2}}} \{\delta_{\tilde{A}_1} + \delta_{\tilde{A}_2} - \delta_{\tilde{A}_1} \times \delta_{\tilde{A}_2}\}, \bigcup_{\substack{\eta_{\tilde{A}_1} \in v_{\tilde{A}_1} \\ \eta_{\tilde{A}_2} \in v_{\tilde{A}_2}}} \{\eta_{\tilde{A}_1} + \eta_{\tilde{A}_2} - \eta_{\tilde{A}_1} \times \eta_{\tilde{A}_2}\} \right), \tag{9.53}$$

$$kh_{\tilde{A}} = \left(\bigcup_{\gamma_{\tilde{A}} \in \mu_{\tilde{A}}} \left\{ 1 - (1 - \gamma_{\tilde{A}})^k \right\}, \bigcup_{\delta_{\tilde{A}} \in w_{\tilde{A}}} \left\{ (\delta_{\tilde{A}})^k \right\}, \bigcup_{\eta_{\tilde{A}} \in \nu_{\tilde{A}}} \left\{ (\eta_{\tilde{A}})^k \right\} \right), \quad (9.54)$$

$$(h_{\tilde{A}})^k = \left(\bigcup_{\gamma_{\tilde{A}} \in \mu_{\tilde{A}}} \left\{ (\gamma_{\tilde{A}})^k \right\}, \bigcup_{\delta_{\tilde{A}} \in w_{\tilde{A}}} \left\{ 1 - (1 - \delta_{\tilde{A}})^k \right\}, \bigcup_{\eta_{\tilde{A}} \in \nu_{\tilde{A}}} \left\{ 1 - (1 - \eta_{\tilde{A}})^k \right\} \right). \quad (9.55)$$

9.6.1 Interval Neutrosophic Hesitant Fuzzy Sets

Definition 9.31 [17] An interval neutrosophic hesitant fuzzy set (INHFS) on the reference set X is defined as below,

$$\tilde{A} = \left\{ \left(x, \mu_{\tilde{A}}(x), w_{\tilde{A}}(x), \nu_{\tilde{A}}(x)\right) \middle| x \in X \right\} \quad (9.56)$$

where, $\mu_{\tilde{A}}(x) = \bigcup_{\left[\gamma^L_{\tilde{A}}, \gamma^U_{\tilde{A}}\right] \in \mu_{\tilde{A}}(x)} \left\{ \left[\gamma^L_{\tilde{A}}, \gamma^U_{\tilde{A}} \right] \right\}$, $w_{\tilde{A}}(x) = \bigcup_{\left[\delta^L_{\tilde{A}}, \delta^U_{\tilde{A}}\right] \in w_{\tilde{A}}(x)} \left\{ \left[\delta^L_{\tilde{A}}, \delta^U_{\tilde{A}} \right] \right\}$, and $\nu_{\tilde{A}}(x) = \bigcup_{\left[\eta^L_{\tilde{A}}, \eta^U_{\tilde{A}}\right] \in \nu_{\tilde{A}}(x)} \left\{ \left[\eta^L_{\tilde{A}}, \eta^U_{\tilde{A}} \right] \right\}$ are respectively the truth-membership function, indeterminacy-membership function, and falsity-membership function of $x \in X$, and are intervals from [0, 1].

Therefore, a INHFS is shown as below too.

$$\tilde{A} = \left\{ \left(x, \bigcup_{\left[\gamma^L_{\tilde{A}}, \gamma^U_{\tilde{A}}\right] \in \mu_{\tilde{A}}(x)} \left\{ \left[\gamma^L_{\tilde{A}}, \gamma^U_{\tilde{A}} \right] \right\}, \bigcup_{\left[\delta^L_{\tilde{A}}, \delta^U_{\tilde{A}}\right] \in w_{\tilde{A}}(x)} \left\{ \left[\delta^L_{\tilde{A}}, \delta^U_{\tilde{A}} \right] \right\}, \bigcup_{\left[\eta^L_{\tilde{A}}, \eta^U_{\tilde{A}}\right] \in \nu_{\tilde{A}}(x)} \left\{ \left[\eta^L_{\tilde{A}}, \eta^U_{\tilde{A}} \right] \right\} \right) \middle| x \in X \right\}. \quad (9.57)$$

For more simplicity, an interval neutrosophic hesitant fuzzy element (INHFE) is shown as below.

$$h_{\tilde{A}} = \left(\mu_{\tilde{A}}(x), w_{\tilde{A}}(x), \nu_{\tilde{A}}(x)\right)$$
$$= \left(\bigcup_{\left[\gamma^L_{\tilde{A}}, \gamma^U_{\tilde{A}}\right] \in \mu_{\tilde{A}}(x)} \left\{ \left[\gamma^L_{\tilde{A}}, \gamma^U_{\tilde{A}} \right] \right\}, \bigcup_{\left[\delta^L_{\tilde{A}}, \delta^U_{\tilde{A}}\right] \in w_{\tilde{A}}(x)} \left\{ \left[\delta^L_{\tilde{A}}, \delta^U_{\tilde{A}} \right] \right\}, \bigcup_{\left[\eta^L_{\tilde{A}}, \eta^U_{\tilde{A}}\right] \in \nu_{\tilde{A}}(x)} \left\{ \left[\eta^L_{\tilde{A}}, \eta^U_{\tilde{A}} \right] \right\} \right),$$

Example 9.12 Assume the reference set $X = \{x_1, x_2\}$, and the INHFEs $h_{\tilde{A}}(x_1) = (\{[0.1, 0.5], [0.6, 0.7]\}, \{[0.2, 0.5], [0.6, 0.8]\}, \{[0.4, 0.6]\})$ and $h_{\tilde{A}}(x_2) = (\{[0.2, 0.2]\}, \{[0.5, 0.7], [0.8, 0.9]\}, \{[0.6, 0.8]\})$. Therefore, $\tilde{A}$ is an INHFS shown as below.

$$\tilde{A} = \left\{ \begin{array}{l} (x_1, \{[0.1, 0.5], [0.6, 0.7]\}, \{[0.2, 0.5], [0.6, 0.8]\}, \{[0.4, 0.6]\}), \\ (x_2, \{[0.2, 0.2]\}, \{[0.5, 0.7], [0.8, 0.9]\}, \{[0.6, 0.8]\}) \end{array} \right\}.$$

Definition 9.32 [8] Assume the INHFEs $h_{\tilde{A}} = \left(\bigcup_{\left[\gamma^L_{\tilde{A}}, \gamma^U_{\tilde{A}}\right] \in \mu_{\tilde{A}}(x)} \left\{\left[\gamma^L_{\tilde{A}}, \gamma^U_{\tilde{A}}\right]\right\}, \bigcup_{\left[\delta^L_{\tilde{A}}, \delta^U_{\tilde{A}}\right] \in w_{\tilde{A}}(x)} \left\{\left[\delta^L_{\tilde{A}}, \delta^U_{\tilde{A}}\right]\right\}, \bigcup_{\left[\eta^L_{\tilde{A}}, \eta^U_{\tilde{A}}\right] \in v_{\tilde{A}}(x)} \left\{\left[\eta^L_{\tilde{A}}, \eta^U_{\tilde{A}}\right]\right\} \right)$, $h_{\tilde{A}_1} = \left(\bigcup_{\left[\gamma^L_{\tilde{A}_1}, \gamma^U_{\tilde{A}_1}\right] \in \mu_{\tilde{A}_1}(x)} \left\{\left[\gamma^L_{\tilde{A}_1}, \gamma^U_{\tilde{A}_1}\right]\right\}, \bigcup_{\left[\delta^L_{\tilde{A}_1}, \delta^U_{\tilde{A}_1}\right] \in w_{\tilde{A}_1}(x)} \left\{\left[\delta^L_{\tilde{A}_1}, \delta^U_{\tilde{A}_1}\right]\right\}, \bigcup_{\left[\eta^L_{\tilde{A}_1}, \eta^U_{\tilde{A}_1}\right] \in v_{\tilde{A}_1}(x)} \left\{\left[\eta^L_{\tilde{A}_1}, \eta^U_{\tilde{A}_1}\right]\right\} \right)$, and $h_{\tilde{A}_2} = \left(\bigcup_{\left[\gamma^L_{\tilde{A}}, \gamma^U_{\tilde{A}_2}\right] \in \mu_{\tilde{A}_2}(x)} \left\{\left[\gamma^L_{\tilde{A}_2}, \gamma^U_{\tilde{A}_2}\right]\right\}, \bigcup_{\left[\delta^L_{\tilde{A}_2}, \delta^U_{\tilde{A}_2}\right] \in w_{\tilde{A}_2}(x)} \left\{\left[\delta^L_{\tilde{A}_2}, \delta^U_{\tilde{A}_2}\right]\right\}, \bigcup_{\left[\eta^L_{\tilde{A}_2}, \eta^U_{\tilde{A}_2}\right] \in v_{\tilde{A}_2}(x)} \left\{\left[\eta^L_{\tilde{A}_2}, \eta^U_{\tilde{A}_2}\right]\right\} \right)$. Then, for $k > 0$,

$$h_{\tilde{A}_1} \cup h_{\tilde{A}_2} = \left(\begin{array}{l} \bigcup_{\substack{\left[\gamma^L_{\tilde{A}_1}, \gamma^U_{\tilde{A}_1}\right] \in \mu_{\tilde{A}_1} \\ \left[\gamma^L_{\tilde{A}_2}, \gamma^U_{\tilde{A}_2}\right] \in \mu_{\tilde{A}_2}}} \left[\max\left\{\gamma^L_{\tilde{A}_1}, \gamma^L_{\tilde{A}_2}\right\}, \max\left\{\gamma^U_{\tilde{A}_1}, \gamma^U_{\tilde{A}_2}\right\}\right], \\ \bigcup_{\substack{\left[\delta^L_{\tilde{A}_1}, \delta^U_{\tilde{A}_1}\right] \in w_{\tilde{A}_1} \\ \left[\delta^L_{\tilde{A}_2}, \delta^U_{\tilde{A}_2}\right] \in w_{\tilde{A}_2}}} \left[\min\left\{\delta^L_{\tilde{A}_1}, \delta^L_{\tilde{A}_2}\right\}, \min\left\{\delta^U_{\tilde{A}_1}, \delta^U_{\tilde{A}_2}\right\}\right], \\ \bigcup_{\substack{\left[\eta^L_{\tilde{A}_1}, \eta^U_{\tilde{A}_1}\right] \in v_{\tilde{A}_1} \\ \left[\eta^L_{\tilde{A}_3}, \eta^U_{\tilde{A}_3}\right] \in v_{\tilde{A}_2}}} \left[\min\left\{\eta^L_{\tilde{A}_1}, \eta^L_{\tilde{A}_2}\right\}, \min\left\{\eta^U_{\tilde{A}_1}, \eta^U_{\tilde{A}_2}\right\}\right] \end{array} \right), \tag{9.58}$$

$$h_{\tilde{A}_1} \cap h_{\tilde{A}_2} = \left(\begin{array}{l} \bigcup_{\substack{\left[\gamma^L_{\tilde{A}_1}, \gamma^U_{\tilde{A}_1}\right] \in \mu_{\tilde{A}_1} \\ \left[\gamma^L_{\tilde{A}_2}, \gamma^U_{\tilde{A}_2}\right] \in \mu_{\tilde{A}_2}}} \left[\min\left\{\gamma^L_{\tilde{A}_1}, \gamma^L_{\tilde{A}_2}\right\}, \min\left\{\gamma^U_{\tilde{A}_1}, \gamma^U_{\tilde{A}_2}\right\}\right], \\ \bigcup_{\substack{\left[\delta^L_{\tilde{A}_1}, \delta^U_{\tilde{A}_1}\right] \in w_{\tilde{A}_1} \\ \left[\delta^L_{\tilde{A}_2}, \delta^U_{\tilde{A}_2}\right] \in w_{\tilde{A}_2}}} \left[\max\left\{\delta^L_{\tilde{A}_1}, \delta^L_{\tilde{A}_2}\right\}, \max\left\{\delta^U_{\tilde{A}_1}, \delta^U_{\tilde{A}_2}\right\}\right], \\ \bigcup_{\substack{\left[\eta^L_{\tilde{A}_1}, \eta^U_{\tilde{A}_1}\right] \in v_{\tilde{A}_1} \\ \left[\eta^L_{\tilde{A}_1}, \eta^U_{\tilde{\tau}}\right] \in v_{\tilde{A}_2}}} \left[\max\left\{\eta^L_{\tilde{A}_1}, \eta^L_{\tilde{A}_2}\right\}, \max\left\{\eta^U_{\tilde{A}_1}, \eta^U_{\tilde{A}_2}\right\}\right] \end{array} \right), \tag{9.59}$$

$$h_{\tilde{A}_1} \oplus h_{\tilde{A}_2} = \begin{pmatrix} \bigcup\limits_{\substack{\left[\gamma^L_{\tilde{A}_1}, \gamma^U_{\tilde{A}_1}\right] \in \mu_{\tilde{A}_1} \\ \left[\gamma^L_{\tilde{A}_2}, \gamma^U_{\tilde{A}_2}\right] \in \mu_{\tilde{A}_2}}} \left\{\left[\left\{\gamma^L_{\tilde{A}_1} + \gamma^L_{\tilde{A}_2} - \gamma^L_{\tilde{A}_1}\gamma^L_{\tilde{A}_2}\right\}, \left\{\gamma^U_{\tilde{A}_1} + \gamma^U_{\tilde{A}_2} - \gamma^U_{\tilde{A}_1}\gamma^U_{\tilde{A}_2}\right\}\right]\right\}, \\ \bigcup\limits_{\substack{\left[\delta^L_{\tilde{A}_1}, \delta^U_{\tilde{A}_1}\right] \in w_{\tilde{A}_1} \\ \left[\delta^L_{\tilde{A}_2}, \delta^U_{\tilde{A}_2}\right] \in w_{\tilde{A}_2}}} \left\{\left[\left\{\delta^L_{\tilde{A}_1} \times \delta^L_{\tilde{A}_2}\right\}, \left\{\delta^U_{\tilde{A}_1} \times \delta^U_{\tilde{A}_2}\right\}\right]\right\}, \\ \bigcup\limits_{\substack{\left[\eta^L_{\tilde{A}_1}, \eta^U_{\tilde{A}_1}\right] \in v_{\tilde{A}_1} \\ \left[\eta^L_{\tilde{A}_2}, \eta^U_{\tilde{A}_2}\right] \in v_{\tilde{A}_2}}} \left\{\left[\left\{\eta^L_{\tilde{A}_1} \times \eta^L_{\tilde{A}_2}\right\}, \left\{\eta^U_{\tilde{A}_1} \times \eta^U_{\tilde{A}_2}\right\}\right]\right\} \end{pmatrix}, \tag{9.60}$$

$$h_{\tilde{A}_1} \otimes h_{\tilde{A}_2} = \begin{pmatrix} \bigcup\limits_{\substack{\left[\gamma^L_{\tilde{A}_1}, \nu^U_{\tilde{A}_1}\right] \in \mu_{\tilde{A}_1} \\ \left[\gamma^L_{\tilde{A}_2}, \gamma^U_{\tilde{A}_2}\right] \in \mu_{\tilde{A}_2}}} \left\{\left[\left\{\gamma^L_{\tilde{A}_1} \times \gamma^L_{\tilde{A}_2}\right\}, \left\{\gamma^U_{\tilde{A}_1} \times \gamma^U_{\tilde{A}_2}\right\}\right]\right\}, \\ \bigcup\limits_{\substack{\left[\delta^L_{\tilde{A}_1}, \delta^U_{\tilde{A}_1}\right] \in w_{\tilde{A}_1} \\ \left[\delta^L_{\tilde{A}_2}, \delta\tilde{A}_{\tilde{A}_2}\right] \in w_{\tilde{A}_2}}} \left\{\left[\left\{\delta^L_{\tilde{A}_1} + \delta^L_{\tilde{A}_2} - \delta^L_{\tilde{A}_1} \times \delta^L_{\tilde{A}_2}\right\}, \right.\right. \\ \left.\left.\left\{\delta^U_{\tilde{A}_1} + \delta^U_{\tilde{A}_2} - \delta^U_{\tilde{A}_1} \times \delta^U_{\tilde{A}_2}\right\}\right]\right\} \\ \bigcup\limits_{\substack{\left[\eta^L_{\tilde{A}_1}, \eta^U_{\tilde{A}_1}\right] \in v_{\tilde{A}_1} \\ \left[\eta^L_{\tilde{A}_2}, \eta^U_{\tilde{A}_2}\right] \in v_{\tilde{A}_2}}} \left\{\left[\left\{\eta^L_{\tilde{A}_1} + \eta^L_{\tilde{A}_2} - \eta^L_{\tilde{A}_1} \times \eta^L_{\tilde{A}_2}\right\}, \right.\right. \\ \left.\left.\left\{\eta^U_{\tilde{A}_1} + \eta^U_{\tilde{A}_2} - \eta^U_{\tilde{A}_1} \times \eta^U_{\tilde{A}_2}\right\}\right]\right\} \end{pmatrix}, \tag{9.61}$$

$$kh_{\tilde{A}} = \begin{pmatrix} \bigcup\limits_{\left[\gamma^L_{\tilde{A}}, \gamma^U_{\tilde{A}}\right] \in \mu_{\tilde{A}}(x)} \left\{\left[1 - \left(1 - \gamma^L_{\tilde{A}}\right)^k, 1 - \left(1 - \gamma^U_{\tilde{A}}\right)^k\right]\right\}, \\ \bigcup\limits_{\left[\delta^L_{\tilde{A}}, \delta^U_{\tilde{A}}\right] \in w_{\tilde{A}}(x)} \left\{\left[\left(\delta^L_{\tilde{A}}\right)^k, \left(\delta^U_{\tilde{A}}\right)^k\right]\right\}, \\ \bigcup\limits_{\left[\eta^L_{\tilde{A}}, \eta^U_{\tilde{A}}\right] \in v_{\tilde{A}}(x)} \left\{\left[\left(\eta^L_{\tilde{A}}\right)^k, \left(\eta^U_{\tilde{A}}\right)^k\right]\right\} \end{pmatrix}, \tag{9.62}$$

$$\left(h_{\tilde{A}}\right)^k = \begin{pmatrix} \bigcup\limits_{\left[\gamma^L_{\tilde{A}}, \gamma^U_{\tilde{A}}\right] \in \mu_{\tilde{A}}(x)} \left\{\left[\left(\gamma^L_{\tilde{A}}\right)^k, \left(\gamma^U_{\tilde{A}}\right)^k\right]\right\}, \\ \bigcup\limits_{\left[\delta^L_{\tilde{A}}, \delta^U_{\tilde{A}}\right] \in w_{\tilde{A}}(x)} \left\{\left[1 - \left(1 - \delta^L_{\tilde{A}}\right)^k, 1 - \left(1 - \delta^U_{\tilde{A}}\right)^k\right]\right\}, \\ \bigcup\limits_{\left[\eta^L_{\tilde{A}}, \eta^U_{\tilde{A}}\right] \in v_{\tilde{A}}(x)} \left\{\left[1 - \left(1 - \eta^L_{\tilde{A}}\right)^k, 1 - \left(1 - \eta^U_{\tilde{A}}\right)^k\right]\right\} \end{pmatrix}. \tag{9.63}$$

9.7 Pythagorean Hesitant Fuzzy Set

Definition 9.33 [18] Assume the reference set X. A Pythagorean hesitant fuzzy set (PHFS) such as $\tilde{A}$ on X based on the membership and non-membership functions $\mu_{\tilde{A}}(x)$ and $\nu_{\tilde{A}}(x)$ where $\mu_{\tilde{A}}(x), \nu_{\tilde{A}}(x) \in [0, 1]$, is defined as below.

$$\tilde{A} = \left\{ \left(x, \mu_{\tilde{A}}(x), \nu_{\tilde{A}}(x)\right) \middle| x \in X \right\} \tag{9.64}$$

Here, for $x \in X$ if we assume $\mu_{\tilde{A}}(x) = \bigcup_{\gamma_{\tilde{A}} \in \mu_{\tilde{A}}(x)} \{\gamma_{\tilde{A}}\}$, $\nu_{\tilde{A}}(x) = \bigcup_{\eta_{\tilde{A}} \in \nu_{\tilde{A}}(x)} \{\eta_{\tilde{A}}\}$, $\gamma_{\tilde{A}}^{+} \in \mu_{\tilde{A}}^{+} = \bigcup_{x \in X} \max_{\gamma_{\tilde{A}} \in \mu_{\tilde{A}}(x)} \{\gamma_{\tilde{A}}\}$, $\eta_{\tilde{A}}^{+} \in \nu_{\tilde{A}}^{+} = \bigcup_{x \in X} \max_{\eta_{\tilde{A}} \in \nu_{\tilde{A}}(x)} \{\eta_{\tilde{A}}\}$, then $0 \le \gamma_{\tilde{A}}, \eta_{\tilde{A}} \le 1$ and $0 \le \left(\gamma_{\tilde{A}}^{+}\right)^2 + \left(\eta_{\tilde{A}}^{+}\right)^2 \le 1$. Therefore, a PHFS can be shown as below too.

$$\tilde{A} = \left\{ \left(x, \bigcup_{\gamma_{\tilde{A}} \in \mu_{\tilde{A}}} \{\gamma_{\tilde{A}}\}, \bigcup_{\eta_{\tilde{A}} \in \nu_{\tilde{A}}} \{\eta_{\tilde{A}}\} \right) \middle| x \in X \right\} \tag{9.65}$$

For more simplicity, in continue the term $h_{\tilde{A}} = \left(\mu_{\tilde{A}}(x), \nu_{\tilde{A}}(x)\right) = \left(\bigcup_{\gamma_{\tilde{A}} \in \mu_{\tilde{A}}} \{\gamma_{\tilde{A}}\}, \bigcup_{\eta_{\tilde{A}} \in \nu_{\tilde{A}}} \{\eta_{\tilde{A}}\}\right)$ represents an intuitionistic hesitant fuzzy element (PHFE).

Example 9.13 [8] Assume the reference set $X = \{x_1, x_2\}$, and the PHFEs $h_{\tilde{A}}(x_1) = (\{0.2, 0.5\}, \{0.3\})$ and $h_{\tilde{A}}(x_2) = (\{0.3, 0.4\}, \{0.1, 0.6\})$. As,

$$\gamma_{\tilde{A}}^{+}(x_1) = 0.5, \eta_{\tilde{A}}^{+}(x_1) = 0.3, \gamma_{\tilde{A}}^{+}(x_2) = 0.4, \eta_{\tilde{A}}^{+}(x_2) = 0.6,$$

$$\left(\gamma_{\tilde{A}}^{+}\right)^2(x_1) + \left(\eta_{\tilde{A}}^{+}\right)^2(x_1) = 0.5^2 + 0.3^2 = 0.34 \le 1,$$
$$\left(\gamma_{\tilde{A}}^{+}\right)^2(x_2) + \left(\eta_{\tilde{A}}^{+}\right)^2(x_2) = 0.4^2 + 0.6^2 = 0.52 \le 1,$$

then, $\tilde{A}$ is a PHFS shown as below.

$$\tilde{A} = \{(x_1, \{0.2, 0.5\}, \{0.3\}), (x_2, \{0.3, 0.4\}, \{0.1, 0.6\})\}.$$

Definition 9.34 [8] Assume the PHFEs $h_{\tilde{A}} = \left(\mu_{\tilde{A}}(x), \nu_{\tilde{A}}(x)\right)$, $h_{\tilde{A}_1} = \left(\mu_{\tilde{A}_1}(x), \nu_{\tilde{A}_1}(x)\right)$, and $h_{\tilde{A}_2} = \left(\mu_{\tilde{A}_2}(x), \nu_{\tilde{A}_2}(x)\right)$. Then,

$$h_{\tilde{A}}^{c} = \left(\nu_{\tilde{A}}(x), \mu_{\tilde{A}}(x)\right), \tag{9.66}$$

$$h_{\tilde{A}_1} \cup h_{\tilde{A}_2} = \left(\mu_{\tilde{A}_1} \cup \mu_{\tilde{A}_2}, \nu_{\tilde{A}_1} \cap \nu_{\tilde{A}_2}\right), \tag{9.67}$$

$$h_{\tilde{A}_1} \cap h_{\tilde{A}_2} = \left(\mu_{\tilde{A}_1} \cap \mu_{\tilde{A}_2}, \nu_{\tilde{A}_1} \cup \nu_{\tilde{A}_2}\right), \tag{9.68}$$

$$h_{\tilde{A}_1} \oplus h_{\tilde{A}_2} = \left(\mu_{\tilde{A}_1} \oplus \mu_{\tilde{A}_2}, \nu_{\tilde{A}_1} \otimes \nu_{\tilde{A}_2}\right), \tag{9.69}$$

$$h_{\tilde{A}_1} \otimes h_{\tilde{A}_2} = \left(\mu_{\tilde{A}_1} \otimes \mu_{\tilde{A}_2}, \nu_{\tilde{A}_1} \oplus \nu_{\tilde{A}_2}\right), \tag{9.70}$$

where $0 \leq \left(\gamma_{\tilde{A}}^{+}\right)^2 + \left(\eta_{\tilde{A}}^{+}\right)^2 \leq 1, 0 \leq \left(\gamma_{\tilde{A}_1}^{+}\right)^2 + \left(\eta_{\tilde{A}_1}^{+}\right)^2 \leq 1, 0 \leq \left(\gamma_{\tilde{A}_2}^{+}\right)^2 + \left(\eta_{\tilde{A}_2}^{+}\right)^2 \leq 1$.

Definition 9.35 [8] Assume the PHFSs $\tilde{A}_1$ and $\tilde{A}_2$. Then,

$$\begin{aligned}\tilde{A}_1 \cup \tilde{A}_2 &= \bigcup_{h_{\tilde{A}_1} \in \tilde{A}_1, h_{\tilde{A}_2} \in \tilde{A}_2} h_{\tilde{A}_1} \cup h_{\tilde{A}_2} \\ &= \left\{\left(x, \mu_{\tilde{A}_1}(x) \cup \mu_{\tilde{A}_2}(x), \nu_{\tilde{A}_1}(x) \cap \nu_{\tilde{A}_2}(x)\right) \middle| x \in X\right\},\end{aligned} \tag{9.71}$$

$$\begin{aligned}\tilde{A}_1 \cap \tilde{A}_2 &= \bigcup_{h_{\tilde{A}_1} \in \tilde{A}_1, h_{\tilde{A}_2} \in \tilde{A}_2} h_{\tilde{A}_1} \cap h_{\tilde{A}_2} \\ &= \left\{\left(x, \mu_{\tilde{A}_1}(x) \cap \mu_{\tilde{A}_2}(x), \nu_{\tilde{A}_1}(x) \cup \nu_{\tilde{A}_2}(x)\right) \middle| x \in X\right\},\end{aligned} \tag{9.72}$$

$$\begin{aligned}\tilde{A}_1 \oplus \tilde{A}_2 &= \bigcup_{h_{\tilde{A}_1} \in \tilde{A}_1, h_{\tilde{A}_2} \in \tilde{A}_2} h_{\tilde{A}_1} \oplus h_{\tilde{A}_2} \\ &= \left\{\left(x, \mu_{\tilde{A}_1}(x) \oplus \mu_{\tilde{A}_2}(x), \nu_{\tilde{A}_1}(x) \otimes \nu_{\tilde{A}_2}(x)\right) \middle| x \in X\right\},\end{aligned} \tag{9.73}$$

$$\begin{aligned}\tilde{A}_1 \otimes \tilde{A}_2 &= \bigcup_{h_{\tilde{A}_1} \in \tilde{A}_1, h_{\tilde{A}_2} \in \tilde{A}_2} h_{\tilde{A}_1} \otimes h_{\tilde{A}_2} \\ &= \left\{\left(x, \mu_{\tilde{A}_1}(x) \otimes \mu_{\tilde{A}_2}(x), \nu_{\tilde{A}_1}(x) \oplus \nu_{\tilde{A}_2}(x)\right) \middle| x \in X\right\}.\end{aligned} \tag{9.74}$$

9.7.1 *Trapezoidal Pythagorean Hesitant Fuzzy Number*

Considering the reference set X, a trapezoidal Pythagorean hesitant fuzzy number (TrPHFN) such as $\tilde{A}$ is defined as $\tilde{A} = \left\{\left(\left((a, b, c, d), \mu_{\tilde{A}}(x), \nu_{\tilde{A}}(x)\right); w_1, w_2\right) : x \in X\right\}$ where $\mu_{\tilde{A}}(x), \nu_{\tilde{A}}(x) \in [0, 1]$ are the membership and non-membership degrees of $x \in X$ in $\tilde{A}$. Here, $a \leq b \leq c \leq d$, and $w_1, w_2 \in [0, 1]$ are the sets of hesitant values for the membership and non-membership degrees (also $0 \leq w_1^2 + w_2^2 \leq 1$). The membership and non-membership functions are defined as below.

$$\mu_{\tilde{A}}(x) = \begin{cases} w_1\left(\frac{x-a}{b-a}\right), a \le x \le b, \\ w_1, b \le x \le c, \\ w_1\left(\frac{d-x}{d-c}\right), c \le x \le d, \\ 0, \textit{otherwise} \end{cases} \tag{9.75}$$

$$\nu_{\tilde{A}}(x) = \begin{cases} w_2\sqrt{1-\left(\frac{x-a}{b-a}\right)^2}, a \le x \le b, \\ w_2, b \le x \le c, \\ w_2\sqrt{1-\left(\frac{d-x}{d-c}\right)^2}, c \le x \le d, \\ 1, \textit{otherwise} \end{cases} \tag{9.76}$$

Definition 9.36 [19] Assume the TrPHFNs $\tilde{A}_1 = \left\{\left(\left((a_1, b_1, c_1, d_1), \mu_{\tilde{A}_1}, \nu_{\tilde{A}_1}\right); w_1, w_2\right)\right\}$ and $\tilde{A}_2 = \left\{\left(\left((a_2, b_2, c_2, d_2), \mu_{\tilde{A}_2}, \nu_{\tilde{A}_2}\right); w_3, w_4\right)\right\}$. The values $w_1 \in \mu_{\tilde{A}_1}$, $w_2 \in \nu_{\tilde{A}_1}$, $w_3 \in \mu_{\tilde{A}_2}$, and $w_4 \in \nu_{\tilde{A}_2}$ are from the interval [0, 1] where $0 \le w_1^2 + w_2^2 \le 1$ and $0 \le w_3^2 + w_4^2 \le 1$. Then, the below arithmetic operators are done on these numbers.

$$\text{Addition: } \tilde{A}_1 \oplus \tilde{A}_2 = \{(((a_1 + a_2, b_1 + b_2, c_1 + c_2, d_1 + d_2)\ , \\ \min\left\{\mu_{\tilde{A}_1}, \mu_{\tilde{A}_2}\right\}, \max\left\{\nu_{\tilde{A}_1}, \nu_{\tilde{A}_2}\right\}); \theta_1, \theta_2)\right\}$$

where, $\theta_1 \in \min\left\{\mu_{\tilde{A}_1}, \mu_{\tilde{A}_2}\right\}$, $\theta_2 \in \max\left\{\nu_{\tilde{A}_1}, \nu_{\tilde{A}_2}\right\}$, and $0 \le \theta_1^2 + \theta_2^2 \le 1$.

$$\text{Subtraction: } \tilde{A}_1 \ominus \tilde{A}_2 = \{(((a_1 - d_2, b_1 - c_2, c_1 - b_2, d_1 - a_2), \\ \min\left\{\mu_{\tilde{A}_1}, \mu_{\tilde{A}_2}\right\}, \max\left\{\nu_{\tilde{A}_1}, \nu_{\tilde{A}_2}\right\}); \theta_3, \theta_4)\right\}$$

where, $\theta_3 \in \min\left\{\mu_{\tilde{A}_1}, \mu_{\tilde{A}_2}\right\}$, $\theta_4 \in \max\left\{\nu_{\tilde{A}_1}, \nu_{\tilde{A}_2}\right\}$, and $0 \le \theta_3^2 + \theta_4^2 \le 1$.

$$\text{Multiplication: } \tilde{A}_1 \otimes \tilde{A}_2 = \{(((a_1 a_2, b_1 b_2, c_1 c_2, d_1 d_2), \\ \min\left\{\mu_{\tilde{A}_1}, \mu_{\tilde{A}_2}\right\}, \max\left\{\nu_{\tilde{A}_1}, \nu_{\tilde{A}_2}\right\}); \theta_5, \theta_6)\right\}$$

where, $\theta_5 \in \min\left\{\mu_{\tilde{A}_1}, \mu_{\tilde{A}_2}\right\}$, $\theta_6 \in \max\left\{\nu_{\tilde{A}_1}, \nu_{\tilde{A}_2}\right\}$, and $0 \le \theta_5^2 + \theta_6^2 \le 1$.

$$\text{Scalar multiplication: } k\tilde{A}_1 = \begin{cases} \left\{\left(\left((ka_1, kb_1, kc_1, kd_1), \mu_{\tilde{A}_1}, \nu_{\tilde{A}_1}\right); w_1, w_2\right)\right\}, k > 0, \\ \left\{\left(\left((kd_1, kc_1, kb_1, ka_1), \mu_{\tilde{A}_1}, \nu_{\tilde{A}_1}\right); w_1, w_2\right)\right\}, k < 0. \end{cases}$$

Definition 9.37 [19] Assume the TrPHFN $\tilde{A} = \left\{\left(\left((a, b, c, d), \mu_{\tilde{A}}, \nu_{\tilde{A}}\right); w_1, w_2\right)\right\}$. A ranking function for this number is defined as,

$$\Re(\tilde{A}) = \frac{\int_a^d x\mu_{\tilde{A}}^2(x)dx + \int_a^d x\nu_{\tilde{A}}^2(x)dx}{\int_a^d \mu_{\tilde{A}}^2(x)dx + \int_a^d \nu_{\tilde{A}}^2(x)dx}$$

$$= \frac{1}{4}\left[\frac{\begin{array}{c} w_1^2(-a^2 - 3b^2 + 3c^2 + d^2 - 2ab + 2cd) \\ + w_2^2(-5a^2 + 3b^2 - 3c^2 + 5d^2 + 2ab - 2cd) \end{array}}{w_1^2(-a - 2b + 2c + d) + w_2^2(-2a + 2b - 2c + 2d)}\right] \tag{9.77}$$

where, $\Re(\tilde{A}) : F(\tilde{A}) \to \mathbb{R}$ and $F(\tilde{A})$ is the set of all TrPHFNs.

Based on this definition, two TrPHFNs are ranked as below.

$$\Re(\tilde{A}_1) > \Re(\tilde{A}_2) \Rightarrow \tilde{A}_1 \succ \tilde{A}_2,$$

$$\Re(\tilde{A}_1) < \Re(\tilde{A}_2) \Rightarrow \tilde{A}_1 \prec \tilde{A}_2,$$

$$\Re(\tilde{A}_1) = \Re(\tilde{A}_2) \Rightarrow \tilde{A}_1 \approx \tilde{A}_2,$$

Lemma 9.1 [19] *considering the ranking function of* Definition 9.37, *for any* $k \geq 0$ *there is* $\Re(k\tilde{A}) = k\Re(\tilde{A})$.

9.7.2 *Interval-Valued Pythagorean Hesitant Fuzzy Sets*

Wang et al. [20] for the first time extended PHFSs to the form of interval-valued Pythagorean fuzzy sets (IVPHFSs). The concepts of this type of fuzzy sets are presented in the rest of this sub-section.

Definition 9.38 [21] Assume the reference set X. An interval-valued intuitionistic hesitant fuzzy set (IVPHFS) such as $\tilde{A}$ on X based on the membership and non-membership functions $\mu_{\tilde{A}}(x)$ and $\nu_{\tilde{A}}(x)$ is defined as below.

$$\tilde{A} = \left\{ \left(x, \mu_{\tilde{A}}(x), \nu_{\tilde{A}}(x)\right) \middle| x \in X \right\} \tag{9.78}$$

Here, $\mu_{\tilde{A}}(x)$ and $\nu_{\tilde{A}}(x)$ are interval values from [0, 1].

In this definition, if for any $x \in X$ is assumed that,

$$\mu_{\tilde{A}}(x) = \bigcup_{\left[\gamma_{\tilde{A}}^L, \gamma_{\tilde{A}}^U\right] \in \mu_{\tilde{A}}(x)} \left\{\left[\gamma_{\tilde{A}}^L, \gamma_{\tilde{A}}^U\right]\right\}, \nu_{\tilde{A}}(x) = \bigcup_{\left[\eta_{\tilde{A}}^L, \eta_{\tilde{A}}^U\right] \in \nu_{\tilde{A}}(x)} \left\{\left[\eta_{\tilde{A}}^L, \eta_{\tilde{A}}^U\right]\right\},$$

$$\gamma_{\tilde{A}}^{U+} \in \bigcup_{x \in X} \max_{\gamma_{\tilde{A}} \in \mu_{\tilde{A}}(x)} \left\{\gamma_{\tilde{A}}^U(x)\right\}, \eta_{\tilde{A}}^{U+} \in \bigcup_{x \in X} \max_{\eta_{\tilde{A}} \in \nu_{\tilde{A}}(x)} \left\{\eta_{\tilde{A}}^U(x)\right\},$$

then, $0 \leq \gamma_{\tilde{A}}^L, \gamma_{\tilde{A}}^U, \eta_{\tilde{A}}^L, \eta_{\tilde{A}}^U \leq 1$ and $0 \leq \left(\gamma_{\tilde{A}}^{U+}\right)^2 + \left(\eta_{\tilde{A}}^{U+}\right)^2 \leq 1$. Therefore, an IVPHFS is shown as,

$$\tilde{A}=\left\{\left(x,\bigcup_{\left[\gamma_{\tilde{A}}^{L},\gamma_{\tilde{A}}^{U}\right]\in\mu_{\tilde{A}}(x)}\{[\gamma_{\tilde{A}}^{L},\gamma_{\tilde{A}}^{U}]\},\bigcup_{\left[\eta_{\tilde{A}}^{L},\eta_{\tilde{A}}^{U}\right]\in\nu_{\tilde{A}}(x)}\{[\eta_{\tilde{A}}^{L},\eta_{\tilde{A}}^{U}]\}\right)\middle|x\in X\right\}. \tag{9.79}$$

In addition, for more simplicity, an interval-valued Pythagorean hesitant fuzzy element (IVPHFE) is shown as below.

$$\begin{aligned}h_{\tilde{A}}&=\left(\mu_{\tilde{A}}(x),\nu_{\tilde{A}}(x)\right)\\&=\left(\bigcup_{\left[\gamma_{\tilde{A}}^{L},\gamma_{\tilde{A}}^{U}\right]\in\mu_{\tilde{A}}(x)}\{[\gamma_{\tilde{A}}^{L},\gamma_{\tilde{A}}^{U}]\},\bigcup_{\left[\eta_{\tilde{A}}^{L},\eta_{\tilde{A}}^{U}\right]\in\nu_{\tilde{A}}(x)}\{[\eta_{\tilde{A}}^{L},\eta_{\tilde{A}}^{U}]\}\right),\end{aligned}$$

Example 9.14 Assume the reference set $X=\{x_1,x_2\}$, and the IVPHFEs $h_{\tilde{A}}(x_1)=(\{[0.2,0.5],[0.3,0.3]\},\{[0.3,0.6]\})$ and $h_{\tilde{A}}(x_2)=(\{[0.3,0.4],[0.1,0.6]\},\{[0.2,0.5]\})$. As, $\gamma_{\tilde{A}}^{U+}(x_1)=0.5$, $\eta_{\tilde{A}}^{U+}(x_1)=0.6$, $\gamma_{\tilde{A}}^{U+}(x_2)=0.6$, $\eta_{\tilde{A}}^{U+}(x_2)=0.5$, then $\left(\gamma_{\tilde{A}}^{U+}\right)^2(x_1)+\left(\eta_{\tilde{A}}^{U+}\right)^2(x_1)=0.61\leq 1$ and $\left(\gamma_{\tilde{A}}^{U+}\right)^2(x_2)+\left(\eta_{\tilde{A}}^{U+}\right)^2(x_2)=0.61\leq 1$. Therefore, $\tilde{A}$ is an IVPHFS shown as below.

$$\begin{aligned}\tilde{A}=\{&(x_1,\{[0.2,0.5],[0.3,0.3]\},\{[0.3,0.6]\}),\\&(x_2,\{[0.3,0.4],[0.1,0.6]\},\{[0.2,0.5]\})\}.\end{aligned}$$

Definition 9.39 [8] Assume the IVPHFEs $h_{\tilde{A}}=\left(\bigcup_{\left[\gamma_{\tilde{A}}^{L},\gamma_{\tilde{A}}^{U}\right]\in\mu_{\tilde{A}}(x)}\left\{\left[\gamma_{\tilde{A}}^{L},\gamma_{\tilde{A}}^{U}\right]\right\},\bigcup_{\left[\eta_{\tilde{A}}^{L},\eta_{\tilde{A}}^{U}\right]\in\nu_{\tilde{A}}(x)}\left\{\left[\eta_{\tilde{A}}^{L},\eta_{\tilde{A}}^{U}\right]\right\}\right)$, $h_{\tilde{A}_1}=\left(\bigcup_{\left[\gamma_{\tilde{A}_1}^{L},\gamma_{\tilde{A}_1}^{U}\right]\in\mu_{\tilde{A}_1}(x)}\left\{\left[\gamma_{\tilde{A}_1}^{L},\gamma_{\tilde{A}_1}^{U}\right]\right\},\bigcup_{\left[\eta_{\tilde{A}_1}^{L},\eta_{\tilde{A}_1}^{U}\right]\in\nu_{\tilde{A}_1}(x)}\left\{\left[\eta_{\tilde{A}_1}^{L},\eta_{\tilde{A}_1}^{U}\right]\right\}\right)$, and $h_{\tilde{A}_2}=\left(\bigcup_{\left[\gamma_{\tilde{A}}^{L},\gamma_{\tilde{A}_2}^{U}\right]\in\mu_{\tilde{A}_2}(x)}\left\{\left[\gamma_{\tilde{A}_2}^{L},\gamma_{\tilde{A}_2}^{U}\right]\right\},\bigcup_{\left[\eta_{\tilde{A}_2}^{L},\eta_{\tilde{A}_2}^{U}\right]\in\nu_{\tilde{A}_2}(x)}\left\{\left[\eta_{\tilde{A}_2}^{L},\eta_{\tilde{A}_2}^{U}\right]\right\}\right)$. Then, for $k>0$,

$$h_{\tilde{A}_1}\oplus h_{\tilde{A}_2}=\left(\begin{array}{c}\bigcup_{\substack{\left[\gamma_{\tilde{A}_1}^{L},\gamma_{\tilde{A}_1}^{U}\right]\in\mu_{\tilde{A}_1}\\\left[\gamma_{\tilde{A}_2}^{L},\gamma_{\tilde{A}_2}^{U}\right]\in\mu_{\tilde{A}_2}}}\left\{\left[\begin{array}{c}\sqrt{\left(\gamma_{\tilde{A}_1}^{L}\right)^2+\left(\gamma_{\tilde{A}_2}^{L}\right)^2-\left(\gamma_{\tilde{A}_1}^{L}\gamma_{\tilde{A}_2}^{L}\right)^2},\\\sqrt{\left(\gamma_{\tilde{A}_1}^{U}\right)^2+\left(\gamma_{\tilde{A}_2}^{U}\right)^2-\left(\gamma_{\tilde{A}_1}^{U}\gamma_{\tilde{A}_2}^{U}\right)^2}\end{array}\right]\right\},\\\bigcup_{\substack{\left[\eta_{\tilde{A}_1}^{L},\eta_{\tilde{A}_1}^{U}\right]\in\nu_{\tilde{A}_1}\\\left[\eta_{\tilde{A}_2}^{L},\eta_{\tilde{A}_2}^{U}\right]\in\nu_{\tilde{A}_2}}}\left\{\left[\eta_{\tilde{A}_1}^{L}\times\eta_{\tilde{A}_2}^{L},\eta_{\tilde{A}_1}^{U}\times\eta_{\tilde{A}_2}^{U}\right]\right\}\end{array}\right), \tag{9.80}$$

$$h_{\tilde{A}_1} \otimes h_{\tilde{A}_2} = \left(\begin{array}{l} \bigcup_{\substack{\left[\gamma^L_{\tilde{A}_1}, \gamma^U_{\tilde{A}_1}\right] \in \mu_{\tilde{A}_1} \\ \left[\gamma^L_{\tilde{A}_2}, \gamma^U_{\tilde{A}_2}\right] \in \mu_{\tilde{A}_2}}} \left\{ \left[\gamma^L_{\tilde{A}_1} \gamma^L_{\tilde{A}_2}, \gamma^U_{\tilde{A}_1} \gamma^U_{\tilde{A}_2} \right] \right\}, \\ \bigcup_{\substack{\left[\eta^L_{\tilde{A}_1}, \eta^U_{\tilde{A}_1}\right] \in \nu_{\tilde{A}_1} \\ \left[\eta^L_{\tilde{A}_1}, \eta^U_{\tilde{A}_2}\right] \in \nu_{\tilde{A}_2}}} \left\{ \left[\sqrt{\left(\eta^L_{\tilde{A}_1}\right)^2 + \left(\eta^L_{\tilde{A}_2}\right)^2 - \left(\eta^L_{\tilde{A}_1} \eta^L_{\tilde{A}_2}\right)^2}, \sqrt{\left(\eta^U_{\tilde{A}_1}\right)^2 + \left(\eta^U_{\tilde{A}_2}\right)^2 - \left(\eta^U_{\tilde{A}_1} \eta^U_{\tilde{A}_2}\right)^2} \right] \right\} \end{array} \right), \tag{9.81}$$

$$kh_{\tilde{A}} = \left(\begin{array}{l} \bigcup_{\left[\gamma^L_{\tilde{A}}, \gamma^U_{\tilde{A}}\right] \in \mu_{\tilde{A}}(x)} \left\{ \left[\sqrt{1 - \left(1 - \left(\gamma^L_{\tilde{A}}\right)^2\right)^k}, \sqrt{1 - \left(1 - \left(\gamma^U_{\tilde{A}}\right)^2\right)^k} \right] \right\}, \\ \bigcup_{\left[\eta^L_{\tilde{A}}, \eta^U_{\tilde{A}}\right] \in \nu_{\tilde{A}}(x)} \left\{ \left[\left(\eta^L_{\tilde{A}}\right)^k, \left(\eta^U_{\tilde{A}}\right)^k \right] \right\} \end{array} \right), \tag{9.82}$$

$$\left(h_{\tilde{A}}\right)^k = \left(\begin{array}{l} \bigcup_{\left[\gamma^L_{\tilde{A}}, \gamma^U_{\tilde{A}}\right] \in \mu_{\tilde{A}}(x)} \left\{ \left[\left(\gamma^L_{\tilde{A}}\right)^k, \left(\gamma^U_{\tilde{A}}\right)^k \right] \right\}, \\ \bigcup_{\left[\eta^L_{\tilde{A}}, \eta^U_{\tilde{A}}\right] \in \nu_{\tilde{A}}(x)} \left\{ \left[\sqrt{1 - \left(1 - \left(\eta^L_{\tilde{A}}\right)^2\right)^k}, \sqrt{1 - \left(1 - \left(\eta^U_{\tilde{A}}\right)^2\right)^k} \right] \right\} \end{array} \right). \tag{9.83}$$

9.8 Picture Hesitant Fuzzy Sets

Definition 9.40 [22] Assume the reference set X. A picture hesitant fuzzy set (PIHFS) such as $\tilde{A}$ on X based on the positive, neutral, and negative membership functions $\mu_{\tilde{A}}(x)$, $w_{\tilde{A}}(x)$, and $\nu_{\tilde{A}}(x)$ is defined as below.

$$\tilde{A} = \left\{ \left(x, \mu_{\tilde{A}}(x), w_{\tilde{A}}(x), \nu_{\tilde{A}}(x)\right) \middle| x \in X \right\} \tag{9.84}$$

Here, $\mu_{\tilde{A}}(x), w_{\tilde{A}}(x), \nu_{\tilde{A}}(x) \in [0, 1]$.

In this definition, if for any $x \in X$ is assumed that,

$$\mu_{\tilde{A}}(x) = \bigcup_{\gamma_{\tilde{A}} \in \mu_{\tilde{A}}(x)} \left\{\gamma_{\tilde{A}}\right\}, w_{\tilde{A}}(x) = \bigcup_{\delta_{\tilde{A}} \in \mu_{\tilde{A}}(x)} \left\{\delta_{\tilde{A}}\right\}, \nu_{\tilde{A}}(x)$$
$$= \bigcup_{\eta_{\tilde{A}} \in \mu_{\tilde{A}}(x)} \left\{\eta_{\tilde{A}}\right\}, \gamma^+_{\tilde{A}} \in \mu^+_{\tilde{A}} = \bigcup_{x \in X} \max_{\gamma_{\tilde{A}} \in \mu_{\tilde{A}}(x)} \left\{\gamma_{\tilde{A}}\right\},$$

$\delta^+_{\tilde{A}} \in w^+_{\tilde{A}} = \bigcup_{x \in X} \max_{\delta_{\tilde{A}} \in w_{\tilde{A}}(x)} \{\delta_{\tilde{A}}\}$, and $\eta^+_{\tilde{A}} \in \nu^+_{\tilde{A}} = \bigcup_{x \in X} \max_{\eta_{\tilde{A}} \in \nu_{\tilde{A}}(x)} \{\eta_{\tilde{A}}\}$,

then, $0 \le \gamma_{\tilde{A}}, \delta_{\tilde{A}}, \eta_{\tilde{A}} \le 1$ and $0 \le \gamma^+_{\tilde{A}} + \delta^+_{\tilde{A}} + \eta^+_{\tilde{A}} \le 1$. Therefore, a PIHFS is shown as,

$$\tilde{A} = \left\{ \left(x, \bigcup_{\gamma_{\tilde{A}} \in \mu_{\tilde{A}}} \{\gamma_{\tilde{A}}\}, \bigcup_{\delta_{\tilde{A}} \in w_{\tilde{A}}} \{\delta_{\tilde{A}}\}, \bigcup_{\eta_{\tilde{A}} \in \nu_{\tilde{A}}} \{\eta_{\tilde{A}}\} \right) \middle| x \in X \right\}. \tag{9.85}$$

In addition, for more simplicity, a picture hesitant fuzzy element (PIHFE) is shown as $h_{\tilde{A}} = \left(\mu_{\tilde{A}}(x), w_{\tilde{A}}(x), \nu_{\tilde{A}}(x)\right)$.

Example 9.15 Assume the reference set $X = \{x_1, x_2\}$, and the PIHFEs $h_{\tilde{A}}(x_1) = (\{0.1, 0.2\}, \{0.3\}, \{0.4, 0.5\})$ and $h_{\tilde{A}}(x_2) = (\{0.3\}, \{0.4\}, \{0.1, 0.2\})$. As,

$$\gamma^+_{\tilde{A}}(x_1) = 0.2, \delta^+_{\tilde{A}}(x_1) = 0.3, \eta^+_{\tilde{A}}(x_1) = 0.5, \gamma^+_{\tilde{A}}(x_2) \\ = 0.3, \delta^+_{\tilde{A}}(x_2) = 0.4, \eta^+_{\tilde{A}}(x_2) = 0.2,$$

$$\gamma^+_{\tilde{A}}(x_1) + \delta^+_{\tilde{A}}(x_1) + \eta^+_{\tilde{A}}(x_1) = 1 \le 1, \gamma^+_{\tilde{A}}(x_2) + \delta^+_{\tilde{A}}(x_2) + \eta^+_{\tilde{A}}(x_2) \\ = 0.9 \le 1,$$

then, $\tilde{A}$ is a PIHFS shown as below.

$$\tilde{A} = \{(x_1, \{[0.2, 0.5], [0.3, 0.3]\}, \{[0.3, 0.6]\}), \\ (x_2, \{[0.3, 0.4], [0.1, 0.6]\}, \{[0.2, 0.5]\})\}.$$

Definition 9.41 [8] Assume the PIHFEs $h_{\tilde{A}} = \left(\mu_{\tilde{A}}(x), w_{\tilde{A}}(x), \nu_{\tilde{A}}(x)\right)$, $h_{\tilde{A}_1} = \left(\mu_{\tilde{A}_1}(x), w_{\tilde{A}_1}(x), \nu_{\tilde{A}_1}(x)\right)$, and $h_{\tilde{A}_2} = \left(\mu_{\tilde{A}_2}(x), w_{\tilde{A}_2}(x), \nu_{\tilde{A}_2}(x)\right)$. Then,

$$h^c_{\tilde{A}} = \left(\nu_{\tilde{A}}(x), w_{\tilde{A}}(x), \mu_{\tilde{A}}(x)\right), \tag{9.86}$$

$$h_{\tilde{A}_1} \cup h_{\tilde{A}_2} = \left(\mu_{\tilde{A}_1} \cup \mu_{\tilde{A}_2}, w_{\tilde{A}_1} \cap w_{\tilde{A}_2}, \nu_{\tilde{A}_1} \cap \nu_{\tilde{A}_2}\right); \tag{9.87}$$

$$h_{\tilde{A}_1} \cap h_{\tilde{A}_2} = \left(\mu_{\tilde{A}_1} \cap \mu_{\tilde{A}_2}, w_{\tilde{A}_1} \cup w_{\tilde{A}_2}, \nu_{\tilde{A}_1} \cup \nu_{\tilde{A}_2}\right), \tag{9.88}$$

$$h_{\tilde{A}_1} \oplus h_{\tilde{A}_2} = \left(\mu_{\tilde{A}_1} \oplus \mu_{\tilde{A}_2}, w_{\tilde{A}_1} \otimes w_{\tilde{A}_2}, \nu_{\tilde{A}_1} \otimes \nu_{\tilde{A}_2}\right), \tag{9.89}$$

$$h_{\tilde{A}_1} \otimes h_{\tilde{A}_2} = \left(\mu_{\tilde{A}_1} \otimes \mu_{\tilde{A}_2}, w_{\tilde{A}_1} \oplus w_{\tilde{A}_2}, \nu_{\tilde{A}_1} \oplus \nu_{\tilde{A}_2}\right), \tag{9.90}$$

where, $0 \le \gamma^+_{\tilde{A}} + \delta^+_{\tilde{A}} + \eta^+_{\tilde{A}} \le 1, 0 \le \gamma^+_{\tilde{A}_1} + \delta^+_{\tilde{A}_1} + \eta^+_{\tilde{A}_1} \le 1$, and $0 \le \gamma^+_{\tilde{A}_1} + \delta^+_{\tilde{A}_1} + \eta^+_{\tilde{A}_1} \le 1$.

Definition 9.42 [8] Assume the PIHFSs $\tilde{A}_1$ and $\tilde{A}_2$. Then,

$$\tilde{A}_1 \cup \tilde{A}_2 = \bigcup_{h_{\tilde{A}_1} \in \tilde{A}_1, h_{\tilde{A}_2} \in \tilde{A}_2} h_{\tilde{A}_1} \cup h_{\tilde{A}_2}$$
$$= \left\{ \left(x, \mu_{\tilde{A}_1} \cup \mu_{\tilde{A}_2}, w_{\tilde{A}_1} \cap w_{\tilde{A}_2}, \nu_{\tilde{A}_1} \cap \nu_{\tilde{A}_2}\right) \middle| x \in X \right\}, \tag{9.91}$$

$$\tilde{A}_1 \cap \tilde{A}_2 = \bigcup_{h_{\tilde{A}_1} \in \tilde{A}_1, h_{\tilde{A}_2} \in \tilde{A}_2} h_{\tilde{A}_1} \cap h_{\tilde{A}_2}$$
$$= \left\{ \left(x, \mu_{\tilde{A}_1} \cap \mu_{\tilde{A}_2}, w_{\tilde{A}_1} \cup w_{\tilde{A}_2}, \nu_{\tilde{A}_1} \cup \nu_{\tilde{A}_2}\right) \middle| x \in X \right\}, \tag{9.92}$$

$$\tilde{A}_1 \oplus \tilde{A}_2 = \bigcup_{h_{\tilde{A}_1} \in \tilde{A}_1, h_{\tilde{A}_2} \in \tilde{A}_2} h_{\tilde{A}_1} \oplus h_{\tilde{A}_2}$$
$$= \left\{ \left(x, \mu_{\tilde{A}_1} \oplus \mu_{\tilde{A}_2}, w_{\tilde{A}_1} \otimes w_{\tilde{A}_2}, \nu_{\tilde{A}_1} \otimes \nu_{\tilde{A}_2}\right) \middle| x \in X \right\}, \tag{9.93}$$

$$\tilde{A}_1 \otimes \tilde{A}_2 = \bigcup_{h_{\tilde{A}_1} \in \tilde{A}_1, h_{\tilde{A}_2} \in \tilde{A}_2} h_{\tilde{A}_1} \otimes h_{\tilde{A}_2}$$
$$= \left\{ \left(x, \mu_{\tilde{A}_1} \otimes \mu_{\tilde{A}_2}, w_{\tilde{A}_1} \oplus w_{\tilde{A}_2}, \nu_{\tilde{A}_1} \oplus \nu_{\tilde{A}_2}\right) \middle| x \in X \right\}. \tag{9.94}$$

9.8.1 Interval-Valued Picture Hesitant Fuzzy Set

Definition 9.43 [8] Assume the reference set X. An interval-valued picture hesitant fuzzy set (IVPIHFS) such as $\tilde{A}$ on X based on the positive, neutral, and negative membership functions $\mu_{\tilde{A}}(x)$, $w_{\tilde{A}}(x)$, and $\nu_{\tilde{A}}(x)$ (from interval type values) is defined as below.

$$\tilde{A} = \left\{ \left(x, \mu_{\tilde{A}}(x), w_{\tilde{A}}(x), \nu_{\tilde{A}}(x)\right) \middle| x \in X \right\} \tag{9.95}$$

Here, $\mu_{\tilde{A}}(x) = \bigcup_{\left[\gamma_{\tilde{A}}^L, \gamma_{\tilde{A}}^U\right] \in \mu_{\tilde{A}}(x)} \left\{\left[\gamma_{\tilde{A}}^L, \gamma_{\tilde{A}}^U\right]\right\}$, $w_{\tilde{A}}(x) = \bigcup_{\left[\delta_{\tilde{A}}^L, \delta_{\tilde{A}}^U\right] \in w_{\tilde{A}}(x)} \left\{\left[\delta_{\tilde{A}}^L, \delta_{\tilde{A}}^U\right]\right\}$, and $\nu_{\tilde{A}}(x) = \bigcup_{\left[\eta_{\tilde{A}}^L, \eta_{\tilde{A}}^U\right] \in \nu_{\tilde{A}}(x)} \left\{\left[\eta_{\tilde{A}}^L, \eta_{\tilde{A}}^U\right]\right\}$, where $0 \leq \gamma_{\tilde{A}}^L, \gamma_{\tilde{A}}^U, \delta_{\tilde{A}}^L, \delta_{\tilde{A}}^U, \eta_{\tilde{A}}^L, \eta_{\tilde{A}}^U \leq 1$ and $0 \leq \gamma_{\tilde{A}}^U + \delta_{\tilde{A}}^U + \eta_{\tilde{A}}^U \leq 1$. Therefore, an IVPIHFS is shown as,

$$\tilde{A} = \left\{ \left(x, \bigcup_{\left[\gamma_{\tilde{A}}^L, \gamma_{\tilde{A}}^U\right] \in \mu_{\tilde{A}}(x)} \{[\gamma_{\tilde{A}}^L, \gamma_{\tilde{A}}^U]\}, \bigcup_{\left[\delta_{\tilde{A}}^L, \delta_{\tilde{A}}^U\right] \in w_{\tilde{A}}(x)} \{[\delta_{\tilde{A}}^L, \delta_{\tilde{A}}^U]\}, \right.\right.$$

$$\left.\left.\bigcup_{\left[\eta_{\tilde{A}}^L,\eta_{\tilde{A}}^U\right]\in\nu_{\tilde{A}}(x)}\{[\eta_{\tilde{A}}^L,\eta_{\tilde{A}}^U]\}\right)\right| x\in X\Bigg\}. \tag{9.96}$$

In addition, for more simplicity, an interval-valued picture hesitant fuzzy element (IVPIHFE) is shown as below.

$$\begin{aligned} h_{\tilde{A}} &= \left(\mu_{\tilde{A}}(x), w_{\tilde{A}}(x), \nu_{\tilde{A}}(x)\right)\\ &= \left(\bigcup_{\left[\gamma_{\tilde{A}}^L,\gamma_{\tilde{A}}^U\right]\in\mu_{\tilde{A}}(x)}\{[\gamma_{\tilde{A}}^L,\gamma_{\tilde{A}}^U]\}, \bigcup_{\left[\delta_{\tilde{A}}^L,\delta_{\tilde{A}}^U\right]\in w_{\tilde{A}}(x)}\{[\delta_{\tilde{A}}^L,\delta_{\tilde{A}}^U]\}, \bigcup_{\left[\eta_{\tilde{A}}^L,\eta_{\tilde{A}}^U\right]\in\nu_{\tilde{A}}(x)}\{[\eta_{\tilde{A}}^L,\eta_{\tilde{A}}^U]\}\right)\end{aligned}$$

Example 9.16 Assume the reference set $X = \{x_1, x_2\}$, and the IVPIHFEs $h_{\tilde{A}}(x_1) = (\{[0.1, 0.2], [0.2, 0.3]\}, \{[0.2, 0.4], [0.3, 0.3]\}, \{[0.1, 0.3]\})$ and $h_{\tilde{A}}(x_2) = (\{[0.1, 0.2]\}, \{[0.2, 0.3], [0.1, 0.4]\}, \{[0.1, 0.2]\})$. As,

$$\begin{aligned}\gamma_{\tilde{A}}^+(x_1)+\delta_{\tilde{A}}^+(x_1)+\eta_{\tilde{A}}^+(x_1) = 1 \le 1, \gamma_{\tilde{A}}^+(x_2)+\delta_{\tilde{A}}^+(x_2)+\eta_{\tilde{A}}^+(x_2)\\ = 0.8 \le 1,\end{aligned}$$

then, $\tilde{A}$ is an IVPIHFS shown as below.

$$\tilde{A} = \left\{\begin{array}{l}(x_1, \{[0.1, 0.2], [0.2, 0.3]\}, \{[0.2, 0.4], [0.3, 0.3]\}, \{[0.1, 0.3]\}),\\ (x_2, \{[0.1, 0.2]\}, \{[0.3, 0.4], [0.1, 0.4]\}, \{[0.1, 0.3]\})\end{array}\right\}.$$

Definition 9.44 [8] Assume the IVPIHFEs $h_{\tilde{A}} = \left(\bigcup_{\left[\gamma_{\tilde{A}}^L,\gamma_{\tilde{A}}^U\right]\in\mu_{\tilde{A}}(x)}\left\{\left[\gamma_{\tilde{A}}^L,\gamma_{\tilde{A}}^U\right]\right\}, \bigcup_{\left[\delta_{\tilde{A}}^L,\delta_{\tilde{A}}^U\right]\in w_{\tilde{A}}(x)}\left\{\left[\delta_{\tilde{A}}^L,\delta_{\tilde{A}}^U\right]\right\}, \bigcup_{\left[\eta_{\tilde{A}}^L,\eta_{\tilde{A}}^U\right]\in\nu_{\tilde{A}}(x)}\left\{\left[\eta_{\tilde{A}}^L,\eta_{\tilde{A}}^U\right]\right\}\right)$, $h_{\tilde{A}_1} = \left(\bigcup_{\left[\gamma_{\tilde{A}_1}^L,\gamma_{\tilde{A}_1}^U\right]\in\mu_{\tilde{A}_1}(x)}\left\{\left[\gamma_{\tilde{A}_1}^L,\gamma_{\tilde{A}_1}^U\right]\right\}, \bigcup_{\left[\delta_{\tilde{A}_1}^L,\delta_{\tilde{A}_1}^U\right]\in w_{\tilde{A}_1}(x)}\left\{\left[\delta_{\tilde{A}_1}^L,\delta_{\tilde{A}_1}^U\right]\right\}, \bigcup_{\left[\eta_{\tilde{A}_1}^L,\eta_{\tilde{A}_1}^U\right]\in\nu_{\tilde{A}_1}(x)}\left\{\left[\eta_{\tilde{A}_1}^L,\eta_{\tilde{A}_1}^U\right]\right\}\right)$, and $h_{\tilde{A}_2} = \left(\bigcup_{\left[\gamma_{\tilde{A}}^L,\gamma_{\tilde{A}_2}^U\right]\in\mu_{\tilde{A}_2}(x)}\left\{\left[\gamma_{\tilde{A}_2}^L,\gamma_{\tilde{A}_2}^U\right]\right\}, \bigcup_{\left[\delta_{\tilde{A}_2}^L,\delta_{\tilde{A}_2}^U\right]\in w_{\tilde{A}_2}(x)}\left\{\left[\delta_{\tilde{A}_2}^L,\delta_{\tilde{A}_2}^U\right]\right\}\right.$, $h_{\tilde{A}_2} = \left(\bigcup_{\left[\gamma_{\tilde{A}}^L,\gamma_{\tilde{A}_2}^U\right]\in\mu_{\tilde{A}_2}(x)}\left\{\left[\gamma_{\tilde{A}_2}^L,\gamma_{\tilde{A}_2}^U\right]\right\}, \bigcup_{\left[\delta_{\tilde{A}_2}^L,\delta_{\tilde{A}_2}^U\right]\in w_{\tilde{A}_2}(x)}\left\{\left[\delta_{\tilde{A}_2}^L,\delta_{\tilde{A}_2}^U\right]\right\}\right.$. Then, for $k > 0$,

$$h_{\tilde{A}_1} \oplus h_{\tilde{A}_2} = \begin{pmatrix} \bigcup_{\substack{\gamma^L_{\tilde{A}_1}, \gamma^U_{\tilde{A}_1} \in \mu_{\tilde{A}_1} \\ \left[\gamma^L_{\tilde{A}_2}, \gamma^U_{\tilde{A}_2}\right] \in \mu_{\tilde{A}_2}}} \left\{\left[\left\{\gamma^L_{\tilde{A}_1} + \gamma^L_{\tilde{A}_2} - \gamma^L_{\tilde{A}_1}\gamma^L_{\tilde{A}_2}\right\}, \left\{\gamma^U_{\tilde{A}_1} + \gamma^U_{\tilde{A}_2} - \gamma^U_{\tilde{A}_1}\gamma^U_{\tilde{A}_2}\right\}\right]\right\}, \\ \bigcup_{\substack{\left[\delta^L_{\tilde{A}_1}, \delta\tilde{A}_1\right] \in w_{\tilde{A}_1} \\ \left[\delta^L_{\tilde{A}_2}, \delta\tilde{A}_2\right] \in w_{\tilde{A}_2}}} \left\{\left[\left\{\delta^L_{\tilde{A}_1} \times \delta^L_{\tilde{A}_2}\right\}, \left\{\delta^U_{\tilde{A}_1} \times \delta^U_{\tilde{A}_2}\right\}\right]\right\}, \\ \bigcup_{\substack{\eta^L_{\tilde{A}_1}, \eta^U_{\tilde{A}_1}\right] \in v_{\tilde{A}_1} \\ \left[\eta^L_{\tilde{A}_2}, \eta_{\tilde{A}_2}\right] \in v_{\tilde{A}_2}}} \left\{\left[\left\{\eta^L_{\tilde{A}_1} \times \eta^L_{\tilde{A}_2}\right\}, \left\{\eta^U_{\tilde{A}_1} \times \eta^U_{\tilde{A}_2}\right\}\right]\right\} \end{pmatrix} \tag{9.97}$$

$$h_{\tilde{A}_1} \otimes h_{\tilde{A}_2} = \begin{pmatrix} \bigcup_{\substack{\gamma^L_{\tilde{A}_1}, \gamma^U_{\tilde{A}_1}\right] \in \mu_{\tilde{A}_1} \\ \left[\gamma^L_{\tilde{A}_2}, \gamma^U_{\tilde{A}_2}\right] \in \mu_{\tilde{A}_2}}} \left\{\left[\left\{\gamma^L_{\tilde{A}_1} \times \gamma^L_{\tilde{A}_2}\right\}, \left\{\gamma^U_{\tilde{A}_1} \times \gamma^U_{\tilde{A}_2}\right\}\right]\right\}, \\ \bigcup_{\substack{\left[\delta^L_{\tilde{A}_1}, \delta^U_{\tilde{A}_1}\right] \in w_{\tilde{A}_1} \\ \left[\delta^L_{\tilde{A}_2}, \delta^U_{\tilde{A}_2}\right] \in w_{\tilde{A}_2}}} \left\{\left[\left\{\delta^L_{\tilde{A}_1} + \delta^L_{\tilde{A}_2} - \delta^L_{\tilde{A}_1} \times \delta^L_{\tilde{A}_2}\right\},\right.\right. \\ \left.\left.\left\{\delta^U_{\tilde{A}_1} + \delta^U_{\tilde{A}_2} - \delta^U_{\tilde{A}_1} \times \delta^U_{\tilde{A}_2}\right\}\right]\right\} \\ \bigcup_{\substack{\left[\eta^L_{\tilde{A}_1}, \eta^U_{\tilde{A}_1}\right] \in v_{\tilde{A}_1} \\ \left[\eta^L_{\tilde{A}_2}, \eta^U_{\tilde{A}_2}\right] \in v_{\tilde{A}_2}}} \left\{\left[\left\{\eta^L_{\tilde{A}_1} + \eta^L_{\tilde{A}_2} - \eta^L_{\tilde{A}_1} \times \eta^L_{\tilde{A}_2}\right\},\right.\right. \\ \left.\left.\left\{\eta^U_{\tilde{A}_1} + \eta^U_{\tilde{A}_2} - \eta^U_{\tilde{A}_1} \times \eta^U_{\tilde{A}_2}\right\}\right]\right\} \end{pmatrix}, \tag{9.98}$$

$$kh_{\tilde{A}} = \begin{pmatrix} \bigcup_{\left[\gamma^L_{\tilde{A}}, \gamma^U_{\tilde{A}}\right] \in \mu_{\tilde{A}}(x)} \left\{\left[1 - \left(1 - \gamma^L_{\tilde{A}}\right)^k, 1 - \left(1 - \gamma^U_{\tilde{A}}\right)^k\right]\right\}, \\ \bigcup_{\left[\delta^L_{\tilde{A}}, \delta^U_{\tilde{A}}\right] \in w_{\tilde{A}}(x)} \left\{\left[\left(\delta^L_{\tilde{A}}\right)^k, \left(\delta^U_{\tilde{A}}\right)^k\right]\right\}, \\ \bigcup_{\left[\eta^L_{\tilde{A}}, \eta^U_{\tilde{A}}\right] \in v_{\tilde{A}}(x)} \left\{\left[\left(\eta^L_{\tilde{A}}\right)^k, \left(\eta^U_{\tilde{A}}\right)^k\right]\right\} \end{pmatrix}, \tag{9.99}$$

$$\left(h_{\tilde{A}}\right)^k = \begin{pmatrix} \bigcup_{\left[\gamma^L_{\tilde{A}}, \gamma^U_{\tilde{A}}\right] \in \mu_{\tilde{A}}(x)} \left\{\left[\left(\gamma^L_{\tilde{A}}\right)^k, \left(\gamma^U_{\tilde{A}}\right)^k\right]\right\}, \\ \bigcup_{\left[\delta^L_{\tilde{A}}, \delta^U_{\tilde{A}}\right] \in w_{\tilde{A}}(x)} \left\{\left[1 - \left(1 - \delta^L_{\tilde{A}}\right)^k, 1 - \left(1 - \delta^U_{\tilde{A}}\right)^k\right]\right\}, \\ \bigcup_{\left[\eta^L_{\tilde{A}}, \eta^U_{\tilde{A}}\right] \in v_{\tilde{A}}(x)} \left\{\left[1 - \left(1 - \eta^L_{\tilde{A}}\right)^k, 1 - \left(1 - \eta^U_{\tilde{A}}\right)^k\right]\right\} \end{pmatrix}. \tag{9.100}$$

9.9 Spherical Hesitant Fuzzy Sets

Definition 9.45 [22] Assume the reference set X. A spherical hesitant fuzzy set (SHFS) such as $\tilde{A}$ on X based on the positive, neutral, and negative membership functions $\mu_{\tilde{A}}(x)$, $w_{\tilde{A}}(x)$, and $\nu_{\tilde{A}}(x)$ is defined as below.

$$\tilde{A} = \left\{ \left(x, \mu_{\tilde{A}}(x), w_{\tilde{A}}(x), \nu_{\tilde{A}}(x)\right) \middle| x \in X \right\} \tag{9.101}$$

Here, $\mu_{\tilde{A}}(x), w_{\tilde{A}}(x), \nu_{\tilde{A}}(x) \in [0, 1]$.

In this definition, if for any $x \in X$ is assumed that,

$$\mu_{\tilde{A}}(x) = \bigcup_{\gamma_{\tilde{A}} \in \mu_{\tilde{A}}(x)} \{\gamma_{\tilde{A}}\}, w_{\tilde{A}}(x) = \bigcup_{\delta_{\tilde{A}} \in w_{\tilde{A}}(x)} \{\delta_{\tilde{A}}\}, \nu_{\tilde{A}}(x) = \bigcup_{\eta_{\tilde{A}} \in \nu_{\tilde{A}}(x)} \{\eta_{\tilde{A}}\},$$

$$\gamma^+_{\tilde{A}} \in \mu^+_{\tilde{A}} = \bigcup_{x \in X} \max_{\gamma_{\tilde{A}} \in \mu_{\tilde{A}}(x)} \{\gamma_{\tilde{A}}\}, \delta^+_{\tilde{A}} \in w^+_{\tilde{A}} = \bigcup_{x \in X} \max_{\delta \tilde{A} \in w_{\tilde{A}}(x)} \{\delta_{\tilde{A}}\},$$
$$\eta^+_{\tilde{A}} \in \nu^+_{\tilde{A}} = \bigcup_{x \in X} \max_{\eta \tilde{A} \in \nu_{\tilde{A}}(x)} \{\eta_{\tilde{A}}\},$$

then, $0 \le \gamma_{\tilde{A}}, \delta_{\tilde{A}}, \eta_{\tilde{A}} \le 1$ and $0 \le \left(\gamma^+_{\tilde{A}}\right)^2 + \left(\delta^+_{\tilde{A}}\right)^2 + \left(\eta^+_{\tilde{A}}\right)^2 \le 1$. Therefore, a SHFS is shown as,

$$\tilde{A} = \left\{ \left(x, \bigcup_{\gamma_{\tilde{A}} \in \mu_{\tilde{A}}} \{\gamma_{\tilde{A}}\}, \bigcup_{\delta_{\tilde{A}} \in w_{\tilde{A}}} \{\delta_{\tilde{A}}\}, \bigcup_{\eta_{\tilde{A}} \in \nu_{\tilde{A}}} \{\eta_{\tilde{A}}\}\right) \middle| x \in X \right\}. \tag{9.102}$$

In addition, for more simplicity, a spherical hesitant fuzzy element (SHFE) is shown as $h_{\tilde{A}} = \left(\mu_{\tilde{A}}(x), w_{\tilde{A}}(x), \nu_{\tilde{A}}(x)\right) = \left(\bigcup_{\gamma_{\tilde{A}} \in \mu_{\tilde{A}}} \{\gamma_{\tilde{A}}\}, \bigcup_{\delta_{\tilde{A}} \in w_{\tilde{A}}} \{\delta_{\tilde{A}}\}, \bigcup_{\eta_{\tilde{A}} \in \nu_{\tilde{A}}} \{\eta_{\tilde{A}}\}\right)$.

Example 9.17 [8] Assume the reference set $X = \{x_1, x_2\}$, and the SHFEs $h_{\tilde{A}}(x_1) = (\{0.2, 0.3\}, \{0.1, 0.4\}, \{0.3\})$ and $h_{\tilde{A}}(x_2) = (\{0.3, 0.4\}, \{0.2\}, \{0.1, 0.2\})$. As,

$$\gamma^+_{\tilde{A}}(x_1) = 0.3, \delta^+_{\tilde{A}}(x_1) = 0.4, \eta^+_{\tilde{A}}(x_1) = 0.3, \gamma^+_{\tilde{A}}(x_2) = 0.4, \delta^+_{\tilde{A}}(x_2)$$
$$= 0.2, \eta^+_{\tilde{A}}(x_2) = 0.2,$$

$$\left(\gamma^+_{\tilde{A}}\right)^2(x_1) + \left(\delta^+_{\tilde{A}}\right)^2(x_1) + \left(\eta^+_{\tilde{A}}\right)^2(x_1) = 0.3^2 + 0.4^2 + 0.3^2 = 0.98 \le 1,$$

$$\left(\gamma^+_{\tilde{A}}\right)^2(x_2) + \left(\delta^+_{\tilde{A}}\right)^2(x_2) + \left(\eta^+_{\tilde{A}}\right)^2(x_2) = 0.4^2 + 0.2^2 + 0.2^2 = 0.24 \le 1,$$

then, $\tilde{A}$ is a SHFS shown as below.

$$\tilde{A} = \{(x_1, \{0.2, 0.3\}, \{0.1, 0.4\}, \{0.3\}), (x_2, \{0.3, 0.4\}, \{0.2\}, \{0.1, 0.2\})\}.$$

Definition 9.46 [8] Assume the SHFEs $h_{\tilde{A}} = \left(\mu_{\tilde{A}}(x), w_{\tilde{A}}(x), \nu_{\tilde{A}}(x)\right)$, $h_{\tilde{A}_1} = \left(\mu_{\tilde{A}_1}(x), w_{\tilde{A}_1}(x), \nu_{\tilde{A}_1}(x)\right)$, and $h_{\tilde{A}_2} = \left(\mu_{\tilde{A}_2}(x), w_{\tilde{A}_2}(x), \nu_{\tilde{A}_2}(x)\right)$. Then,

$$h_{\tilde{A}}^c = \left(\nu_{\tilde{A}}(x), w_{\tilde{A}}(x), \mu_{\tilde{A}}(x)\right), \tag{9.103}$$

$$h_{\tilde{A}_1} \cup h_{\tilde{A}_2} = \left(\mu_{\tilde{A}_1} \cup \mu_{\tilde{A}_2}, w_{\tilde{A}_1} \cap w_{\tilde{A}_2}, \nu_{\tilde{A}_1} \cap \nu_{\tilde{A}_2}\right), \tag{9.104}$$

$$h_{\tilde{A}_1} \cap h_{\tilde{A}_2} = \left(\mu_{\tilde{A}_1} \cap \mu_{\tilde{A}_2}, w_{\tilde{A}_1} \cup w_{\tilde{A}_2}, \nu_{\tilde{A}_1} \cup \nu_{\tilde{A}_2}\right), \tag{9.105}$$

$$h_{\tilde{A}_1} \oplus h_{\tilde{A}_2} = \left(\mu_{\tilde{A}_1} \oplus \mu_{\tilde{A}_2}, w_{\tilde{A}_1} \otimes w_{\tilde{A}_2}, \nu_{\tilde{A}_1} \otimes \nu_{\tilde{A}_2}\right), \tag{9.106}$$

$$h_{\tilde{A}_1} \otimes h_{\tilde{A}_2} = \left(\mu_{\tilde{A}_1} \otimes \mu_{\tilde{A}_2}, w_{\tilde{A}_1} \oplus w_{\tilde{A}_2}, \nu_{\tilde{A}_1} \oplus \nu_{\tilde{A}_2}\right), \tag{9.107}$$

where, $0 \le \left(\gamma_{\tilde{A}}^+\right)^2 + \left(\delta_{\tilde{A}}^+\right)^2 + \left(\eta_{\tilde{A}}^+\right)^2 \le 1$, $0 \le \left(\gamma_{\tilde{A}_1}^+\right)^2 + \left(\delta_{\tilde{A}_1}^+\right)^2 + \left(\eta_{\tilde{A}_1}^+\right)^2 \le 1$, and $0 \le \left(\gamma_{\tilde{A}_2}^+\right)^2 + \left(\delta_{\tilde{A}_2}^+\right)^2 + \left(\eta_{\tilde{A}_2}^+\right)^2 \le 1$.

Definition 9.47 [8] Assume the PIHFSs $\tilde{A}_1$ and $\tilde{A}_2$. Then,

$$\tilde{A}_1 \cup \tilde{A}_2 = \bigcup_{h_{\tilde{A}_1} \in \tilde{A}_1, h_{\tilde{A}_2} \in \tilde{A}_2} h_{\tilde{A}_1} \cup h_{\tilde{A}_2}, \tag{9.108}$$

$$\tilde{A}_1 \cap \tilde{A}_2 = \bigcup_{h_{\tilde{A}_1} \in \tilde{A}_1, h_{\tilde{A}_2} \in \tilde{A}_2} h_{\tilde{A}_1} \cap h_{\tilde{A}_2}, \tag{9.109}$$

$$\tilde{A}_1 \oplus \tilde{A}_2 = \bigcup_{h_{\tilde{A}_1} \in \tilde{A}_1, h_{\tilde{A}_2} \in \tilde{A}_2} h_{\tilde{A}_1} \oplus h_{\tilde{A}_2}, \tag{9.110}$$

$$\tilde{A}_1 \otimes \tilde{A}_2 = \bigcup_{h_{\tilde{A}_1} \in \tilde{A}_1, h_{\tilde{A}_2} \in \tilde{A}_2} h_{\tilde{A}_1} \otimes h_{\tilde{A}_2}. \tag{9.111}$$

9.10 Application: Hesitant Fuzzy TOPSIS

In this sub-section, the hesitant fuzzy TOPSIS proposed by Keikha [23] is presented. In this method, m alternatives are to be ranked according to n criteria (including a set of benefit criteria (B) and a set of cost criteria (C)) with weight vector of $w = (w_1, \ldots, w_n)$ and hesitant fuzzy decision matrix of $\tilde{D} = \left(d_{ij}, \left\{\gamma_{ij1}, \ldots, \gamma_{ijk_i}\right\}\right)_{m \times n}$. The steps of this method are as below.

Step 1. Obtain the adjusted hesitant fuzzy decision matrix $A\tilde{D} = \left(d_{ij}, \left\{\gamma_{ij}^{\sigma(1)}, \ldots, \gamma_{ij}^{\sigma(k)}\right\}\right)_{m\times n}$ where $\gamma_{ij}^{\sigma(1)} \leq \ldots \leq \gamma_{ij}^{\sigma(k)}$.

Step 2. Obtain the normalized decision matrix $N\tilde{D} = \left(n\tilde{h}\right)_{m\times n}$ as below,

$$\begin{aligned} n\tilde{h}_{ij} &= \left(nd_{ij}, \left\{\gamma_{ij}^{\sigma(1)}, \ldots, \gamma_{ij}^{\sigma(k)}\right\}\right) \\ &= \left\{\left(w_i \times \frac{d_j^{\max_{ij}}}{d_j^{\max_j^{\min}}, \left\{\gamma_{ij}^{\sigma(1)}, \ldots, \gamma_{ij}^{\sigma(k)}\right\}()}\right|\right. \\ &\quad \left.\left(w_i \times \frac{d_{ij} - d_j^{\min}}{d_j^{\max_j^{\min}}, \left\{\gamma_{ij}^{\sigma(1)}, \ldots, \gamma_{ij}^{\sigma(k)}\right\}()}\right)\right) \end{aligned}$$

Step 3. Determine the positive and negative ideal solutions respectively as below,

$$\begin{aligned} A^+ &= \left\{d_i^+ | i = 1, \ldots, m\right\} \\ &= \left\{\left(\max_i nd_{ij}, \{1, \ldots, 1 | j \in B\}\right) \& \left(\min_i nd_{ij} \bigcup_{r=1}^{k} \left\{\min_i \left\{\gamma_{ij}^{\sigma(r)}, 0.5 | j \in C\right\}\right\}\right)\right\}, \end{aligned}$$

$$\begin{aligned} A^- &= \left\{d_i^- | i = 1, \ldots, m\right\} \\ &= \left\{\left(\min_i nd_{ij}, \bigcup_{r=1}^{k} \left\{\min_i \left\{\gamma_{ij}^{\sigma(r)}, 0.5 | j \in B\right\}\right\}\right) \& \left(\max_i nd_{ij}, \{1, \ldots, 1 | j \in C\}\right)\right\}. \end{aligned}$$

Step 4. Determine the distance of each alternative from the positive and negative solutions of Step 3 using the Euclidean distance formula (9.6) as below.

$$S_i^+ = \sqrt{\sum_{j=1}^{n} \left(d_E\left(d_j^+, nd_{ij}\right)\right)^2} \quad i = 1, 2, \ldots, m,$$

$$S_i^- = \sqrt{\sum_{j=1}^{n} \left(d_E\left(d_j^-, nd_{ij}\right)\right)^2} \quad i = 1, 2, \ldots, m,$$

Step 5. Calculate the relative score of each alternative as $C_i = \frac{S_i^-}{S_i^+ + S_i^-}$.

Step 6. Rank the alternatives according to the increasing order of C_i values.

References

1. Torra, V.: Hesitant fuzzy sets. Int. J. Intell. Syst. **25**(6), 529–539 (2010)
2. Yu, D., Merigó, J.M., Xu, Y.: Group decision making in information systems security assessment using dual hesitant fuzzy set. Int. J. Intell. Syst. **31**(8), 786–812 (2016)
3. Xia, M., Xu, Z.: Hesitant fuzzy information aggregation in decision making. Int. J. Approximate Reasoning **52**(3), 395–407 (2011)
4. Liao, H., Xu, Z.: Subtraction and division operations over hesitant fuzzy sets. J. Intell. Fuzzy Syst. **27**(1), 65–72 (2014)
5. Chen, N., Xu, Z., Xia, M.: Correlation coefficients of hesitant fuzzy sets and their applications to clustering analysis. Appl. Math. Model. **37**(4), 2197–2211 (2013)
6. Farhadinia, B.: A novel method of ranking hesitant fuzzy values for multiple attribute decision-making problems. Int. J. Intell. Syst. **28**(8), 752–767 (2013)
7. Peng, J.J., Wang, J.Q., Wang, J., Yang, L.J., Chen, X.H.: An extension of ELECTRE to multi-criteria decision-making problems with multi-hesitant fuzzy sets. Inform. Sci. **307**, 113–126 (2015)
8. Farhadinia, B.: Hesitant fuzzy set. In: Hesitant Fuzzy Set: Theory and Extension. Springer Singapore, Singapore (2021)
9. Garg, H., Keikha, A., Mishmast Nehi, H.: Multiple-attribute decision-making problem using TOPSIS and Choquet integral with hesitant fuzzy number information. Math. Probl. Eng. **2020**(1), 9874951 (2020)
10. Keikha, A.: Generalized hesitant fuzzy numbers: introducing, arithmetic operations, aggregation operators, and an application. Int. J. Intell. Syst. **36**(12), 7709–7730 (2021)
11. Deli, I.: A TOPSIS method by using generalized trapezoidal hesitant fuzzy numbers and application to a robot selection problem. J. Intell. Fuzzy Syst. **38**(1), 779–793 (2020)
12. Deli, I., Karaaslan, F.: Generalized trapezoidal hesitant fuzzy numbers and their applications to multi criteria decision-making problems. Soft. Comput. **25**(2), 1017–1032 (2021)
13. Chen, N., Xu, Z., Xia, M.: Interval-valued hesitant preference relations and their applications to group decision making. Knowl.-Based Syst. **37**, 528–540 (2013)
14. Xu, Z.S., Da, Q.L.: The uncertain OWA operator. Int. J. Intell. Syst. **17**(6), 569–575 (2002)
15. Ye, J.: Multiple-attribute decision-making method under a single-valued neutrosophic hesitant fuzzy environment. J. Intell. Syst. **24**(1), 23–36 (2015)
16. Akram, M., Naz, S., Smarandache, F.: Generalization of maximizing deviation and TOPSIS method for MADM in simplified neutrosophic hesitant fuzzy environment. Symmetry **11**(8), 1058 (2019)
17. Liu, P., Shi, L.: The generalized hybrid weighted average operator based on interval neutrosophic hesitant set and its application to multiple attribute decision making. Neural Comput. Appl. **26**(2), 457–471 (2015)
18. Khan, M.S.A., Abdullah, S., Ali, A., Siddiqui, N., Amin, F.: Pythagorean hesitant fuzzy sets and their application to group decision making with incomplete weight information. J. Intell. Fuzzy Syst. **33**(6), 3971–3985 (2017)
19. Ghosh, S., Küfer, K.H., Roy, S.K., Weber, G.W.: Carbon mechanism on sustainable multi-objective solid transportation problem for waste management in Pythagorean hesitant fuzzy environment. Complex Intell. Syst. **8**(5), 4115–4143 (2022)
20. Wang, L., Wang, H., Xu, Z., Ren, Z.: The interval-valued hesitant Pythagorean fuzzy set and its applications with extended TOPSIS and Choquet integral-based method. Int. J. Intell. Syst. **34**(6), 1063–1085 (2019)
21. Zhang, M., Zheng, T., Zheng, W., Zhou, L.: Interval-valued pythagorean hesitant fuzzy set and its application to multiattribute group decision-making. Complexity **2020**(1), 1724943 (2020)
22. Wang, R., Li, Y.: Picture hesitant fuzzy set and its application to multiple criteria decision-making. Symmetry **10**(7), 295 (2018)
23. Keikha, A.: New extension of TOPSIS method for solving inaccurate MADM problems modeled with hesitant fuzzy numbers. J. Decisions Oper. Res. **7**(1), 1–16 (2022)

Chapter 10
Bipolar Fuzzy Sets

10.1 Introduction

The basic concept of bipolar fuzzy set (BFS) was introduced by Zhang [1] for the first time. It is an extension of classical fuzzy sets where instead of one membership function, two membership functions are used. In a bipolar fuzzy set, there are two membership functions to represent positive and negative aspects of a fuzzy set separately. This property can help to reflect the nature and concept of a fuzzy set more efficiently.

Todays, bipolar fuzzy sets are widely applied in decision-making, expert systems, data mining, scoring functions, etc. In this section the concepts of bipolar fuzzy sets and their operations are presented.

10.2 Bipolar Fuzzy Set

In this sub-section some basic concepts of bipolar fuzzy sets are introduced.

Definition 10.1 A bipolar fuzzy set (BFS) such as $\tilde{B}$ on the non-empty reference set X is defined as below.

$$\tilde{B} = \left\{ \left(x, \mu_{\tilde{B}}^{+}(x), \mu_{\tilde{B}}^{-}(x)\right) \middle| x \in X \right\}, \tag{10.1}$$

Here, $\mu_{\tilde{B}}^{+}(x) : X \rightarrow [0, 1]$ is the satisfaction degree (positive membership degree) of x and $\mu_{\tilde{B}}^{-}(x) : X \rightarrow [-1, 0]$ is the dissatisfaction degree (negative membership degree) of element x. Satisfaction degree is between 0 and 1, and its greater value reflects more satisfaction. On the other hand, dissatisfaction degree is between 0 and − 1, and its less value (close to − 1) reflects more dissatisfaction.

A. Mahmoodirad et al., *Recent Advances on Fuzzy Sets*, Studies in Fuzziness and Soft Computing 440, https://doi.org/10.1007/978-3-032-08776-8_10

Example 10.1 Assume the reference set $X = \{x_1, x_2, x_3\}$ and the BFS $\tilde{B} = \{(x_1, 0.8, -0.2), (x_2, 0.4, -0.5), (x_3, 0, -1)\}$ defined on this set. For example, for x_1 the positive membership degree of 0.8 and negative membership degree of – 0.2 is considered by this BFS.

Definition 10.2 [2] Consider the BFSs $\tilde{B}_1 = \left\{\left(x, \mu^+_{\tilde{B}_1}(x), \mu^-_{\tilde{B}_1}(x)\right)\middle| x \in X\right\}$, and $\tilde{B}_2 = \left\{\left(x, \mu^+_{\tilde{B}_2}(x), \mu^-_{\tilde{B}_2}(x)\right)\middle| x \in X\right\}$, on the reference set X. Then,

- $\tilde{B}^c_1 = \left\{\left(x, 1 - \mu^+_{\tilde{B}_1}(x), \left|\mu^-_{\tilde{B}_1}(x)\right| - 1\right)\middle| x \in X\right\}$,
- $\tilde{B}_1 \cup \tilde{B}_2 = \left\{\left(x, \max\left\{\mu^+_{\tilde{B}_1}(x), \mu^+_{\tilde{B}_2}(x)\right\}, \min\left\{\mu^-_{\tilde{B}_1}(x), \mu^-_{\tilde{B}_2}(x)\right\}\right)\middle| x \in X\right\}$,
- $\tilde{B}_1 \cap \tilde{B}_2 = \left\{\left(x, \min\left\{\mu^+_{\tilde{B}_1}(x), \mu^+_{\tilde{B}_2}(x)\right\}, \max\left\{\mu^-_{\tilde{B}_1}(x), \mu^-_{\tilde{B}_2}(x)\right\}\right)\middle| x \in X\right\}$,
- $\tilde{B}_1 \oplus \tilde{B}_2 = \left\{\left(x, \mu^+_{\tilde{B}_1}(x) + \mu^+_{\tilde{B}_2}(x) - \mu^+_{\tilde{B}_1}(x) \times \mu^+_{\tilde{B}_2}(x), -\left|\mu^-_{\tilde{B}_1}(x)\right| \times \left|\mu^-_{\tilde{B}_2}(x)\right|\right)\middle| x \in X\right\}$,
- $\tilde{B}_1 \otimes \tilde{B}_2 = \left\{\left(x, \mu^+_{\tilde{B}_1}(x) \times \mu^+_{\tilde{B}_2}(x), \mu^-_{\tilde{B}_1}(x) + \mu^-_{\tilde{B}_2}(x) - \mu^-_{\tilde{B}_1}(x) \times \mu^-_{\tilde{B}_2}(x)\right)\middle| x \in X\right\}$,
- $k\tilde{B}_1 = \left\{\left(x, 1 - \left(1 - \mu^+_{\tilde{B}_1}(x)\right)^k, -\left|\mu^-_{\tilde{B}_1}(x)\right|^k\right)\middle| x \in X\right\}, k > 0$,
- $\tilde{B}^k_1 = \left\{\left(x, \left(\mu^+_{\tilde{B}_1}(x)\right)^k, -1 + \left|1 + \mu^-_{\tilde{B}_1}(x)\right|^k\right)\middle| x \in X\right\}, k > 0$.

Definition 10.3 [3] A BFS on the reference set X is a bipolar fuzzy number (BFN) if it meets the below conditions.

- $\mu^+_{\tilde{B}}(x)$ and $\mu^-_{\tilde{B}}(x)$ are piece-wise continuous functions.
- $\tilde{B}$ is convex on X.
- $\tilde{B}$ is a normal BFS. Meaning that there exist $x_1, x_2 \in X$ such that $\mu^+_{\tilde{B}}(x_1) = 1$ and $\mu^-_{\tilde{B}}(x_2) = -1$.

Definition 10.4 [4] A trapezoidal bipolar fuzzy number (TrBFN) is shown as $\tilde{B} = (\langle a_1, a_2, a_3, a_4\rangle, \langle b_1, b_2, b_3, b_4\rangle)$, and its satisfaction and dissatisfaction degrees are determined as below.

$$\mu^+_{\tilde{B}}(x) = \begin{cases} \frac{x-a_1}{a_2-a_1}, a_1 \le x \le a_2, \\ 1, a_2 \le x \le a_3, \\ \frac{a_4-x}{a_4-a_3}, a_3 \le x \le a_4, \\ 0, O.W. \end{cases}$$

$$\mu^-_{\tilde{B}}(x) = \begin{cases} \frac{b_1-x}{b_2-b_1}, b_1 \le x \le b_2, \\ -1, b_2 \le x \le b_3, \\ \frac{x-b_4}{b_4-b_3}, b_3 \le x \le b_4, \\ 0, O.W. \end{cases}$$

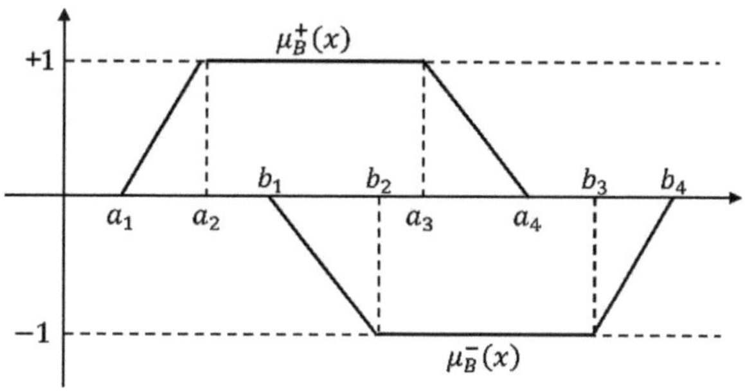

Fig. 10.1 Schematic representation of a TrBFN

A TrBFN is represented by Fig. 10.1.

In the TrBFN $\tilde{B} = (\langle a_1, a_2, a_3, a_4\rangle, \langle b_1, b_2, b_3, b_4\rangle)$ if $a_2 = a_3$ and $b_2 = b_3$, it is a triangular bipolar fuzzy number (TBFN).

Definition 10.5 [4] Assume $H = \{\tilde{h}_1, \ldots, \tilde{h}_n\}$ as a set of BFNs. Then, for $\tilde{h}_i$ and $\tilde{h}_j$ from H, the ranking function $\Re : H \to \mathbb{R}$ results in the below rankings,

- $\Re(\tilde{h}_i) < \Re(\tilde{h}_j) \Rightarrow \tilde{h}_i \prec \tilde{h}_j$,
- $\Re(\tilde{h}_i) > \Re(\tilde{h}_j) \Rightarrow \tilde{h}_i \succ \tilde{h}_j$,
- $\Re(\tilde{h}_i) = \Re(\tilde{h}_j) \Rightarrow \tilde{h}_i \approx \tilde{h}_j$,

where, this ranking function for the BFN $\tilde{B} = (\langle a_1, a_2, a_3, a_4\rangle, \langle b_1, b_2, b_3, b_4\rangle)$ is defined as below.

$$\Re(\tilde{B}) = \left(\frac{a_1 + a_2 + a_3 + a_4}{4} + \frac{-a_1 - a_2 + a_3 + a_4}{2}\right) - \left(\frac{b_1 + b_2 + b_3 + b_4}{4} + \frac{-b_1 - b_2 + b_3 + b_4}{2}\right). \tag{10.2}$$

Example 10.2 [4] Assume the TrBFNs $\tilde{B}_1 = (\langle 20, 40, 50, 70\rangle, \langle 40, 60, 70, 90\rangle)$ and $\tilde{B}_1 = (\langle 20, 40, 50, 70\rangle, \langle 40, 60, 70, 90\rangle)$. Using the ranking function (10.2), the values of $\Re(\tilde{B}_1) = -20$ and $\Re(\tilde{B}_2) = -12.5$ are calculated. Therefore, as $\Re(\tilde{B}_1) = -20 < \Re(\tilde{B}_2) = -12.5$, then $\tilde{B}_1 \prec \tilde{B}_2$.

Definition 10.6 [5] Assume E as the set of all triangular bipolar fuzzy numbers, and $\tilde{B} = (\langle a_1, a_2, a_3\rangle, \langle b_1, b_2, b_3\rangle)$ as a TBFN. The function $\varphi : E \to \{-1, 1\}$ is defined as below.

$$\varphi(\tilde{B}) = \begin{cases} 1, a_1 + a_2 + a_3 + b_1 + b_2 + b_3 \geq 0, \\ -1, a_1 + a_2 + a_3 + b_1 + b_2 + b_3 < 0, \end{cases} \tag{10.3}$$

In addition, the NFR index is defined as below and is used to rank triangular bipolar fuzzy numbers.

$$NFR(\tilde{B}) = \left(\frac{a_1 + a_2 + a_3 + b_1 + b_2 + b_3}{3}, a_2 + b_2, \frac{\varphi(\tilde{B}) \times (a_3 - a_1 + b_3 - b_1)}{2}\right) \in \mathbb{R}^3 \quad (10.4)$$

As defuzzification based ranking methods lose the fuzzy information of the number in their ranking procedure, applying the NFR index, Babakordi and Taghi-Nezhad [5] proposed the below logics in order to overcome this shortcoming. Therefore, for ranking the TBFNs $\tilde{B} = (\langle a_1, a_2, a_3\rangle, \langle b_1, b_2, b_3\rangle)$ and $\tilde{D} = (\langle c_1, c_2, c_3\rangle, \langle d_1, d_2, d_3\rangle)$, the below logics are proposed by Babakordi and Taghi-Nezhad [5].

- If $NFR(\tilde{B}) <_L NFR(\tilde{D})$, then $\tilde{B} \prec \tilde{D}$.
- If $NFR(\tilde{B}) =_L NFR(\tilde{D})$, then $\tilde{B} \approx \tilde{D}$.

Here, the sign $<_L$ is the alphabetical order in comparing the NFRs.

Theorem 10.1 [5] *Assume the TBFNs* $\tilde{A}$ and $\tilde{B}$. *Then,*

- $NFR(-\tilde{B}) = -NFR(\tilde{B})$,
- *if* $\tilde{A} < \tilde{B}$ *then* $-\tilde{B} < -\tilde{A}$,
- *the index NFR is linear, meaning that if* $\alpha > 0$, *then* $NFR(\alpha\tilde{A} \oplus \tilde{B}) = \alpha NFR(\tilde{A}) \oplus NFR(\tilde{B})$.

Example 10.3 [5] Consider the below TBFNs.

$$\begin{aligned}
\tilde{A} &= (30, 50, 60, 30, 50, 60)\\
\tilde{B} &= (40, 60, 70, 40, 60, 70)\\
\tilde{C} &= (10, 20, 30, 10, 20, 30)\\
\tilde{D} &= (1, 2, 3, 1, 2, 3)\\
\tilde{E} &= (50, 70, 90, 50, 70, 90)\\
\tilde{F} &= (20, 50, 70, 20, 50, 70)
\end{aligned}$$

The ranking function values of Table 10.1 are obtained for these numbers.

According to the values of Table 10.1, the NFR index and the ranking logic of Definition 10.6 result in the ranking of $\tilde{D} \prec \tilde{C} \prec \tilde{A} \prec \tilde{F} \prec \tilde{B} \prec \tilde{E}$. But, the method of Akram and Arshad [4] cannot introduce a ranking as all the ranking function values are equal to zero.

Definition 10.7 An intuitionistic bipolar fuzzy set such as $\tilde{B}$ on the reference set X is defined as below.

$$\tilde{B} = \left\{ \left(x, \mu^+_{\tilde{B}}(x), \nu^+_{\tilde{B}}(x), \mu^-_{\tilde{B}}(x), \nu^-_{\tilde{B}}(x)\right) \middle| x \in X \right\} \quad (10.5)$$

Table 10.1 The ranking function values obtained for the TBFNs of Example 10.3

TBFN	NFR index	Akram and Arshad [4]
$\tilde{A}$	(93.33, 100, 30)	$\Re(\tilde{A}) = 0$
$\tilde{B}$	(113.33, 120, 30)	$\Re(\tilde{B}) = 0$
$\tilde{C}$	(40, 40, 20)	$\Re(\tilde{C}) = 0$
$\tilde{D}$	(4, 4, 2)	$\Re(\tilde{D}) = 0$
$\tilde{E}$	(140, 140, 40)	$\Re(\tilde{E}) = 0$
$\tilde{F}$	(93.33, 100, 50)	$\Re(\tilde{F}) = 0$

Here, $\mu_{\tilde{B}}^{+}(x), \nu_{\tilde{B}}^{+}(x) : X \to [0, 1]$, $\mu_{\tilde{B}}^{-}(x), \nu_{\tilde{B}}^{-}(x) : X \to [-1, 0]$, $0 \le \mu_{\tilde{B}}^{+}(x) + \nu_{\tilde{B}}^{+}(x) \le 1$, and $-1 \le \mu_{\tilde{B}}^{-}(x) + \nu_{\tilde{B}}^{-}(x) \le 0$.

Definition 10.8 [6] A Pythagorean bipolar fuzzy set such as $\tilde{P}$ on the reference set X is defined as below.

$$\tilde{P} = \left\{ \left(x, \mu_{\tilde{P}}^{+}(x), \nu_{\tilde{P}}^{+}(x), \mu_{\tilde{P}}^{-}(x), \nu_{\tilde{P}}^{-}(x)\right) \middle| x \in X \right\} \tag{10.6}$$

Here, $\mu_{\tilde{B}}^{+}(x), \nu_{\tilde{B}}^{+}(x) : X \to [0, 1]$, $\mu_{\tilde{B}}^{-}(x), \nu_{\tilde{B}}^{-}(x) : X \to [-1, 0]$, $0 \le \left(\mu_{\tilde{B}}^{+}(x)\right)^2 + \left(\nu_{\tilde{B}}^{+}(x)\right)^2 \le 1$, and $0 \le \left(\mu_{\tilde{B}}^{-}(x)\right)^2 + \left(\nu_{\tilde{B}}^{-}(x)\right)^2 \le 1$.

Example 10.4 [6] Assume the reference set $X = \{x_1, x_2, x_3\}$. Then the set $\tilde{P}$ represented as below is a Pythagorean bipolar fuzzy set.

$$\begin{aligned}\tilde{P} = \{&(x_1, 0.6, 0.3, -0.6, -0.4), (x_2, 0.3, 0.3, -0.2, -0.3)\\ &(x_3, 0.5, 0.6, -0.5, -0.2)\}.\end{aligned}$$

Definition 10.9 A spherical bipolar fuzzy set such as $\tilde{B}$ on the reference set X is defined as below.

$$\tilde{B} = \left\{ \left(x, \mu_{\tilde{B}}^{+}(x), \nu_{\tilde{B}}^{+}(x), w_{\tilde{B}}^{+}(x), \mu_{\tilde{B}}^{-}(x), \nu_{\tilde{B}}^{-}(x), w_{\tilde{B}}^{-}(x)\right) \middle| x \in X \right\} \tag{10.7}$$

Here, $\mu_{\tilde{B}}^{+}(x), \nu_{\tilde{B}}^{+}(x), w_{\tilde{B}}^{+}(x) : \mathbb{R} \to [0, 1]$, $\mu_{\tilde{B}}^{-}(x), \nu_{\tilde{B}}^{-}(x), w_{\tilde{B}}^{-}(x) : \mathbb{R} \to [-1, 0]$, $0 \le \left(\mu_{\tilde{B}}^{+}(x)\right)^2 + \left(\nu_{\tilde{B}}^{+}(x)\right)^2 + \left(w_{\tilde{B}}^{+}(x)\right)^2 \le 1$, and $0 \le \left(\mu_{\tilde{B}}^{-}(x)\right)^2 + \left(\nu_{\tilde{B}}^{-}(x)\right)^2 + \left(w_{\tilde{B}}^{-}(x)\right)^2 \le 1$. For more information about spherical bipolar fuzzy sets, the study of Princy and Mohana [7] can be referred to.

10.3 Neutrosophic Bipolar Fuzzy Set

Definition 10.10 [8] A neutrosophic bipolar fuzzy set such as $\tilde{B}$ on the reference set X is defined as below.

$$\tilde{B} = \left\{ \left(x, T_{\tilde{B}}^{+}(x), I_{\tilde{B}}^{+}(x), F_{\tilde{B}}^{+}(x), T_{\tilde{B}}^{-}(x), I_{\tilde{B}}^{-}(x), F_{\tilde{B}}^{-}(x)\right) \middle| x \in X \right\} \tag{10.8}$$

Here, $T_{\tilde{B}}^{+}(x), I_{\tilde{B}}^{+}(x), F_{\tilde{B}}^{+}(x) : \mathbb{R} \to [0, 1]$ and $T_{\tilde{B}}^{-}(x), I_{\tilde{B}}^{-}(x), F_{\tilde{B}}^{-}(x) : \mathbb{R} \to [-1, 0]$.

Example 10.5 [8] Considering the reference set $X = \{x_1, x_2, x_3\}$, the below set is a neutrosophic bipolar fuzzy set.

$$\tilde{B} = \left\{ \begin{array}{l} (x_1, 0.5, 0.3, 0.1, -0.6, -0.4, -0.01), \\ (x_2, 0.3, 0.2, 0.7, -0.02, -0.03, -0.5), \\ (x_3, 0.8, 0.05, 0.4, -0.1, -0.5, -0.06) \end{array} \right\}$$

Definition 10.11 [8] Assume the neutrosophic bipolar fuzzy sets $\tilde{B}_1 = \left\{ \left(x, T_{\tilde{B}_1}^{+}(x), I_{\tilde{B}_1}^{+}(x), F_{\tilde{B}_1}^{+}(x), T_{\tilde{B}_1}^{-}(x), I_{\tilde{B}_1}^{-}(x), F_{\tilde{B}_1}^{-}(x)\right) \middle| x \in X \right\}$ and $\tilde{B}_2 = \left\{ \left(x, T_{\tilde{B}_2}^{+}(x), I_{\tilde{B}_2}^{+}(x), F_{\tilde{B}_2}^{+}(x), T_{\tilde{B}_2}^{-}(x), I_{\tilde{B}_2}^{-}(x), F_{\tilde{B}_2}^{-}(x)\right) \middle| x \in X \right\}$. Then,

- $\tilde{B}_1^c = \left\{ \left(\begin{array}{l} x, \{1^+\} - T_{\tilde{B}_1}^{+}(x), \{1^+\} - I_{\tilde{B}_1}^{+}(x), \{1^+\} - F_{\tilde{B}_1}^{+}(x), \\ \{1^-\} - T_{\tilde{B}_1}^{-}(x), \{1^-\} - I_{\tilde{B}_1}^{-}(x), \{1^-\} - F_{\tilde{B}_1}^{-}(x) \end{array} \right) \middle| x \in X \right\}$,
- $\tilde{B}_1 \subseteq \tilde{B}_2$ if and only if $T_{\tilde{B}_1}^{+}(x) \le T_{\tilde{B}_2}^{+}(x), I_{\tilde{B}_1}^{+}(x) \ge I_{\tilde{B}_2}^{+}(x), F_{\tilde{B}_1}^{+}(x) \ge F_{\tilde{B}_2}^{+}(x)$ and $T_{\tilde{B}_1}^{+}(x) \ge T_{\tilde{B}_2}^{+}(x), I_{\tilde{B}_1}^{+}(x) \le I_{\tilde{B}_2}^{+}(x), F_{\tilde{B}_1}^{+}(x) \le F_{\tilde{B}_2}^{+}(x)$, for all $x \in X$,
- $\tilde{B}_1 = \tilde{B}_2$ if and only if $T_{\tilde{B}_1}^{+}(x) = T_{\tilde{B}_2}^{+}(x), I_{\tilde{B}_1}^{+}(x) = I_{\tilde{B}_2}^{+}(x), F_{\tilde{B}_1}^{+}(x) = F_{\tilde{B}_2}^{+}(x)$ and $T_{\tilde{B}_1}^{+}(x) = T_{\tilde{B}_2}^{+}(x), I_{\tilde{B}_1}^{+}(x) = I_{\tilde{B}_2}^{+}(x), F_{\tilde{B}_1}^{+}(x) = F_{\tilde{B}_2}^{+}(x)$, for all $x \in X$,
- $\tilde{B}_1 \cup \tilde{B}_2 = \left\{ \left(\begin{array}{l} x, \max\left\{T_{\tilde{B}_1}^{+}(x), T_{\tilde{B}_2}^{+}(x)\right\}, \\ \frac{T_{\tilde{B}_1}^{+}(x)+T_{\tilde{B}_2}^{+}(x)}{2}, \min\left\{F_{\tilde{B}_1}^{+}(x), F_{\tilde{B}_2}^{+}(x)\right\}, \\ \min\left\{T_{\tilde{B}_1}^{-}(x), T_{\tilde{B}_2}^{-}(x)\right\}, \\ \frac{T_{\tilde{B}_1}^{-}(x)+T_{\tilde{B}_2}^{-}(x)}{2}, \max\left\{F_{\tilde{B}_1}^{-}(x), F_{\tilde{B}_2}^{-}(x)\right\} \end{array} \right) \middle| x \in X \right\}$,
- $\tilde{B}_1 \cap \tilde{B}_2 = \left\{ \left(\begin{array}{l} x, \min\left\{T_{\tilde{B}_1}^{+}(x), T_{\tilde{B}_2}^{+}(x)\right\}, \\ \frac{T_{\tilde{B}_1}^{+}(x)+T_{\tilde{B}_2}^{+}(x)}{2}, \max\left\{F_{\tilde{B}_1}^{+}(x), F_{\tilde{B}_2}^{+}(x)\right\}, \\ \max\left\{T_{\tilde{B}_1}^{-}(x), T_{\tilde{B}_2}^{-}(x)\right\}, \\ \frac{T_{\tilde{B}_1}^{-}(x)+T_{\tilde{B}_2}^{-}(x)}{2}, \min\left\{F_{\tilde{B}_1}^{-}(x), F_{\tilde{B}_2}^{-}(x)\right\} \end{array} \right) \middle| x \in X \right\}$

Example 10.6 [8] Consider the below neutrosophic bipolar fuzzy sets on the reference set $X = \{x_1, x_2, x_3\}$.

$$\tilde{B}_1 = \left\{ \begin{array}{l} (x_1, 0.5, 0.3, 0.1, -0.6, -0.4, -0.01), \\ (x_2, 0.3, 0.2, 0.7, -0.02, -0.03, -0.5), \\ (x_3, 0.8, 0.05, 0.4, -0.1, -0.5, -0.06) \end{array} \right\}$$

$$\tilde{B}_2 = \left\{ \begin{array}{l} (x_1, 0.4, 0.6, 0.3, -0.3, -0.5, -0.1), \\ (x_2, 0.5, 0.1, 0.4, -0.2, -0.3, -0.3), \\ (x_3, 0.2, 0.5, 0.6, -0.4, -0.6, -0.7) \end{array} \right\}$$

Then,

- $\tilde{B}_1^c = \left\{ \begin{array}{l} (x_1, 0.5, 0.7, 0.9, -0.4, -0.6, -0.99), \\ (x_2, 0.7, 0.8, 0.3, -0.98, -0.97, -0.5), \\ (x_3, 0.2, 0.95, 0.6, -0.9, -0.5, -0.94) \end{array} \right\}$,
- $\tilde{B}_1 \cup \tilde{B}_2 = \left\{ \begin{array}{l} (x_1, 0.5, 0.45, 0.2, -0.6, -0.45, -0.1), \\ (x_2, 0.5, 0.15, 0.7, -0.2, -0.152, -0.5), \\ (x_3, 0.8, 0.47, 0.6, -0.4, -0.55, -0.7) \end{array} \right\}$,
- $\tilde{B}_1 \cap \tilde{B}_2 = \left\{ \begin{array}{l} (x_1, 0.4, 0.65, 0.9, -0.6, -0.45, -0.01), \\ (x_2, 0.3, 0.45, 0.7, -0.2, -0.152, -0.3), \\ (x_3, 0.2, 0.5, 0.6, -0.4, -0.55, -0.06) \end{array} \right\}$.

For more simplification, a neutrosophic bipolar fuzzy number (NBFN) is represented as $\tilde{b} = \left(T^+, I^+, F^+, T^-, I^-, F^-\right)$.

Definition 10.12 [8] Consider the NBFNs $\tilde{b}_1 = \left(T_1^+, I_1^+, F_1^+, T_1^-, I_1^-, F_1^-\right)$ and $\tilde{b}_2 = \left(T_2^+, I_2^+, F_2^+, T_2^-, I_2^-, F_2^-\right)$. Further assume $k > 0$. Then,

- $k\tilde{b}_1 = \left(1 - \left(1 - T_1^+\right)^k, \left(I_1^+\right)^k, \left(F_1^+\right)^k, -\left(-T_1^-\right)^k, -\left(-I_1^-\right)^k, -\left(1 - \left(1 - \left(-F_1^-\right)\right)^k\right)\right)$,
- $\left(\tilde{b}_1\right)^k = \left(\left(T_1^+\right)^k, 1 - \left(1 - I_1^+\right)^k, 1 - \left(1 - F_1^+\right)^k, -\left(1 - \left(1 - \left(-T_1^-\right)\right)^k\right), -\left(-I_1^-\right)^k, -\left(-F_1^-\right)^k\right)$,
- $\tilde{b}_1 \oplus \tilde{b}_2 = \left(T_1^+ + T_2^+ - T_1^+T_2^+, I_1^+I_2^+, F_1^+F_2^+, -T_1^-T_2^-, -\left(-I_1^- - I_2^- - I_1^-I_2^-\right), -\left(-F_1^- - F_2^- - F_1^-F_2^-\right)\right)$,
- $\tilde{b}_1 \otimes \tilde{b}_2 = \left(T_1^+T_2^+, I_1^+ + I_2^+ - I_1^+I_2^+, F_1^+ + F_2^+ - F_1^+F_2^+, -\left(-T_1^- - T_2^- - T_1^-T_2^-\right), -I_1^-I_2^-, -F_1^-F_2^-\right)$.

Definition 10.13 [8] Assume the NBFN $\tilde{b} = \left(T^+, I^+, F^+, T^-, I^-, F^-\right)$. For this number, the below score, accuracy, and certainty functions are defined respectively.

$$S\left(\tilde{b}\right) = \frac{3 + T^+ - I^+ - F^+ + T^- - I^- - F^-}{6} \tag{10.9}$$

$$A\left(\tilde{b}\right) = T^+ - F^+ + T^- - F^-, \tag{10.10}$$

$$C\left(\tilde{b}\right) = T^+ - F^-, \tag{10.11}$$

Definition 10.14 [8] Consider the NBFNs $\tilde{b}_1 = \left(T_1^+, I_1^+, F_1^+, T_1^-, I_1^-, F_1^-\right)$ and $\tilde{b}_2 = \left(T_2^+, I_2^+, F_2^+, T_2^-, I_2^-, F_2^-\right)$. In order to compare and rank these numbers, the below rules are considered.

- If $S(\tilde{b}_1) > S(\tilde{b}_2) \Rightarrow \tilde{b}_1 \succ \tilde{b}_2$
- If $S(\tilde{b}_1) = S(\tilde{b}_2)$ and $A(\tilde{b}_1) > A(\tilde{b}_2)$, then $\tilde{b}_1 \prec \tilde{b}_2$.
- If $S(\tilde{b}_1) = S(\tilde{b}_2), A(\tilde{b}_1) = A(\tilde{b}_2)$, and $C(\tilde{b}_1) > C(\tilde{b}_2)$, then $\tilde{b}_1 \succ \tilde{b}_2$.
- If $S(\tilde{b}_1) = S(\tilde{b}_2), A(\tilde{b}_1) = A(\tilde{b}_2)$, and $C(\tilde{b}_1) = C(\tilde{b}_2)$, then $\tilde{b}_1 \approx \tilde{b}_2$.

Definition 10.15 [8] For the set of NBFNs $\tilde{a}_j = \left(T_j^+, I_j^+, F_j^+, T_j^-, I_j^-, F_j^-\right), j = 1, \ldots, n$, the below neutrosophic bipolar fuzzy weighted arithmetic (NBFWA) and geometric (NBFWG) aggregators are defined.

$$BNFWA(\tilde{a}_1, \ldots, \tilde{a}_n) = \oplus_{j=1}^n w_j \tilde{a}_j = \begin{pmatrix} 1 - \prod_{j=1}^n \left(1 - T_j^+\right)^{w_j}, \prod_{j=1}^n \left(I_j^+\right)^{w_j}, \\ \prod_{j=1}^n \left(F_j^+\right)^{w_j}, -\prod_{j=1}^n \left(-T_j^-\right)^{w_j}, \\ -\left(1 - \prod_{j=1}^n \left(1 - \left(-I_j^-\right)\right)^{w_j}\right), -\left(1 - \prod_{j=1}^n \left(1 - \left(-F_j^-\right)\right)^{w_j}\right) \end{pmatrix} \tag{10.12}$$

$$BNFWG(\tilde{a}_1, \ldots, \tilde{a}_n) = \prod_{j=1}^n w_j \tilde{a}_j = \begin{pmatrix} \prod_{j=1}^n \left(T_j^+\right)^{w_j}, 1 - \prod_{j=1}^n \left(1 - I_j^+\right)^{w_j}, \\ 1 - \prod_{j=1}^n \left(1 - F_j^+\right)^{w_j}, \\ -\left(1 - \prod_{j=1}^n \left(1 - T_j^-\right)^{w_j}\right), -\left(\prod_{j=1}^n \left(-I_j^-\right)^{w_j}\right), -\left(\prod_{j=1}^n \left(-F_j^-\right)^{w_j}\right) \end{pmatrix} \tag{10.13}$$

Here, $w_j \in [0, 1]$ is the weight of NBFN $\tilde{a}_j$, and $\sum_{j=1}^n w_j = 1$.

Definition 10.16 [9] A single-valued neutrosophic bipolar fuzzy number (SVNBFN) is shown as $\tilde{b} = \left\{(b_1, b_2, b_3), \mu_{\tilde{b}}, \nu_{\tilde{b}}, w_{\tilde{b}}\right\}$. It follows the conditions $b_1, b_2, b_3 \in \mathbb{R}$, $b_1 \le b_2 \le b_3$, and $\mu_{\tilde{b}}, \nu_{\tilde{b}}, w_{\tilde{b}} \in [0, 1]$. This number is shown by Fig. 10.2 and its truth, indeterminacy, and falsity membership functions are defined as below.

$$T_{\tilde{b}}^+(x) = \begin{cases} \frac{x-b_1}{b_2-b_1}\mu_{\tilde{b}}, b_1 \le x \le b_2, \\ \mu_{\tilde{b}}, x = b_2, \\ \frac{b_3-x}{b_3-b_2}\mu_{\tilde{b}}, b_2 \le x \le b_3, \\ 0, O.W. \end{cases} \quad T_{\tilde{b}}^-(x) = \begin{cases} \frac{b_1-x}{b_2-b_1}\mu_{\tilde{b}}, b_1 \le x \le b_2, \\ -\mu_{\tilde{b}}, x = b_2, \\ \frac{x-b_3}{b_3-b_2}\mu_{\tilde{b}}, b_2 \le x \le b_3, \\ 0, O.W. \end{cases}$$

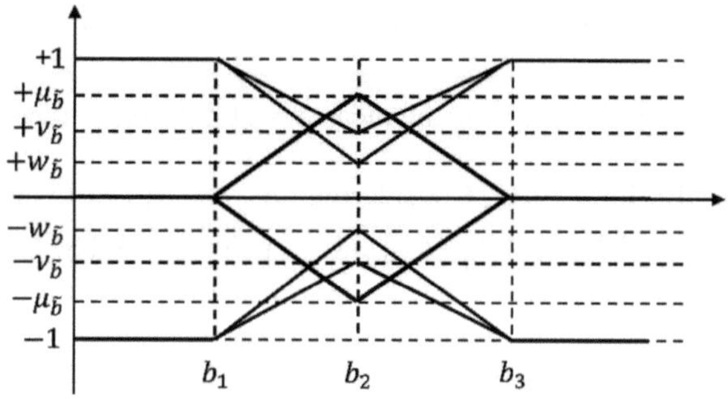

Fig. 10.2 The schematic of a single-valued triangular neutrosophic bipolar fuzzy number

$$I_{\tilde{b}}^{+}(x)=\begin{cases}\frac{b_2-x+(x-b_1)\nu_{\tilde{b}}}{b_2-b_1}, b_1\le x\le b_2,\\ \nu_{\tilde{b}}, x=b_2,\\ \frac{x-b_2+(b_3-x)\nu_{\tilde{b}}}{b_3-b_2}, b_2\le x\le b_3,\\ 1, O.W.\end{cases}\qquad I_{\tilde{b}}^{-}(x)=\begin{cases}\frac{x-b_2+(b_1-x)\nu_{\tilde{b}}}{b_2-b_1}, b_1\le x\le b_2,\\ -\nu_{\tilde{b}}, x=b_2,\\ \frac{b_2-x+(x-b_3)\nu_{\tilde{b}}}{b_3-b_2}, b_2\le x\le b_3,\\ -1, O.W.\end{cases}$$

$$F_{\tilde{b}}^{+}(x)=\begin{cases}\frac{b_2-x+(x-b_1)w_{\tilde{b}}}{b_2-b_1}, b_1\le x\le b_2,\\ w_{\tilde{b}}, x=b_2,\\ \frac{x-b_2+(b_3-x)w_{\tilde{b}}}{b_3-b_2}, b_2\le x\le b_3,\\ 1, O.W.\end{cases}\qquad F_{\tilde{b}}^{-}(x)=\begin{cases}\frac{x-b_2+(b_1-x)w_{\tilde{b}}}{b_2-b_1}, b_1\le x\le b_2,\\ -w_{\tilde{b}}, x=b_2,\\ \frac{b_2-x+(x-b_3)w_{\tilde{b}}}{b_3-b_2}, b_2\le x\le b_3,\\ -1, O.W.\end{cases}$$

Here, $\mu_{\tilde{b}}$, $\nu_{\tilde{b}}$, and $w_{\tilde{b}}$ are the maximum values for the truth, indeterminacy, and falsity membership functions respectively.

Definition 10.17 [10] Assume the SVTBFNs $\tilde{a}=\{(a_1,a_2,a_3),\mu_{\tilde{a}},\nu_{\tilde{a}},w_{\tilde{a}}\}$ and $\tilde{b}=\{(b_1,b_2,b_3),\mu_{\tilde{b}},\nu_{\tilde{b}},w_{\tilde{b}}\}$, and $k>0$. Then,

- $\tilde{a}\oplus\tilde{b}=\{(a_1+b_1,a_2+b_2,a_3+b_3),\min\{\mu_{\tilde{a}},\mu_{\tilde{b}}\},\max\{\nu_{\tilde{a}},\nu_{\tilde{b}}\},\max\{w_{\tilde{a}},w_{\tilde{b}}\}\}$,
- $k\tilde{a}=\begin{cases}\{(ka_1,ka_2,ka_3),\mu_{\tilde{a}},\nu_{\tilde{a}},w_{\tilde{a}}\}, k>0,\\ \{(ka_3,ka_2,ka_1),\mu_{\tilde{a}},\nu_{\tilde{a}},w_{\tilde{a}}\}, k<0,\end{cases}$,
- $\tilde{a}\otimes\tilde{b}=\{(a_1b_1,a_2b_2,a_3b_3),\min\{\mu_{\tilde{a}},\mu_{\tilde{b}}\},\max\{\nu_{\tilde{a}},\nu_{\tilde{b}}\},\max\{w_{\tilde{a}},w_{\tilde{b}}\}\}, a_1>0, b_1>0.$

Definition 10.18 [9] Assume the SVTBFN $\tilde{a}=\{(a_1,a_2,a_3),\mu_{\tilde{a}},\nu_{\tilde{a}},w_{\tilde{a}}\}$. The positive and negative sets of (α,β,γ) cuts of this number are shown by $Q_{\tilde{a}}^{+}$ and $Q_{\tilde{a}}^{-}$ respectively and defined as below.

$$Q_{\tilde{a}}^{+}=\{x\in\mathbb{R}|T_{\tilde{a}}^{+}(x)\ge\alpha, I_{\tilde{a}}^{+}(x)\le\beta, F_{\tilde{a}}^{+}(x)\le\gamma,\},$$

$$Q_{\tilde{a}}^{-}=\{x\in\mathbb{R}|T_{\tilde{a}}^{-}(x)\le-\alpha, I_{\tilde{a}}^{-}(x)\ge-\beta, F_{\tilde{a}}^{-}(x)\ge-\gamma,\},$$

Here, $0\le\alpha\le\mu_{\tilde{a}}$, $\nu_{\tilde{a}}\le\beta\le1$, $w_{\tilde{a}}\le\gamma\le1$, and $0\le\alpha+\beta+\gamma\le3$.

10.4 Possibility Measures for SVTBFN

This section presents the measures of possibility theory for SVTBFNs.

Definition 10.19 Considering $\tilde{A} = \left(T_{\tilde{A}}^{+}, I_{\tilde{A}}^{+}, F_{\tilde{A}}^{+}, T_{\tilde{A}}^{-}, I_{\tilde{A}}^{-}, F_{\tilde{A}}^{-}\right)$ and $\tilde{B} = \left(T_{\tilde{B}}^{+}, I_{\tilde{B}}^{+}, F_{\tilde{B}}^{+}, T_{\tilde{B}}^{-}, I_{\tilde{B}}^{-}, F_{\tilde{B}}^{-}\right)$ as a pair of SVTBFNs, the positive and negative possibility measure of a fuzzy event for the truth-membership, indeterminacy-membership, and falsity-membership functions (shown as $Pos_{T^+}, Pos_{I^+}, Pos_{F^+}$) are defined by the following equations, respectively.

$$Pos_{T^+}\left(\tilde{A} * \tilde{B}\right) = sup\left\{\min\left(T_{\tilde{A}}^{+}(x), T_{\tilde{B}}^{+}(y)\right), x, y \in \mathbb{R}, x * y\right\},$$

$$Pos_{I^+}\left(\tilde{A} * \tilde{B}\right) = sup\left\{\min\left(I_{\tilde{A}}^{+}(x), I_{\tilde{B}}^{+}(y)\right), x, y \in \mathbb{R}, x * y\right\},$$

$$Pos_{F^+}\left(\tilde{A} * \tilde{B}\right) = sup\left\{\min\left(F_{\tilde{A}}^{+}(x), F_{\tilde{B}}^{+}(y)\right), x, y \in \mathbb{R}, x * y\right\},$$

$$Pos_{T^-}\left(\tilde{A} * \tilde{B}\right) = inf\left\{\max\left(T_{\tilde{A}}^{-}(x), T_{\tilde{B}}^{-}(y)\right), x, y \in \mathbb{R}, x * y\right\},$$

$$Pos_{I^-}\left(\tilde{A} * \tilde{B}\right) = inf\left\{\max\left(I_{\tilde{A}}^{-}(x), I_{\tilde{B}}^{-}(y)\right), x, y \in \mathbb{R}, x * y\right\},$$

$$Pos_{F^-}\left(\tilde{A} * \tilde{B}\right) = sup\left\{\min\left(F_{\tilde{A}}^{-}(x), F_{\tilde{B}}^{-}(y)\right), x, y \in \mathbb{R}, x * y\right\},$$

In these equations, the notation * can be of the signs $<, >, \leq, \geq, =$.

Definition 10.20 Let $\tilde{A} = \left\{(a_1, a_2, a_3); \mu_{\tilde{A}}, \nu_{\tilde{A}}, w_{\tilde{A}}\right\}$ and $\tilde{B} = \left\{(b_1, b_2, b_3); \mu_{\tilde{B}}, \nu_{\tilde{B}}, w_{\tilde{B}}\right\}$ are two SVTBFNs. From Definition 10.19, the possibility measures of the event $\tilde{A} \leq \tilde{B}$ for all of the membership functions are as below.

$$Pos_{T^+}\left(\tilde{A} \leq \tilde{B}\right) = \begin{cases} \min\{\mu_{\tilde{A}}, \mu_{\tilde{B}}\}, & a_2 \leq b_3, \\ \frac{(b_3-a_1)\min\{\mu_{\tilde{A}}, \mu_{\tilde{B}}\}}{(b_2-b_1)\mu_{\tilde{A}}+(a_2-a_1)\mu_{\tilde{B}}}, & a_1 < b_3, a_2 > b_2, \\ 0, & b_4 \leq a_1 \end{cases} \tag{10.14}$$

$$Pos_{I^+}\left(\tilde{A} \leq \tilde{B}\right) = \begin{cases} \max\{\nu_{\tilde{A}}, \nu_{\tilde{B}}\}, & a_2 \leq b_3 \\ \frac{(a_2-a_1\nu_{\tilde{A}})\left(1-\nu_{\tilde{B}}\right)+\left(b_1\nu_{\tilde{B}}-b_3\right)\left(1-\nu_{\tilde{A}}\right)}{(a_2-a_1)\left(1-\nu_{\tilde{B}}\right)+(b_1-b_2)\left(1-\nu_{\tilde{A}}\right)}, & a_1 < b_3, a_2 > b_2 \\ 1, & b_3 \leq a_1 \end{cases} \tag{10.15}$$

$$Pos_{F^+}\left(\tilde{A} \leq \tilde{B}\right) = \begin{cases} \max\{w_{\tilde{A}}, w_{\tilde{B}}\}, & a_2 \leq b_2 \\ \frac{(a_2-a_1w_{\tilde{A}})\left(1-w_{\tilde{B}}\right)+\left(b_1w_{\tilde{B}}-b_2\right)\left(1-w_{\tilde{A}}\right)}{(a_2-a_1)\left(1-w_{\tilde{B}}\right)+(b_1-b_2)\left(1-w_{\tilde{A}}\right)}, & a_1 < b_3, a_2 > b_2 \\ 1, & b_3 \leq a_1 \end{cases} \tag{10.16}$$

$$Pos_{T^-}\left(\tilde{A} \le \tilde{B}\right) = \begin{cases} -\max\{\mu_{\tilde{A}}, \mu_{\tilde{B}}\}, & b_2 \le a_2 \\ \frac{(b_1-a_3)\max\{\mu_{\tilde{A}},\mu_{\tilde{B}}\}}{(b_2-b_1)\mu_{\tilde{A}}+(a_3-a_2)\mu_{\tilde{B}}}, & b_1 < a_3, b_2 > a_2 \\ 0, & O.W. \end{cases} \tag{10.17}$$

$$Pos_{I^-}\left(\tilde{A} \le \tilde{B}\right) = \begin{cases} -\min\{\nu_{\tilde{A}}, \nu_{\tilde{B}}\}, & b_2 \ge a_2 \\ \frac{(a_2-a_3\nu_{\tilde{A}})(1-\nu_{\tilde{B}})+(b_1\nu_{\tilde{B}}-b_2)(1-\nu_{\tilde{A}})}{(a_3-a_2)(1-\nu_{\tilde{B}})+(b_2-b_1)(1-\nu_{\tilde{A}})}, & b_2 < a_2, b_1 > a_3 \\ 1, & O.W. \end{cases} \tag{10.18}$$

$$Pos_{F^-}\left(\tilde{A} \le \tilde{B}\right) = \begin{cases} -\min\{w_{\tilde{A}}, w_{\tilde{B}}\}, & b_2 \le a_2 \\ \frac{(a_2-a_3w_{\tilde{A}})(1-w_{\tilde{B}})+(b_1w_{\tilde{B}}-b_2)(1-w_{\tilde{A}})}{(a_3-a_2)(1-w_{\tilde{B}})+(b_2-b_1)(1-w_{\tilde{A}})}, & b_2 < a_2, b_1 > a_3 \\ -1, & O.W. \end{cases} \tag{10.19}$$

For the particular case of $\tilde{A} \le x$, according to Definition 10.20, the following possibility, measures are obtained.

$$Pos_{T^+}\left(\tilde{A} \le x\right) = \begin{cases} \mu_{\tilde{A}} & x \ge a_2, \\ \frac{\mu_{\tilde{A}}(x-a_1)}{a_2-a_1} & a_1 \le x \le a_2, \\ 0 & x \le a_1, \end{cases} \tag{10.20}$$

$$Pos_{I^+}\left(\tilde{A} \le x\right) = \begin{cases} \nu_{\tilde{A}} & x \ge a_2, \\ \frac{(a_2-x)+\nu_{\tilde{A}}(x-a_1)}{a_2-a_1} & a_1 \le x \le a_2, \\ 1, & x \le a_1. \end{cases} \tag{10.21}$$

$$Pos_{F^+}\left(\tilde{A} \le x\right) = \begin{cases} w_{\tilde{A}}, & x \ge a_2, \\ \frac{(a_2-x)+w_{\tilde{A}}(x-a_1)}{a_2-a_1}, & a_1 \le x \le a_2, \\ 1, & x \le a_1. \end{cases} \tag{10.22}$$

$$Pos_{T^-}\left(\tilde{A} \le x\right) = \begin{cases} -\mu_{\tilde{A}}, & x \le a_2, \\ \frac{\mu_{\tilde{A}}(x-a_3)}{a_3-a_2}, & a_2 \le x \le a_3, \\ 0, & x \ge a_3, \end{cases} \tag{10.23}$$

$$Pos_{I^-}\left(\tilde{A} \le x\right) = \begin{cases} -\nu_{\tilde{A}}, & x \le a_2, \\ \frac{(a_2-a_3\nu_{\tilde{A}})+x(\nu_{\tilde{A}}-1)}{a_3-a_2}, & a_2 \le x \le a_3, \\ -1, & x \ge a_3. \end{cases} \tag{10.24}$$

$$Pos_{F^-}\left(\tilde{A} \le x\right) = \begin{cases} -w_{\tilde{A}}, & x \le a_2, \\ \frac{(a_2-a_3w_{\tilde{A}})+x(w_{\tilde{A}}-1)}{a_2-a_1}, & a_2 \le x \le a_3, \\ -1, & x \ge a_3. \end{cases} \tag{10.25}$$

Lemma 10.1 *Consider the SVTBFN* $\tilde{A} = \left\{(a_1, a_2, a_3); \mu_{\tilde{A}}, \nu_{\tilde{A}}, w_{\tilde{A}}\right\}$. *Then,*

$$Pos_{T^+}(\tilde{A} \le x) \ge \xi_1 \Leftrightarrow \frac{\mu_{\tilde{A}}(x - a_1)}{a_2 - a_1} \ge \xi_1 \Leftrightarrow x \ge a_1 + \frac{(a_2 - a_1)\xi_1}{\mu_{\tilde{A}}}, \quad (10.26)$$

$$\begin{aligned} Pos_{I^+}(\tilde{A} \le x) \le \xi_2 &\Leftrightarrow \frac{(a_2 - x) + \nu_{\tilde{A}}(x - a_1)}{a_2 - a_1} \le \xi_2 \\ &\Leftrightarrow x \ge \frac{(\xi_2 + \nu_{\tilde{A}})a_1 + (1 - \xi_2)a_2}{1 - \nu_{\tilde{A}}}, \end{aligned} \quad (10.27)$$

$$\begin{aligned} Pos_{F^+}(\tilde{A} \le x) \le \xi_3 &\Leftrightarrow \frac{(a_2 - x) + w_{\tilde{A}}(x - a_1)}{a_2 - a_1} \\ &\le \xi_3 \Leftrightarrow x \ge \frac{(\xi_3 - w_{\tilde{A}})a_1 + (1 - \xi_3)a_2}{1 - w_{\tilde{A}}}. \end{aligned} \quad (10.28)$$

$$Pos_{T^-}(\tilde{A} \le x) \le \xi_1 \Leftrightarrow \frac{\mu_{\tilde{A}}(x - a_3)}{a_3 - a_2} \le \xi_1 \Leftrightarrow x \le a_3 + \frac{(a_3 - a_2)\xi_1}{\mu_{\tilde{A}}}, \quad (10.29)$$

$$\begin{aligned} Pos_{I^-}(\tilde{A} \le x) \ge \xi_2 &\Leftrightarrow \frac{(a_2 - a_3\nu_{\tilde{A}}) + x(\nu_{\tilde{A}} - 1)}{a_3 - a_2} \\ &\ge \xi_2 \Leftrightarrow x \le \frac{(1 + \xi_2)a_2 - (\nu_{\tilde{A}} + \xi_2)a_3}{1 - \nu_{\tilde{A}}}, \end{aligned} \quad (10.30)$$

$$\begin{aligned} Pos_{F^-}(\tilde{A} \le x) \le \xi_3 &\Leftrightarrow \frac{(a_2 - a_3 w_{\tilde{A}}) + x(w_{\tilde{A}} - 1)}{a_3 - a_2} \\ &\ge \xi_3 \Leftrightarrow x \le \frac{(\xi_3 + w_{\tilde{A}})a_3 - (1 + \xi_3)a_2}{w_{\tilde{A}} - 1}. \end{aligned} \quad (10.31)$$

10.5 Hesitant Bipolar Fuzzy Set

Definition 10.21 [11] Assume the reference set X. A hesitant bipolar fuzzy set (HBFS) for the positive and negative membership functions $\mu_{\tilde{A}}^{+}(x) \in [0, 1]$ (satisfaction function) and $\mu_{\tilde{A}}^{-}(x) \in [-1, 0]$ (dissatisfaction function) is shown as $\tilde{A} = \left\{\left(x, \mu_{\tilde{A}}^{+}(x), \mu_{\tilde{A}}^{-}(x)\right) \middle| x \in X\right\}$. Here, if for any $x \in X$, we assume that $\mu_{\tilde{A}}^{+}(x) = \cup_{\gamma_{\tilde{A}} \in \mu_{\tilde{A}}^{+}(x)}\{\gamma_{\tilde{A}}\}$ and $\mu_{\tilde{A}}^{-}(x) = \cup_{\eta_{\tilde{A}} \in \mu_{\tilde{A}}^{-}(x)}\{\eta_{\tilde{A}}\}$, then $0 \le \gamma_{\tilde{A}} \le 1$ and $-1 \le \eta_{\tilde{A}} \le 0$. In continue, for more simplification, $h_{\tilde{A}} = \left(\mu_{\tilde{A}}^{+}, \mu_{\tilde{A}}^{-}\right)$ is used to show a hesitant bipolar fuzzy element (HBFE). Notably, $h_{\tilde{A}} = \left(\mu_{\tilde{A}}^{+}, \mu_{\tilde{A}}^{-}\right)$ is used instead of $h_{\tilde{A}} = \cup_{\substack{\gamma_{\tilde{A}} \in \mu_{\tilde{A}}^{+}(x), \\ \eta_{\tilde{A}} \in \mu_{\tilde{A}}^{-}(x)}} \left\{\left(\gamma_{\tilde{A}}, \eta_{\tilde{A}}\right)\right\}$.

Example 10.7 Assume the reference set $X = \{x_1, x_2, x_3\}$, and the HBFEs $h_{\tilde{A}}(x_1) = \left(\mu^+_{\tilde{A}}(x_1), \mu^-_{\tilde{A}}(x_1)\right) = \{(0.5, -0.4), (0.5, -0.2)\}$, $h_{\tilde{A}}(x_2) = \left(\mu^+_{\tilde{A}}(x_2), \mu^-_{\tilde{A}}(x_2)\right) = \{(0.7, -0.1), (0.6, -0.4)\}$, and $h_{\tilde{A}}(x_3) = \left(\mu^+_{\tilde{A}}(x_3), \mu^-_{\tilde{A}}(x_3)\right) = \{(0.3, -0.2), (0.6, -0.2)\}$. Then, the below HBFS can be defined.

$$\begin{aligned}\tilde{A} = \{&(x_1, \{(0.5, -0.4), (0.5, -0.2)\}), (x_2, \{(0.7, -0.1), (0.6, -0.4)\}),\\ &(x_3, \{(0.3, -0.2), (0.6, -0.2)\})\}.\end{aligned}$$

Definition 10.22 [11] Assume the HBFEs $h_{\tilde{A}} = \left(\mu^+_{\tilde{A}}, \mu^-_{\tilde{A}}\right)$, $h_{\tilde{A}_1} = \left(\mu^+_{\tilde{A}_1}, \mu^-_{\tilde{A}_1}\right)$, and $h_{\tilde{A}_2} = \left(\mu^+_{\tilde{A}_2}, \mu^-_{\tilde{A}_2}\right)$. Then,

- $h_{\tilde{A}_1} \oplus h_{\tilde{A}_2} = \cup_{\substack{\gamma_{\tilde{A}_1} \in \mu^+_{\tilde{A}_1}, \eta_{\tilde{A}_1} \in \mu^-_{\tilde{A}_1} \\ \gamma_{\tilde{A}_2} \in \mu^+_{\tilde{A}_2}, \eta_{\tilde{A}_2} \in \mu^-_{\tilde{A}_2}}} \left\{\left(\gamma_{\tilde{A}_1} + \gamma_{\tilde{A}_2} - \gamma_{\tilde{A}_1} \times \gamma_{\tilde{A}_2}, -\left|\eta_{\tilde{A}_1}\right| \times \left|\eta_{\tilde{A}_2}\right|\right)\right\}$,
- $h_{\tilde{A}_1} \otimes h_{\tilde{A}_2} = \cup_{\substack{\gamma_{\tilde{A}_1} \in \mu^+_{\tilde{A}_1}, \eta_{\tilde{A}_1} \in \mu^-_{\tilde{A}_1} \\ \gamma_{\tilde{A}_2} \in \mu^+_{\tilde{A}_2}, \eta_{\tilde{A}_2} \in \mu^-_{\tilde{A}_2}}} \left\{\left(\gamma_{\tilde{A}_1} \times \gamma_{\tilde{A}_2}, \eta_{\tilde{A}_1} + \eta_{\tilde{A}_2} - \eta_{\tilde{A}_1} \times \eta_{\tilde{A}_2}\right)\right\}$,
- $kh_{\tilde{A}} = \cup_{\gamma_{\tilde{A}_1} \in \mu^+_{\tilde{A}}, \eta_{\tilde{A}_1} \in \mu^-_{\tilde{A}}} \left\{\left(1 - \left(1 - \gamma_{\tilde{A}}\right)^k, -\left|\eta_{\tilde{A}}\right|^k\right)\right\}$,
- $h^k_{\tilde{A}} = \cup_{\gamma_{\tilde{A}_1} \in \mu^+_{\tilde{A}}, \eta_{\tilde{A}_1} \in \mu^-_{\tilde{A}}} \left\{\left(\left(\gamma_{\tilde{A}}\right)^k, -1 + \left|1 + \eta_{\tilde{A}}\right|^k\right)\right\}$,

where, $0 \le \gamma_{\tilde{A}} \le 1$, $-1 \le \eta_{\tilde{A}} \le 0$, $0 \le \gamma_{\tilde{A}_1} \le 1$, $-1 \le \eta_{\tilde{A}_1} \le 0$, $0 \le \gamma_{\tilde{A}_2} \le 1$, and $-1 \le \eta_{\tilde{A}_2} \le 0$.

10.6 Application: Group Decision Making by Trapezoidal Bipolar Fuzzy TOPSIS

This approach was developed by Akram and Arshad [4] and its steps are described by the rest of this sub-section.

Step 1. The below three-dimensional decision matrix is considered.

$$\begin{array}{c} \\ A_1 \\ A_2 \\ \vdots \\ A_m \end{array} \begin{array}{c} \begin{array}{cccc} C_1 & C_2 & \cdots & C_n \end{array} \\ \begin{pmatrix} x^k_{11} & x^k_{12} & \cdots & x^k_{1n} \\ x^k_{21} & x^k_{22} & \cdots & x^k_{2n} \\ \vdots & \vdots & \vdots & \vdots \\ x^k_{m1} & x^k_{m2} & \cdots & x^k_{mn} \end{pmatrix} \end{array} \tag{10.32}$$

Here, $x_{ij}^{k} = \left(\left(a_{ij}^{k,1}, a_{ij}^{k,2}, a_{ij}^{k,3}, a_{ij}^{k,4}\right), \left(b_{ij}^{k,1}, b_{ij}^{k,2}, b_{ij}^{k,3}, b_{ij}^{k,4}\right)\right)$ (where, $i = 1, \ldots, m$, $j = 1, \ldots, n$, and $k = 1, \ldots, p$), is a trapezoidal bipolar fuzzy number showing the performance of alternative i in criterion j determined by expert k. Furthermore, $w^k = \left(w_1^k, w_2^k, \ldots, w_n^k\right)^T$ is the weight vector of criteria determined by expert k.

Step 2. The aggregated performance of alternative i in criterion j is obtained as below.

$$a_{ij}^{1} = \frac{1}{p}\sum_{k=1}^{p} a_{ij}^{k,1}, a_{ij}^{2} = \frac{1}{p}\sum_{k=1}^{p} a_{ij}^{k,2}, a_{ij}^{3} = \frac{1}{p}\sum_{k=1}^{p} a_{ij}^{k,3}, a_{ij}^{4} = \frac{1}{p}\sum_{k=1}^{p} a_{ij}^{k,4},$$

$$b_{ij}^{1} = \frac{1}{p}\sum_{k=1}^{p} b_{ij}^{k,1}, b_{ij}^{2} = \frac{1}{p}\sum_{k=1}^{p} b_{ij}^{k,2}, b_{ij}^{3} = \frac{1}{p}\sum_{k=1}^{p} b_{ij}^{k,3}, b_{ij}^{4} = \frac{1}{p}\sum_{k=1}^{p} b_{ij}^{k,4},$$

Here, $x_{ij} = \left(\left(a_{ij}^1, a_{ij}^2, a_{ij}^3, a_{ij}^4\right), \left(b_{ij}^1, b_{ij}^2, b_{ij}^3, b_{ij}^4\right)\right)$ is defined. Moreover, the aggregated weight of criterion j is obtained as $w_j = \frac{1}{p}\sum_{k=1}^{p} w_{jk}$.

Step 3. The weighted trapezoidal bipolar fuzzy decision matrix of $E = \left(e_{ij}\right)_{m\times n}$ is obtained, where, $e_{ij} = w_j \times x_{ij} = \left(\left(\eta_{ij}^1, \eta_{ij}^2, \eta_{ij}^3, \eta_{ij}^4\right), \left(\delta_{ij}^1, \delta_{ij}^2, \delta_{ij}^3, \delta_{ij}^4\right)\right)$.

Step 4. The below positive $\left(A^+\right)$ and negative $\left(A^-\right)$ trapezoidal bipolar fuzzy ideal solutions are obtained.

$$A^+ = \left\{e_1^+, e_2^+, \ldots, e_n^+\right\} = \left\{\left(\max e_{ij} \middle| j \in B\right), \left(\min e_{ij} \middle| j \in C\right), i = 1, \ldots, m\right\},$$

$$A^- = \left\{e_1^-, e_2^-, \ldots, e_n^-\right\} = \left\{\left(\min e_{ij} \middle| j \in B\right), \left(\max e_{ij} \middle| j \in C\right), i = 1, \ldots, m\right\}.$$

Step 5. The below distances from the positive and negative ideal solutions of Step 4 is obtained for each alternative.

$$d_i^+ = \sqrt{\frac{1}{2}\sum_{j=1}^{n}\left\{\begin{array}{l}\left(\eta_{ij}^1 - \eta_{j*}^1\right)^2 + \left(\eta_{ij}^2 - \eta_{j*}^2\right)^2 + \left(\eta_{ij}^3 - \eta_{j*}^3\right)^2 + \left(\eta_{ij}^4 - \eta_{j*}^4\right)^2 \\ +\left(\delta_{ij}^1 - \delta_{j*}^1\right)^2 + \left(\delta_{ij}^2 - \delta_{j*}^2\right)^2 + \left(\delta_{ij}^3 - \delta_{j*}^3\right)^2 + \left(\delta_{ij}^4 - \delta_{j*}^4\right)^2\end{array}\right\}},$$

$$d_i^- = \sqrt{\frac{1}{2}\sum_{j=1}^{n}\left\{\begin{array}{l}\left(\eta_{ij}^1 - \eta_{j-}^1\right)^2 + \left(\eta_{ij}^2 - \eta_{j-}^2\right)^2 + \left(\eta_{ij}^3 - \eta_{j-}^3\right)^2 + \left(\eta_{ij}^4 - \eta_{j-}^4\right)^2 \\ +\left(\delta_{ij}^1 - \delta_{j-}^1\right)^2 + \left(\delta_{ij}^2 - \delta_{j-}^2\right)^2 + \left(\delta_{ij}^3 - \delta_{j-}^3\right)^2 + \left(\delta_{ij}^4 - \delta_{j-}^4\right)^2\end{array}\right\}},$$

Step 6. The below closeness score for each alternative is obtained. The alternatives are ranked according to the decreasing order of the closeness score values.

$$C_i = \frac{d_i^-}{d_i^- + d_i^+}, i = 1, \ldots, m,$$

References

1. Zhang, W.R.: Bipolar fuzzy sets and relations: a computational framework for cognitive modeling and multiagent decision analysis. In: NAFIPS/IFIS/NASA'94. Proceedings of the First International Joint Conference of the North American Fuzzy Information Processing Society Biannual Conference. The Industrial Fuzzy Control and Intelligent, pp. 305–309. IEEE (1994)
2. Farhadinia, B.: Hesitant fuzzy set. In: Hesitant Fuzzy Set: Theory and Extension. Springer Singapore, Singapore (2021)
3. Akhtar, N., Islam, S.: Linear fractional transportation problem in bipolar fuzzy environment. Results Control Optim. **17**, 100482 (2024)
4. Akram, M., Arshad, M.: A novel trapezoidal bipolar fuzzy TOPSIS method for group decision-making. Group Decis. Negot. **28**(3), 565–584 (2019)
5. Babakordi, F., Taghi-Nezhad, N.A.: Review and comparison of bipolar fuzzy number types. Fuzzy Syst. Appl. **6**(2), 91–113 (2023)
6. Mandal, W.A.: Bipolar pythagorean fuzzy sets and their application in multi-attribute decision making problems. Ann. Data Sci. **10**(3), 555–587 (2023)
7. Princy, R., Mohana, K.: Spherical bipolar fuzzy sets and its application in multi criteria decision making problem. J. New Theory **32**, 58–70 (2019)
8. Deli, I., Ali, M., Smarandache, F.: Bipolar neutrosophic sets and their application based on multi-criteria decision making problems. In: 2015 International Conference on Advanced Mechatronic Systems (ICAMechS), pp. 249–254. IEEE (2015)
9. Samanta, S., Chakraborty, D., Jana, D.K.: Bipolar neutrosophic multi-item four-dimensional transportation problem with variable routes for breakable items. CMS **22**(1), 7 (2025)
10. Garai, T., Garg, H.: Multi-criteria decision making of COVID-19 vaccines (in India) based on ranking interpreter technique under single valued bipolar neutrosophic environment. Expert Syst. Appl. **208**, 118160 (2022)
11. Wei, G., Alsaadi, F.E., Hayat, T., Alsaedi, A.: Hesitant bipolar fuzzy aggregation operators in multiple attribute decision making. J. Intell. Fuzzy Syst. **33**(2), 1119–1128 (2017)

MIX
Papier aus verantwortungsvollen Quellen
Paper from responsible sources
FSC® C105338

If you have any concerns about our products,
you can contact us on
ProductSafety@springernature.com

In case Publisher is established outside the EU,
the EU authorized representative is:
Springer Nature Customer Service Center GmbH
Europaplatz 3, 69115 Heidelberg, Germany

Printed by Libri Plureos GmbH
in Hamburg, Germany